Nanostructure Science and Technology

Series Editor

David J. Lockwood, FRSC, National Research Council of Canada, Ottawa, ON, Canada

More information about this series at http://www.springer.com/series/6331

Toshimi Shimizu

Smart Soft-Matter Nanotubes

Preparation, Functions, and Applications

Springer

Toshimi Shimizu
Nanomaterials Research Institute (NMRI)
Department of Materials and Chemistry
National Institute of Advanced Industrial Science
and Technology (AIST)
Tsukuba, Ibaraki, Japan

ISSN 1571-5744 ISSN 2197-7976 (electronic)
Nanostructure Science and Technology
ISBN 978-981-16-2684-5 ISBN 978-981-16-2685-2 (eBook)
https://doi.org/10.1007/978-981-16-2685-2

This Springer imprint is published by the registered company Springer Nature Singapore Pte Ltd.
The registered company address is: 152 Beach Road, #21-01/04 Gateway East, Singapore 189721, Singapore

Preface

In 1991, Prof. Dr. Sumio Iijima discovered an eccentric, hollow cylindrical structure made of carbon atom under an electron microscope and then coined it "carbon nanotube." Thereafter, strategic research and development on the carbon nanotubes, which covers from the basic to commercialization level, have been vigorously promoting in all over the world. Speaking of nanotube, every researchers and engineers involved with nanotechnology will be able to imagine the carbon nanotube instantly. The carbon nanotubes are currently very familiar and technical terminology that represents one-dimensional nanomaterials. Self-assembled lipid nanotubes from amphiphilic molecules, however, emerged in liquid media in 1984, i.e., seven years before the astonishing discovery of the carbon nanotubes.

More interestingly, many researchers including chemists, physicists, and materials scientists tackled on the potent applications of the self-assembled tiny lipid tubules at the Defense Advanced Research Projects Agency for three days. The attendees have, at that moment, no way of knowing the existence of the carbon nanotubes with similar dimensionality as the organic tiny tubules. Triggered by this deep and promising discussion, immeasurable and intellectual curiosity regarding the organic tubular architectures with nanometric or micrometric dimension has developed to not only chemical and physical functions but medical applications. In this context, multifarious self-assembly and co-assembly techniques and pathways into the hollow cylindrical architectures have created strikingly talented organic nanotubes from rationally designed and multiple molecular building blocks, providing them with structural complexity and hierarchy for individual applications.

In this book, I define self-assembled organic nanotubes made of single or multiple molecular building blocks with "soft-matter nanotubes" (SMNTs) in contradistinction to hard-matter nanotubes like carbon, metallic, and inorganic nanotubes. I will, then, address the manufacturing method, characteristic tubular morphologies, diverse functions, and potent applications of every possible SMNTs. This book starts with general remarks of the SMNTs in Chap. 1. The SMNTs discussed in detail in this book are lipid nanotubes (Chap. 2), bolaamphiphile-based nanotubes (Chap. 3), Diphenylalanine-based nanotubes (Chap. 4), peptide-based nanotubes (Chap. 5), cyclic peptide-based and cyclic peptide–polymer-based nanotubes (Chap. 6), protein-based nanotubes (Chap. 7), bottlebrush copolymer-based nanotubes (Chap. 8), and rigid–flexible block molecule-based nanotubes (Chap. 9).

The hollow cylindrical structures with high-aspect-ratios can invest them with unique functions and properties, which can be differentiated from those of well-known self-assembled organic nanofibers, nanoribbons, and nanorods. Indeed, encapsulation, stabilization, transportation, release, and their cooperated functions for a certain guest substance lead to pave the way for innovative chemical, physical, biological, and medical applications. I also sincerely hope that the topics and figures should intrigue not only academic researchers but also engineers and university students.

To conclude this preface, I would like to thank my talented and skilled co-workers for their great contribution to the development of bilayered and monolayered soft-matter nanotubes. Their names are found in chapter references throughout this book. I would like to, particularly, thank my colleagues Drs. Naohiro Kameta and Wuxiao Ding for their current and enthusiastic collaboration. Drs. Mitsutoshi Masuda, Hiroyuki Minamikawa, Masaki Kogiso, Masumi Asakawa, and Masaru Aoyagi of AIST are also acknowledged for their fruitful collaboration. Our work was partly supported by the CREST and SORST projects from the Japan Science and Technology Agency (JST). Finally, I am grateful to Dr. Shin'ichi Koizumi and Mr. Ramamoorthy Rajangam, Springer, for taking care of this book patiently from beginning to end.

Tsukuba, Japan
February 2021

Toshimi Shimizu

Contents

Abbreviations

1D	One dimensional
2D	Two dimensional
3D	Three dimensional
5-FU	5-fluorouracil
AAO	Anodic aluminum oxide
AbNT	$(PLAR/HSA)_2$PLAR/PLG/HBsAb nanotube
AFM	Atomic force microscopy
AIE	Aggregation-induced emission
AMP	Adenosine monophosphate
AO	Acridine orange
AP	Amphiphilic peptide
ATP	Adenosine-5′-triphosphate
ATRP	Atom-transfer radical polymerization
Au NP	Gold nanoparticle
Au NR	Gold nanorod
Avi	Avidin
AZNT	Azo derivative-based nanotube
BA	Boronic acid
BAEAM	*Tert*-butyloxycarbonyl-aminoethyl acrylamide
BANT	Bolaamphiphile-based nanotube
BBC	Bottlebrush copolymer
BBCNT	Bottlebrush copolymer-based nanotube
BChl	Bacteriochlorophyll
BCP	Block copolymer
BGA	Biogenic amine
Boc	*Tert*-butyloxycarbonyl
BSPSA	9,10-bis[4-(3-sulfonatopropoxyl)-styryl] anthracene
BTNT	1,3,5-benzenetricarbonyl derivative-based nanotube
BXNT	Boroxine-based nanotube
B-αGluD	Biotinylated α-glucosidase
C60	Fullerene
CAB	Carbonic anhydrase

CalB	Candida antarctica lipase B
c-AMP	Cyclic adenosine monophosphate
Cbz	Benzyloxycarbonyl
CDDP	Cis-dichlorodiamineplatinum (II)
CF	Carboxyfluorescein
CNT	Carbon nanotube
COF	Covalent organic framework
COL	Collagen
Con A	Concanavalin A
CP	Cyclic peptide
CPL	Circularly polarized luminescence
CPNT	Cyclic peptide-based nanotube
CPPNT	Cyclic peptide–polymer-based nanotube
CPT	20-(S)-camptothecin
CT	Computed tomography
Cyt c	Cytochrome c
DA	Drug amphiphile
DBAE	2-(dibutylamino)-ethanol
DCE	1,2-dichlorethane
DDL	12-dodecanolactone
DEX	Dexamethasone
DFT	Density functional theory
DMF	*N,N*-dimethylformamide
DMSO	Dimethyl sulfoxide
DOPC	1,2-dioleoyl-sn-glycero-3-phoshatidylcholine
DOTA	Tetraazacyclododecane tetraacetic acid
DOX	Doxorubicin
DSC	Differential scanning calorimetry
DTC	3,3′-diethylthiacarbocyanine iodide
DTE	Dithienylethene
DWNT	Double-walled nanotube
EB	Eonsin B
ECL	Electrogenerated chemiluminescence
EG	Ethylene glycol
EPD	Electrophoretic deposition
FA	Folic acid
FB	Fuchsin basic
FFFF	Tetra-phenylalanine
FFNT	Di-phenylalanine-based nanotube
Flu-PAAH	Fluorescent poly (allylamine hydrochloride)
Fmoc	*N*-fluorenylmethoxycarbonyl
FR	Folate receptor
FRET	Fluorescence resonance energy transfer
FT	Fourier transform
GFP	Green fluorescence protein

GOD	Glucose oxidase
HBC	Hexa-*peri*-hexabenzocoronene
HBsAb	Anti-HBsAg antibody
HBsAg	Hepatitis B surface antigen
HBV	Hepatitis B virus
HEIZ	*N*-(2-hydroxyethyl)-imidazole
HFIP	1,1,1,3,3,3-hexafluoro-2-propanol
HGN	Hollow graphene nanoshell
HPM	Hierarchically porous material
HRP	Horseradish peroxidase
HSA	Human serum albumin
IR	Infrared spectroscopy
ITO	Indium-tin oxide
LbL	Layer-by-layer
LC	Liquid crystal
LCA	Lithocholic acid
LCANT	Lithocholic acid-based nanotube
LH	Light harvesting
LiTf	Lithium trifluoromethanesulfonate
LNT	Lipid-based nanotube
LUV	Large unilamellar vesicle
Mb	Myoglobin
MBL	Methylene blue
MC	Merocyanine
MCNT	Macrocycle-based nanotube
MD	Molecular dynamics
MLM	Monolayer lipid membrane
MNP	γ-Fe_3O_4 magnetic nanoparticle
MOF	Metal–organic framework
MONF	Microporous organic nanotube framework
MONN	Microporous organic nanotube network
MONWF	Microporous organic nanowire framework
MOP	Microporous organic polymer
MU	4-methyl-umbelliferone
MUB	4-methyl-umbelliferyl butyrate
MUGlc	4-methyl-umbelliferyl-α-D-glucopyranoside
MWCNT	Multi-walled carbon nanotube
NADH	Nicotinamide adenine dinucleotide
NDI	1,4,5,8-naphthalenetetracarboxylic acid
Neu5Ac	*N*-acetyl neuraminic acid
NIPAM	*N*-isopropylacrylamide
PA	Peptide amphiphile
PAA	Poly acrylic acid)
PAAH	Poly(allylamine hydrochloride)
PAH	Polycyclic aromatic hydrocarbon

PBA	Poly(*n*-butyl acrylate)
PBAEAM	Poly(tert-butyloxy-aminoethyl acrylamide)
PBS	Phosphate buffered saline
PC	Polycarbonate
PCEMA	Poly(2-cinnamoylethyl methacrylate)
PCHA	Poly(cyclohexyl acrylate)
PCLEMA	Poly(2-chloroethyl methacrylate)
PCP	Porous coordination polymer
Pd NP	Palladium nanoparticle
PDA	Polydopamine
PDI	Perylene diimide
PDMS	Poly(dimethyl siloxane)
P–E	Polarization–electric field
PEG	Poly(ethylene glycol)
PEI	1-(2-(prop-2-yn-1-yloxy) ethyl)-1H-imidazole
PEO	Poly(ethylene oxide)
PFS	Poly(ferrocenyldimethylsilane)
PGM	Poly(glycidyl methacrylate)
PHEA	poly(2-hydroxyethylacrylate)
PHPMA	Poly(hydroxypropyl methacrylamide)
PI	Poly(isoprene)
PL	Photoluminescence
PLA	Polylactide
PLAR	Poly-L-arginine
PLG	Poly-L-glutamic acid
PMMA	Poly(methyl methacrylate)
PNIPAM	Poly(*N*-isopropylacrylamide)
PNT	Peptide-based nanotube
POM	Porous organic material
PPO	Poly(propylene oxide)
PPQ	Poly(phenylquinoline)
PRNT	Protein-based nanotube
PS	Polystyrene
PSS	Poly(styrene sulfonate)
Pt NP	Pt nanoparticle
PTB	Poly(thiopheneboronic acid)
PtBA	Poly(*tert*-butylacrylate)
PTSA	*P*-toluene sulfonic acid
PVBC/BS	Poly(4-vinylbenzyl chloride-*co*-4-(3-butenyl styrene)
PVBS	Phenyl 4-vinylbenzene sulfonate
PVD	Physical vapor deposition
PVDF	Polyvinylidene difluoride
QC	Quantum confinement
R123	Rhodamine 123
R6G	Rhodamine 6G

RAPTA-C	Ruthenium dichloride (*p*-cymene)-1,3,5-triaza-7-phosphaadamantane
RFBM	Rigid–flexible block molecule
RFBMNT	Rigid–flexible block molecule-based nanotube
RGDG	Arg–Gly–Asp–Gly
ROP	Ring-opening polymerization
SA	Salicylic acid
SA-MONF	Salicylaldehyde-modified microporous organic nanotube framework
SAXS	Small-angle X-ray scattering
SC	Supercapacitor
SDS	Sodium dodecyl sulfate
SEM	Scanning electron microscopy
SHG	Second harmonic generation
SMNT	Soft-matter nanotube
SNP	Superparamagnetic nanoparticle
SNT	Silica nanotube
SPS	Sodium prop-2-yne-1-sulfonate
SRB	Sulforhodamine B
SSNMR	Solid-state nuclear magnetic resonance
ST	Safranine T
STEM	Scanning transmission electron microscopy
SWCNT	Single-walled carbon nanotube
SWNT	Single-walled nanotube
TAM	Tamoxifen
TC	Trithiocarbonate
TEG	Triethylene glycol
TEM	Transmission electron microscopy
TFA	Trifluoroacetic acid
$T_{g\text{-}l}$	Gel-to-liquid crystalline phase transition temperature
ThT	Thioflavin T
TMV	Tobacco mosaic virus
TNF	Trinitrofluorenone
TWNT	Triple-walled nanotube
VO	Oxo-vanadium (IV)
VT	Variable-temperature
VVVV	Tetra-valine
XRD	Powder X-ray diffraction
ZnChl	Zinc chlorin
αGluD	αv-glucosidase

Chapter 1
General Remarks of Soft-Matter Nanotubes

1.1 Introduction

What comes to your mind when you hear nanotube? Carbon nanotubes (CNTs) made of carbon atom will come to mind of readers who are interested in current nanotechnology and nanomaterials (Fig. 1.1) [1]. Those who are related to biology may be reminded of tobacco mosaic virus (TMV), in which tubular architectures self-assembled from identical protein units encapsulate RNA within the hollow cylinder (Fig. 1.1) [2]. Nanotubes have been undoubtedly becoming one of representative buzzwords in nano and nanobiotechnology. Indeed, the numbers of papers and patents dealing with the CNTs have a tendency to increase rapidly since 2000. Moreover, diverse tubular architectures made of organic molecules including amphiphiles, synthetic polymers, proteins, and DNAs have emerged in large numbers and have demonstrated novel application fields that are different from those of CNTs [3–6]. In other words, we enter a new era, in which nanotubular structures can be fabricated from every substances and atomic elements. This book focuses on soft-matter nanotubes (SMNTs) made of organic molecular building blocks as mother components (Fig. 1.1). Targeting classes of the molecules as molecular building blocks are associated with lipids and lipid-like amphiphiles (Chap. 2), bola-form amphiphiles (bolaamphiphiles) (Chap. 3), di-phenyl alanine (Chap. 4), amphiphilic peptides and peptide amphiphiles (Chap. 5), cyclic peptide and cyclic peptide–polymer conjugate (Chap. 6), protein (Chap. 7), bottlebrush copolymers (Chap. 8), and rigid–flexible block molecules (Chap. 9). Historical background, methodologies of fabrication, structures, functions, and applications of each SMNT are also described.

1.2 Comparison with Other Nanoporous Materials

Porous coordination polymers (PCPs) [8], metal–organic frameworks (MOFs) [9], covalent organic frameworks (COFs) [10, 11], and inorganic mesoporous materials

T. Shimizu, *Smart Soft-Matter Nanotubes*, Nanostructure Science and Technology,
https://doi.org/10.1007/978-981-16-2685-2_1

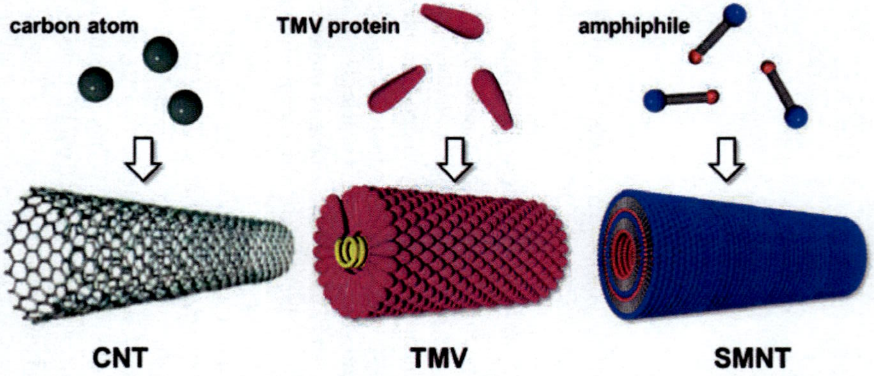

Fig. 1.1 Self-assembly of carbon atom, a TMV protein, and an amphiphile into CNTs, TMV, and SMNTs, respectively. The figure of CNT was reproduced with permission from Ref. [7]. © 2011 American Chemical Society

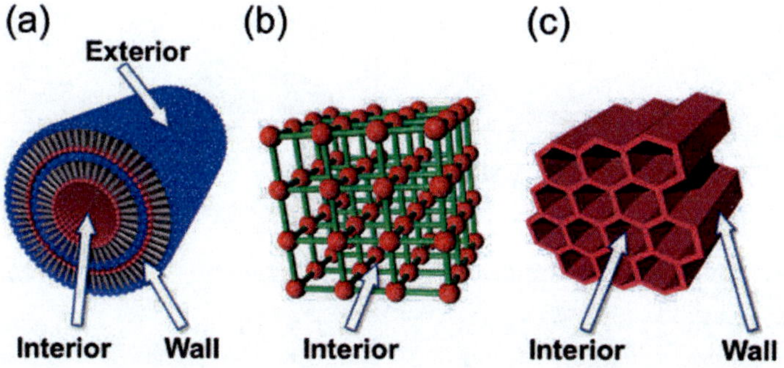

Fig. 1.2 Representative nanoporous materials with controlled dimensions and well-defined hollow cavities: **a** self-assembled SMNT, **b** MOF framework, and **c** inorganic mesoporous material

[12–16] have recently emerged as a booming class of porous materials. The reason for that is due to their specific properties and potent applications ascribable to their unique nanochannels. The surface features for the exterior of those porous materials, however, arouse less attention when compared to the characteristics of their interiors and scaffolds. Meanwhile, one-dimensional (1D), discrete high-aspect-ratio nanotube materials, e.g., SMNTs and CNTs, can exhibit synergetic and cooperative functions by utilizing the internal and external surfaces together with the nanotube wall entity (Fig. 1.2) [1, 6, 17]. Furthermore, by contrast with hard-matter nanotubes including CNTs and inorganic nanotubes, we categorize here organic molecule-derived nanotubes made of single or multiple components into SMNTs. A broad range of tunable sizes of the SMNT channel diameters (1 nm to several μm) can benefit the organic nanotubes to provide ideal accommodation for various guest substances ranging from small molecules to viruses [5, 18]. The interior diameters of the SMNTs can sometimes change dynamically by shrinkage, expansion, and reversible open–closed motion [6]. In addition, tunable lengths of the SMNTs are favorable not only to undergo hierarchization to form gels or liquid crystals (LCs) [19] but also to well disperse in or to hybridize with other matrix substances [20].

1.3 Nanotubes with Inner Diameters of 10 nm

Let us consider the nanostructures and characteristics of the SMNT featuring a 10-nm inner diameter. Well-known multi-walled CNTs (MWCNTs) have commonly the inner diameters of 5–10 nm that are 5 times larger compared with those of single-walled CNTs (SWCNTs). Some proteins like tubulin also self-organize to form a highly elaborate tubular structures called microtubules (Fig. 1.3). A dimer consisting of α- and β-tubulin makes a helical arrangement with 13 dimers to result in the formation of hollow cylindrical structures of 25 nm in outer diameter. While

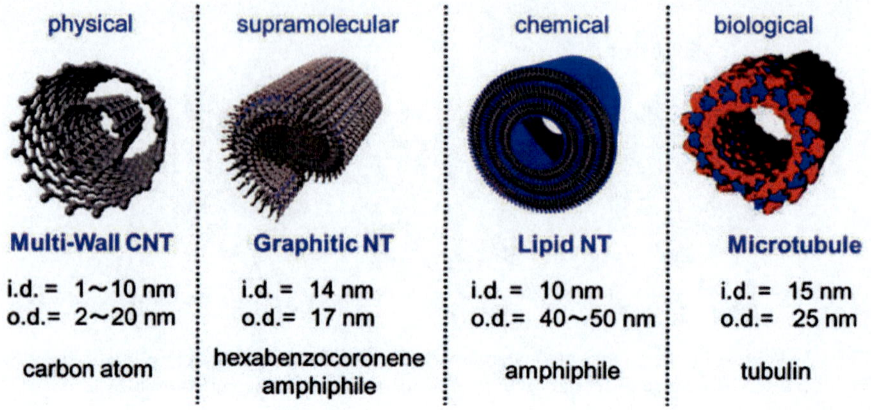

Fig. 1.3 Various nanotube structures with inner diameters of 10 nm, which are classified according to different building blocks. The images of the graphitic NT and microtubule were provided by National Partnership for Advanced Computational Infrastructure (NPACI) and courtesy of Dr. Takanori Fukushima, Aida ERATO project, JST, respectively

CNTs are generally fabricated by physical methodology based on 1D growth mechanism (Fig. 1.3), the microtubule grows via biological process such as polymerization. Chemical process can also produce nanotube structures through the self-assembly technique of rationally designed molecules [5]. Lipid nanotubes (LNTs) are the typical products of the molecular self-assembly (Fig. 1.3). Lipid, i.e., amphiphilic molecules containing both hydrophilic and hydrophobic functional groups self-assemble in water or organic solvents to form the SMNTs. In case of bilayer-based nanotubes in aqueous media, the hydrophilic groups expose to inner and outer environments to give a well aqueous dispersion. The minimum building blocks of the MWCNTs, microtubule, and LNTs differ individually and consist of carbon, tubulin protein, and amphiphile, respectively. Each hollow tubular structure is, however, common in term of the stabilization by helical arrangement of the building blocks. The amphiphile, which has a molecular fragment of graphene sheet, i.e., hexabenzocoronene as a molecular rigid core, was shown to self-assemble into supramolecular graphene-like nanotubes (Fig. 1.3) [21]. This SMNTs give an intriguing nanotube structure having both properties of CNT and LNT. Meanwhile, immersion of functional proteins onto the internal surfaces of porous templates allowed for the preparation of tubular structures as chemically produced artificial nanotubes. For example, Martin et al. reported many examples of SMNTs made of synthetic polymers by utilizing this template method [22].

1.4 Characteristics of Diameters and Internal Volume

Well-documented internal diameters measured for amphiphile- and polymer-derived nanotubes roughly range from 5 to 1000 nm [5, 6]. Neither top-down-type microfabrication nor any other fabrication techniques for carbon nanotubes can distinctively manufacture desired tubular structures with these dimensions. Therefore, self-assembled supramolecular nanotubes with such well-defined 1D hollow channels should serve as novel host nanomaterials in terms of mesoscale host–guest chemistry [23]. The reason is because their internal diameters are 10–100 times larger than those of well-known cyclodextrin- and cyclic peptide (CP)-based tubular host substances [24, 25]. Namely, those nanochannels of SMNTs should be able to encapsulate guest substances 10–100 times as large as conventional small guest, e.g., metal cations, low-molecular weight and aromatic compounds, or a single extended polymer chain.

Figure 1.4 summarizes the distribution of the diameters reported for various self-assembled nanotubes. Molecular nanotubes, e.g., cyclodextrin and cyclic peptide nanotubes can offer the lumen with internal diameters less than 1 nm. The cyclodextrin nanotubes can, therefore, accommodate different kinds of single polymer chains in the inner channel. Meanwhile, cyclic peptide nanotubes can mediate the transport of alkali or alkaline earth metal cations through lipid membranes. SWCNTs, MWCNTs, and inorganic nanotubes including molybdenum disulfide nanotubes [26], and imogolite nanotubes [27] span commonly the inner diameter region from 1 to 10 nm. That region corresponds to the inner diameter sizes that cannot be covered by the molecular nanotubes. Inner diameters of ca. 100 μm provided by polymer hollow fibers cover the other end of the size spectrum. Finely pulled glass

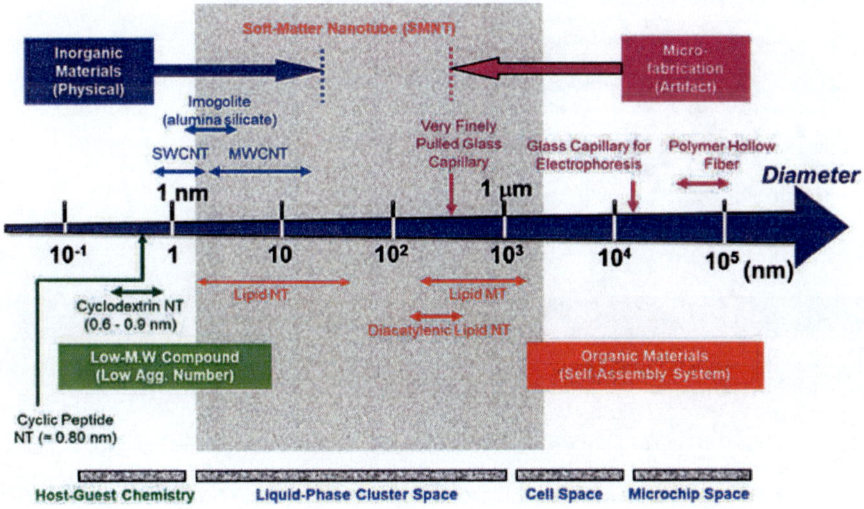

Fig. 1.4 Size distribution of typical nanotube structures. Abbreviations: NT, nanotube; MT, microtube; M.W, molecular weight; Agg., aggregation

capillaries for microinjection use may be the smallest tubular nanomaterials commercially available. For example, the Femtotips™ manufactured by Eppendorf Co. Ltd. have an internal diameter of 500 nm at the tip. Taken together, amphiphile- and polymer-derived nanotubes are the main tubular nanomaterials that can cover the inner diameters in the few nm–1000 nm range (Fig. 1.4).

Assuming that a single piece of cell is shaped by a cubic structure with a width of 10 μm, the content of the included water within the cell corresponds to 1 picoliter (pL $= 10^{-12}$ L) volume. If picoliter or attoliter (aL $= 10^{-18}$ L) volumes of liquids are confined in a micrometeric or nanometric space such as biological cell, capillary for chromatography, and microfluidic device, the liquids undergo a dramatic change in their fluid properties. For example, restricted water in nanometeric geometries including micelles and nanoporous silica often displays unexpected features as compared to bulk water [28, 29]. One dimensional channels with femtoliter (fL $=10^{-15}$ L) or attoliter volumes, such as CNT [30] and silica nanotubes [31], currently intrigue researchers to further miniaturize conventional microchip devices consisting of poly(dimethyl siloxane), silicone, and glass. Those needs have also encouraged many researches on supramolecular assemblies with well-defined morphologies and dimensions [6]. Indeed, self-assembled LNTs of 10 nm in internal diameter and 1 μm in length are able to present the confined volume of water corresponding to1 aL. Notably, the 1 aL volume is 10^3- and 10^6-fold smaller than those of the femtoliter chamber [32] and single cell, respectively (Fig. 1.5). "Attoliter chemistry" features unique chemical events that take place in confined liquid nanospace of attoliter volume [23]. Nanotube architectures self-assembled from a wide variety of molecular building blocks, e.g., peptides, proteins, DNAs, synthetic bottlebrush copolymers, rigid–flexible block molecules and amphiphiles, have recently emerged in large

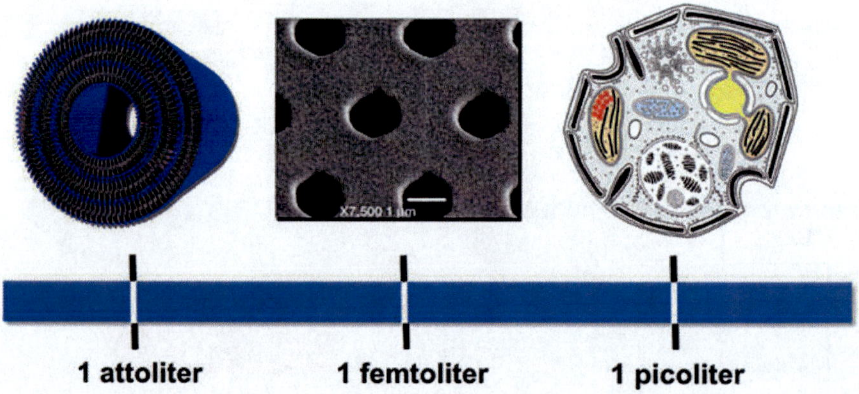

Fig. 1.5 Typical example of ultrasmall amount of liquids confined in various nanospace including SMNT, femtoliter chamber, and single cell (from the left). The images of the chamber and cell are reproduced with permission from Ref. [32], © 2005, Nature Publishing Group and by courtesy of Professor Isao Inouye of University of Tsukuba, respectively

numbers [6]. They also have performed highly attractive activities for practical use in diverse fields from chemistry, physics, biology to medical science.

1.5 Historical Background of Lipid Nanotubes and Carbon Nanotubes

In 1977, the formation of a spherical vesicle made of totally synthetic bilayer membrane, which is self-assembled from didodecyldimethylammonium bromide **1** (Chart 1.1) in water, was reported by Kunitake and Okahata for the first time (Fig. 1.6) [33]. This notable discovery and well-connected molecular design attract researchers, leading to successive development of the creation of supramolecular self-assemblies based on functionalizable membrane structures [34]. For example, Kunitake and Fuhrhop groups devoted their research to extensively investigate a great number of bilayer- [35] and monolayer-based molecular assemblies [36], respectively, with attractive biology-related properties.

Of particular note is that through chiral molecular self-assembly [37, 38], Kunitake and co-workers discovered the spontaneous assembly of hollow tube structures from chiral glutamic acid-derived, double-chain ammonium amphiphiles **2** (Chart 1.1) in

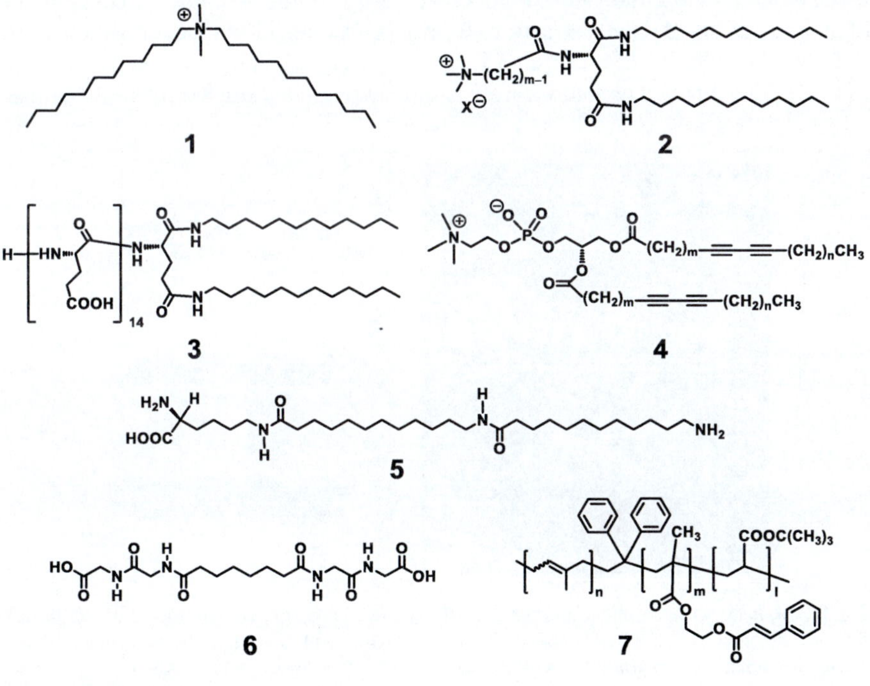

Chart 1.1 Chemical structures of **1–7**

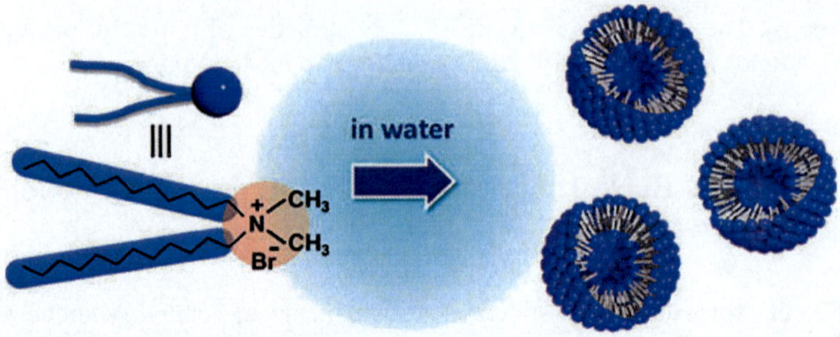

Fig. 1.6 Formation of synthetic bilayer membrane-based vesicular assemblies from a double-chain ammonium salt **1**

1984 (Fig. 1.7) [39, 40]. Both Ihara's group [41] and Yager et al. [42] concurrently reported on the self-assembly of tubular structures from double-chain glutamic acid amphiphiles **3** (Chart 1.1) and diacetylenic phospholipids **4** (Chart 1.1), respectively. In the dawn of organic nanotube history, the dimensions for the self-assembled nanotubes showed 200–500 nm external diameters, 100—200 nm internal diameters, and 50—200 μm length. These values present relatively larger ones than those reported until now. The membrane walls of the tubes are stabilized by multiple bilayer membranes with thicknesses of several tens nanometers. In contrast to the bilayer-based nanotubes, Fuhrhop and co-workers reported for the first time the construction of monolayer-based tubular structures from α-(L-Lysine), ω-(amino) bolaamphiphile **5** (Chart 1.1) with lysine and amino groups as two terminal hydrophilic headgroups [34, 43]. The obtained nanotubes are featured by relatively smaller internal diameter

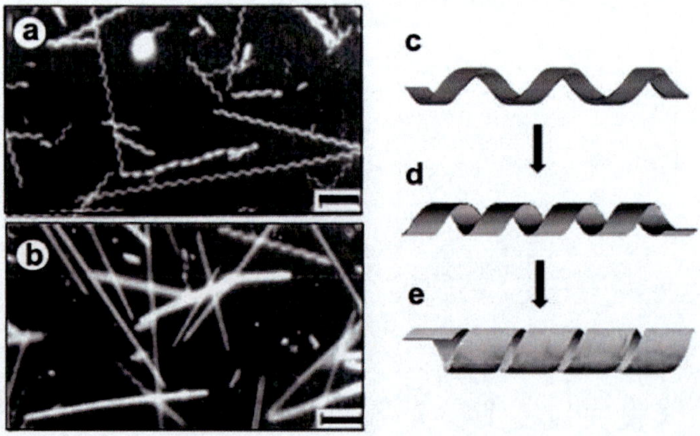

Fig. 1.7 **a, b** Dark-field optical micrographs of aqueous dispersion containing the amphiphile **2**, showing the growth from helically coiled ribbons (**c, d**) to nanotube structures (**e**) (scale bars, 10 μm; the micrographs were reproduced with permission from Ref. [40], © 1985 American Chemical Society)

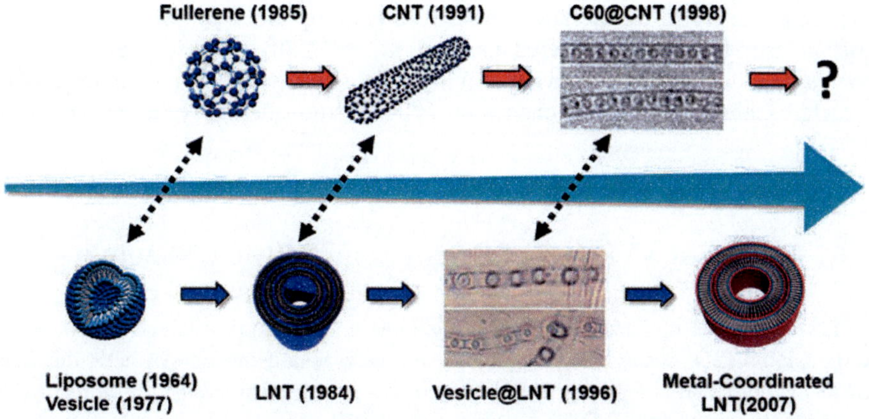

Fig. 1.8 Historical aspect of LNTs and CNTs. The images of fullerene and CNT were given by courtesy of Professor Hitoshi Nakahara of Nagoya University. The image of C60@CNT was reproduced with permission from Ref. [51] , © 2011 American Physical Society

(~50 nm) and wall thickness (~4.4 nm), showing the presence of a single monolayer membrane.

Fullerenes (C60), CNTs, and C60-encapsulated CNTs are well-known as three representative morphologies of carbon nanomaterials. Each structure corresponds to spherical, hollow cylindrical, and sphere-containing tubular structures. Supramolecular assemblies with similar morphological features as those of the carbon nanomaterials are vesicles [33], LNTs [39, 41, 42], and vesicle-encapsulated microtubes [44]. This morphological diversity is ascribable to multiple structures, shapes, and functionalities of constituent amphiphilic molecules. Notably, the first construction of the vesicles, LNTs, and vesicle-encapsulated microtubes were reported prior to that of corresponding C60 [45], CNTs [46], and C60-encapsulated CNTs [47], respectively (Fig. 1.8). Liposomes made of phospholipids and vesicles made of synthetic lipids were first reported in 1964 [48] and 1977 [33], respectively. This event took place at least 7 years earlier than the discovery of C60. As already described above, regarding the LNTs, individual three research groups reported the self-assembly of glutamic acid long-chain dialkyl derivatives **2** and **3** or diacetylene-containing phospholipid derivatives **4** into nanotubes. Those findings were reported 7 years before the historical moment for the discovery of CNT structures by Professor Dr. Iijima. Of particular note is that a great many researchers in different fields gather together to discuss on "What are tiny nanotubes good for?" by using the tiny lipid nanotubule at an American Physical Society Meeting in 1990 [49]. In 1996, we found very interesting 1D molecular assemblies, in which microtubular assembly of 1–2 μm in outer diameters encapsulated vesicles in their lumens, from the diglycine bolaamphiphile **6** (Chart 1.1) [44]. This discovery is indeed 2 years before that of C60-encapsulated CNTs [47], so-called peapod (Fig. 1.8). Eleven years later, we also reported that a metal cation-complexed building block of peptide lipid self-assembles to form a multi-layered nanotube structure [50]. Considering that SMNT system has demonstrated forward-looking topologies and architectures in tubular structures, novel carbon nanotube materials that feature alternative stacked layers of hetero atoms may emerge.

1.6 Fabrication Method of Organic and Polymer Nanotubes

A discrete SMNT means a single piece of nanostructure that is involved with no two-dimensional (2D) assembled or bundled structures. Focusing on synthetic discrete SMNTs, both ends of which are opened to external environment, the preparation methodologies for the SMNTs are overviewed. We can roughly classify into three categories including (1) self-assembly technique, (2) hollowing approach of cylindrical core, and (3) template methods (Fig. 1.9). The first self-assembly technique is exemplified by the scheme, in which self-assembling molecules spontaneously form tubular architectures in water or organic solvents. Driving forces for the self-assembly are van der Waals force, hydrogen bond, coordination bond, and π–π stacking. The molecular building blocks as rationally designed self-assembling molecules are classified into more than ten groups, as described later in **1.11**. One can implement the information on the molecular shape, topological arrangement of functional groups, and localized environment of solvophobic/solvophilic segments into the molecular design. Combination of synthetic strategy with rational molecular design has enabled diverse tubular structures.

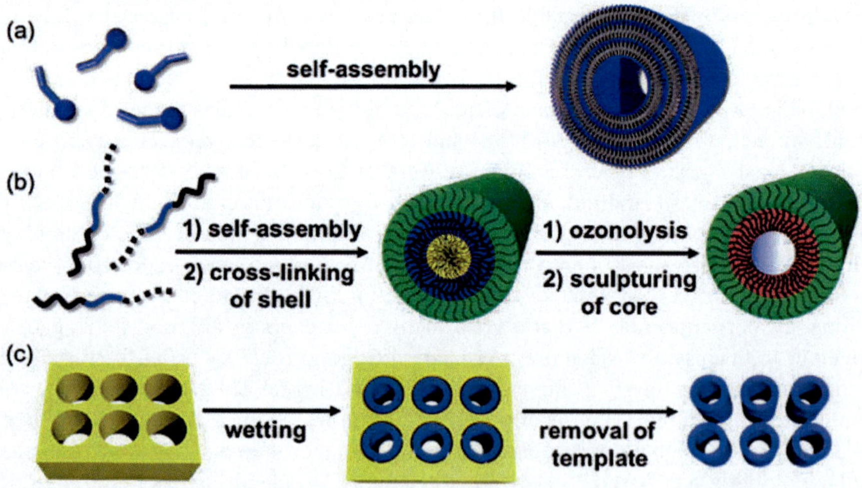

Fig. 1.9 General classification of synthetic methodology to fabricate various SMNTs

The self-assembly shown in Fig. 1.9a is further divided into eleven pathways (Fig. 1.10). For example, current SMNT formation is attributed not only to self-assembly through bilayer or monolayer membrane structures but also to supramolecular stacking and stacking of self-assembled nanorings or cyclic compounds. Figure 1.10 illustrates categorization of nanotube formation through self-assembly, which are based on the structures of building blocks and self-assembly route. Mainly four formation pathways are represented, containing (I) membrane- or sheet-based,

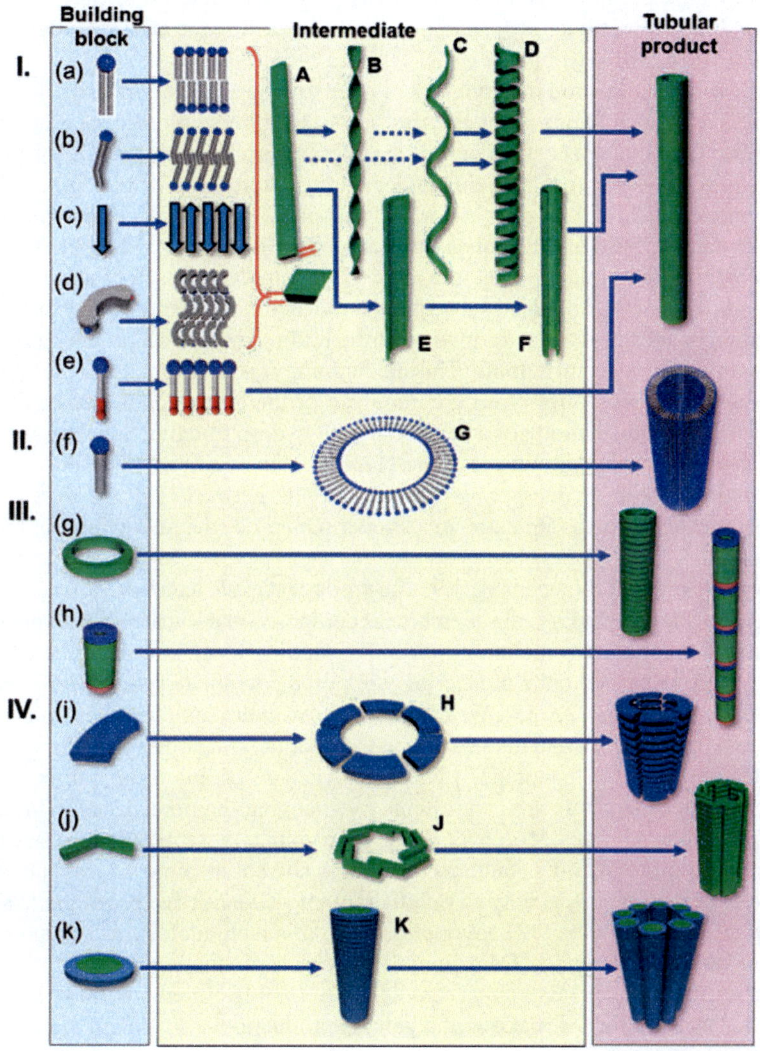

Fig. 1.10 Classification of proposed self-assembly pathways of various building blocks into tubular architectures. Reproduced with permission from Ref. [5], © The Chemical Society of Japan

(II) nanoring- or nanotoroid-based, (III) stacking-based, and (IV) supramolecular stacking-based routes. All the simplified images of the building blocks, intermediates, and tubular products just show representative structures and morphologies. Before completing the nanotube formation, many building blocks pass through various intermediate structures, e.g., 1D sheets (A in Fig. 1.10), twisted ribbons (B), helically coiled ribbons (C and D), 2D curved sheets (E and F), self-assembled nanorings (G), rosettes (H) and macrocycles (J), and vertical stack of discs (K). The self-assembled nanotube structures as major products often coexist with thermodynamically unstable intermediates. Of course, packing-directed self-assembly drives building block directly to form nanotubes without passing through any intermediate structures [Fig. 1.10, I(e)] [4, 37].

Hollowing-out method of a cylindrical fiber structures shown in Fig. 1.9b begins with the self-assembly of nanofibers from self-assembling molecules or amphiphilic polymers. The obtained central cores of the fibers are hollowed out by chemical reaction like ozonolysis and degradation under acidic conditions. For example, tri-block copolymers **7** (Chart 1.1) comprising poly(isoprene) (PI), poly(2-cinnamoylethyl methacrylate) (PCEMA), and poly(*t*-butyl acrylate) (PtBA) (e.g., 130:130:800, wt%) produce cylindrical micelle structures via self-assembly. Photo-crosslinking of the shell part made of PCEMA and subsequent removal of the inner core part made of PI via ozonolysis result into the formation of the hollow cylindrical architectures with outer diameters of several tens nm. This methodology takes advantage of crosslinking ability and solvent affinity of each component of the block copolymers. Obtained inner surfaces of the nanotubes are covered with hydrophilic functional groups such as aldehyde or carboxy groups. It is, therefore, easy to be subjected to further chemical functionalization. Microporous organic nanotube networks (MONNs) developed by Huang et al. are also fabricated in a similar manner as the hollowing-out method [52].

Template method shown in Fig. 1.9c can produce organic nanotubes by moistening the inner surfaces of nanoporous membranes composed of polycarbonate or anodized alumina with solutions containing molecular building blocks of nanotubes. Martin and co-workers carried out a pioneering work on the template synthesis of polymer and metal microtubes [22, 53]. Wendorff and co-workers followed the similar technique as Martin's work and also developed a modular assembly system to fabricate free-standing polymer nanotubes [54]. An advantage of this technique is that the nanotube outer diameters are controllable by changing the pore sizes (15–400 nm) of the porous membranes. Moreover, this gives chemically stable nanotube structures that are tolerant to alkaline solutions or organic solvent as removal reagents of the template membranes. A variety of polymers such as conductive polymers [53] and macromolecular proteins [18] are applicable to this methodology as the molecular building block of nanotubes. Compared with massive production method for organic nanotubes through self-assembly technique, the template methods further need to solve the problem for the massive and general production.

1.7 Membrane- or Sheet-Based

1.7.1 Bilayer-Based

1.7.1.1 Chiral Self-assembly

Bilayer-based nanotube structures are generally based on chiral molecular packing within a membrane wall. Chiral interactions between constituent amphiphilic molecules make the molecules pack at a nonzero angle concerning their nearest neighbors. One particular orientation of the self-assembled bilayer membrane in a solid state should be energetically favorable. This process leads to twisting and/or coiling of the bilayer membranes, thereby resulting in the formation of hollow tubular structures composed of the bilayer membranes (Fig. 1.11).

As already described above, chiral self-assembly of amphiphilic molecules into nanotubes commonly passes through helically twisted and/or coiled ribbon or sheet morphologies as an intermediate structure. As a typical example, when hot aqueous solutions containing tube-forming amphiphiles are allowed to cool slowly, two different self-assembly pathways may proceed. One route is involved with shortening of the helical pitch of the ribbon while keeping the ribbon width constant (Fig. 1.12a) [55, 56]. Meanwhile, the second route proceeds via widening of the ribbon width while keeping the helical pitch length constant (Fig. 1.12b) [40]. The latter "growing width" route is more familiar in the literature than the former "closing pitch" one. A synchronized scheme accompanied with both "growing width" and "closing pitch" route was also reported to explain the nanotube formation from a newly designed lysine amphiphile [57]. Both the helical ribbon and tubular structure are thermodynamically stable only at temperatures ($T < T_{g\text{-}l}$) below the gel-to-liquid crystalline phase transition temperature ($T_{g\text{-}l}$) of the amphiphiles. Heating up the aqueous dispersion of the nanotube to temperatures ($T > T_{g\text{-}l}$) above $T_{g\text{-}l}$ causes instantly a drastic morphological change from tubular into spherical morphology.

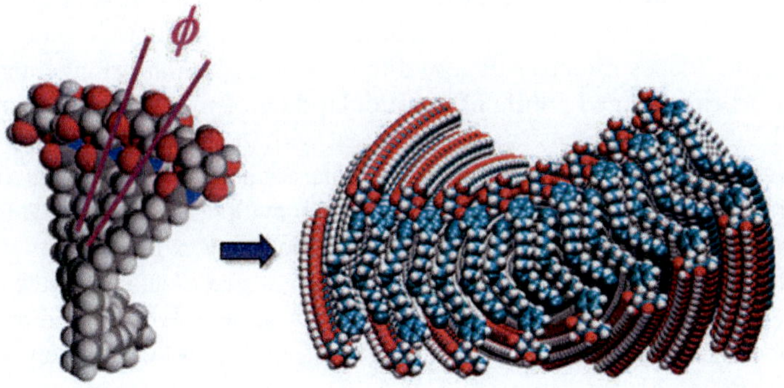

Fig. 1.11 Chiral molecular self-assembly that leads to twisting of bilayer membrane

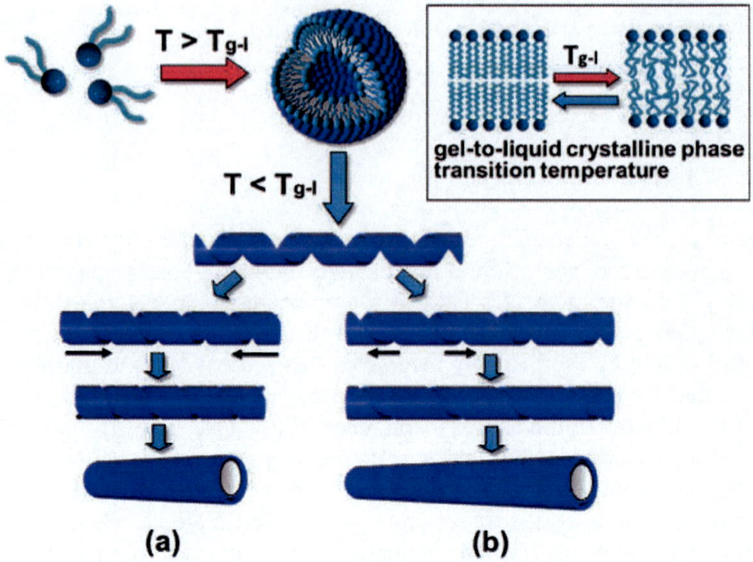

Fig. 1.12 Two different pathways of chiral molecular self-assembly into nanotubes; **a** "closing pitch" and **b** "growing width" routes

1.7.1.2 Bilayer Membrane-Based

Multi-walled organic nanotubes stabilized by bilayer membranes gave the first example of a discrete nanotube through the self-assembly of chiral amphiphiles (Fig. 1.10a) [39, 41, 42]. A great number of practices in constructing tubular structures through chiral self-assembly has been performed through the 36-year history of organic nanotubes [5, 38]. At the early stage of the development, chiral amphiphiles carried a double-tailed chain as a hydrophobic segment. A lot of scanning and transmission electron microscopic (SEM and TEM) observation corroborated that self-assembled nanotubes are composed of helically coiled bilayer-based membranes [5, 38]. Helically arranged adsorption of nanoparticles or polymers onto the nanotube surfaces directly suggested the chiral molecular arrangements in the nanotubes. Mirror-imaged profile of the circular dichroism spectra for enantiomeric amphiphiles-based nanotubes is also consistent with the chiral molecular assembly. The presence of helically coiled ribbons that interconnect with tubular architectures on a single ribbon structure can also provide direct evidence for the chiral molecular self-assembly. Diverse self-assembly pathways that convert from spherical assemblies including micelles or vesicles to completely matured tubular architectures have ever been discussed. As described in the previous section, twisted and/or helically coiled ribbon structures (B, C, and D in Fig. 1.10) are reported to commonly intermediate the nanotube self-assembly (Fig. 1.13). Spontaneous rolling-up and subsequent zipping of a sheet structure is also reported for the self-assembly of a bile acid amphiphile into nanotubes (E and F in Fig. 1.10).

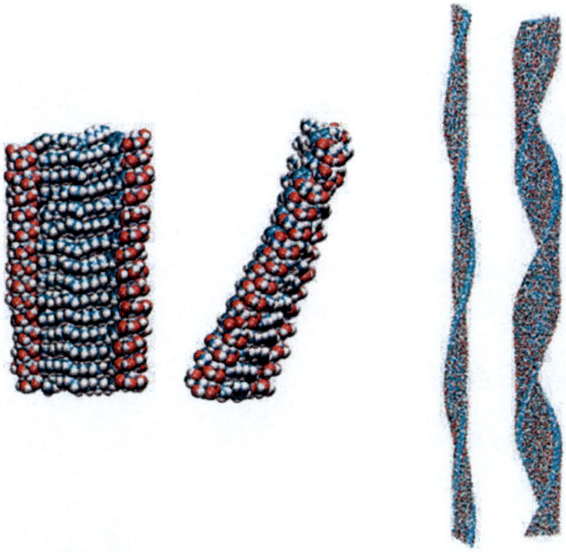

Fig. 1.13 Various morphology of bilayer membranes formed by self-assembly of chiral amphiphiles

1.7.1.3 Interdigitated Bilayer-Based

Phospholipids, main components of biomembranes, self-assemble in water to commonly form so-called liposomes that comprise bilayer-based spherical assemblies. Under high pressure or in the presence of amphiphilic substances, however, the phospholipid forms an interdigitated bilayer membrane as a result of the unusual molecular packing of the acyl chains. The lipid molecules that face each other within bilayer phase interpenetrate their hydrophobic chains and construct a monolayer-thick membrane [58]. This unique nonbilayer structure can be typically observed for the phospholipids having a bulky hydrophilic headgroups. Self-assembled nanotubes from a single-head, single-tailed amphiphile carrying an unsaturated hydrophobic chain have a tendency to organize the interdigitated bilayer membrane (Fig. 1.10b) [59–63]. The hydrocarbon chains in diacetylenic amphiphiles and monoene-appended glycolipids take a kink conformation, resulting in the formation of the interdigitated bilayer membranes (Fig. 1.14). Helically coiled ribbons, self-assembled from a mixture of cardanol glycolipid **8–1** (Chart 1.2), self-assembled in water to yield a tubular structure [59]. By contrast, a twisted ribbon, self-assembled from the saturated glycolipid derivative **8d** (Chart 1.2), produced no tubular structures through self-assembly [59].

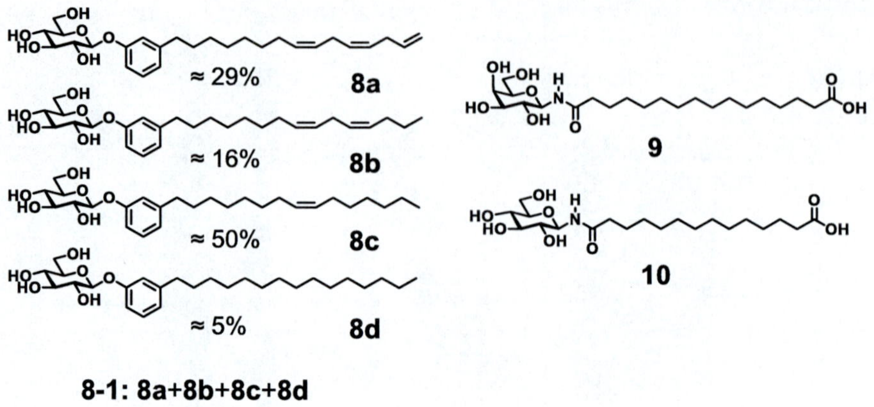

Chart 1.2 Chemical structures of **8-1–10**

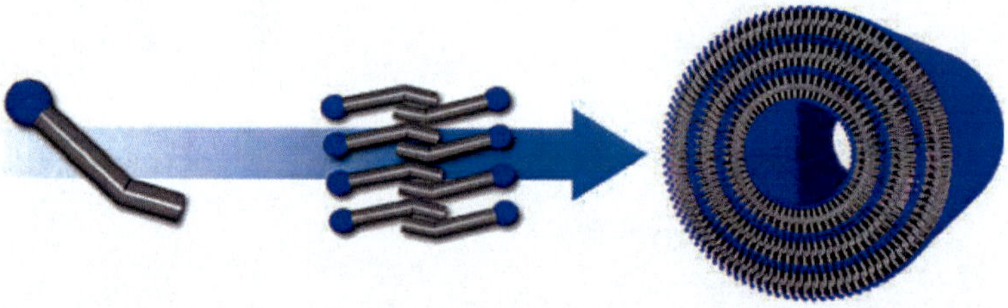

Fig. 1.14 Nanotube structure based on three layers of interdigitated bilayer membranes

1.7.2 Monolayer-Based

1.7.2.1 Packing-Directed Self-assembly

Besides the chiral self-assembly pathway described in Sect. 1.7.1.1, another route called packing-directed self-assembly should be noted. No remarkable chiral structures represented by helically coiled or twisted ribbons emerge in the middle of the self-assembly into tubular structures. Wedge-shaped bolaamphiphiles, in which two different-sized hydrophilic headgroups are linked to both ends of a long hydrophobic chain, tend to self-assemble into hollow cylindrical structures. If the molecular packing of hydrophobic chains in a rationally designed amphiphilic molecule is in a crystalline phase, it is normally difficult or complicated to make a prediction of the resultant self-assembled morphology from the amphiphile. Only when the hydrophobic chains of the amphiphile are in a fluid state, the following guideline for the self-assembly is applicable for the prediction of the final aggregate morphology. Israelachivili proposed a relationship between the structure of an amphiphile and its self-assembled morphology, which can be discussed by the critical packing parameter, $P = v/(a_0 l_c)$. In this equation, v is the volume of the hydrophobic chain, a_0

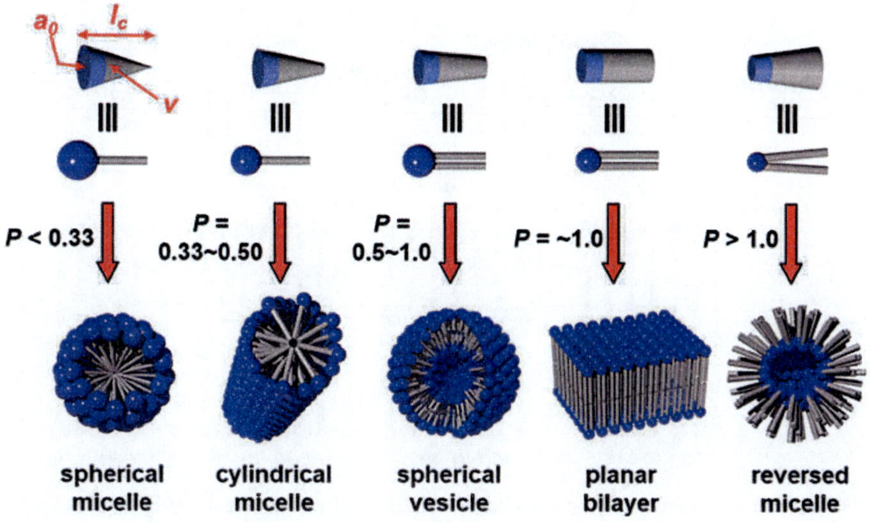

Fig. 1.15 Various self-assembled morphologies depending on the critical packing parameter (P) of each amphiphile, which is defined as $P = v/(a_0 l_c)$

is the polar head surface area at the critical micellar concentration, and l_c is the chain length (Fig. 1.15) [64]. If $P < 1/3$, the amphiphile tends to self-assemble into spherical micelles; if $1/3 < P < 1/2$, cylindrical micelles will be preferred; if $1/2 < P < 1$, bilayer vesicles with a spontaneous curvature are producible; if $P = 1$, planar bilayers will be preferred; and if $P > 1$, reversed micelles will be formed (Fig. 1.15).

1.7.2.2 Control of *Polymorph* and *Polytype*

Exclusive construction of SMNTs with completely different internal and external surfaces through self-assembly has attracted growing attention for further development of novel applications. For that purpose, not only the rational optimization of the building block structures but also control of the *polymorph* and *polytype* within monolayer-based membranes are critical. Here, we discuss an unsymmetrical bolaamphiphiles, in which two polar headgroups with different bulkiness or functionality are connected to both ends of a nonpolar hydrocarbon chain. Figure 1.16 shows the packing profile (*polymorph*) of monolayer lipid membranes (MLMs) comprising an unsymmetrical bolaamphiphile and also illustrates the additional stacking feature (*polytype*) of each MLM. The *polymorph* of the MLM can be divided into two classes, i.e., unsymmetrical and symmetrical MLMs. The classification depends on whether their constituent bolaamphiphiles are packed in a parallel or antiparallel fashion, respectively (Fig. 1.16a, b). Based on the molecular orientation at the MLM interface, two additional kinds of membrane stacking motifs can appear as a *polytype*. That is a head-to-tail or head-to-head interface, where head and tail mean relatively

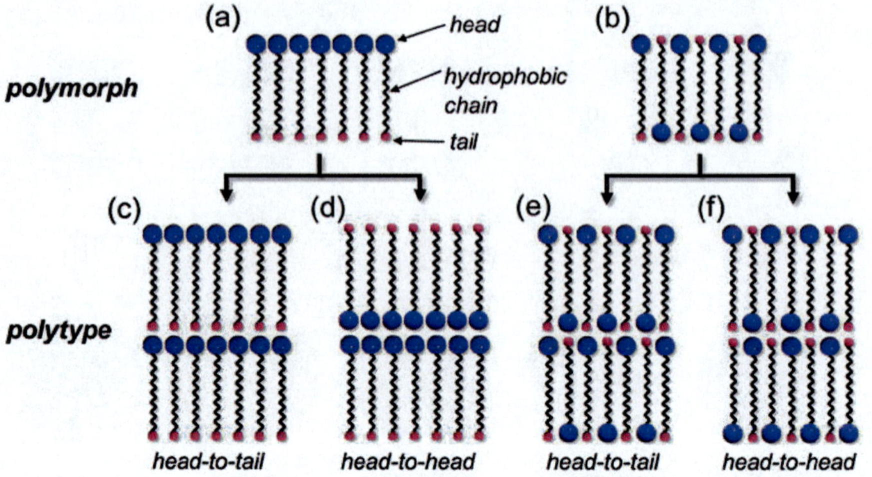

Fig. 1.16 **a** Unsymmetrical, **b** symmetrical MLMs, and **c–f** four different self-assembled MLMs from an unsymmetrical bolaamphiphile. These MLMs are formed through a *polymorph* of a single MLM and a subsequent *polytype* of the stacked MLMs. Reproduced with permission from Ref. [4], © 2016 American Chemical Society

larger and smaller headgroups, respectively (Fig. 1.16c–f). Antiparallel molecular packing can well compensate the spatial void and dipole moment, which are derived from the relatively bulky headgroup. Indeed, all of the *polymorph* in crystal structures reported for unsymmetrical bolaamphiphiles until 2001 corresponds to this symmetrical MLMs (Fig. 1.16b) [65, 66]. No matter how amphiphiles form an unsymmetrical MLM in a solid phase, each MLM stacks with a head-to-head interface as illustrated in Fig. 1.16d [67]. Then, we reported for the first time that in a single-crystal structure the unsymmetrical 1-galactosamide bolaamphiphile **9** (Chart 1.2) self-assembles to form an unsymmetrical MLM with a head-to-tail interface (Fig. 1.16c) [68, 69]. Meanwhile, the 1-glucosamide bolaamphiphile **10** (Chart 1.2) forms a symmetrical MLM in a crystal structure [69]. However, it remains still challenging to selectively form unsymmetrical MLMs with a head-to-tail stack.

1.7.2.3 Monolayer-Based Nanotube Compared with Bilayer-Based One

In nature, the cell membranes of eubacteria and archaea are composed of totally different bilayer- and monolayer-based phospholipid membranes, respectively. For example, the former bilayer membrane is self-assembled from the phospholipid that has aliphatic hydrocarbons with phosphoric acid via an ester linkage (Fig. 1.17a). Meanwhile, the archaea construct monolayer-based membranes comprising a large macrocyclic phospholipid with an ether linkage. This unique structural feature serves to keep biogenic activity of the cell even in highly acidic, high-temperature, and high-pH aqueous media (Fig. 1.17b).

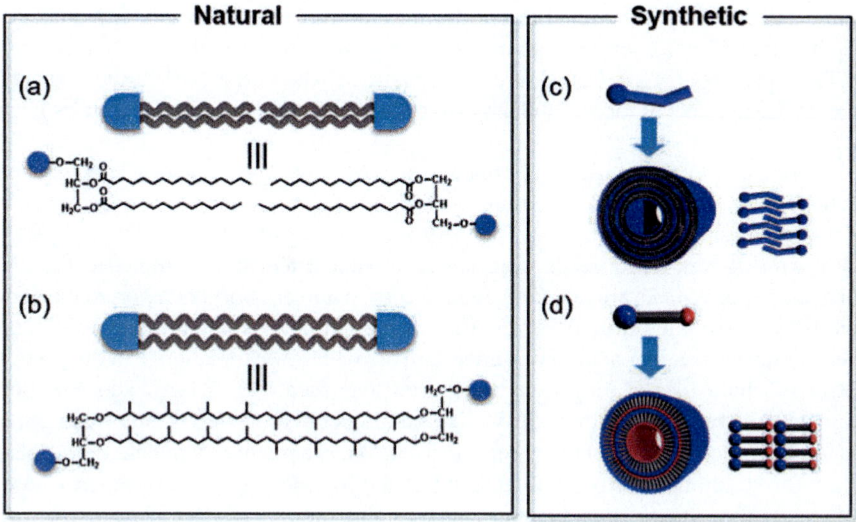

Fig. 1.17 Structures of natural lipids that constitute cell membranes in **a** eukaryotes and eubacteria and **b** archaea. Self-assembly of **c** a synthetic single-head, single-chain amphiphile into an interdigitated bilayer-based SMNT and **d** a synthetic unsymmetrical bolaamphiphile into a monolayer-based counterpart

In a synthetic system, the functionality of the interior and exterior surfaces differs between the bilayer- and monolayer-based nanotubes [4]. For example, it is considerably hard to separately modify both surfaces of the bilayer-based nanotubes with desired functional groups (Fig. 1.17c). Free regulation of the interior diameter of the bilayer-based nanotubes is also difficult even by optimizing the self-assembly conditions [17, 37]. This drawback eventually leads to low efficiency of passive capture for nanoparticles and biopolymers [37]. In contrast, the monolayer-based nanotubes can be utilized for the selective modification of the interior or exterior surfaces of the nanotubes (Fig. 1.17d) [3, 17]. Indeed, we were able to make only nanotube inner surfaces hydrophobic [70, 71], cationic [72, 73], or anionic [74]. Additionally, nanotube structures with very homogeneous inner diameter are formed by rationally designing the structure of bolaamphiphiles. The obtained monolayer-based nanotubes with distinct inner and outer surface properties resulted in the highly efficient encapsulation of diverse nanostructures and biomolecules [4].

1.7.2.4 Strategy and Tactics for Unsymmetrical Monolayer Lipid Membrane Formation

As already mentioned in Sect. 1.1, 2130 pieces of identical TMV proteins as building blocks undergo the self-assembly to form the hollow tubular structure of TMV [75]. The molecule of TMV protein comprising 158 amino acid residues has a wedge shape and helically assembled to construct a hollow cylinder structure. Notably, a single-stranded RNA serves as a long core template to stabilize the helical assembly of TMV proteins (Fig. 1.18). Not only the wedge shape of the TMV protein but the internal hoop effect of the RNA chain evokes some clues to construct nanotubes from bolaamphiphiles.

Learning from this structural concept of TMV, we formulated a strategy and tactics for the formation of unsymmetrical MLMs from rationally designed bolaamphiphiles, as shown in Fig. 1.19. The utilization of unsymmetrical bolaamphiphiles with two different-sized headgroups should be a shortcut to unsymmetrical MLM formation [74, 76]. As already described, Fuhrhop and co-workers reported the self-assembly of nanotube structures from the unsymmetrical bolaamphiphile **5**, although the detailed molecular packing was uncertain [43]. First, we took notice of the preservation of obtained *polymorph* via head-to-tail interface (Fig. 1.19a). This situation leads to the construction of multi-walled nanotubes with different inner and outer surfaces. Immobilization of unsymmetrical MLMs by a polymer template should also allow for additional head-to-tail stacking of the MLMs (Fig. 1.19b). Furthermore, we made other strategies and tactics for the stabilization and selective construction of unsymmetrical MLMs (Fig. 1.19c–g-2). The first tactic is to include diglycine (Gly–Gly) or triglycine residues (Gly–Gly–Gly), which is able to form multiple hydrogen bonds, in the vicinity of the terminal hydrophilic headgroup (Fig. 1.19c). In some instances, both nanotube formation and highly efficient loading of an anti-cancer metallo-drug occur concurrently through *endo*-complexation of the terminal carboxylate group with the drug (Fig. 1.19d). Binary co-assembly of triglycine-appended unsymmetrical bolaamphiphiles with a fluorescent probe- or hydrophobic group-containing analogue enables partial immobilization of the characteristic functionalities on the internal surfaces (Fig. 1.19e). In addition, two-step self-assembly

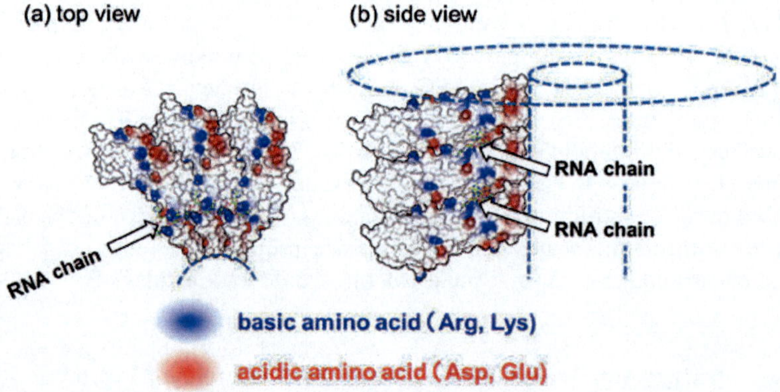

Fig. 1.18 Electrostatic potential of the atomic model based on the cryo-TEM structure of TMV. Identical wedge-shaped protein building blocks self-assemble to form TMV that is stabilized by backing with a single RNA chain as a template. Reproduced with permission from Ref. [2], © 2007 Elsevier Ltd.

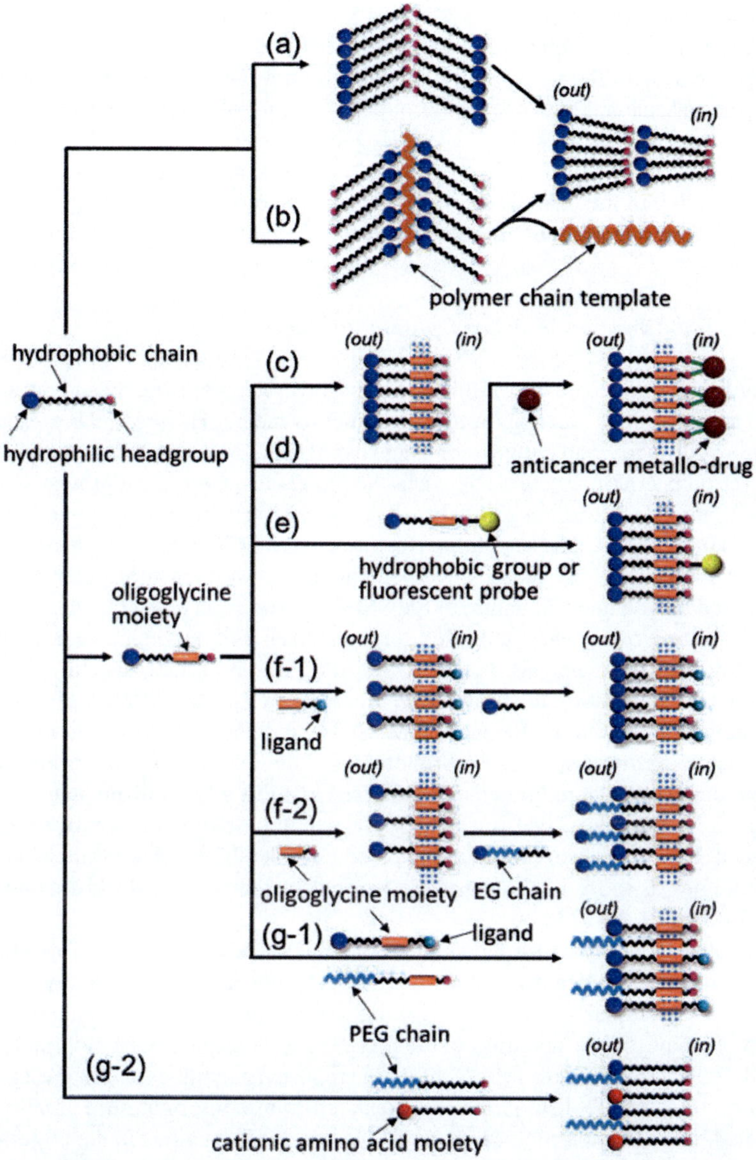

Fig. 1.19 Selective formation of unsymmetrical MLMs from unsymmetrical bolaamphiphiles based on various strategies and tactics. **a** *Polymorph* control, **b** polymer template, **c** multiple hydrogen bonds, **d** *endo*-complexation, **e** binary co-assembly, **f-1**, **f-2** two-step self-assembly, and **g-1**, **g-2** ternary co-assembly. The notation *(out)* and *(in)* means outer and inner surfaces of the curved MLM, respectively. Reproduced with permission from Ref. [4], © 2016 American Chemical Society

approach with three components attains the interior localization of a ligand that can bind an anticancer metallo-drugs (Fig. 1.19f–1). Two of the three components are associated with the networks of hydrogen bonds. With the similar strategy, we were able to anchor a glucose headgroup through an oligo(ethylene glycol) chain onto the external surfaces (Fig. 1.19f-2). Finally, ternary co-assembly of unsymmetrical bolaamphiphiles with other two analogues carrying different functional groups resulted in the formation of notable unsymmetrical MLMs, thereby, leading to controlled localization of additional two functionalities (Fig. 1.19g-1 and g-2).

1.7.2.5 Identification of *Polymorph* and *Polytype*

It is hardly possible to elucidate full details of the molecular packing and orientation within the MLMs of bolaamphiphiles in aqueous media. However, the presence of multiple hydrogen-bond networks through a single MLM of the bolaamphiphile can stabilize the lateral interactions between the molecules and/or headgroups. If that is the case, we can freeze the resultant self-assembly in a high vacuum and store as a dried nanomaterial. The crystalline feature of a SMNT in aqueous media should preserve the orientation and arrangement of the constituent molecules even in a lyophilized state. The solid nature of the SMNT is also in an advantageous condition to use various experimental tools for solid analyses. For example, we can employ single-crystal X-ray analysis, powder X-ray diffraction (XRD), and infrared spectroscopy (IR) in addition to atomic force microscopy (AFM), SEM, and cryo-TEM under aqueous conditions. Indeed, AFM, SEM, and TEM studies have manifested that in water, the dimensions and morphologies of the self-assembled nanotubes from unsymmetrical bolaamphiphiles are preserved even in a lyophilized state [74, 77]. Additionally, the cryo-TEM images for the nanotube hydrogel from a triglycine-appended glucose bolaamphiphile displayed completely the same dimensions and morphologies as those of the nanotube xerogel, which were observed using SEM and TEM (Fig. 1.20) [78].

To identify the *polymorph* and *polytype* of solid MLMs within the self-assemblies, we adopted the analytical tools (TEM, IR, XRD, and single-crystal X-ray analysis) and criteria as summarized in Table 1.1. First, when examining bolaamphiphile-based MLMs, we have to pay attention to the presence of U-shaped bolaamphiphiles [34, 79, 80]. IR spectroscopy measurement for the reported nanotubes herein revealed that all of the constituent bolaamphiphiles take an all-trans conformation. These findings mean that the bolaamphiphiles have no U-shaped conformations in the oligomethylene spacer chains. The $\nu_{as}(CH_2)$ and $\nu_s(CH_2)$ stretching vibration bands for the oligomethylene chains appear around 2916–2928 and 2848–2860 cm^{-1}, respectively. Upon heating to temperatures above the $T_{g\text{-}1}$ of the oligomethylene chain, the ratio of gauche/trans conformers increases in the spacer chains. Then, the $\nu_{as}(CH_2)$ and $\nu_s(CH_2)$ band frequencies shift from 2916 to 2923 cm^{-1} and from 2848 to 2853 cm^{-1}, respectively [81–84]. No major or even shoulder peaks for the $\nu_{as}(CH_2)$ and $\nu_s(CH_2)$ IR bands have been appeared at relatively higher frequencies ascribable to a gauche conformation [74, 77].

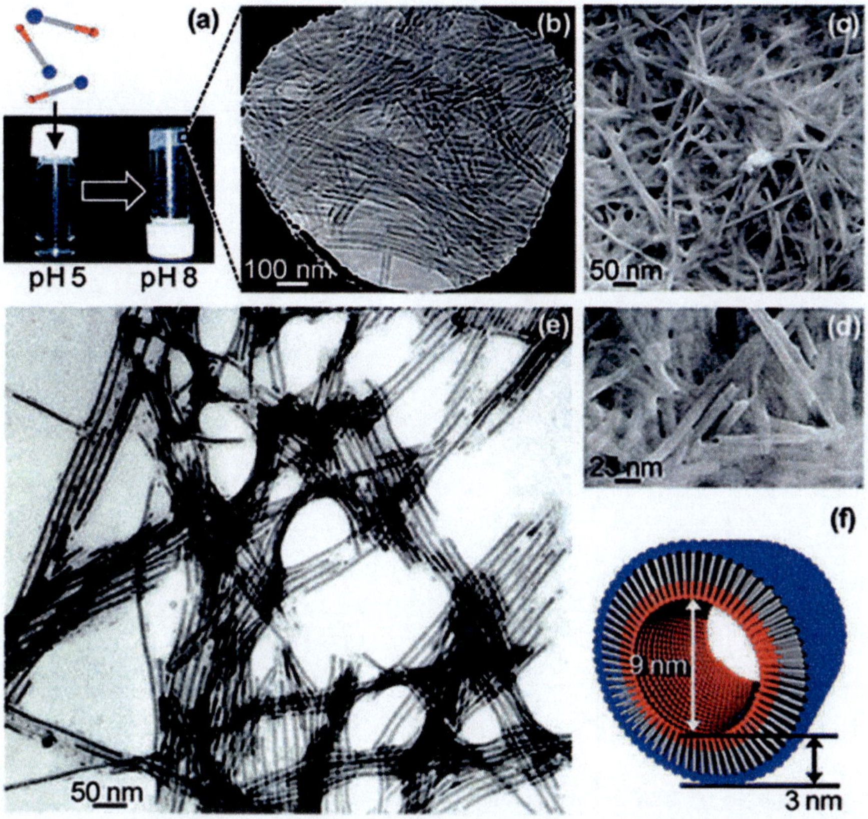

Fig. 1.20 **a** Photograph of the nanotube hydrogel formation that depends on pH condition. **b** Cryo-TEM and **c**, **d** SEM images of the nanotube hydrogel and xerogel, respectively. **e** TEM image of the nanotubes dispersed from the hydrogel. **f** Monolayer-based nanotube self-assembled from an unsymmetrical bolaamphiphile. Reproduced with permission from Ref. [78], © 2009 American Chemical Society

IR absorption bands of dried self-assemblies also gave a decisive proof to confirm a symmetrical or unsymmetrical MLM within the nanotube walls. Figure 1.21 shows typical subcell structures of saturated hydrocarbon chains. For instance, the triclinic ($T_{//}$) and orthorhombic ($O_{\perp}$) subcell structures of an oligomethylene spacer are strongly associated with the unsymmetrical and symmetrical MLMs of the corresponding membranes, respectively. Indeed, the single-crystal structures of unsymmetrical bolaamphiphiles **9** and **10** form an unsymmetrical and symmetrical MLMs, respectively (Fig. 1.22) [68, 69]. The appearance of two separate $\delta(CH_2)$ scissoring bands at around 1463 and 1473 cm^{-1} is consistent with an orthorhombic ($O_{\perp}$) subcell structure, whereas a single sharp peak at around 1464 cm^{-1} of the $\delta(CH_2)$ band is indicative of a triclinic ($T_{//}$) type [85, 86]. A single sharp peak of the $\gamma(CH_2)$ rocking vibration at 719 cm^{-1} also suggests the triclinic ($T_{//}$) subcell structure [85, 86].

Table 1.1 Experimental tools and various criteria utilized for the identification of *polymorph* and *polytype* of solid MLMs

		Polymorph and polytype	TEM	IR $\nu_{as}(CH_2)^a$ $\nu_s(CH_2)^b$ (cm^{-1})	IR $\delta(CH_2)^c$ (cm^{-1})	IR $\gamma(CH_2)^d$ (cm^{-1})	IR CH def.[e] CH skl.[f] (cm^{-1})	XRD	Single cryst. analysis
Ref.				[74, 77, 81–84]	[85, 86]	[85, 86]	[87–91]	[74, 77]	[65, 66, 68]
Bilayer of U-shaped molecules			$t^h \sim L^i$	2923–2928[a] 2853–2860[b]					–
Single monolayer	(a)[g]			2916– 2920[a] 2848–2850[b]	~1464 (single) $T_{//}{}^j$	719 (single) $T_{//}{}^j$	1420[e] 1026[f]	$d^l \sim L^i$ $d^l \leq L^i$	–
	(b)[g]				1460 and 1473 (double) $O_\perp{}^k$			$d^l \geq L^i$	–

(continued)

Table 1.1 (continued)

		Polymorph and polytype	TEM	IR $\nu_{as}(CH_2)^a$ $\nu_s(CH_2)^b$ (cm^{-1})	IR $\delta(CH_2)^c$ (cm^{-1})	IR $\gamma(CH_2)^d$ (cm^{-1})	IR CH def.[e] CH skl.[f] (cm^{-1})	XRD	Single cryst. analysis
Multiple monolayers	(c)[g]		$t^h \gg L^i$	2916–2920 2848–2850	~1464 (single) $T_{//}^j$	719 (single) $T_{//}^j$		$d^l \ll L^i$	Reported
	(d)[g]							$d^l \gg L^i/2$	No report
	(e)[g]				1460 and 1473 (double) $O_{\perp}^k$				Reported

(continued)

Table 1.1 (continued)

		Polymorph and polytype	TEM	IR $\nu_{as}(CH_2)^a$ $\nu_s(CH_2)^b$ (cm^{-1})	IR $\delta(CH_2)^c$ (cm^{-1})	IR $\gamma(CH_2)^d$ (cm^{-1})	IR CH def.[e] CH skl.[f] (cm^{-1})	XRD	Single cryst. analysis
	(f)[g]								No report

Reproduced with permission from Ref. [4], © 2016 American Chemical Society
[a]Antisymmetric stretching band
[b]Symmetric stretching band
[c]Scissoring band
[d]Rocking band
[e]Deformation band
[f]Skeletal band
[g]The same numbering as in Fig. 1.16
[h]Membrane thickness
[i]Extended molecular length
[j]Triclinic
[k]Orthorhombic
[l]*d*-spacing

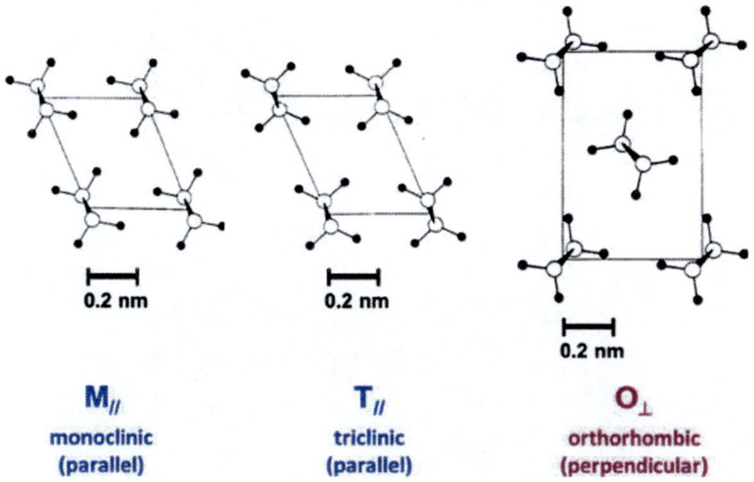

Fig. 1.21 Monoclinic ($M_{//}$), triclinic ($T_{//}$), and orthorhombic ($O_{\perp}$) subcell structures of saturated hydrocarbon chains

Furthermore, the unsymmetrical MLMs can be probed with both the CH deformation at 1420 cm^{-1} and skeletal IR bands at 1026 cm^{-1} for the oligoglycine segment in a bolaamphiphile [87–91]. Notably, the nanofibers self-assembled from unsymmetrical bolaamphiphiles with monoglycine or diglycine terminal headgroup exhibited the presence of a mixture of symmetrical and unsymmetrical MLMs [92].

Further supporting evidence is necessary to corroborate the construction of unsymmetrical nanotubes, the inner and outer surfaces of which are covered with completely different functional groups individually. Powder XRD and, if possible, X-ray crystallography give a conclusive information to discriminate the *polymorph* and *polytype* of MLMs in addition to the criteria of IR band frequencies mentioned above. By using powder XRD technique, we carried out elaborate analyses of the self-assembled nanotubes from a series of unsymmetrical bolaamphiphiles. As a result of detailed analyses, we proposed a prediction rule to differentiate symmetrical or unsymmetrical packing type within the MLMs. That is the relationship between the membrane stacking periodicity (d) of the MLMs and the extended molecular length (L) of the bolaamphiphile obtained by calculation (Fig. 1.23) [74, 77]. If the obtained d values for nanotubes are nearly equal to L or somewhat shorter than L, the MLM type should be unsymmetrical with a head-to-tail interface. Different multiple unsymmetrical MLMs with a head-to-tail interface can also be confirmed by the d values much shorter than L. If the d values are slightly larger than L, a symmetrical MLM motif with an antiparallel molecular packing should be present. When the $2d$ values are larger than L, additional unsymmetrical MLM type having a head-to-head orientation can be derived [74].

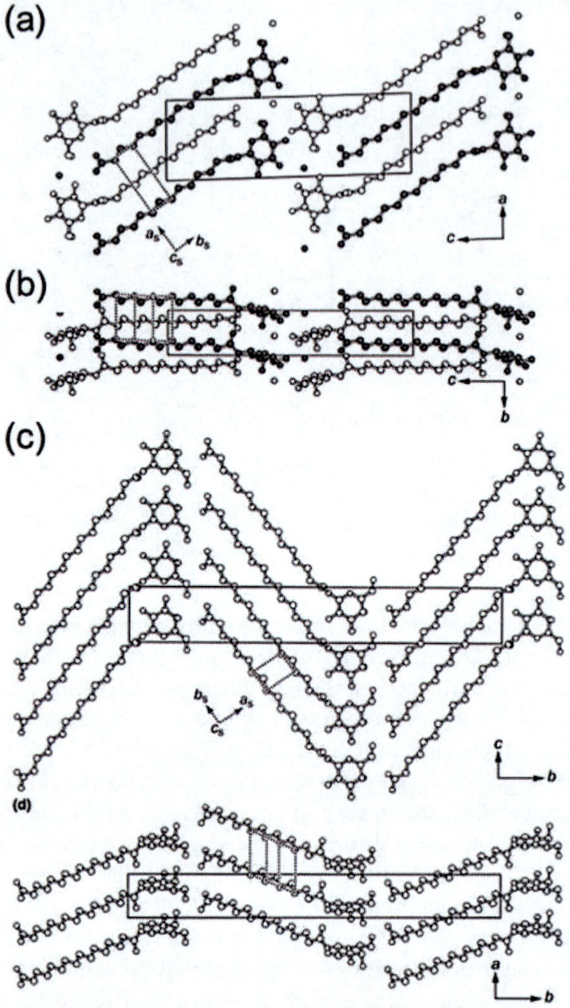

Fig. 1.22 Molecular packing of **9** viewed along **a** the *b*-axis and **b** *a*-axis and that of **10** viewed along **c** the a-axis and **d** *c*-axis. Hydrogen atoms are omitted for clarity. Rectangular frames and arrows with a_s, b_s, and c_s represent the subcells and their axes. Reproduced with permission from Ref. [69], © 2005 Elsevier Ltd.

1.7.3 β-Sheet Structure-Based

Peptides having amphiphilic property can be mainly categorized into two classes [93]. The first class is associated with purely peptide molecules that possess amphiphilic properties attributable to rationally sequenced hydrophilic and hydrophobic amino

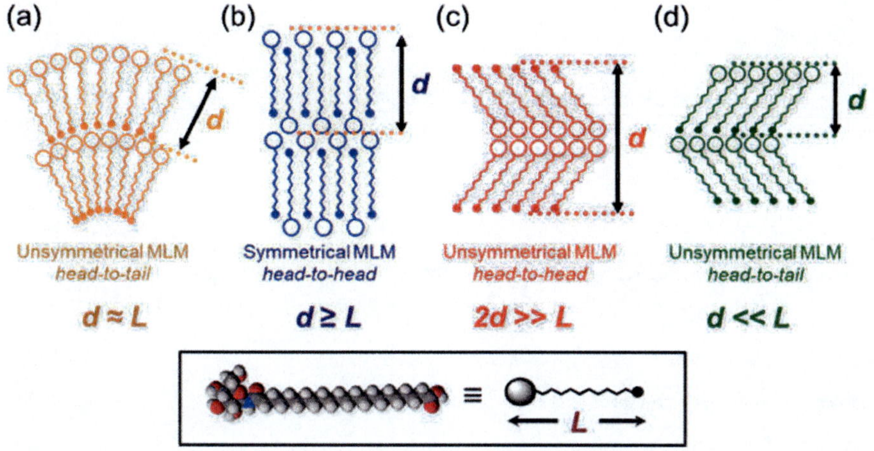

Fig. 1.23 Symmetrical or unsymmetrical packing type within the MLMs, which depend on the relationship between the membrane stacking periodicity (*d*) of the MLMs and the extended molecular length (*L*) of the bolaamphiphile

acid residues. We term herein those peptides "amphiphilic peptides (APs)". Meanwhile, a class of peptides which we term herein "peptide amphiphile (PAs)" are peptide derivatives with long nonpolar hydrocarbon chains attached. Ordered secondary structures such as a β-sheet provide the basis for constructing vesicles, micelles, fibrils, and nanotubes through self-assembly [93]. Fully matured structures of the self-assembled APs or PAs are, however, not always hollow cylindrical morphologies [94]. Few studies proposed the formation mechanism of the self-assembly of those peptide derivatives into tubular structures via helically coiled and/or twisted ribbon morphologies (Figs. 1.10c and 1.24) [57, 94–96]. The PAs **11** (Chart 1.3) having the heptapeptide segment from the amyloid β peptide (Aβ1622) self-assembles to yield the β-sheet structure-based nanotubes as a typical example [97]. The heptapeptide-based AP **12** (Chart 1.3) containing hexaalanine residues as a hydrophobic segment also self-assembles in water to form nanotubes, although the formation pathway is indefinite [98]. Fourier transform (FT)-IR and solid-state nuclear magnetic resonance (SSNMR) measurement confirmed that β-sheet structure-based bilayers of the AP **12** organize the nanotube wall [99].

1.7.4 Bile Acid Membrane-Based

Several research groups demonstrated intriguing works on the nanotube formation based on bile acid membrane, although molecular arrangement and orientation are not always clear. Benedek and co-workers reported that in the crystallization of

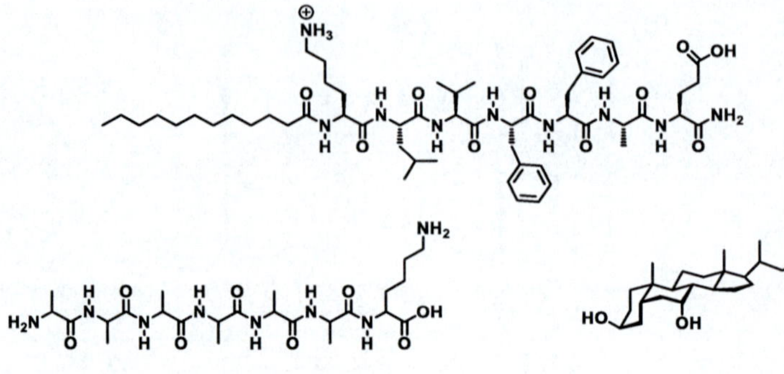

11

12

13

Chart 1.3 Chemical structures of **11–13**

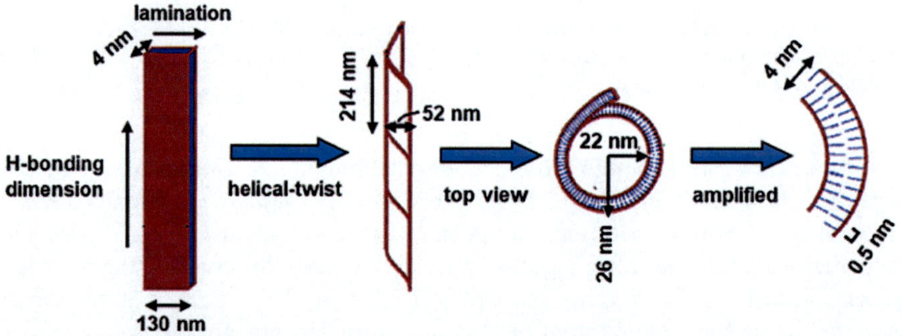

Fig. 1.24 Model for the self-assembly of β-sheet structure-based nanotube. Reproduced with permission from Ref. [96], © 2003 American Chemical Society

cholesterol a series of metastable helical microcoils and microtubular ribbon structures appeared as an intermediate morphology [56]. Collins and co-workers described the self-assembly of lithocholic acid (LCA) **13** (Chart 1.3) into organic nanotubes, which was intermediated by helical ribbon membranes comprising trimers of **13** (Figs. 1.10d and 1.25) [100]. Hydrogen bonding connects the molecules **13** in head-to-tail manner and arranges them in a fashion of alternating concave and convex orientation. The research group performed no detailed molecular packing analyses within the membranes, although the thickness of the nanotube wall is micrometeric scale. For the self-assembly of J-aggregate-based nanotubes, Fang and co-workers proposed an ion complex as a plausible building block, which was composed of a hydrophobic tail segment of LCA **13** and a hydrophilic carbocyanine dye [101].

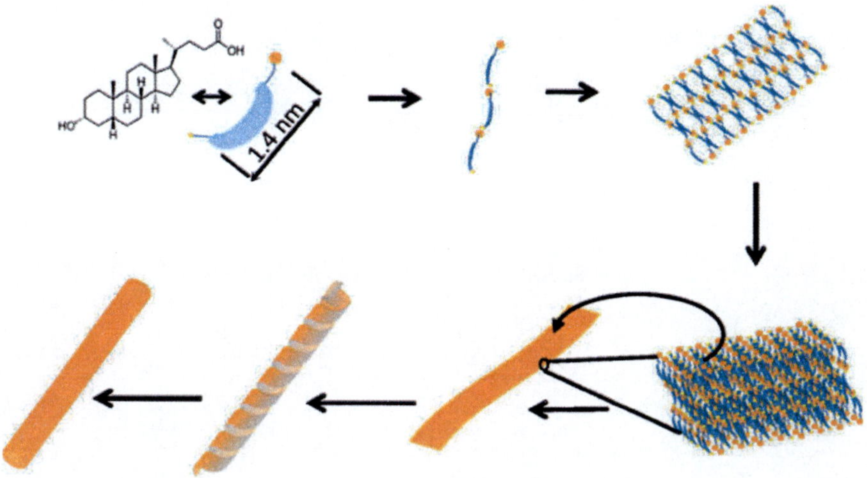

Fig. 1.25 Self-assembly of LCA **13** into SMNTs stabilized by hydrogen-bonding and hydrophobic interactions among the molecules. Reproduced with permission from Ref. [100], © 2015 Elsevier Ltd.

1.8 Nanoring- or Nanotoroid-Based

It is well-known that intrinsically cyclic compounds, e.g., cyclodextrin and cyclic peptides, have potential ability to stack themselves on top of each other. Rational molecular designing and optimization of solvent for self-assembly result in the construction of nanotube structures [25, 94, 102]. It has recently been reported that self-assembled nanorings or nanotoroid structures from noncyclic amphiphilic compounds intermediate additional self-assembly into the tubular structures (Fig. 1.10f) [103–105]. For example, the cone-shaped chiral compound **14a** and **14b** (Chart 1.4) with two alkoxyazobenzene arms (Fig. 1.26) [103], bolaamphiphilic donor/acceptor dyads **15** (Chart 1.4) [104], and amphiphiles composed of diglycine and azobenzene moieties **16** (Chart 1.4) [105] self-assemble to construct stackable nanoring (G in Fig. 1.10) or nanotoroid morphologies with homogeneous dimensions. All the generated ring-shaped intermediates uniaxially overlap each other to yield discrete organic nanotubes.

1.9 Stacking-Based

1.9.1 Cyclic Peptide

Uniaxial stacking approach of inherently ring-shaped amphiphiles or compounds gives a straightforward and rational manner for the formation of tubular structures

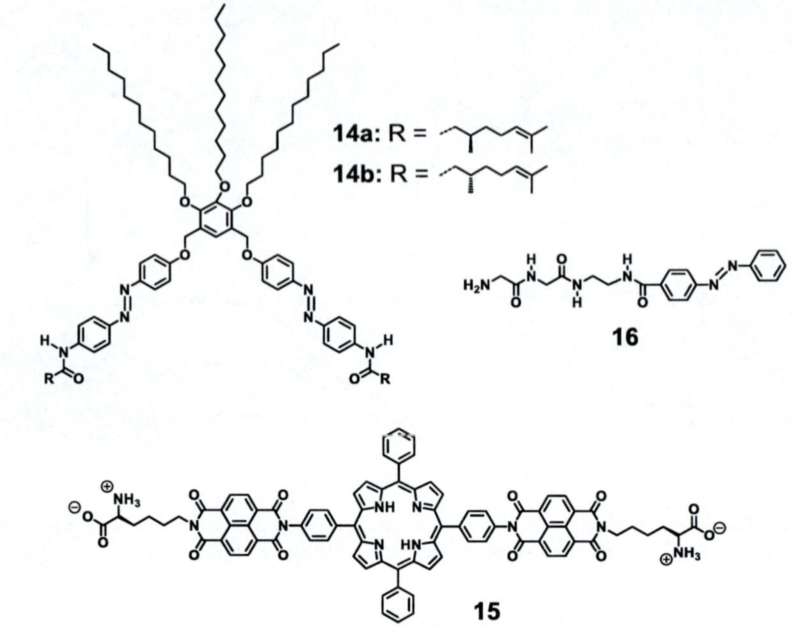

Chart 1.4 Chemical structures of **14–16**

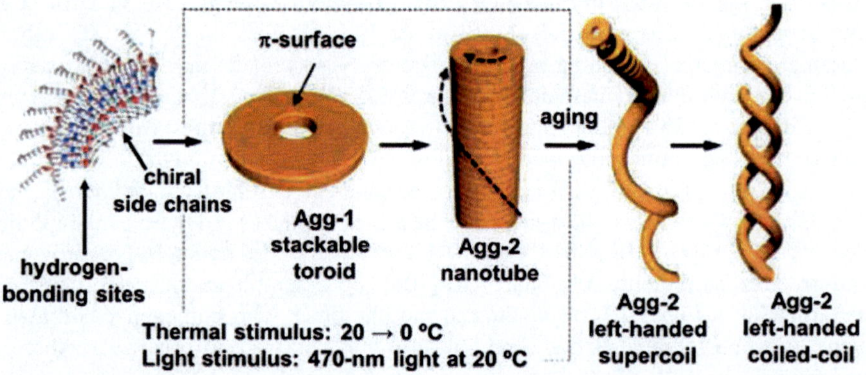

Fig. 1.26 Schematic illustration of hierarchical self-assembly of **14**. Reproduced with permission from Ref. [103], © 2012 American Chemical Society

with two open ends (Fig. 1.10g) [25]. Disulfide bond formation is known to trigger the longitudinal stacking of doughnut-like protein building block, thereby leading to the formation of stable discrete nanotubes [106, 107]. Ghadiri and co-workers carried out a pioneering work on very attractive peptide nanotubes, in which the building block of a cyclic peptide antiparallelly stacks to self-assemble. The molecular design

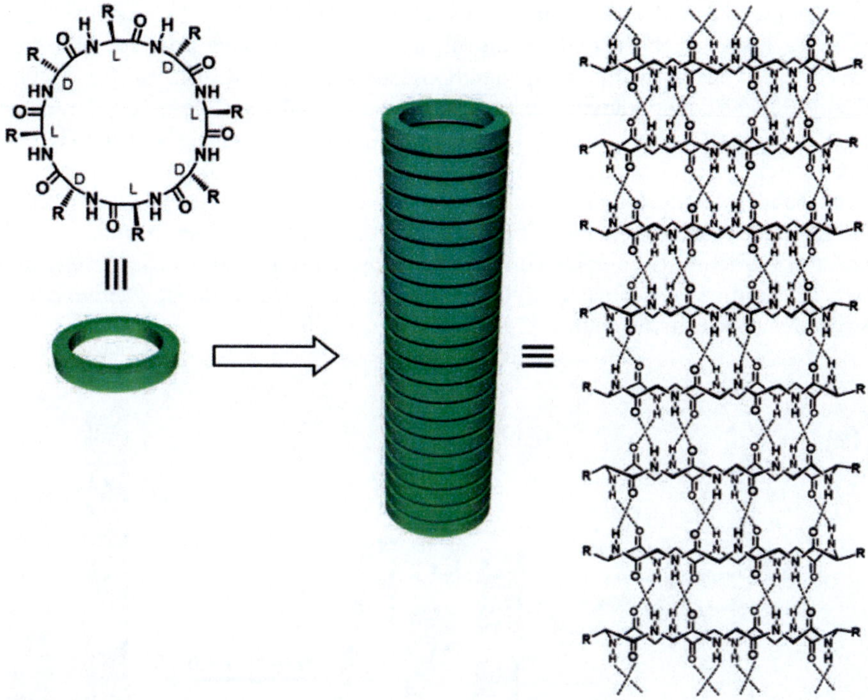

Fig. 1.27 Self-assembly of a cyclic D,L-peptide into nanotube structure. For clarity, most of the side chains of the cyclic peptide are omitted

they adopted is grounded on rationally designed peptide sequence like cyclo-(D-Ala-L-Glu-D-Ala-L-Gln)$_2$ consisting of even numbers of alternating D- and L-amino acids (Fig. 1.27) [108, 109]. The cyclic peptides with such a sequence were assumed to take a flat ring conformation with low energy. As a result of the generated configuration, all backbone amide groups are compelled to lie approximately perpendicular to the ring structure plane. An excellent review of the nanotube self-assembly from various classes of cyclic peptides discusses the design principle for functional nanotubes as well as the incorporation of exterior functionalities into the nanotube scaffold [102].

1.9.2 Helical Biomolecule

Two or more α-helical peptide chains are coiled around each other in a similar fashion as the twisted strand of a rope. A coiled-coil motif, which is a superhelical structure in protein structures, is thus formed [110]. The coiled-coil structures are not always associated with nanotube structures, for example, through self-assembly. If provided with complementary interfaces at both ends, coiled-coil lock-washer structures are, however, able to undergo the self-assembly into tubular structures with high aspect ratios due to the mutual recognition between their end surfaces (Fig. 1.10h). Conticello and collaborative research groups synthesized a well-designed, sequential peptide **7HSAP1** [111] through structurally informed mutagenesis of the **GCN4-pAA** leucine zipper peptide [112]. The obtained peptide **7HSAP1** constructs a 7-helix bundle structure that resembles the shape of a supramolecular lock washer [111]. Electrostatic interactions between glutamic acid and lysine residues, which construct structurally complementary bundle edges, promote the self-assembly into helical nanotubes as the major driving force (Fig. 1.28). No outstanding intermediate structures appear in the course of this stacking assembly.

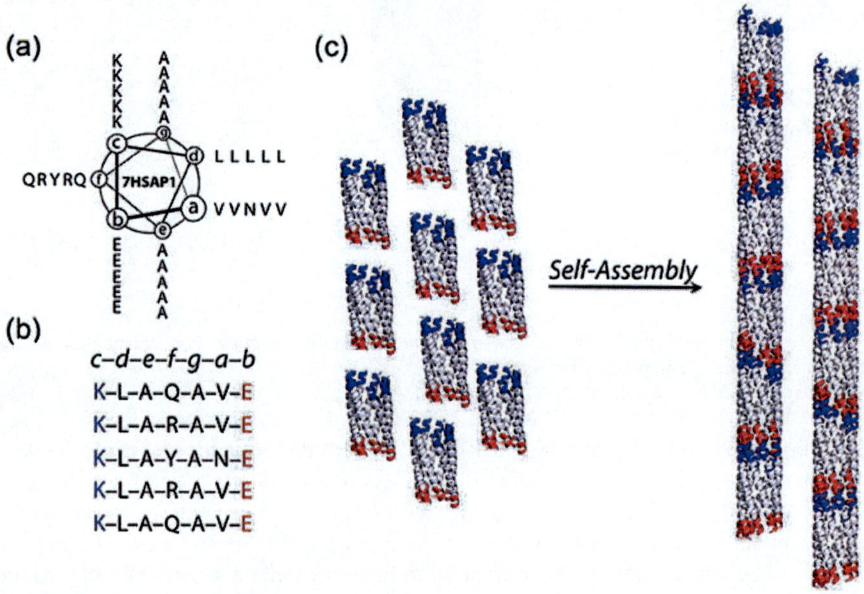

Fig. 1.28 **a** Helical wheel projection and **b** linear description of the amino acid sequence of the peptide **7HSAP1**. **c** A proposed model for the self-assembly of lock-washer structures based on **7HSAP1** into helical nanotubes. Reproduced with permission from [111], © 2013 American Chemical Society

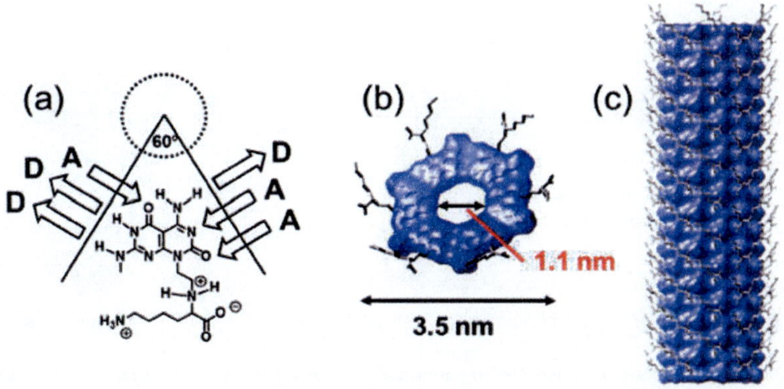

Fig. 1.29 **a** Bicyclic pyrimido[4,5-*d*]pyrimidine building block featuring the hydrogen-bond donors (D) and acceptors (A) of both guanine and cytosine. **b** Self-organization of six building blocks into a hexameric rosette. **c** A nanotube structure formed by stacking of the rosettes. Reproduced with permission from Ref. [115], © The Royal Society of Chemistry 2016

1.10 Supramolecular Stacking-Based

1.10.1 Fan-Shaped Molecule

Well-designed hydrogen-bond-mediated assembly of a fan-shaped molecule should lead to the formation of supramolecular rosette-like cyclic structures. For instance, melamine/dicarboximide complexes [113], fan-shaped dendric polymers [114], and bicyclic pyrimido-[4,5-*d*]pyrimidine [115] undergo self-organization to yield supramolecular macrocycle (rosette) structures (H in Fig. 1.10) through well-designed noncovalent intermolecular interactions. The generated rosette can further π–π stack into hollow cylindrical structures such as discrete nanofibers or nanotubes (Fig. 1.10i and 1.29) [115–117]. This self-assembly strategy into nanotube structures has a noteworthy advantage. The structures and dimensions of each fan-like molecule can precisely control the internal and external diameters of the resultant nanotubes [115]. Conversely, the disadvantage of this approach is that sophisticated attention should be paid to the molecular design in consideration of optimized selection of required functional groups and their most suitable orientation and arrangement [118].

1.10.2 Bent-Shaped Aromatic Amphiphile

Lee and co-workers have carried out pioneering work on the hierarchical assembly of bent-shaped aromatic amphiphiles into fascinating tubular nanostructures (Fig. 1.10j) [119, 120]. For example, as the first step of two successive assembly processes, bent-shaped aromatic amphiphiles self-organize into hexameric macrocycles in aqueous

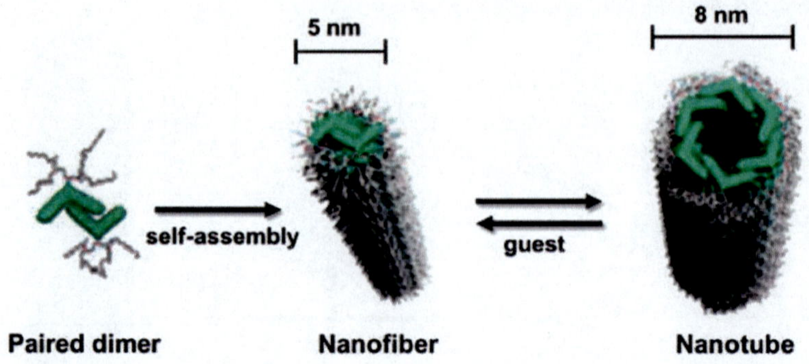

Fig. 1.30 Inflation of the self-assembled nanofiber into a hollow tubular architecture upon adding a guest molecule. Reproduced with permission from Ref. [120], © 2014 American Chemical Society

media (J in Fig. 1.10) [119]. Next, uniaxial stacking of the resultant macrocyclic units spontaneously yields chiral nanotubes as a result of the mutual rotation of the macrocycles. Representative structure of the building blocks following this scheme is composed of a bent-shaped aromatic moiety with an internal angle of 120° and hydrophilic four-armed oligoether dendron at its apex [119]. In an analogous fashion, bent-shaped aromatic rod amphiphiles bearing a *m*-pyridine moiety and a four-armed hydrophilic dendron at its apex are involved with guest-triggered nanotube formation (Fig. 1.30) [120]. The aromatic molecules firstly assemble into a pair of dimer. The resultant dimers, then, stack on top of each other to hierarchically form nanofibers through self-assembly. Addition of a guest molecule with hydrogen-bonding property induces the inflation of the nanofibers and eventually results in the formation of helical tubular structures.

1.10.3 Aromatic Macrocycle Amphiphile

Lee's research group have also explored dynamic and novel nanochannel formed by the self-assembly of aromatic macrocycle amphiphile. Interestingly, the hollow cylindrical nanochannel exhibits rapid open–closed switching in water by a thermal signal as a trigger [121]. They designed amphiphiles with a disk shape, which comprise a hydrophobic ring of cyclohexa-*m*-phenylene and a four-armed hydrophilic oligoether dendron at the periphery. The disk-shaped building blocks stack to form fibrillar morphologies as an intermediate structure during self-assembly (K in Figs. 1.10 and 1.31). The resultant fibrils then assemble laterally to construct high-aspect-ratio nanotube structures (Fig. 1.10k). The lateral aggregation of each fibril should be originated from asymmetric bilayer packings, which is generated by the mutual rotation of the aromatic discs concerning the neighboring discs [122].

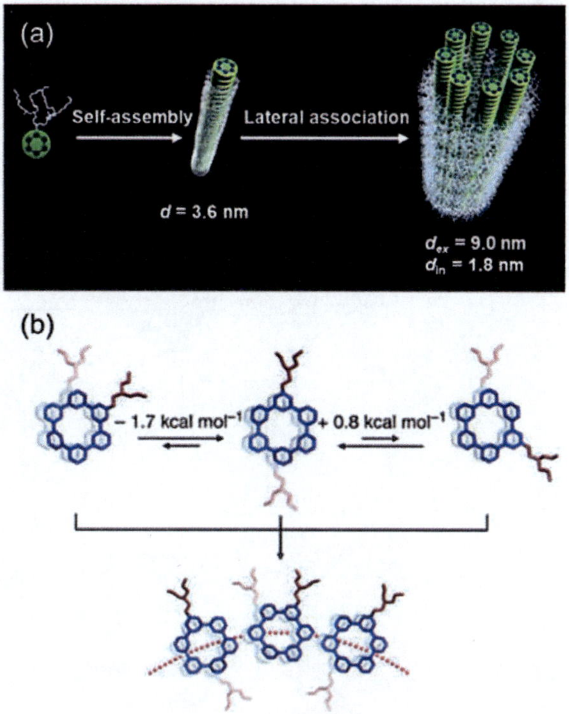

Fig. 1.31 **a** Self-assembly of a disk-shaped amphiphile into a fiber structure and subsequent lateral association of the fiber into a tubular nanostructure. **b** Lateral association-induced curvature of the nanotube wall. Density functional theory (DFT) calculation shows the low energy barrier of the mutual rotation of stacked aromatic discs. Reproduced with permission from Ref. [121], © 2015 Springer Nature

1.11 Different Fields of Action for Diverse Applications

In this book, we intend to make reference to mainly fifteen different classes of SMNTs. In the Chap. 2, we discuss mainly LNTs [5], lithocholic acid-based (LCANTs) [123], and azo derivative-based nanotubes (AZNTs) [124]. Chapter 3 addresses the formation of bolaamphiphile-based nanotubes (BANTs) [4]. Chap. 4 discusses di-phenylalanine-based nanotubes (FFNTs) [125]. Chapter 5 describes the self-assembly of peptide-based nanotubes (PNTs) [93, 94]. Chapter 6 refers to cyclic peptide-based (CPNTs) [102, 126] and cyclic peptide–polymer-based nanotubes (CPPNTs) [102, 127]. Chapter 7 mentions protein-based nanotubes (PRNTs) [18]. Chapter 8 addresses SMNTs self-assembled from bottlebrush copolymer (BBC)-based nanotubes (BBCNTs) [128] and microporous organic nanotube networks (MONNs) [129]. Finally, Chap. 9 describes rigid–flexible block molecule-based (RFBMNTs) [130–132], macrocycle-based (MCNTs), boroxine-based (BXNTs) [133], and 1,3,5-benzenetricarbonyl derivative-based nanotubes (BTNTs) [134]. Figure 1.32 represents the molecular structures of typical building blocks that can self-assemble into individual family of SMNTs.

The SMNTs display various fascinating characteristics and functions that can lead to potential applications in chemical, physical, biological, and medical fields. We have to, however, take notice of the action sites and fields (i.e., tube interior,

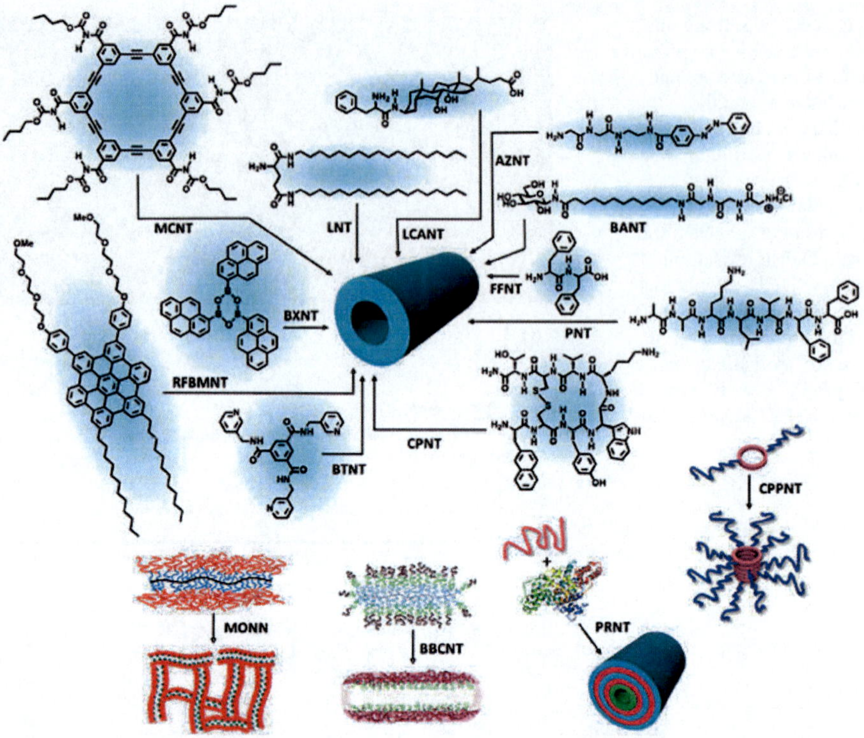

Fig. 1.32 Fifteen families of SMNTs and representative molecular structure that assemble to form each SMNT as a building block. Reproduced with permission from Ref. , © 2020 American Chemical Society [6]

exterior, wall, and whole of nanotubes), where each function and property of the nanotubes really emerge. Here, we remark four different action fields associated with the nanotube body, i.e., the (1) interior 1D nanospace, (2) nanotube wall and membrane, (3) exterior surface exposed to bulk environment, and (4) whole of nanostructures and their ensembles (Fig. 1.33). We define the fourth category as the action field, where either collective actions of the interior, wall, and exterior of the nanotubes or an ensemble of a great many nanotubes are associated with the expression of the functions. For instance, unless otherwise noted, the whole structures of the nanotubes or the nanotube ensembles are involved in the manifestation of gelation, liquid crystal formation, superhydrophobicity, detection, sensing, and the expression of physical properties. In the next **1.11.1–1.11.4**, we overview the features of the functions and potent applications that are expressed in the individual action field.

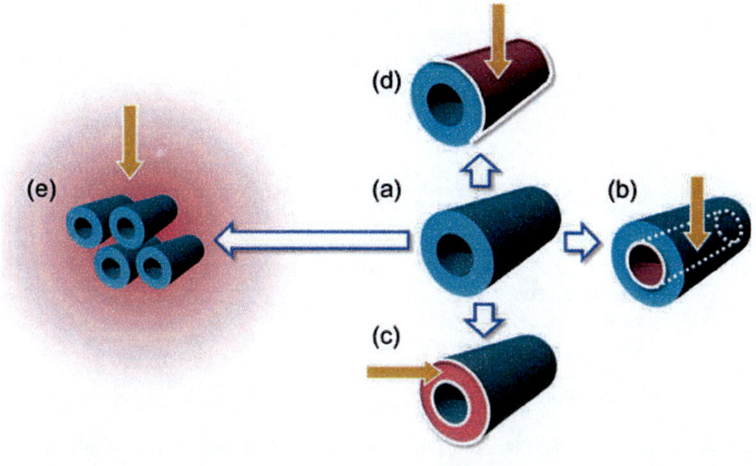

Fig. 1.33 Four different action fields represented by filled arrows, in which SMNTs are involved with the expression of their functions. **a** original nanotube, **b** interior nanospace, **c** nanotube wall, **d** exterior surface, and **e** whole of nanostructures and ensembles of many nanotubes

1.11.1 Interior One-Dimensional Nanospace

The SMNTs undoubtedly have a 1D well-defined, cylindrical tubular nanospace with a high aspect ratio in the central core part of rod or fibrous nanostructures. The SMNT with this unique hollowness, unlike pores, cavities, or holes of short length strictly differs from any other 1D nanostructures such as crystalline nanorods, rod-like micelles, self-assembled nanoribbons, nanowires, and nanofibers [19, 135–139]. The most suitable uses of the high-aspect-ratio lumens drive themselves to express unprecedented functions and properties, which any other 1D nanomaterials cannot exhibit. Figure 1.34 shows typical functions and properties that can be expressed in the interior 1D nanospaces of SMNTs. By utilizing the interior 1D nanospaces as the action field, the SMNT can encapsulate, capture, trap, and release a wide range of guest substances, e.g., low-molecular-weight molecules, synthetic polymers, proteins, metal nanoparticles, and viruses. Those unique functions directly relate to promising applications, such as sensing and separation of the substance. Electrostatic or specific host–guest interaction, capillary force, or other noncovalent interactions including hydrophobic and π–π interactions play a significant role in serving as a driving force. The SMNT thus encapsulates and sometimes holds the substances in the nanochannel. Moreover, the confinement of the encapsulated substances in the 1D nanochannel causes the stabilization of, for example, the three-dimensional (3D) structures of native proteins. Accordingly, this confinement effect results in the prevention of thermal or chemical denaturation of the proteins from heat and denaturant. The relationship between the nanotube inner diameter and the size of a guest substance has a strong effect on the ability of encapsulation, stabilization, release, refolding, and diffusion of the guest substances including proteins.

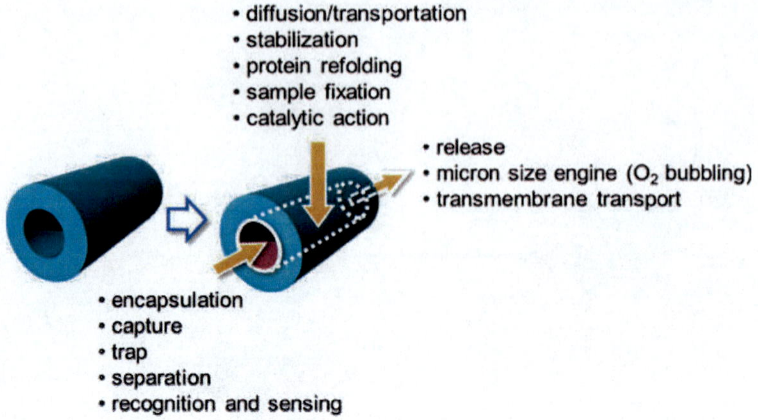

Fig. 1.34 Potent properties and functions that have ever been demonstrated mainly in the interior nanospaces of various SMNTs

Functioning as artificial molecular chaperone gives a representative example that ingeniously utilizes the nanotube channel. Optimization of interior functionalities toward the external environment of a guest protein enhances the refolding efficiency of the SMNT [71, 124]. One more remarkable advance in novel functions that appear in the interior nanospace is the employment of a glycolipid-based nanotube as a sample immobilization substrate for TEM computed tomography (CT) [140, 141]. As another example, confined nanospaces, created by metal-coordinated nanotubes [142–146] and MONNs constructed from BBC [147], partially provide an action field for some catalytic reactions. Similarly, chemical nanoreactors are also realized in the confined nanospace of some organic nanotubes [23, 148–150]. Protein-based nanotubes with internal surfaces composed of α-glucosidase or lipase B also exhibited the effective activity as an enzymatic reactor [151, 152]. Of particular note is the fact that self-propelling of the nanotube itself in water is promoted by O_2 bubbles that jets from the interior nanospace of a protein- and polymer-based nanotube (Fig. 1.35a) [153, 154]. Studies on biomimicry of transmembrane protein channels have made remarkable progress by utilizing the interior nanochannel of SMNTs. For instance, PNTs [155], CPPNTs [156], and MCNTs [157] were shown to exhibit the activity of transmembrane transport for ion, water, anticancer drugs, and dye. Meanwhile, as a novel medical application, protein-based nanotubes, the internal surfaces of which are covered with an appropriate protein, were demonstrated to trap infectious hepatitis B virus (HBV)[158] or influenza virus A PR8 (Fig. 1.35b) [159].

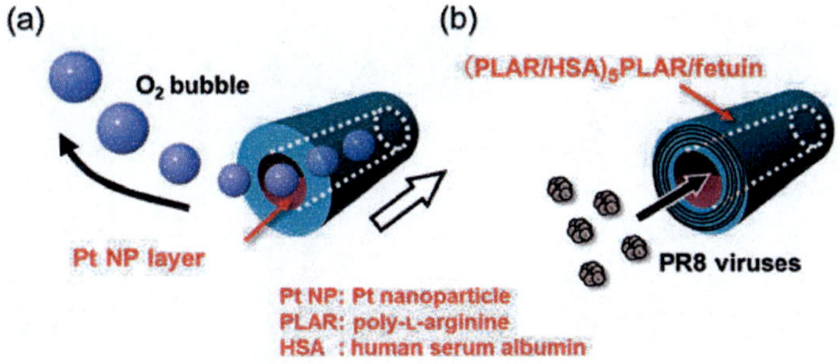

Fig. 1.35 **a** Self-propelling of a human serum albumin (HSA)–Pt nanoparticle (Pt NP) microtube, which is promoted by O_2 bubble ejection from the interior nanospace. **b** Trapping of influenza virus A PR8 into the interior nanospace

1.11.2 Nanotube Wall and Membrane

The backbone and network of cyclic peptide, copolymers, or other molecular building block organize nanotube walls, whereas lipid monolayer and bilayer, and alternatively arranged proteins nanotube membranes. Those nanotube walls and membranes construct well-defined molecular boundaries that play a role in dividing the internal nanochannels from the external environment [4, 5]. The applications associated with the nanotube walls and membranes are, thus, strongly influenced by the chemical, physical, biological, and medical functions and properties of the wall materials (Fig. 1.36). For example, interdigitated solid bilayer membranes, which constitute an organic nanotube, function as an organic scaffold that can immobilize hydrophobic guest molecules such as 8-anilinonaphthalene-1-sulfonate and Zn-phthalocyanine [160]. Interestingly, temperature-dependent release behavior displayed different results for encapsulated and immobilized guest substances in the nanotube channel and membrane wall, respectively [160]. Alternating protein layers of a solid protein-based nanotube serve as an ideal scaffold that can embed gold nanoparticles (Au NPs) and iron(III) oxide (α-Fe_2O_3) NPs within them [161, 162]. Template method using a layer-by-layer (LbL) lamination of more than two different proteins, therefore, allows for the easy and controlled incorporation of desired proteins and functional NPs into the nanotube walls.

Fig. 1.36 Potent properties and functions that have ever been demonstrated mainly in the nanotube walls and membranes of various SMNTs

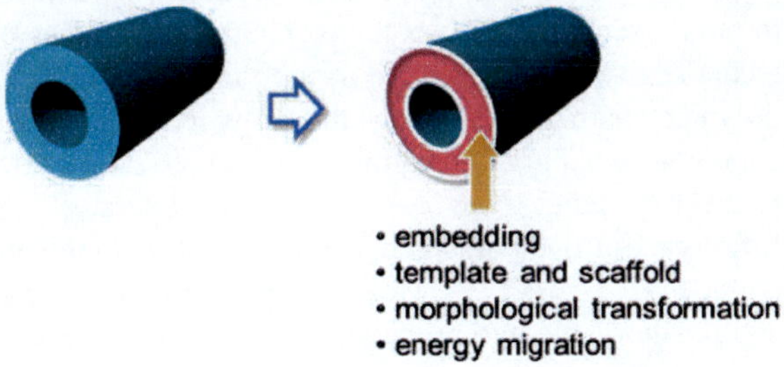

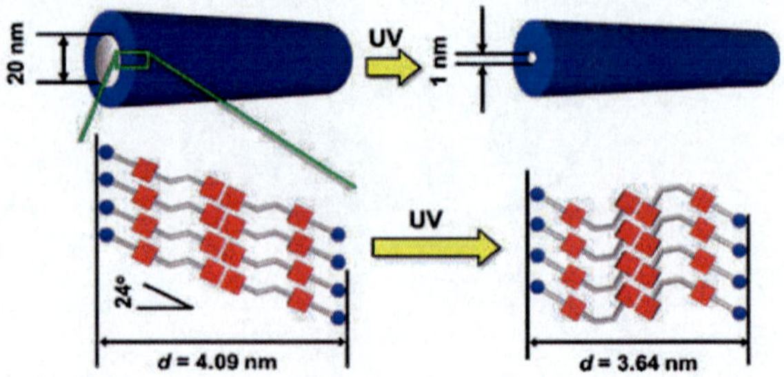

Fig. 1.37 Photoresponsive shrinkage of the hollow cylinder of a SMNT by UV light

Stimuli-responsive and drastic morphological change in SMNTs improves the quality of their drug delivery function. In some cases, the dynamic feature of the nanotube morphology can add an additional value to the intrinsic function of the nanotubes. As an example, UV photoirradiation onto a nanotube was shown to cause dynamic shrinkage of the hollow cylindrical shape (Fig. 1.37) [105, 163]. The trans–cis photoisomerization of the azobenzene unit in a bilayer-forming amphiphile is responsible for the drastic morphological transformation in this case. The change in molecular packing of macrocyclic discs and aromatic cores also trigger reversible open—closed motion of a nanotube wall [119, 121] and guest-driven inflation of the wall, respectively [120]. Meanwhile, the walls of n- and p-type nanotubes [164–166] and donor (D)–acceptor (A) nanotubes [104, 167] play a vital role in the occurrence of energy migration in physical applications.

1.11.3 Exterior Surface

As a matter of course, all the well-defined 1D nanostructures including tubular, ribbon-like, rod-like, and wire-like self-assemblies furnish with external surfaces. The surfaces are covered or partially modified with some functionalities, exposing to external spaces and environment. The properties of the external surfaces of the nanotubes play an important role in improving the quality of the nanotube characteristics (Fig. 1.38). In this context, selective detection and recognition of low-molecular-weight substances [168], heavy metal ions [169], proteins [170], and viruses [171] can be demonstrated on the nanotube exterior surfaces. Antigen–antibody interaction, host–guest interaction, chirality matching, and so on function as driving forces to target, sense, or deliver them. The outer surfaces of the nanotubes also serve as a chemically active template for polymerization of a functional monomer. A single polymer chain generated by the polymerization sometimes propagates along the nanotube axis [172]. Free radical polymerization creates another type of polymer chain that shapes a nanotube shell. The generated polymer chain by this divergent (grafting-from) approach covalently connects to the nanotube core based on a cyclic peptide [102, 173–175]. Meanwhile, a nanotube exterior serves not only as an interaction site of the pyrene paddle of CPNTs to prepare dual composites comprising

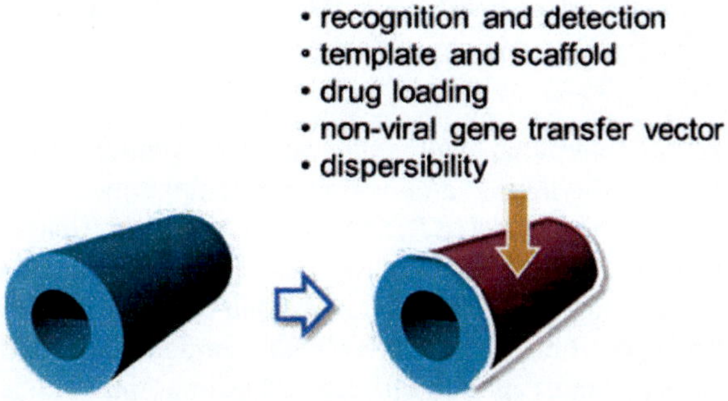

Fig. 1.38 Potent properties and functions that have ever been demonstrated mainly on the exterior surface of various SMNTs

CNTs and CPNTs [176] but also as a nanotube template for the immobilization of Au NPs [177]. With regards to medical applications, the external surfaces of SMNTs play an important role as scaffolds that assist sensing of endogenous opioid peptides [178], trapping of cells [179, 180], as well as appending of anticancer drugs [181] and genes (Fig. 1.39) [182–185]. Commonly, drug or gene nanocarriers, the exterior of which is covered with poly(ethylene glycol)s (PEGs), display an enhanced in vitro and in vivo stability and stability [19].

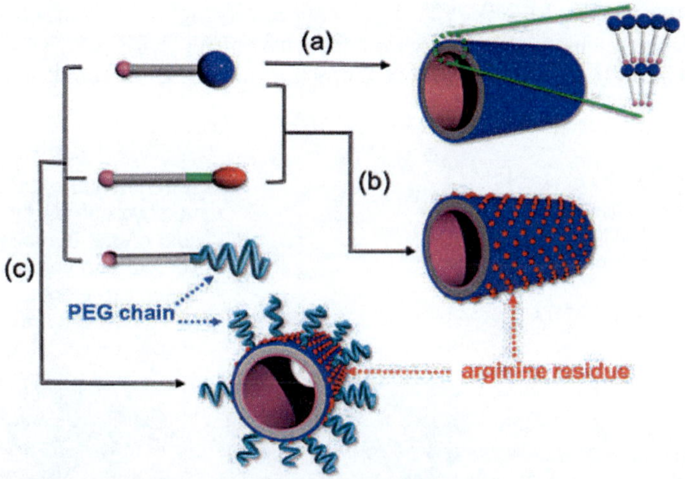

Fig. 1.39 Formation of exterior-functionalized nanotubes as nonviral gene transfer vector through **a** self-assembly of a mother bolaamphiphile, **b** binary co-assembly of the mother and arginine-appended bolaamphiphiles, and **c** ternary co-assembly of the mother, arginine-appended, and PEG-attended bolaamphiphiles

1.11.4 *Whole of Nanostructure and Ensemble of Many Nanotubes*

A well-prepared technique for the synthesis of MONNs is involved with hyper-crosslinking and following templating of three-arm branched core–shell BBC molecules [147]. Notably, three different-sized pores, i.e., micropores, meso/macropores, and mesopores can be constructed through intrabrush, interbrush crosslinking of the polymer shell, and selective removal of the polymer core, respectively. The obtained MONNs by the similar approach display an excellent adsorption ability of dyes and proteins as well as the separation ability of dyes [186]. However, nobody will be able to identify the action sites of those functions due to the existence of the hybrid pores. Namely, the pores are all implicated in the realization of their specific functions (Fig. 1.40). Ultrasensitive environmental sensing gives another example, in which whole of nanotube structures and their assemblies contribute to express a unique function. The highest sensitivity for the detection of phenol was achieved by using the functional electrode, the surface area of which was highly enhanced by coating with vertically arrayed PNTs [125, 187]. Superhydrophobicity was also realized by applying vertically aligned PNT film and peptide nanowires [188, 189]. SMNTs with high aspect ratios are also associated with the creation of different types of LC phases, e.g., hexagonal [190], columnar [191], nematic [98], and lyotropic phases [20, 170, 192]. As a hierarchically higher-ordered self-assembly system, some of nanotube structures can form hydrogel [71, 78, 170, 193–196] and organogel [134, 168, 195, 197–201] in a similar manner as self-assembled nanofibers do.

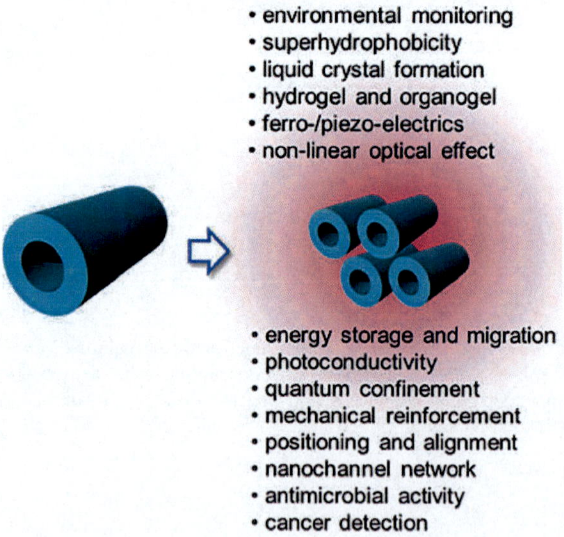

Fig. 1.40 Potent properties and functions that have ever been demonstrated by applying the whole of nanostructures and ensembles of many SMNTs

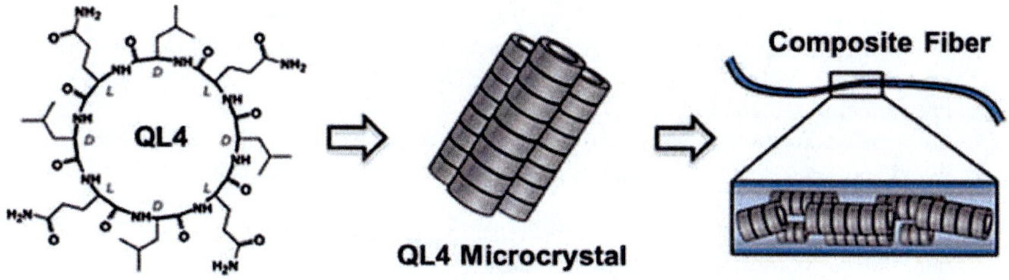

Fig. 1.41 Mechanical reinforcement of polymer fibers by incorporating a synthetic cyclic peptide (QL4) nanotube bundles as a structural filler. Reproduced with permission from [219], © 2013 American Chemical Society

Interestingly, the whole structures of FFNTs are concerned with the display of a variety of physical functions, e.g., ferroelectric [202, 203], piezoelectric [204–208], nonlinear optical effect [207, 209–211], energy storage [212, 213], energy migration [164–167], photoconductivity [214, 215], and quantum confinement properties [206, 216, 217]. Mechanical reinforcement of composite materials with SMNTs or their assemblies as incorporating fillers [218–220], fiber mats [196], or additives [221] gives other examples that demonstrate the utilization of whole nanotube structures (Fig. 1.41). In this context, it is essential to measure the mechanical properties of each nanotube structure, i.e., elasticity, rigidity, and stiffness in a liquid medium or in a dried state [222–224]. Each species of the whole nanotube structure is subjected to various manipulation techniques including high-frequency alternating current electric field [225, 226], vapor deposition [188, 227], micro-extrusion [228–230], and inkjet printing [231, 232].

Biological applications that utilize nanotube-based nanochannels deal with the whole structures of the nanotubes, especially both the interior and exterior functionalities [233–236]. Controlling nanotube lengths, angles between related nanotubes, and connectivity of each organic nanotube and spherical vesicles enables the creation of a variety of nanotube- and vesicle-based network systems [237–240]. Meanwhile, in a few membrane permeation modes involving from a single to multiple pieces of the CPNT, the whole structures strongly affect the antimicrobial activity on gram-positive and negative bacterial [241, 242]. As a medical application, a graphene electrode functionalized with a newly prepared PNT–folic acid composite enables high detection of human cancer cells [243]. Rational immobilization of the whole nanotube composites on the graphene electrode holds the key to perform this electrical function.

References

1. Hayden O, Nielsch K (2011) Molecular- and nano-tubes. Springer, Heidelberg. https://doi.org/10.1007/978-1-4419-9443-1
2. Sachse C, Chen JZ, Coureux PD, Stroupe ME, Fandrich M, Grigorieff N (2007) High-resolution electron microscopy of helical specimens: a fresh look at Tobacco mosaic virus. J

Mol Biol 371:812–835. https://doi.org/10.1016/j.jmb.2007.05.088
3. Shimizu T, Minamikawa H, Kogiso M, Aoyagi M, Kameta N, Ding W, Masuda M (2014) Self-organized nanotube materials and their application in bioengineering. Polym J 46:831–858. https://doi.org/10.1038/Pj.2014.72
4. Shimizu T, Kameta N, Ding W, Masuda M (2016) Supramolecular self-assembly into biofunctional soft nanotubes: from bilayers to monolayers. Langmuir 32:12242–12264. https://doi.org/10.1021/acs.langmuir.6b01632
5. Shimizu T (2018) Self-assembly of discrete organic nanotubes. Bull Chem Soc Jpn 91:623–668. https://doi.org/10.1246/bcsj.20170424
6. Shimizu T, Ding W, Kameta N (2020) Soft-matter nanotubes: a platform for diverse functions and applications. Chem Rev 120:2347–2407. https://doi.org/10.1021/acs.chemrev.9b00509
7. Meyer F, Raquez JM, Verge P, de Arenaza IM, Coto B, Van Der Voort P, Meaurio E, Dervaux B, Sarasua JR, Du Prez F, Dubois P (2011) Poly(ethylene oxide)-b-poly(L-lactide) Diblock copolymer/carbon nanotube-based nanocomposites: LiCl as supramolecular structure-directing agent. Biomacromol 12:4086–4094. https://doi.org/10.1021/bm201149g
8. Kitao T, Zhang Y, Kitagawa S, Wang B, Uemura T (2017) Hybridization of MOFs and polymers. Chem Soc Rev 46:3108–3133. https://doi.org/10.1039/c7cs00041c
9. Schoedel A, Li M, Li D, O'Keeffe M, Yaghi OM (2016) Structures of metal-organic frameworks with rod secondary building units. Chem Rev 116:12466–12535. https://doi.org/10.1021/acs.chemrev.6b00346
10. Diercks CS, Yaghi OM (2017) The atom, the molecule, and the covalent organic framework. Science **355**, eaal1585. https://doi.org/10.1126/science.aal1585
11. Lohse MS, Bein T (2018) Covalent organic frameworks: structures, synthesis, and applications. Adv Funct Mater 28:1705553. https://doi.org/10.1002/adfm.201705553
12. Dusselier M, Davis ME (2018) Small-pore zeolites: synthesis and catalysis. Chem Rev 118:5265–5329. https://doi.org/10.1021/acs.chemrev.7b00738
13. Snyder BER, Bols ML, Schoonheydt RA, Sels BF, Solomon EI (2018) Iron and copper active sites in zeolites and their correlation to metalloenzymes. Chem Rev 118:2718–2768. https://doi.org/10.1021/acs.chemrev.7b00344
14. Benzigar MR, Talapaneni SN, Joseph S, Ramadass K, Singh G, Scaranto J, Ravon U, Al-Bahily K, Vinu A (2018) Recent Advances in functionalized micro and mesoporous carbon materials: synthesis and applications. Chem Soc Rev 47:2680–2721. https://doi.org/10.1039/c7cs00787f
15. Wang J, Ma Q, Wang Y, Li Z, Li Z, Yuan Q (2018) New insights into the structure-performance relationships of mesoporous materials in analytical science. Chem Soc Rev 47:8766–8803. https://doi.org/10.1039/c8cs00658j
16. Shamzhy M, Opanasenko M, Concepcion P, Martinez A (2019) New trends in tailoring active sites in zeolite-based catalysts. Chem Soc Rev 48:1095–1149. https://doi.org/10.1039/c8cs00887f
17. Kameta N, Minamikawa H, Masuda M (2011) Supramolecular organic nanotubes: how to utilize the inner nanospace and the outer space. Soft Matter 7:4539–4561. https://doi.org/10.1039/c0sm01559h
18. Komatsu T (2012) Protein-based nanotubes for biomedical applications. nanoscale 4:1910–1918. https://doi.org/10.1039/c1nr11224d
19. Babu SS, Praveen VK, Ajayaghosh A (2014) Functional π-gelators and their applications. Chem Rev 114:1973–2129. https://doi.org/10.1021/cr400195e
20. Ding W, Minamikawa H, Kameta N, Wada M, Masuda M, Shimizu T (2015) Spontaneous nematic alignment of a lipid nanotube in aqueous solutions. Langmuir 31:1150–1154. https://doi.org/10.1021/la5042772
21. Hill JP, Jin W, Kosaka A, Fukushima T, Ichihara H, Shimomura T, Ito K, Hashizume T, Ishii N, Aida T (2004) Self-assembled hexa-*peri*-hexabenzocoronene graphitic nanotube. Science 304:1481–1483. https://doi.org/10.1126/science.1097789
22. Martin CR (1994) Nanomaterials: a membrane-based synthetic approach. Science 266:1961–1966. https://doi.org/10.1126/science.266.5193.1961

23. Shimizu T (2008) Self-assembled organic nanotubes: toward attoliter chemistry. J Polym Sci Part A: Polym Chem 46:2601–2611. https://doi.org/10.1002/Pola.22652
24. Harada A, Li J, Kamachi M (1993) Synthesis of a tubular polymer from threaded cyclodextrins. Nature 364:516–518. https://doi.org/10.1038/364516a0
25. Bong DT, Clark TD, Granja JR, Ghadiri MR (2001) Self-assembling organic nanotubes. Angew Chem Int Ed 40:988–1011. https://doi.org/10.1002/1521-3773(20010316)40:6%3c988::AID-ANIE9880%3e3.0.CO;2-N
26. Remskar M, Mrzel A, Skraba Z, Jesih A, Ceh M, Demsar J, Stadelmann P, Levy F, Mihailovic D (2001) Self-assembly of subnanometer-diameter single-wall MoS_2 nanotubes. Science 292:479–481. https://doi.org/10.1126/science.1059011
27. Amara MS, Paineau E, Rouziere S, Guiose B, Krapf MEM, Tache O, Launois P, Thill A (2015) Hybrid, tunable-diameter, metal oxide nanotubes for trapping of organic molecules. Chem Mater 27:1488–1494. https://doi.org/10.1021/cm503428q
28. Ricci MA, Bruni F, Gallo P, Rovere M, Soper AK (2000) Water in confined geometries: experiments and simulations. J Phys Condens Matter 12:A345–A350. https://doi.org/10.1088/0953-8984/12/8a/346
29. Scodinu A, Fourkas JT (2002) Comparison of the orientational dynamics of water confined in hydrophobic and hydrophilic nanopores. J Phys Chem B 106:10292–10295. https://doi.org/10.1021/jp0263491
30. Ito T, Sun L, Henriquez RR, Crooks RM (2004) A carbon nanotube-based Coulter nanoparticle counter. Acc Chem Res 37:937–945. https://doi.org/10.1021/ar040108+
31. Goldberger J, Fan R, Yang PD (2006) Inorganic nanotubes: a novel platform for nanofluidics. Acc Chem Res 39:239–248. https://doi.org/10.1021/ar040274h
32. Rondelez Y, Tresset G, Tabata KV, Arata H, Fujita H, Takeuchi S, Noji H (2005) Microfabricated arrays of femtoliter chambers allow single molecule enzymology. Nat Biotechnol 23:361–365. https://doi.org/10.1038/nbt1072
33. Kunitake T, Okahata Y (1977) A totally synthetic bilayer membrane. J Am Chem Soc 99:3860–3861. https://doi.org/10.1021/ja00453a066
34. Fuhrhop JH, Koening J (1994) Membranes and molecular assemblies: the synkinetic approach. Monographs in supramolecular chemistry. The Royal Society of Chemistry, Cambridge. https://doi.org/10.1039/9781847551368
35. Kunitake T (1992) Synthetic bilayer membrane: molecular design, self-organization, and application. Angew Chem Int Ed 31:709–726. https://doi.org/10.1002/anie.199207091
36. Fuhrhop JH, Helfrich W (1993) Fluid and solid fibers made of lipid molecular bilayers. Chem Rev 93:1565–1582. https://doi.org/10.1021/cr00020a008
37. Shimizu T, Masuda M, Minamikawa H (2005) Supramolecular nanotube architectures based on amphiphilic molecules. Chem Rev 105:1401–1443. https://doi.org/10.1021/cr030072j
38. Barclay TG, Constantopoulos K, Matisons J (2014) Nanotubes self-assembled from amphiphilic molecules via helical intermediates. Chem Rev 114:10217–10291. https://doi.org/10.1021/Cr400085m
39. Nakashima N, Asakuma S, Kim JM, Kunitake T (1984) Helical superstructures are formed from chiral ammonium bilayers. Chem Lett 13:1709–1712. https://doi.org/10.1246/cl.1984.1709
40. Nakashima N, Asakuma S, Kunitake T (1985) Optical microscopic study of helical superstructures of chiral bilayer membranes. J Am Chem Soc 107:509–510. https://doi.org/10.1021/ja00288a043
41. Yamada K, Ihara H, Ide T, Fukumoto T, Hirayama C (1984) Formation of helical super structure from single-walled bilayers by amphiphiles with oligo-l-glutamic acid-head group. Chem Lett 13:1713–1716. https://doi.org/10.1246/cl.1984.1713
42. Yager P, Schoen PE (1984) Formation of tubules by polymerizable surfactant. Mol Cryst Liq Cryst 106:371–381. https://doi.org/10.1080/00268948408071454
43. Fuhrhop JH, Spiroski D, Boettcher C (1993) Molecular monolayer rods and tubules made of a-(L-Lysine),ω-(Amino) bolaamphiphiles. J Am Chem Soc 115:1600–1601. https://doi.org/10.1021/ja00057a069

44. Shimizu T, Kogiso M, Masuda M (1996) Vesicle assembly in microtubes. Nature 383:487–488. https://doi.org/10.1038/383487b0
45. Kroto HW, Heath JR, Obrien SC, Curl RF, Smalley RE (1985) C-60—Buckminsterfullerene. Nature 318:162–163. https://doi.org/10.1038/318162a0
46. Iijima S (1991) Helical microtubules of graphitic carbon. Nature 354:56–58. https://doi.org/10.1038/354056a0
47. Smith BW, Monthioux M, Luzzi DE (1998) Encapsulated C-60 in carbon nanotubes. Nature 396:323–324. https://doi.org/10.1038/24521
48. Bangham AD, Horne RW (1964) Negative staining of phospholipids + their structural modification by-surface active agents as observed in electron microscope. J Mol Biol 8:660–668. https://doi.org/10.1016/S0022-2836(64)80115-7
49. Pool R (1990) Physicists tackle theory, tubes, and temperature. Science 247:1410–1412. https://doi.org/10.1126/science.247.4949.1410-a
50. Kogiso M, Zhou Y, Shimizu T (2007) Instant preparation of self-assembled metal-complexed lipid nanotubes that act as templates to produce metal oxide nanotubes. Adv Mater 19:242–246. https://doi.org/10.1002/adma.200601117
51. Hirahara K, Suenaga K, Bandow S, Kato H, Okazaki T, Shinohara H, Iijima S (2000) One-dimensional metallofullerene crystal generated inside single-walled carbon nanotubes. Phys Rev Lett 85:5384–5387. https://doi.org/10.1103/PhysRevLett.85.5384
52. Huang K, Rzayev J (2009) Well-defined organic nanotubes from multicomponent bottlebrush copolymers. J Am Chem Soc 131:6880–6885. https://doi.org/10.1021/ja901936g
53. Martin CR (1995) Template synthesis of electronically conductive polymer nanostructures. Acc Chem Res 28:61–68. https://doi.org/10.1021/ar00050a002
54. Steinhart M, Wehrspohn RB, Gosele U, Wendorff JH (2004) Nanotubes by template wetting: a modular assembly system. Angew Chem Int Ed 43:1334–1344. https://doi.org/10.1002/anie.200300614
55. Shimizu T, Hato M (1993) Self-assembling propertise of synthetic peptidic lipids. Biochim Biophys Acta 1147:50–58. https://doi.org/10.1016/0005-2736(93)90315-Q
56. Chung DS, Benedek GB, Konikoff FM, Donovan JM (1993) Elastic free energy of anisotropic helical ribbons as metastable intermediates in the crystallization of cholesterol. Proc Natl Acad Sci USA 90:11341–11345. https://doi.org/10.1073/pnas.90.23.11341
57. Ziserman L, Lee HY, Raghavan SR, Mor A, Danino D (2011) Unraveling the mechanism of nanotube formation by chiral self-assembly of amphiphiles. J Am Chem Soc 133:2511–2517. https://doi.org/10.1021/Ja107069f
58. Slater JL, Huang CH (1988) Interdigitated bilayer-membranes. Prog Lipid Res 27:325–359. https://doi.org/10.1016/0163-7827(88)90010-0
59. John G, Masuda M, Okada Y, Yase K, Shimizu T (2001) Nanotube formation from renewable resources via coiled nanofibers. Adv Mater 13:715–718. https://doi.org/10.1002/1521-4095(200105)13:10%3c715::AID-ADMA715%3e3.0.CO;2-Z
60. John G, Jung JH, Minamikawa H, Yoshida K, Shimizu T (2002) Morphological control of helical solid bilayers in high-axial-ratio nanostructures through binary self-assembly. Chem Eur J 8:5494–5500. https://doi.org/10.1002/1521-3765(20021202)8:23%3c5494::AID-CHEM5494%3e3.0.CO;2-P
61. John G, Mason M, Ajayan PM, Dordick JS (2004) Lipid-based nanotubes as functional architectures with embedded fluorescence and recognition capabilities. J Am Chem Soc 126:15012–15013. https://doi.org/10.1021/ja0446449
62. Kamiya S, Minamikawa H, Jung JH, Yang B, Masuda M, Shimizu T (2005) Molecular structure of glucopyranosylamide lipid and nanotube morphology. Langmuir 21:743–750. https://doi.org/10.1021/la047765v
63. Yui H, Minamikawa H, Danev R, Nagayama K, Kamiya S, Shimizu T (2008) Growth process and molecular packing of a self-assembled lipid nanotube: phase-contrast transmission electron microscopy and XRD analyses. Langmuir 24:709–713. https://doi.org/10.1021/La702488u

64. Israelachvili JN (1985) Intrmolecular and surface forces: with application to colloidal and biological systems . Academic, London
65. Aigouy PT, Costeseque P, Sempere R, Senac T (1995) Caractérisation Structurale et Évolution Thermiquue de I'Acide Amino-11-undécanoïque($C_{11}H_{23}NO_2$.1,5H_2O). Acta Cystallogr B51:55–61. https://doi.org/10.1107/S010876819400159X
66. Szafran M, Dega-Szafran Z, Katrusiak A, Buczak G, Glowiak T, Sitkowski J, Stefaniak L (1998) Electrostatic interactions and conformations of zwitterionic pyridinium alkanoates. J Org Chem 63:2898–2908. https://doi.org/10.1021/jo9720694
67. Sim GA (1955) The crystal structure of 11-amino-undecanoic acid hydrobromide hemihydrate. Acta Crystallogr 8:833–840. https://doi.org/10.1107/S0365110x5500248x
68. Masuda M, Shimizu T (2001) Multilayer structure of unsymmetrical monolayer lipid membrane with head-to-tail interface. Chem Commun 2001:2442–2443. https://doi.org/10.1039/B106581P
69. Masuda M, Yoza K, Shimizu T (2005) Polymorphism of monolayer lipid membrane structures made from unsymmetrical bolaamphiphiles. Carbohydr Res 340:2502–2509. https://doi.org/10.1016/j.carres.2005.08.005
70. Ding W, Kameta N, Minamikawa H, Wada M, Shimizu T, Masuda M (2012) Hybrid organic nanotubes with dual functionalities localized on cylindrical nanochannels control the release of doxorubicin. Adv Healthcare Mater 1:699–706. https://doi.org/10.1002/adhm.201200133
71. Kameta N, Masuda M, Shimizu T (2012) Soft nanotube hydrogels functioning as artificial chaperones. ACS Nano 6:5249–5258. https://doi.org/10.1021/Nn301041y
72. Kameta N, Masuda M, Minamikawa H, Goutev NV, Rim JA, Jung JH, Shimizu T (2005) Selective construction of supramolecular nanotube hosts with cationic inner surfaces. Adv Mater 17:2732–2736. https://doi.org/10.1002/adma.200501092
73. Kameta N, Masuda M, Minamikawa H, Mishima Y, Yamashita I, Shimizu T (2007) Functionalizable organic nanochannels based on lipid nanotubes: encapsulation and nanofluidic behavior of biomacromolecules. Chem Mater 19:3553–3560. https://doi.org/10.1021/Cm070626p
74. Masuda M, Shimizu T (2004) Lipid nanotubes and microtubes: experimental evidence for unsymmetrical monolayer membrane formation from unsymmetrical bolaamphiphiles. Langmuir 20:5969–5977. https://doi.org/10.1021/la049085y
75. Namba K, Stubbs G (1986) Structure of tobacco mosaic-virus at 3.6-a resolution—implications for assembly. Science 231:1401–1406. https://doi.org/10.1126/science.3952490
76. Fuhrhop JH, Wang T (2004) Bolaamphiphiles. Chem Rev 104:2901–2938. https://doi.org/10.1021/cr030602b
77. Kameta N, Masuda M, Minamikawa H, Shimizu T (2007) Self-assembly and thermal phase transition behavior of unsymmetrical bolaamphiphiles having glucose- and amino-hydrophilic headgroups. Langmuir 23:4634–4641. https://doi.org/10.1021/La063542o
78. Kameta N, Yoshida K, Masuda M, Shimizu T (2009) Supramolecular nanotube hydrogels: remarkable resistance effect of confined proteins to denaturants. Chem Mater 21:5892–5898. https://doi.org/10.1021/Cm903108h
79. Gu Q, Zou AH, Yuan CW, Guo R (2003) Effects of a bolaamphiphile on the structure of phosphatidylcholine liposomes. J Colloid Interface Sci 266:442–447. https://doi.org/10.1016/S0021-9797(03)00649-0
80. Baek K, Kim Y, Kim H, Yoon M, Hwang I, Ko YH, Kim K (2010) Unconventional U-shaped conformation of a bolaamphiphile embedded in a synthetic host. Chem Commun 46:4091–4093. https://doi.org/10.1039/c0cc00752h
81. Masuda M, Vill V, Shimizu T (2000) Conformational and thermal phase behavior of oligomethylene chains constrained by carbohydrate hygrogen-bond networks. J Am Chem Soc 122:12327–12333. https://doi.org/10.1021/ja001884p
82. Snyder RG, Hsu SL, Krimm S (1978) Vibrational-spectra in C-H stretching region and structure of polymethylene chain. Spectrochim Acta Part A Mol Biomol Spectr 34:395–406. https://doi.org/10.1016/0584-8539(78)80167-6

83. Snyder RG, Strauss HL, Elliger CA (1982) C-H stretching modes and the structure of normal-alkyl chains. 1. Long disordered chains. J Phys Chem 86:5145–5150. https://doi.org/10.1021/J100223a018
84. Mantsch HH, Mcelhaney RN (1991) Phospholipid phase-transitions in model and biological-membranes as studied by infrared-spectroscopy. Chem Phys Lipids 57:213–226. https://doi.org/10.1016/0009-3084(91)90077-O
85. Garti N, Sato K (1988) Crystallization and Polymorphism of Fats and Fatty Acids. Marcel Dekker
86. Yamada N, Okuyama K, Serizawa T, Kawasaki M, Oshima S (1996) Periodic change in absorption maxima due to different chain packing between the bilayers of amphiphiles possessing even and odd numbers of carbons in the hydrophobic chain. J Chem Soc, Perkin Trans 2:2707–2714. https://doi.org/10.1039/P29960002707
87. Blout ER, Linsley SG (1952) Infrared spectra and the structure of glycine and leucine peptides. J Am Chem Soc 74:1946–1951. https://doi.org/10.1021/ja01128a023
88. Elliott A, Malcolm BR (1956) Infra-red studies of polypeptides related to silk. Trans Faraday Soc 52:528–536. https://doi.org/10.1039/tf9565200528
89. Crick FHC, Rich A (1955) Structure of polyglycine-II. Nature 176:780–781. https://doi.org/10.1038/176780a0
90. Bamford CH, Brown L, Cant EM, Elliott A, Hanby WE, Malcolm BR (1955) Structure of polyglycine. Nature 176:396–397. https://doi.org/10.1038/176396a0
91. Shimizu T, Kogiso M, Masuda M (1997) Noncovalent formation of polyglycine II-type structure by hexagonal self-assembly of linear polymolecular chains. J Am Chem Soc 119:6209–6210. https://doi.org/10.1021/ja970844r
92. Kameta N, Mizuno G, Masuda M, Minamikawa H, Kogiso M, Shimizu T (2007) Molecular monolayer nanotubes having 7–9 nm inner diameters covered with different inner and outer surfaces. Chem Lett 36:896–897. https://doi.org/10.1246/Cl.2007.896
93. Hamley IW (2011) Self-assembly of amphiphilic peptides. Soft Matter 7:4122–4138. https://doi.org/10.1039/c0sm01218a
94. Hamley IW (2014) Peptide nanotubes. Angew Chem Int Ed 53:6866–6881. https://doi.org/10.1002/anie.201310006
95. Boettcher C, Schade B, Fuhrhop J-H (2001) Comparative cryo-electron microscopy of noncovalent N-dodecanoyl-(D- and L-) serine assemblies in vitreous toluene and water. Langmuir 17:873–877. https://doi.org/10.1021/la001054p
96. Lu K, Jacob J, Thiyagarajan P, Conticello VP, Lynn DG (2003) Exploiting amyloid fibril lamination for nanotube self-assembly. J Am Chem Soc 125:6391–6393. https://doi.org/10.1021/ja0341642
97. Ni R, Childers WS, Hardcastle KI, Mehta AK, Lynn DG (2012) Remodeling cross-beta nanotube surfaces with peptide/lipid chimeras. Angew Chem Int Ed 51:6635–6638. https://doi.org/10.1002/anie.201201173
98. Bucak S, Cenker C, Nasir I, Olsson U, Zackrisson M (2009) Peptide nanotube nematic phase. Langmuir 25:4262–4265. https://doi.org/10.1021/la804175h
99. Middleton DA, Madine J, Castelletto V, Hamley IW (2013) Insights into the molecular architecture of a peptide nanotube using FTIR and solid-state NMR spectroscopic measurements on an aligned sample. Angew Chem Int Ed 52:10537–10540. https://doi.org/10.1002/anie.201301960
100. Lu Q, Kim Y, Bassim N, Collins GE (2015) Impact of confinement on proteins concentrated in lithocholic acid based organic nanotubes. J Colloid Interface Sci 454:97–104. https://doi.org/10.1016/j.jcis.2015.05.004
101. Liang W, He S, Fang J (2014) Self-assembly of J-aggregate nanotubes and their applications for sensing dopamine. Langmuir 30:805–811. https://doi.org/10.1021/la404022q
102. Chapman R, Danial M, Koh ML, Jolliffe KA, Perrier S (2012) Design and properties of functional nanotubes from the self-assembly of cyclic peptide templates. Chem Soc Rev 41:6023–6041. https://doi.org/10.1039/C2cs35172b

103. Yagai S, Yamauchi M, Kobayashi A, Karatsu T, Kitamura A, Ohba T, Kikkawa Y (2012) Control over hierarchy levels in the self-assembly of stackable nanotoroids. J Am Chem Soc 134:18205–18208. https://doi.org/10.1021/Ja308519b
104. Tu S, Kim SH, Joseph J, Modarelli DA, Parquette JR (2011) Self-assembly of a donor-acceptor nanotube. a strategy to create bicontinuous arrays. J Am Chem Soc 133:19125–19130. https://doi.org/10.1021/ja205868b
105. Kameta N, Masuda M, Shimizu T (2015) Photoinduced morphological transformations of soft nanotubes. Chem Eur J 21:8832–8839. https://doi.org/10.1002/chem.201500430
106. Ballister ER, Lai AH, Zuckermann RN, Cheng Y, Mougous JD (2008) In vitro self-assembly from a simple protein of tailorable nanotubes building block. Proc Natl Acad Sci USA 105:3733–3738. https://doi.org/10.1073/pnas.0712247105
107. Miranda FF, Iwasaki K, Akashi S, Sumitomo K, Kobayashi M, Yamashita I, Tame JRH, Heddle JG (2009) A Self-Assembled protein nanotube with high aspect ratio. Small 5:2077–2084. https://doi.org/10.1002/smll.200900667
108. Ghadiri MR, Granja JR, Milligan RA, Mcree DE, Khazanovich N (1993) Self-assembling organic nanotubes based on a cyclic peptide architecture. Nature 366:324–327. https://doi.org/10.1038/366324a0
109. Ghadiri MR (1995) Self-assembled nanoscale tubular ensembles. Adv Mater 7:675–677. https://doi.org/10.1002/adma.19950070718
110. Wolf E, Kim PS, Berger B (1997) MultiCoil: a program for predicting two- and three-stranded coiled coils. Protein Sci 6:1179–1189. https://doi.org/10.1002/pro.5560060606
111. Xu C, Liu R, Mehta AK, Guerrero-Ferreira RC, Wright ER, Dunin-Horkawicz S, Morris K, Serpell LC, Zuo X, Wall JS, Conticello VP (2013) Rational design of helical nanotubes from self-assembly of coiled-coil lock washers. J Am Chem Soc 135:15565–15578. https://doi.org/10.1021/ja4074529
112. Liu J, Zheng Q, Deng Y, Cheng CS, Kallenbach NR, Lu M (2006) A seven-helix coiled coil. Proc Natl Acad Sci USA 103:15457–15462. https://doi.org/10.1073/pnas.0604871103
113. Kimizuka N, Kawasaki T, Hirata K, Kunitake K (1995) Tube-like nanostructures composed of networks of complementary hydrogen bonds. J Am Chem Soc 117:6360–6361. https://doi.org/10.1021/ja00128a027
114. Percec V, Ahn CH, Unger G, Yeardley DJP, Moeller M, Sheiko SS (1998) Controlling polymer shape through the self-assembly of dendritic side-groups. Nature 391:161–164. https://doi.org/10.1038/34384
115. Beingessner RL, Fan YW, Fenniri H (2016) Molecular and supramolecular chemistry of rosette nanotubes. RSC Adv 6:75820–75838. https://doi.org/10.1039/c6ra16315g
116. Jonkheijm P, Miura A, Zdanowska M, Hoeben FJ, De Feyter S, Schenning AP, De Schryver FC, Meijer EW (2004) π-conjugated oligo-(*p*-phenylenevinylene) rosettes and their tubular self-assembly. Angew Chem Int Ed 43:74–78. https://doi.org/10.1002/anie.200352790
117. Yagai S, Nakajima T, Kishikawa K, Kohmoto S, Karatsu T, Kitamura A (2005) Hierarchical organization of photoresponsive hydrogen-bonded rosettes. J Am Chem Soc 127:11134–11139. https://doi.org/10.1021/ja052645a
118. Block MAB, Kaiser C, Khan A, Hecht S (2005) Discrete organic nanotubes based on a combination of covalent and non-covalent approaches. Top Curr Chem 245:89–150. https://doi.org/10.1007/b98167
119. Huang Z, Kang SK, Banno M, Yamaguchi T, Lee D, Seok C, Yashima E, Lee M (2012) Pulsating tubules from noncovalent macrocycles. Science 337:1521–1526. https://doi.org/10.1126/science.1224741
120. Wang Y, Huang Z, Kim Y, He Y, Lee M (2014) Guest-driven inflation of self-assembled nanofibers through hollow channel formation. J Am Chem Soc 136:16152–16155. https://doi.org/10.1021/ja510182x
121. Kim Y, Kang J, Shen B, Wang Y, He Y, Lee M (2015) Open-closed switching of synthetic tubular pores. Nat Commun 6:8650. https://doi.org/10.1038/ncomms9650
122. Pisula W, Kastler M, Yang C, Enkelmann V, Mullen K (2007) Columnar mesophase formation of cyclohexa-m-phenylene-based macrocycles. Chem Asian J 2:51–56. https://doi.org/10.1002/asia.200600338

123. Gubitosi M, D'Annibale A, Schillen K, Olsson U, Pavel NV, Galantini L (2017) On the stability of lithocholate derivative supramolecular tubules. Rsc Adv 7:512–517. https://doi.org/10.1039/c6ra26092f
124. Kameta N, Akiyama H, Masuda M, Shimizu T (2016) Effect of photoinduced size changes on protein refolding and transport abilities of soft nanotubes. Chem Eur J 22:7198–7205. https://doi.org/10.1002/chem.201504613
125. Adler-Abramovich L, Gazit E (2014) The physical properties of supramolecular peptide assemblies: from building block association to technological applications. Chem Soc Rev 43:6881–6893. https://doi.org/10.1039/c4cs00164h
126. Brea RJ, Reiriz C, Granja JR (2010) Towards functional bionanomaterials based on self-assembling cyclic peptide nanotubes. Chem Soc Rev 39:1448–1456. https://doi.org/10.1039/b805753m
127. Chapman R, Jolliffe KA, Perrier S (2011) Modular design for the controlled production of polymeric nanotubes from polymer/peptide conjugates. Polym Chem 2:1956–1963. https://doi.org/10.1039/c1py00202c
128. Huang K, Rzayev J (2011) Charge and size selective molecular transport by amphiphilic organic nanotubes. J Am Chem Soc 133:16726–16729. https://doi.org/10.1021/ja204296v
129. Xu Y, Wang T, He Z, Zhong A, Huang K (2016) Carboxyl-containing microporous organic nanotube networks as a platform for Pd catalysts. RSC Adv 6:39933–39939. https://doi.org/10.1039/c6ra05753e
130. Kim Y, Li H, He Y, Chen X, Ma X, Lee M (2017) Collective helicity switching of a DNA-coat assembly. Nat Nanotechnol 12:551–558. https://doi.org/10.1038/Nnano.2017.42
131. Kim HJ, Kim T, Lee M (2011) Responsive nanostructures from aqueous assembly of rigid-flexible block molecules. Acc Chem Res 44:72–82. https://doi.org/10.1021/ar100111n
132. Prasanthkumar S, Zhang W, Jin W, Fukushima T, Aida T (2015) Selective synthesis of single- and multi-walled supramolecular nanotubes by using solvophobic/solvophilic controls: step-wise radial growth via "coil-on-tube" intermediates. Angew Chem Int Ed 54:11168–11172. https://doi.org/10.1002/anie.201505806
133. Ishikawa K, Kameta N, Masuda M, Asakawa M, Shimizu T (2014) Boroxine nanotubes: moisture-sensitive morphological transformation and guest release. Adv Funct Mater 24:603–609. https://doi.org/10.1002/adfm.201302005
134. Cao H, Duan P, Zhu X, Jiang J, Liu M (2012) Self-assembled organic nanotubes through instant gelation and universal capacity for guest molecule encapsulation. Chem Eur J 18:5546–5550. https://doi.org/10.1002/chem.201103654
135. Yagai S (2015) Supramolecularly engineered functional π-assemblies based on complementary hydrogen-bonding interactions. Bull Chem Soc Jpn 88:28–58. https://doi.org/10.1246/bcsj.20140261
136. Zhang J, Chen XF, Wei HB, Wan XH (2013) Tunable assembly of amphiphilic rod-coil block copolymers in solution. Chem Soc Rev 42:9127–9154. https://doi.org/10.1039/c3cs60192g
137. Qian Z, Wang Z, Zhao N, Xu J (2018) Aerogels derived from polymer nanofibers and their applications. Macromol Rapid Commun 39:1700724. https://doi.org/10.1002/marc.201700724
138. Su B, Wu Y, Jiang L (2012) The art of aligning one-dimensional (1D) nanostructures. Chem Soc Rev 41:7832–7856. https://doi.org/10.1039/c2cs35187k
139. Otter R, Klinker K, Spitzer D, Schinnerer M, Barz M, Besenius P (2018) Folding induced supramolecular assembly into pH-responsive nanorods with a protein repellent shell. Chem Commun 54:401–404. https://doi.org/10.1039/c7cc08127h
140. Furusho H, Mishima Y, Kameta N, Masuda M, Yamashita I, Shimizu T (2008) Lipid nanotube encapsulating method for two- and three-dimensional transmission electron microscopy analyses of cage-shaped proteins. Jpn J Appl Phys 47:394–399. https://doi.org/10.1143/Jjap.47.394
141. Furusho H, Mishima Y, Kameta N, Yamane M, Masuda M, Asakawa M, Yamashita I, Mori H, Takaoka A, Shimizu T (2009) Lipid nanotube encapsulating method in low-energy scanning transmission electron microscopy analyses. Jpn J Appl Phys **48**, 097001. https://doi.org/10.1143/Jjap.48.097001

142. Chattopadhyay T, Kogiso M, Asakawa M, Shimizu T, Aoyagi M (2010) Copper(II)-coordinated organic nanotube: a novel heterogeneous catalyst for various oxidation reactions. Catal Commun 12:9–13. https://doi.org/10.1016/j.catcom.2010.07.013
143. Chattopadhyay T, Kogiso M, Aoyagi M, Yui H, Asakawa M, Shimizu T (2011) Single bilayered organic nanotubes: anchors for production of a reusable catalyst with nickel ions. Green Chem 13:1138–1140. https://doi.org/10.1039/C1gc00005e
144. Jin Q, Zhang L, Cao H, Wang T, Zhu X, Jiang J, Liu M (2011) Self-assembly of Copper(II) ion-mediated nanotube and its supramolecular chiral catalytic behavior. Langmuir 27:13847–13853. https://doi.org/10.1021/La203110z
145. Jiang J, Meng Y, Zhang L, Liu M (2016) Self-assembled single-walled metal-helical nanotube (M-HN): creation of efficient supramolecular catalysts for asymmetric reaction. J Am Chem Soc 138:15629–15635. https://doi.org/10.1021/jacs.6b08808
146. Wu S, Li Y, Xie SY, Ma C, Lim J, Zhao J, Kim DS, Yang M, Yoon DK, Lee M, Kim SO, Huang Z (2017) Supramolecular nanotubules as a catalytic regulator for palladium cations: applications in selective catalysis. Angew Chem Int Ed 56:11511–11514. https://doi.org/10.1002/anie.201706373
147. He Z, Zhong A, Zhang H, Xiong L, Xu Y, Wang T, Zhou M, Huang K (2016) Three-arm branched microporous organic nanotube networks. Macromol Rapid Commun 37:1566–1572. https://doi.org/10.1002/marc.201600327
148. Xiong L, Zhang H, He Z, Wang T, Xu Y, Zhou M, Huang K (2018) Acid-base bifunctional amphiphilic organic nanotubes as a catalyst for one-pot cascade reactions in water. New J Chem 42:1368–1372. https://doi.org/10.1039/c7nj04209d
149. Lee KS, Parquette JR (2015) A self-assembled nanotube for the direct aldol reaction in water. Chem Commun 51:15653–15656. https://doi.org/10.1039/c5cc06142c
150. Wang T, Xu Y, He Z, Zhang H, Xiong L, Zhou M, Yu W, Shi B, Huang K (2017) Acid-base bifunctional microporous organic nanotube networks for cascade reactions. Macromol Chem Phys 218:1600431. https://doi.org/10.1002/macp.201600431
151. Amano Y, Komatsu T (2015) Nanotube reactor with a lipase wall interior for enzymatic ring-opening oligomerization of lactone. Chem Lett 44:1646–1648. https://doi.org/10.1246/cl.150789
152. Komatsu T, Terada H, Kobayashi N (2011) Protein nanotubes with an enzyme interior surface. Chem Eur J 17:1849–1854. https://doi.org/10.1002/chem.201001937
153. Kobayakawa S, Nakai Y, Akiyama M, Komatsu T (2017) Self-propelled soft protein microtubes with a Pt nanoparticle interior surface. Chem Eur J 23:5044–5050. https://doi.org/10.1002/chem.201605055
154. Wu Z, Wu Y, He W, Lin X, Sun J, He Q (2013) Self-propelled polymer-based multilayer nanorockets for transportation and drug release. Angew Chem Int Ed 52:7000–7003. https://doi.org/10.1002/anie.201301643
155. Chen J, Zhang B, Xia F, Xie Y, Jiang S, Su R, Lu Y, Wu W (2016) Transmembrane delivery of anticancer drugs through self-assembly of cyclic peptide nanotubes. Nanoscale 8:7127–7136. https://doi.org/10.1039/c5nr06804e
156. Danial M, Tran CMN, Young PG, Perrier S, Jolliffe KA (2013) Janus cyclic peptide-polymer nanotubes. Nat Commun 4:2780. https://doi.org/10.1038/ncomms3780
157. Gong B, Shao Z (2013) Self-assembling organic nanotubes with precisely defined, sub-nanometer pores: formation and mass transport characteristics. Acc Chem Res 46:2856–2866. https://doi.org/10.1021/ar400030e
158. Komatsu T, Qu X, Ihara H, Fujihara M, Azuma H, Ikeda H (2011) Virus trap in human serum albumin nanotube. J Am Chem Soc 133:3246–3248. https://doi.org/10.1021/Ja1096122
159. Yuge S, Akiyama M, Ishii M, Namkoong H, Yagi K, Nakai Y, Adachi R, Komatsu T (2017) Glycoprotein nanotube traps influenza virus. Chem Lett 46:95–97. https://doi.org/10.1246/cl.160805
160. Kameta N, Asakawa M, Masuda M, Shimizu T (2011) Self-assembled organic nanotubes embedding hydrophobic molecules within solid bilayer membranes. Soft Matter 7:85–90. https://doi.org/10.1039/C0sm00375a

161. Goto S, Amano Y, Akiyama M, Bottcher C, Komatsu T (2013) Gold nanoparticle inclusion into protein nanotube as a layered wall component. Langmuir 29:14293–14300. https://doi.org/10.1021/La403283x
162. Qu X, Kobayashi N, Komatsu T (2010) Solid nanotubes comprising α-Fe_2O_3 nanoparticles prepared from ferritin protein. ACS Nano 4:1732–1738. https://doi.org/10.1021/Nn901879d
163. Kameta N, Tanaka A, Akiyama H, Minamikawa H, Masuda M, Shimizu T (2011) Photoresponsive soft nanotubes for controlled guest release. Chem Eur J 17:5251–5255. https://doi.org/10.1002/chem.201100179
164. Ma X, Zhang Y, Zheng Y, Zhang Y, Tao X, Che Y, Zhao J (2015) Highly fluorescent one-handed nanotubes assembled from a chiral asymmetric perylene diimide. Chem Commun 51:4231–4233. https://doi.org/10.1039/c5cc00365b
165. Gao M, Paul S, Schwieters CD, You ZQ, Shao H, Herbert JM, Parquette JR, Jaroniec CP (2015) A structural model for a self-assembled nanotube provides insight into its exciton dynamics. J Phys Chem C 119:13948–13956. https://doi.org/10.1021/acs.jpcc.5b03398
166. Uji H, Kim H, Imai T, Mitani S, Sugiyama J, Kimura S (2016) Electronic properties of tetrathiafulvalene-modified cyclic-β-peptide nanotube. Biopolymers 106:275–282. https://doi.org/10.1002/bip.22850
167. Tu S, Kim SH, Joseph J, Modarelli DA, Parquette JR (2013) Proton-coupled self-assembly of a porphyrin-naphthalenediimide dyad. ChemPhysChem 14:1609–1617. https://doi.org/10.1002/cphc.201300023
168. Wang X, Duan P, Liu M (2014) Self-assembly of π-conjugated gelators into emissive chiral nanotubes: emission enhancement and chiral detection. Chem Asian J 9:770–778. https://doi.org/10.1002/asia.201301518
169. de la Rica R, Mendoza E, Matsui H (2010) Bioinspired target-specific crystallization on peptide nanotubes for ultrasensitive pb ion detection. Small 6:1753–1756. https://doi.org/10.1002/smll.201000489
170. Kameta N, Masuda M, Shimizu T (2015) Two-step naked-eye detection of lectin by hierarchical organization of soft nanotubes into liquid crystal and gel phases. Chem Commun 51:6816–6819. https://doi.org/10.1039/C5cc01464f
171. de la Rica R, Mendoza E, Lechuga LM, Matsui H (2008) Label-free pathogen detection with sensor chips assembled from peptide nanotubes. Angew Chem Int Ed 47:9752–9755. https://doi.org/10.1002/anie.200804299
172. Ishihara Y, Kimura S (2012) Peptide nanotube composed of cyclic tetra-β-peptide having polydiacetylene. Biopolymers 98:155–160. https://doi.org/10.1002/Bip.22029
173. Catrouillet S, Brendel JC, Larnaudie S, Barlow T, Jolliffe KA, Perrier S (2016) Tunable length of cyclic peptide-polymer conjugate self-assemblies in water. ACS Macro Lett 5:1119–1123. https://doi.org/10.1021/acsmacrolett.6b00586
174. Koh ML, FitzGerald PA, Warr GG, Jolliffe KA, Perrier S (2016) Study of (cyclic peptide)–polymer conjugate assemblies by small-angle neutron scattering. Chem Eur J 22:18419–18428. https://doi.org/10.1002/chem.201603091
175. Larnaudie SC, Brendel JC, Jolliffe KA, Perrier S (2016) Cyclic peptide-polymer conjugates: grafting-to vs grafting-from. J Polym Sci Part A: Polym Chem 54:1003–1011. https://doi.org/10.1002/pola.27937
176. Montenegro J, Vazquez-Vazquez C, Kalinin A, Geckeler KE, Granja JR (2014) Coupling of carbon and peptide nanotubes. J Am Chem Soc 136:2484–2491. https://doi.org/10.1021/ja410901r
177. Kim JU, Haberkorn N, Theato P, Zentel R (2011) Controlled fabrication of organic nanotubes via self-assembly of non-symmetric bis-acylurea. Colloid Polym Sci 289:1855–1862. https://doi.org/10.1007/s00396-011-2512-y
178. Kameta N, Dong J, Yui H (2018) Thermoresponsive PEG-coated nanotubes as chiral selectors of amino acids and peptides. Small 14:1800030. https://doi.org/10.1002/smll.201800030
179. Liu X, Chen L, Liu H, Yang G, Zhang P, Han D, Wang S, Jiang L (2013) Bio-inspired soft polystyrene nanotube substrate for rapid and highly efficient breast cancer-cell capture. Npg Asia Mater **5**, e63. https://doi.org/10.1038/am.2013.43

180. Jin H, Ding YH, Wang M, Song Y, Liao Z, Newcomb CJ, Wu X, Tang XQ, Li Z, Lin Y, Yan F, Jian T, Mu P, Chen CL (2018) Designable and dynamic single-walled stiff nanotubes assembled from sequence-defined peptoids. Nat Commun 9:270. https://doi.org/10.1038/s41467-017-02059-1
181. Blunden BM, Chapman R, Danial M, Lu H, Jolliffe KA, Perrier S, Stenzel MH (2014) Drug conjugation to cyclic peptide-polymer self-assembling nanotubes. Chem Eur J 20:12745–12749. https://doi.org/10.1002/chem.201403130
182. Ding W, Wada M, Kameta N, Minamikawa H, Shimizu T, Masuda M (2011) Functionalized organic nanotubes as tubular nonviral gene transfer vector. J Controlled Rel 156:70–75. https://doi.org/10.1016/j.jconrel.2011.07.007
183. Hsieh WH, Chang SF, Chen HM, Chen JH, Liaw J (2012) Oral gene delivery with cyclo-(D-Trp-Tyr) peptide nanotubes. Mol Pharm 9:1231–1249. https://doi.org/10.1021/Mp200523n
184. Lee YH, Chang SF, Liaw J (2015) Anti-apoptotic gene delivery with cyclo-(D-Trp-Tyr) peptide nanotube via eye drop following corneal epithelial debridement. Pharmaceutics 7:122–136. https://doi.org/10.3390/pharmaceutics7030122
185. Li M, Ehlers M, Schlesiger S, Zellermann E, Knauer SK, Schmuck C (2016) Incorporation of a non-natural arginine analogue into a cyclic peptide leads to formation of positively charged nanofibers capable of gene transfection. Angew Chem Int Ed 55:598–601. https://doi.org/10.1002/anie.201508714
186. Wang TQ, Xu Y, He ZD, Zhou MH, Huang K (2018) Microporous organic nanotube networks from hyper cross-linking core-shell bottlebrush copolymers for selective adsorption study. Chin J Polym Sci 36:98–105. https://doi.org/10.1007/s10118-018-2007-0
187. Adler-Abramovich L, Badihi-Mossberg M, Gazit E, Rishpon J (2010) Characterization of peptide-nanostructure-modified electrodes and their application for ultrasensitive environmental monitoring. Small 6:825–831. https://doi.org/10.1002/smll.200902186
188. Adler-Abramovich L, Aronov D, Beker P, Yevnin M, Stempler S, Buzhansky L, Rosenman G, Gazit E (2009) Self-Assembled Arrays of Peptide Nanotubes by Vapour Deposition. Nat Nanotechnol 4:849–854. https://doi.org/10.1038/nnano.2009.298
189. Lee JS, Ryu J, Park CB (2009) Bio-inspired fabrication of superhydrophobic surfaces through peptide self-assembly. Soft Matter 5:2717–2720. https://doi.org/10.1039/b906783c
190. Terech P, Jean B, Ne F (2006) Hexagonally ordered ammonium lithocholate self-assembled nanotubes with highly monodisperse sections. Adv Mater 18:1571–1574. https://doi.org/10.1002/adma.200502358
191. Amorin M, Perez A, Barbera J, Ozores HL, Serrano JL, Granja JR, Sierra T (2014) Liquid crystal organization of self-assembling cyclic peptides. Chem Commun 50:688–690. https://doi.org/10.1039/c3cc47400c
192. Kameta N, Shiroishi H (2018) PEG-nanotube liquid crystals as templates for construction of surfactant-free gold nanorods. Chem Commun 54:4665–4668. https://doi.org/10.1039/c8cc02013b
193. Liu Y, Wang T, Li Z, Liu M (2013) Copper(II) ion selective and strong acid-tolerable hydrogels formed by an L-histidine ester terminated bolaamphiphile: from single molecular thick nanofibers to single-wall nanotubes. Chem Commun 49:4767–4769. https://doi.org/10.1039/c3cc41786g
194. Lalitha K, Prasad YS, Maheswari CU, Sridharan V, John G, Nagarajan S (2015) Stimuli responsive hydrogels derived from a renewable resource: synthesis, self-assembly in water and application in drug delivery. J Mater Chem B 3:5560–5568. https://doi.org/10.1039/c5tb00864f
195. Mukai M, Minamikawa H, Aoyagi M, Asakawa M, Shimizu T, Kogiso M (2013) A hydro/organo/hybrid gelator: a peptide lipid with turning aspartame head groups. J Colloid Interface Sci 395:154–160. https://doi.org/10.1016/j.jcis.2012.12.060
196. Bernet A, Behr M, Schmidt HW (2011) Supramolecular nanotube-based fiber mats by self-assembly of a tailored amphiphilic low molecular weight hydrogelator. Soft Matter 7:1058–1065. https://doi.org/10.1039/C0sm00456a

197. Kameta N, Ishikawa K, Masuda M, Asakawa M, Shimizu T (2012) Soft nanotubes acting as a light-harvesting antenna system. Chem Mater 24:209–214. https://doi.org/10.1021/Cm2030526
198. Kameta N, Aoyagi M, Asakawa M (2017) Enhancement of the photocatalytic activity of Rhenium(I) complexes by encapsulation in light-harvesting soft nanotubes. Chem Commun 53:10116–10119. https://doi.org/10.1039/c7cc05337a
199. Zhu X, Li Y, Duan P, Liu M (2010) Self-assembled ultralong chiral nanotubes and tuning of their chirality through the mixing of enantiomeric components. Chem Eur J 16:8034–8040. https://doi.org/10.1002/chem.201000595
200. Zhang L, Liu C, Jin Q, Zhu X, Liu M (2013) Pyrene-functionalized organogel and spacer effect: from emissive nanofiber to nanotube and inversion of supramolecular chirality. Soft Matter 9:7966–7973. https://doi.org/10.1039/C3sm51204e
201. Ni W, Liang F, Liu J, Qu X, Zhang C, Li J, Wang Q, Yang Z (2011) Polymer nanotubes toward gelating organic chemicals. Chem Commun 47:4727–4729. https://doi.org/10.1039/c1cc10900f
202. Gan Z, Wu X, Zhu X, Shen J (2013) Light-induced ferroelectricity in bioinspired self-assembled diphenylalanine nanotubes/microtubes. Angew Chem Int Ed 52:2055–2059. https://doi.org/10.1002/anie.201207992
203. Handelman A, Beker P, Mishina E, Semin S, Amdursky N, Rosenman G (2012) Ferroelectric properties and phase transition in dipeptide nanotubes. Ferroelectrics 430:84–91. https://doi.org/10.1080/00150193.2012.677721
204. Zelenovskiy PS, Koryukova TA, Yuzhakov VV, Vasilev SG, Nuraeva AS, Gunina EV, Chezganov DS, Kholkin AL, Shur VY (2018) Piezoelectric properties and young's moduli of diphenylalanine microtubes-oxide nanoparticles composites. Ferroelectrics 525:146–155. https://doi.org/10.1080/00150193.2018.1432826
205. Bosne ED, Heredia A, Kopyl S, Karpinsky DV, Pinto AG, Kholkin AL (2013) Piezoelectric resonators based on self-assembled diphenylalanine microtubes. Appl Phys Lett 102:073504. https://doi.org/10.1063/1.4793417
206. Handelman A, Beker P, Amdursky N, Rosenman G (2012) Physics and engineering of peptide supramolecular nanostructures. PCCP 14:6391–6408. https://doi.org/10.1039/c2cp40157f
207. Rosenman G, Beker P, Koren I, Yevnin M, Bank-Srour B, Mishina E, Semin S (2011) Bioinspired peptide nanotubes: deposition technology, basic physics and nanotechnology applications. J Pept Sci 17:75–87. https://doi.org/10.1002/psc.1326
208. Kholkin A, Amdursky N, Bdikin I, Gazit E, Rosenman G (2010) Strong piezoelectricity in bioinspired peptide nanotubes. ACS Nano 4:610–614. https://doi.org/10.1021/nn901327v
209. Handelman A, Apter B, Turko N, Rosenman G (2016) Linear and nonlinear optical waveguiding in bio-inspired peptide nanotubes. Acta Biomater 30:72–77. https://doi.org/10.1016/j.actbio.2015.11.004
210. Handelman A, Lavrov S, Kudryavtsev A, Khatchatouriants A, Rosenberg Y, Mishina E, Rosenman G (2013) Nonlinear optical bioinspired peptide nanostructures. Adv Opt Mater 1:875–884. https://doi.org/10.1002/adom.201300282
211. Yan X, Su Y, Li J, Fruh J, Mohwald H (2011) Uniaxially oriented peptide crystals for active optical waveguiding. Angew Chem Int Ed 50:11186–11191. https://doi.org/10.1002/anie.201103941
212. Xu X, Chen S, Chen Y, Sun H, Song L, He W, Wang X (2016) Polyoxometalate cluster-incorporated metal-organic framework hierarchical nanotubes. Small 12:2982–2990. https://doi.org/10.1002/smll.201503695
213. Ryu J, Kim SW, Kang K, Park CB (2010) Mineralization of self-assembled peptide nanofibers for rechargeable lithium ion batteries. Adv Mater 22:5537–5541. https://doi.org/10.1002/adma.201000669
214. Saeki A, Yamamoto Y, Koizumi Y, Fukushima T, Aida T, Seki S (2011) Photoconductivity of self-assembled hexabenzocoronene nanotube: insight into the charge carrier mobilities on local and long-range scales. J Phys Chem Lett 2:2549–2554. https://doi.org/10.1021/Jz201223e

215. Yamamoto Y (2011) Electroactive nanotubes from π-conjugated discotic molecules. Bull Chem Soc Jpn 84:17–25. https://doi.org/10.1246/bcsj.20100272
216. Amdursky N, Gazit E, Rosenman G (2010) Quantum confinement in self-assembled bioinspired peptide hydrogels. Adv Mater 22:2311–2315. https://doi.org/10.1002/adma.200904034
217. Amdursky N, Molotskii M, Aronov D, Adler-Abramovich L, Gazit E, Rosenman G (2009) Blue luminescence based on quantum confinement at peptide nanotubes. Nano Lett 9:3111–3115. https://doi.org/10.1021/nl9008265
218. Rubin DJ, Amini S, Zhou F, Su HB, Miserez A, Joshi NS (2015) Structural, nanomechanical, and computational characterization of D, L-cyclic peptide assemblies. ACS Nano 9:3360–3368. https://doi.org/10.1021/acsnano.5b00672
219. Rubin DJ, Nia HT, Desire T, Nguyen PQ, Gevelber M, Ortiz C, Joshi NS (2013) Mechanical reinforcement of polymeric fibers through peptide nanotube incorporation. Biomacromol 14:3370–3375. https://doi.org/10.1021/bm4008293
220. Even N, Adler-Abramovich L, Buzhansky L, Dodiuk H, Gazit E (2011) Improvement of the mechanical properties of epoxy by peptide nanotube fillers. Small 7:1007–1011. https://doi.org/10.1002/smll.201001940
221. Goldshtein K, Golodnitsky D, Peled E, Adler-Abramovich L, Gazit E, Khatun S, Stallworth P, Greenbaum S (2012) Effect of peptide nanotube filler on structural and ion-transport properties of solid polymer electrolytes. Solid State Ionics 220:39–46. https://doi.org/10.1016/j.ssi.2012.05.028
222. Azuri I, Adler-Abramovich L, Gazit E, Hod O, Kronik L (2014) Why are diphenylalanine-based peptide nanostructures so rigid? insights from first principles calculations. J Am Chem Soc 136:963–969. https://doi.org/10.1021/ja408713x
223. Adler-Abramovich L, Kol N, Yanai I, Barlam D, Shneck RZ, Gazit E, Rousso I (2010) Self-assembled organic nanostructures with metallic-like stiffness. Angew Chem Int Ed 49:9939–9942. https://doi.org/10.1002/anie.201002037
224. Niu L, Chen X, Allen S, Tendler SJB (2007) Using the bending beam model to estimate the elasticity of diphenylalanine nanotubes. Langmuir 23:7443–7446. https://doi.org/10.1021/la7010106
225. Frusawa H, Manabe T, Kagiyama E, Hirano K, Kameta N, Masuda M, Shimizu T (2013) Electric moulding of dispersed lipid nanotubes into a nanofluidic device. Sci Rep 3:2165. https://doi.org/10.1038/Srep02165
226. Hirano K, Aoyagi M, Ishido T, Ooie T, Frusawa H, Asakawa M, Shimizu T, Ishikawa M (2009) Measuring the length distribution of self-assembled lipid nanotubes by orientation control with a high-frequency alternating current electric field in aqueous solutions. Anal Chem 81:1459–1464. https://doi.org/10.1021/Ac8022795
227. Reches M, Gazit E (2006) Controlled patterning of aligned self-assembled peptide nanotubes. Nat Nanotechnol 1:195–200. https://doi.org/10.1038/nnano.2006.139
228. Mahajan N, Zhao Y, Du T, Fang J (2006) Nanoscale ripples in self-assembled lipid tubules. Langmuir 22:1973–1975. https://doi.org/10.1021/la051751n
229. Fang J (2007) Ordered arrays of self-assembled lipid tubules: fabrication and applications. J Mater Chem 17:3479–3484. https://doi.org/10.1039/B705350a
230. Zhao Y, Fang J (2008) Direct printing of self-assembled lipid tubules on substrates. Langmuir 24:5113–5117. https://doi.org/10.1021/la703634t
231. Safaryan S, Slabov V, Kopyl S, Romanyuk K, Bdikin I, Vasilev S, Zelenovskiy P, Shur VY, Uslamin EA, Pidko EA, Vinogradov AV, Kholkin AL (2018) Diphenylalanine-based microribbons for piezoelectric applications via inkjet printing. ACS Appl Mater Interfaces 10:10543–10551. https://doi.org/10.1021/acsami.7b19668
232. Adler-Abramovich L, Gazit E (2008) Controlled patterning of peptide nanotubes and nanospheres using inkjet printing technology. J Pept Sci 14:217–223. https://doi.org/10.1002/psc.963
233. Garcia-Fandino R, Amorin M, Castedo L, Granja JR (2012) Transmembrane ion transport by self-assembling α,γ-peptide nanotubes. Chem Sci 3:3280–3285. https://doi.org/10.1039/c2sc21068a

234. Ruiz L, Benjamin A, Sullivan M, Keten S (2015) Regulating ion transport in peptide nanotubes by tailoring the nanotube lumen chemistry. J Phys Chem Lett 6:1514–1520. https://doi.org/10.1021/acs.jpclett.5b00252
235. Montenegro J, Ghadiri MR, Granja JR (2013) Ion Channel models based on self-assembling cyclic peptide nanotubes. Acc Chem Res 46:2955–2965. https://doi.org/10.1021/ar400061d
236. Rodriguez-Vazquez N, Amorin M, Granja JR (2017) Recent advances in controlling the internal and external properties of self-assembling cyclic peptide nanotubes and dimers. Org Biomol Chem 15:4490–4505. https://doi.org/10.1039/c7ob00351j
237. Karlsson A, Karlsson R, Karlsson M, Cans A-S, Stroemberg A, Ryttsen F, Orwar O (2001) Networks of nanotubes and containers. Nature 409:150–152. https://doi.org/10.1038/35051656
238. Hurtig J, Orwar O (2008) Injection and transport of bacteria in nanotube-vesicle networks. Soft Matter 4:1515–1520. https://doi.org/10.1039/b800333e
239. Lizana L, Bauer B, Orwar O (2008) Controlling the rates of biochemical reactions and signaling networks by shape and volume changes. Proc Natl Acad Sci USA 105:4099–4104. https://doi.org/10.1073/pnas.0709932105
240. Sekine Y, Abe K, Shimizu A, Sasaki Y, Sawada S, Akiyoshi K (2012) Shear flow-induced nanotubulation of surface-immobilized liposomes. Rsc Adv 2:2682–2684. https://doi.org/10.1039/C2ra00629d
241. Fernandez-Lopez S, Kim HS, Choi EC, Delgado M, Granja JR, Khasanov A, Kraehenbuehl K, Long G, Weinberger DA, Wilcoxen KM, Ghadiri MR (2001) Antibacterial agents based on the cyclic D, L-α-peptide architecture. Nature 412:452–455. https://doi.org/10.1038/35086601
242. Motiei L, Rahimipour S, Thayer DA, Wong CH, Ghadiri MR (2009) Antibacterial cyclic D,L-α-Glycopeptides. Chem Commun 3693–3695. https://doi.org/10.1039/b902455g
243. Castillo JJ, Svendsen WE, Rozlosnik N, Escobar P, Martineza F, Castillo-Leon J (2013) Detection of cancer cells using a peptide nanotube-folic acid modified graphene electrode. Analyst 138:1026–1031. https://doi.org/10.1039/c2an36121c

Chapter 2
Lipid Nanotubes

2.1 Glutamic Acid-Based Nanotube

A typical example of amino acid-based amphiphile is glutamic acid derived lipids, *N,N′*-bis(octadecyl)-L-glutamic diamide **1** (Chart 2.1) [1]. Amidation of the *tert*-butyloxycarbonyl (Boc)-glutamic acid with octadecylamine and subsequent elimination of the protecting Boc group result in the targeted product. Usage of L- or D-form amino acid gives enantiomeric long-chain amino acid derivatives. Slow cooling of boiled ethanol solution of the lipid **1** to room temperature produced insoluble precipitate (< 1 mg/mL) or opaque organogels (> 19.5 mg/mL) that depend on the lipid concentration. TEM revealed that the dimensions of the nanotubes self-assembled in the ethanol solution (concentration: 20 mg/mL) are outer (116.7 ± 1.3 nm), inner (23.7 ± 1.2 nm) diameters and lengths (several hundred μm). The measured membrane thicknesses are equivalent to ten bilayers. Notably, right-handed nanotubes self-assemble from the L-enantiomer, whereas left-handed nanotubes from the D-form lipid. This finding clearly shows that each assembly is based on chiral molecular packing of the molecular building block.

Changing the mixing ratios of the L- and D-enantiomer of **1** continuously was found to enable the tuning of resultant self-assembled morphologies including one-handed nanotubes, twisted ribbons, platelets, and the opposite-handed nanotubes. Importantly, the chirality of the self-assembled nanostructures conforms to the majority rules principle [2, 3]. The morphological control in the self-assembly of mixed enantiomers expresses from one-handed seamless nanotubes through twisted ribbons and nanosheets to opposite-handed seamless nanotubes. Balancing of the chiral interactions between two enantiomers plays an essential role in controlling over the shape.

Aiming at the self-assembly of simple π-conjugated nanotubes, Liu and co-workers designed novel gelator molecules **1a, 1b, 1c**, and **1d** (Chart 2.1), in which four different phenyl, 1-naphthyl, 2-naphthyl, and 9-anthryl aromatic rings are attached to one end of the *N,N*'-bis(octadecyl)-L-glutamic diamide, respectively [4]. The phenyl derivative **1a** self-assembles in polar solvents, e.g., dimethyl sulfoxide (DMSO),

T. Shimizu, *Smart Soft-Matter Nanotubes*, Nanostructure Science and Technology,
https://doi.org/10.1007/978-981-16-2685-2_2

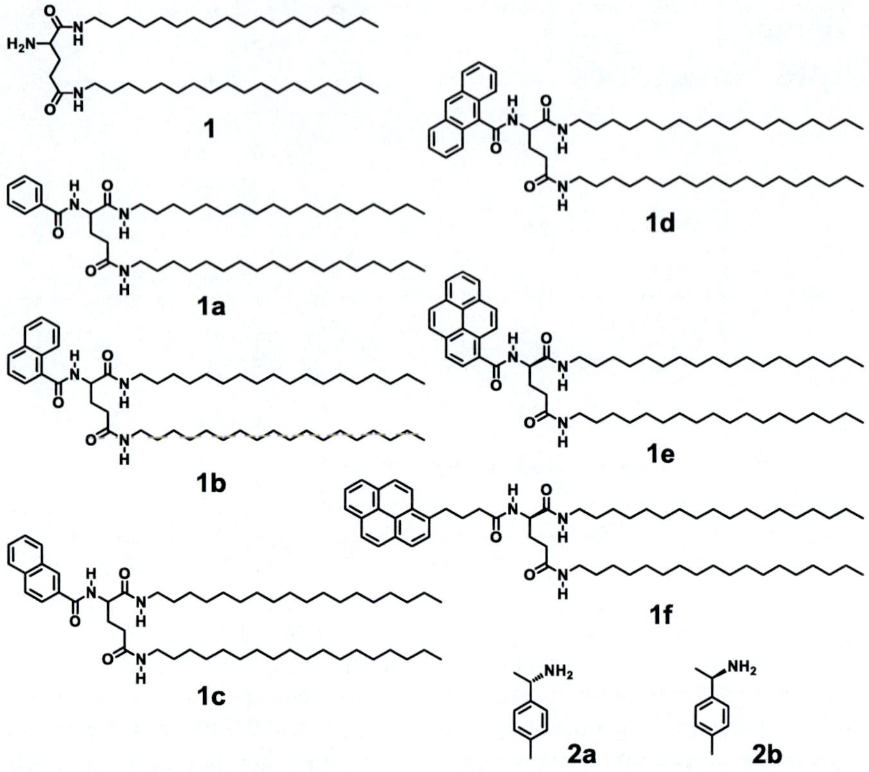

Chart 2.1 Chemical structures of **1**–**2**

N,*N*-dimethylformamide (DMF), and acetonitrile to form nanotubular structures that further hierarchically assemble into organogels. Meanwhile, the naphthyl derivatives **1b** and **1c** form gels in polar solvents, e.g., DMSO, DMF, acetonitrile, ethyl acetate, acetone, tetrahydrofuran (THF), and ethanol, the fundamental unit structure of which is nanotubes. No outstanding tubular structures are obtained in the self-assembled gels from larger anthryl derivative **1d**. The formation pathway of all the nanotubes is based on the closing of helically rolled up nanobelts composed of interdigitated bilayer membranes. UV irradiation on the gels undergoes strong blue emission due to the gelation-enhanced fluorescence emission, whereas the same irradiation induced no fluorescence in solution [4]. Interestingly, the self-assembled nanotubes can exhibit enantioselective fluorescence of chiral amines (Fig. 2.1). For example, the L-glutamide-based gelator **1b** self-assembles in acetonitrile to yield nanotubular architectures. When the injection volume of aniline vapor onto the fluorescent gel increased, the fluorescence quenched step-by-step. In particular, the quenching efficiency of the fluorescence of **1b** nanotube shows higher value for chiral organic amine (*S*)-(−)-1-(*p*-tolyl)ethylamine **2a** (Chart 2.1) than that for its enantiomeric amine (*R*)-(+)-1-(*p*-tolyl)ethylamine **2b** (Chart 2.1). With regard to related nanotube

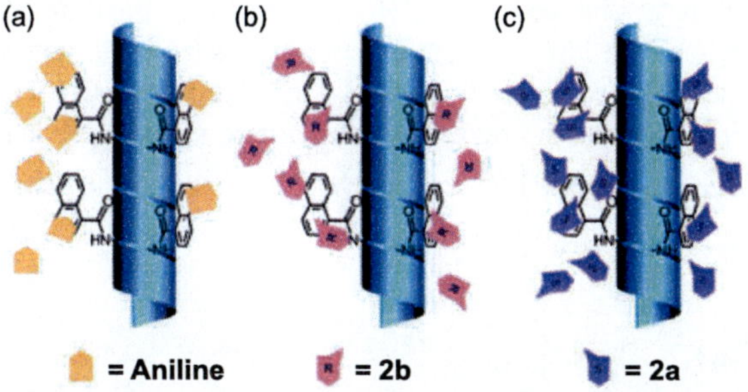

Fig. 2.1 **a** Detection of aniline by the self-assembled nanotube from 1-naphthyl derivative **1b**. Detection of chiral amine by the same nanotube; **b** R enantiomer **2b** and **c** S enantiomer **2a**. Reproduced with permission from Ref. [4], © 2014 WILEY–VCH Verlag GmbH & Co. KGaA, Weinheim

formation, the pyrene-bearing L-glutamide **1e** and **1f** (Chart 2.1) were prepared to examine the gelation activity toward polar or nonpolar solvents [5]. Interestingly, the derivative **1e**, in which the pyrene segment is directly connected with the glutamide unit, can yield nanotube architectures through the self-assembly in DMSO. No tube formation was observed for the derivative **1f** with trimethylene spacer. The research group assumes that the close location of the pyrene segment nearby the chiral center makes the rolling-up of nanosheets into tubular structures easier.

Liu and co-workers demonstrated that a self-assembled light-harvesting chiral nanotube from the cyanostilbene-bearing glutamate amphiphile (**3-L** or **3-D**) (Chart 2.2) can transfer chirality cooperatively and energy sequentially in an aqueous phase [6]. The compounds **3-L** and **3-D** self-assemble in water to form right- and left-handed helical nanotube, respectively, thus, resulting in the formation of hydrogel. The nanotube formation mediates the rolling of bilayer-based ribbons into helical hollow cylinders. A light-harvesting antenna system of organic nanotube was fabricated through binary or ternary assembly of the helical nanotubes from **3-L** or **3-D** with water-soluble achiral acceptors, thioflavin T (ThT) **4** (Chart 2.2) and acridine orange (AO) **5** (Chart 2.2). Fortunately, the ThT molecule **4** could serve as an appropriate receptor for **3-L** or **3-D** and further employed as an intermediating donor for the AO molecule **5**. Indeed, the ternary assembly system comprising **3-L** or **3-D**, **4**, and **5** displays only circularly polarized luminescence (CPL) signal ascribable to the AO acceptor [6]. This observation represents that the acceptor AO efficiently

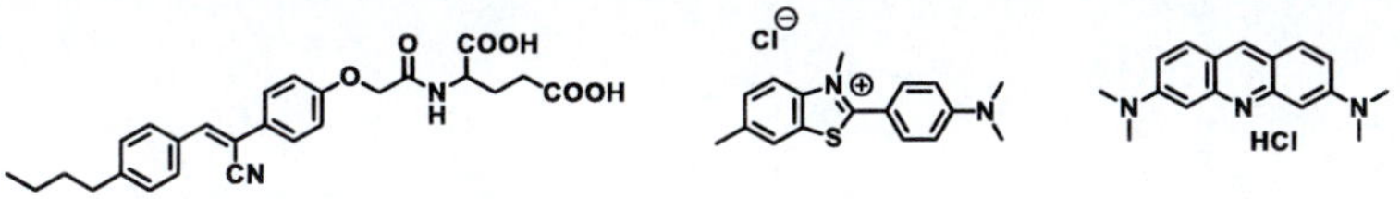

3-L or **3-D** **4** **5**

Chart 2.2 Chemical structures of **3**–**5**

collects the circularly polarized energy sequentially from the donor (**3-L** or **3-D**) and intermediate the donor **4**. The result, therefore, differs from the strong CPL induction of nonchiral organic fluorophores by helical nanofiber templates [7]. Thus, this work should lead to deep understanding of the cooperative chirality and sequential energy transfer in a light-harvesting system based on naturally occurring supramolecular assemblies [8].

2.2 Bile Acid

2.2.1 Cholic Acid

Amphiphilic bile acid derivatives have been widely known to self-assemble in liquid media to produce well-defined helical ribbon and tubular structures [9, 10]. Interestingly, slight chemically modified bile acids often undergo dramatical changes in their self-assembly behavior that largely differs from those of original compounds. Reaction of *p-tert*-butylbenzoyl chloride with the 3β-amino derivative of cholic acid **6** (Chart 2.3) gave the anionic derivative of sodium cholate **7** (Chart 2.3) [11]. Similar manner to the reaction procedure developed by Maitra et al. resulted in the production of the cationic derivative **8** (Chart 2.3) [12]. Galantini and co-workers reported that novel catanionic nanotubes based on rolling of sheets were shown to self-assemble in the mixture solutions of anionic **7** and cationic derivatives **8** of the sodium cholate **6** [13]. TEM images clearly showed that in some cases membrane sheets with stair-shaped borders or triangle edges are rolling-up to form tubular structures. Electrophoretic mobility measurement by using laser Doppler velocimetry corroborated that the compositions and charges of the nanostructures are controllable by the mixture stoichiometry. The Galantini's group also synthesized a novel

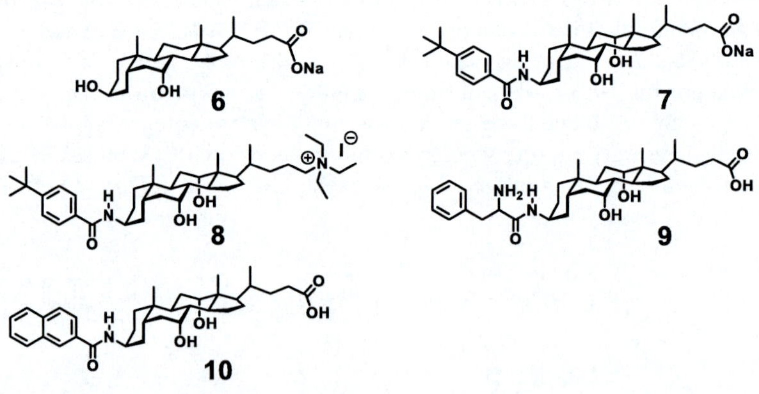

Chart 2.3 Chemical structures of **6**–**10**

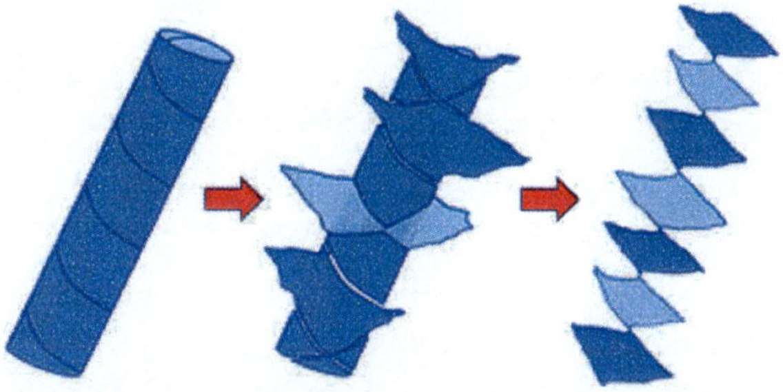

Fig. 2.2 Schematic illustration of the opening mechanism of a nanotube structure. Reproduced with permission from Ref. [15], © The Owner Societies 2013

amphoteric bile acid derivative **9** (Chart 2.3) by linking an L-Phe residue to the cholic acid. The amphiphile **9** gives transparent solutions under acidic conditions (36 mM, pH = 1.1) and self-assemble into hydrogels by the addition of NaCl (0.15 M) [14]. Cryo-TEM and AFM measurement revealed that extremely narrow nanotubes with 3-nm inner and 6-nm outer diameters organize the gel structures. The nanotube wall is based on monolayer membrane stabilized by the wedge-shaped molecular structure. The same research group also synthesized the cholic acid derivative **10** (Chart 2.3) by introducing 2-naphthamide group into the C3 position of the cholic acid. The obtained molecule self-assembles into smart nanotubes that exhibit a pH-sensitive opening of the nanotube in aqueous solutions (Fig. 2.2) [15]. The nanotubes comprise monolayer membrane and have outer diameters of 60 nm at pH 8–9, whereas they split open to give twisted or untwisted ribbons upon an increase in pH to 9–10. Such a pH-responsive morphological change strikingly differs from slow rolling and unrolling of ribbons so far reported [16]. The same molecule **10** is responsible to several stimuli of temperature and salt concentration in addition to pH, thus resulting in the formation of several self-assembled architectures such as nanotubes, ribbons, scrolls, or rolled lamellae [17].

2.2.2 *Lithocholic Acid*

Controlling the dimensions of the nanotube cross section, for example, in the range of 1–50 nm is crucial to maximize the functions of confined nanospaces that are applicable to a wide spectrum of nanobiofields. Very few approaches for tailoring the inner and outer diameters have been reported until now [18]. Galantini and co-workers found that the outer diameter of co-assembled organic nanotubes from a mixture of sodium lithocholate **6** and its bolaamphiphilic derivative **11** (Chart 2.4) is tunable by optimizing the mixture stoichiometry in aqueous solutions at pH 12 [19]. Individual homo-assembly of each component yields the nanotubes with different outer diameters of 46 and 18.5 nm for **6** and **11**, respectively. Detailed analysis of the small-angle X-ray scattering (SAXS) profile for the mixtures displayed that the outer diameters increase slowly from 18.5 to 46 nm with the decrease in the molar ratio of the bolaamphiphilic derivative **11** from 1.00 to 0.50.

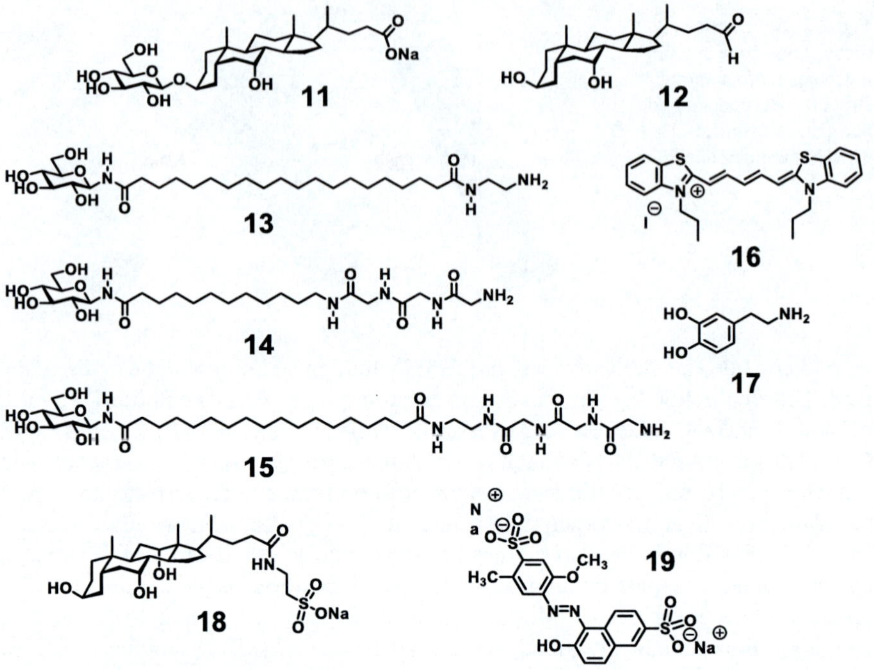

Chart 2.4 Chemical structures of **11**–**19**

By employing LCA **12** (Chart 2.4) as a molecular building block, Fang and co-workers found an interesting result. In alkaline solution (pH 12.0) at room temperature the molecules self-assemble into vesicles with diameters of ca. 1.5 μm after 2 h, which subsequently assemble linearly to yield microtubes with time [20]. Notably, the pathway of the microtube formation is quite different from conventional one that mediates helically coiled or twisted ribbon structures [16]. The generated tubes further coil to form 3D spirals over a few days. More interestingly, acidification of the microtube alkaline solution to pH 7.4 with HCl induced a pH-switchable spiral-to-straight morphological transition.

By utilizing **12**-based organic nanotubes in aqueous solutions, Collins and co-workers described the encapsulation behavior of a pair of enhanced green fluorescent protein (eGFP, donor) and its acceptor (mCherry) in a fluorescent resonance energy transfer (FRET) [21]. Meanwhile, self-assembled organic nanotubes from bolaamphiphiles **13**–**15** (Chart 2.4), having glucose and oligoglycine moieties at each end, displayed a characteristic diffusion and release behavior of the encapsulated GFP, which are highly dependent on the inner diameter sizes of the nanotubes [22]. Conversely, the thermal and chemical stability of GFP in the **12**-based nanotubes were enhanced irrespective of the inner diameter sizes from 20 to 40 nm. Horseradish peroxidase (HRP) is also trapped in the interior nanospace (inner diameters: 40 nm) of the self-assembled organic nanotubes from **12** [23]. Note worthy is that the encapsulated HRP in the nanotube can preserve its catalytic activity despite thermal treatment at 55 °C for 8 h. Meanwhile, free HRP dramatically loses the catalytic activity under same conditions. The loading amount of HRP into the interior nanospace of the LCA nanotube at 55 °C also increased with time, thus reaching an almost constant level after 4 h treatment. The encapsulated amount of HRP attains four times as much as that by the treatment at 4 °C.

Fang and co-workers found that the co-assembly of 3,3′-dipropylthiadicarbocyanine iodide **16** (Chart 2.4) and LCA **12** yield unique LCA nanotubes stabilized by J-aggregate interaction [24]. Absorption spectra of the J-aggregate nanotube showed an intense and sharp J-band ascribable to the delocalized excitations at 711 nm in phosphate buffered saline (PBS) solution. In the same solutions, the LCA nanotubes can behave as an selective and sensitive probe that can detect dopamine **17** (Chart 2.4) with the sensing limit of ~0.4 nM [24]. The J-band intensity changes, if effective photoinduced electron transfer occurs from the nanotube to the adsorbed dopamine, indicate the detected concentration of dopamine.

Fang's research group also reported smart organic nanotubes through the co-assembly of taurolithocholic acid **18** (Chart 2.4) and LCA **12** in aqueous solution [25]. The nanotubes undergo reversible tube-to-sheet morphological change that is induced by longitudinal zipping–unzipping motion of the membrane sheet (Fig. 2.3). Of great interest in its release action is that unzipping from the tubes to flat sheets by capillary force after dehydration enables the ejection of red dye #40 **19** (Chart 2.4) that is encapsulated in advance onto the substrate. Conversely, the resultant sheets are hydrated with aqueous solution to enable the zipping process into the nanotube formation.

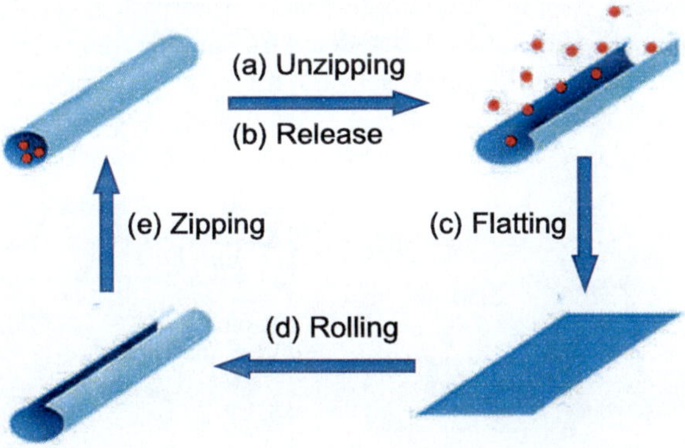

Fig. 2.3 **a** Longitudinal unzipping into a flat sheet by capillary force after being dehydrated and **b** subsequent release of encapsulated guest molecules. **c** Flatting into a sheet, **d** rolling, and **e** subsequent zipping back of the flat sheet into a hollow tube upon hydration with aqueous solution. Reproduced with permission from Ref. [25], © 2011 American Chemical Society)

2.3 Carbohydrate

2.3.1 Glycolipid

Glucopyranoside and glucopyranosylamide amphiphiles with saturated or unsaturated long hydrocarbon chains are well-known to self-assemble into supramolecular helical and tubular fibers in aqueous solutions [26]. Taking notice of long-chain phenols derived from cashew nut shell liquid, so-called cardanol, as a plant-derived renewable hydrophobic segment, we aimed to combine the principle of green chemistry with supramolecular chemistry (Fig. 2.4) [27]. The starting material cardanol, obtained from *Anacardium occidentale* L., is coupled with β-D-glucopyranose to give a mixture of 1-O-3′-*n*-(8′(Z),11′(Z),14′-pentadecatrienyl)-phenyl-β-D-glucopyranoside **20a** (Chart 2.5), 1-O-3′-*n*-(8′(Z),-11′(Z)-pentadecadienyl)phenyl-β-D-glucopyranoside **20b** (Chart 2.5), 1-O-3′-*n*-(8′(Z)-pentadecenyl)phenyl-β-D-glucopyranoside **20c** (Chart 2.5), and 1-O-3′-n-(pentadecyl)phenyl-β-D-glucopyranoside **20d**. The mixture of renewable resource-derived glycolipids **20-1** (**20a**:**20b**:**20c**:**20d**, 29:16:50:5 wt%) (Chart 2.5) undergo the self-assembly in water to yield uniform nanotubes with relatively smaller inner and outer diameters among the nanotubes so far obtained [27]. The dimensions can be featured by inner and outer diameters of 10–15 nm and 40–50 nm, respectively. Significantly, unsaturation of the hydrophobic chains has a striking effect on the fibrous self-assembled morphologies from **20-1** or **20c**. For example, helically coiled nanoribbons self-assembled from **20-1** slowly transform into a hollow cylindrical shape with opened two ends. Conversely, the saturated homologue **20d** gives no tubular architectures even after 1 year. This finding provides the first example about the unsaturation effect of the long hydrocarbon chains on the control over the self-assembled helical morphologies. During the self-assembly of the phenolic glycopyranoside **20-1**, π–π interactions between the aromatic rings play a crucial role in stabilizing the nanotubes. Indeed, octadecyl β-D-glucopyranoside **21** (Chart 2.5), which lacks aromatic phenolic rings, self-assembles to produce no fiber structures [27].

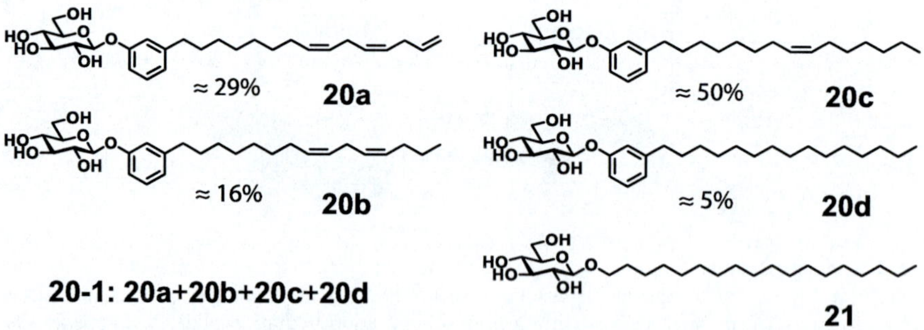

Chart 2.5 Chemical structures of **20-1**–**21**

Fig. 2.4 Photographs of cashew nut trees, cashew apples, and the cross section of the cashew nut [cited from http://www.bolacashew.com/more.htm for the cashew nut tree (left) and https://cardanol.vn/technology.html for the cashew apple and its cross section (right)]

Only a few systematic studies were performed to investigate the effect of unsaturation, i.e., the effect of the number and position of the double bond segment on the self-assembly of synthetic glucoside amphiphiles into helically twisted or coiled, and tubular nanostructures. Reverse-phase medium-pressure column chromatography allowed for the precise fractionation of the cardanyl glucopyranoside **20-1** into the four constituent molecules **20a–20d**. Upon self-assembly, the obtained monoene glucopyranoside **20c** proved to yield nanotube structures, whereas the saturated homologue **20d** gave twisted ribbons (Fig. 2.5) [28]. Binary assembly using the combination of the monoene **20c** and saturated **20d** derivatives enabled free control of resultant self-assembled morphologies ranging from twisted ribbons and helical

Fig. 2.5 EF-TEM images of a self-assembled **a** nanotube from 20 °C, **b** twisted ribbon from **20d**, and **c** helically coiled ribbon from a binary composition of **20c** and **20d**

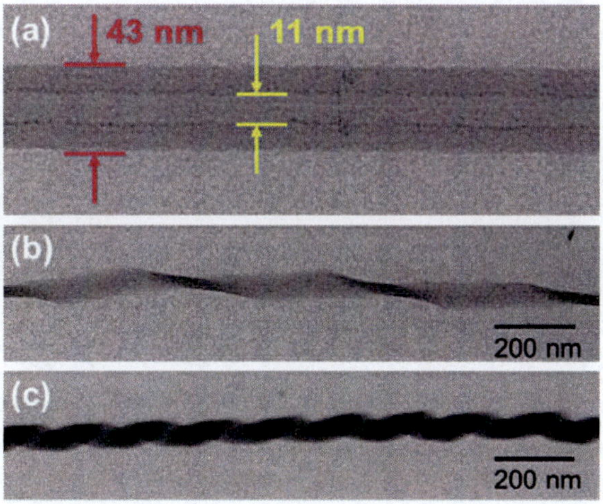

ribbons to nanotube structures [28]. Importantly, this approach is promising from the viewpoints that the assembly processes not only produce nanotubes with and without helical marking separately, but also regulate the pitch length of the helical ribbons.

2.3.1.1 Unsaturation Effect

Following the above-mentioned studies, we investigated the self-assembly behavior of a series of long hydrocarbon-chain-containing phenyl glucosides **22a–22d** (Chart 2.6) [29]. The four glucopyranoside amphiphiles differ in the number of cis double bonds, i.e., saturated, monoene, diene, and triene, in the long hydrophobic chains. With the increase in the number of cis double bonds, the glycolipids **22a**, **22b**, and **22c** self-assemble in aqueous solution to form twisted fiber, helical ribbon, and nanotube structures, respectively. Careful CD studies of **22a** and **22b** evidenced the formation of tubular assemblies through chiral molecular assembly from the existence of strong negative bands at 225 and 237 nm, respectively.

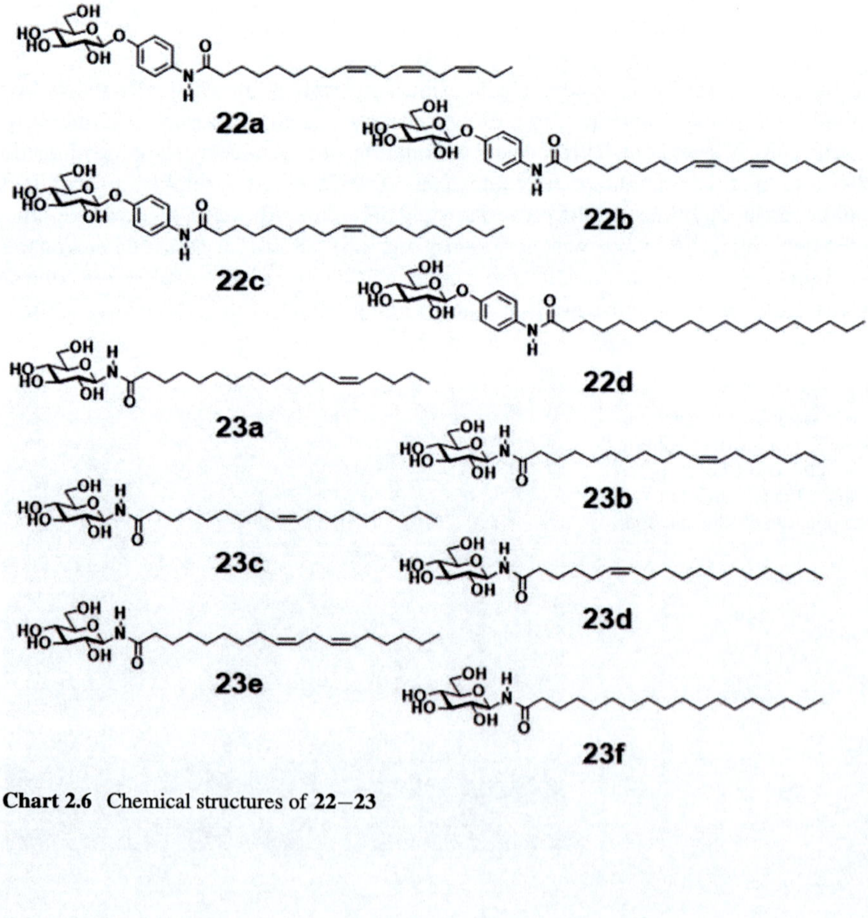

Chart 2.6 Chemical structures of **22–23**

The insert position of the cis double bond within the long hydrophobic chains of glycolipids also sensitively affects the yields and uniformity of the dimensions, particularly the outside diameters of resultant self-assembled nanotubes. We synthesized a series of *N*-gluconamide derivatives **23a–23f** (Chart 2.6) to optimize the most favorable structure as a nanotube self-assembling molecule [30]. Elaborated analyses of the resultant dimensions were carried out for the self-assembled nanotubes from six gluconamides. As a result, the gluconamide **23b** with a cis double bond at the C11 position underwent the self-assembly in water to homogeneous nanotubes having the narrowest distribution of outer diameters, giving excellent yields of more than 98%. CD spectroscopy of the aqueous nanotube solutions revealed that the intensity of the CD band at around 235 nm increased in the order of **23c**, **23a**, and **23b**. The obtained tendency is well compatible with the order of the uniformity observed for the outer diameters. An enforced bent conformation by the introduction of the 11-cis double bond should be responsible for the most favorable chiral molecular packing in the solid bilayer membranes, resulting in the narrowest diameter distribution.

2.3.1.2 Stiffness, Rigidity, and Elasticity

Young's modulus measured for a single piece of the self-assembled nanotube from the glycolipid **20-1** is $E = 720$ MPa in an aqueous solution [31]. This value is mostly the same as that measured for naturally occurring microtubules ($E = 1000$ MPa) featured by the outer and inner diameters of the same order [32]. By exploiting this moderate stiffness of the single glycolipid nanotubes from **20-1**, Ito and co-workers explored a new microinjection technique [31]. We can thus extrude a single piece of nanotube from a superfine glass capillary with internal diameters of 500 nm onto a glass slide (Fig. 2.6). The nanotubes, the long hydrocarbon chains of which take a crystalline (rigid) state, can be freely arranged and aligned. This micromanipulation approach of an organic nanotube is totally different from the electroinjection technique [33] that is associated with fluid lipid bilayer systems.

The external diameters of glycolipid nanotubes are generally larger than those of single-walled carbon nanotubes (1 nm ~ few tens of nm), but smaller than those of superfine glass capillaries (ca. 500 nm). The diameter range of the glycolipid nanotubes, thus, cover unobtainable dimension from any other nano- and microscale hollow cylindrical materials. Moreover, if the length is the same, the glycolipid nanotubes with inner diameters of 10 nm are smaller by a factor of 10^8 in volume

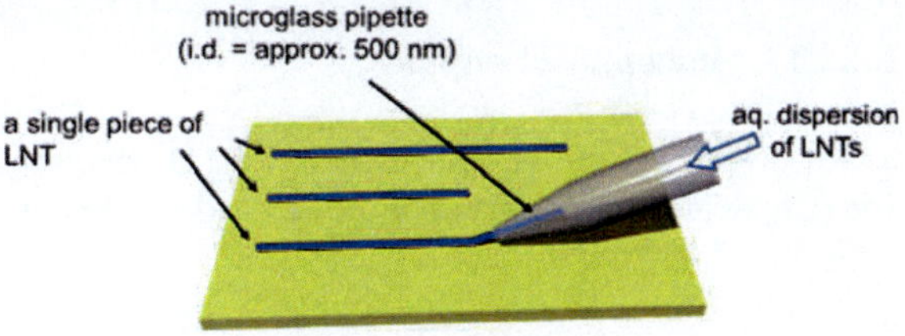

Fig. 2.6 Schematic image of a "dip pen" by using a single piece of LNT in place of an ink

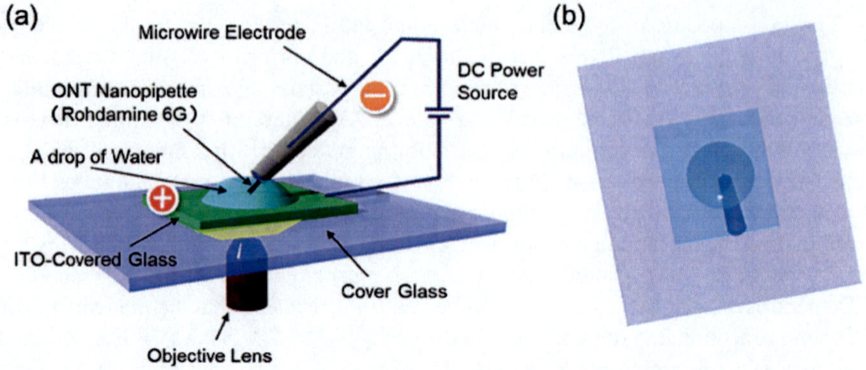

Fig. 2.7 **a** Schematic illustration for the spout of a solution of R6G from the SMNT nanopipette by voltage loading. **b** A view from the side of objective lens

than conventional microchannels used for DNA chips (inner diameter: ≈100 μm). It remains still challenging to miniaturize the internal diameters of the conventional high-axial-ratio microchannels even with advanced top-down nanotechnology. Ito and co-workers developed a resist resin-made nanochannel device, the channel diameters of which are in the 50–100 nm range [26]. The aforementioned microinjection technique [31] enabled the deposition of ten discrete linear arrays of the glycolipid nanotubes of **20-1** on a glass substrate. These nanotube arrays serve as a removable template after UV crosslinking reaction of the photosensitive resin.

We also developed a glycolipid-based nanopipette that can eject attoliter volume solutions. A 3D micromanipulation procedure allowed for the immobilization of a self-assembled glycolipid from **23b** with an inner diameter of 50 nm [30] onto the inner surface of a glass micropipette with an inner diameter of 1.8 μm [34]. The research group employed a photo-crosslinkable resin as a glue to seal the interspace between the external nanotube surface and the internal micropipette surface. Electroosmotic force enabled the ejection of an aqueous solution of fluorescent Rhodamine 6G (R6G). Fluorescent optical microscopy confirmed the initial ejection of the solution from the nanotube tip when the voltage of 426 V was applied (Fig. 2.7) [34]. The ejected volume of the solution is attoliter-scale, being largely different from that from nanotube-free glass micropipette. The volume coincides with from one-hundredth to one ten-thousandth of the amount that can be injected from commercially available pipette systems.

2.3.1.3 Glucophospholipid

Krafft, Riess, and co-workers report a new class of anionic glucophospholipids **24a**–**24c** (Chart 2.7) that can form hollow cylindrical nanotubes in aqueous solutions [35].

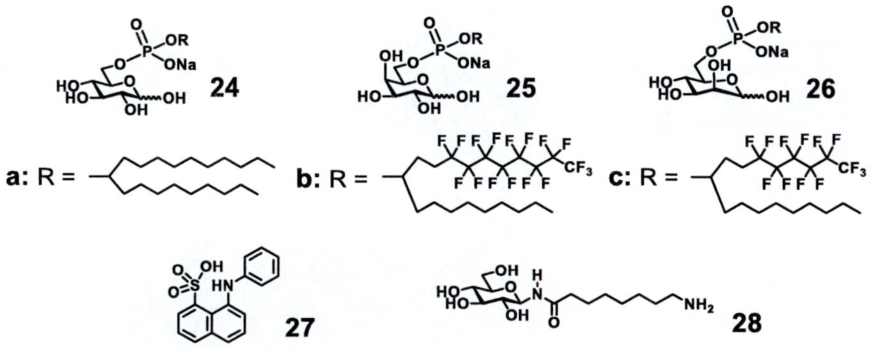

Chart 2.7 Chemical structures of **24**–**28**

These glycolipids carry hydrophobic long double chains that connect to the O-6 position of a hydrophilic glucose headgroup via a phosphate linkage. The nanotube formation takes place depending on pH of the aqueous solutions and proceeds favorably under alkaline conditions (pH $\approx$ 11), because the glucophospholipids have a negative charge. Notably, the galactose **25a**–**25c** (Chart 2.7) and mannose homologues **26a**–**26c** (Chart 2.7) produce no organic nanotubes through self-assembly. The self-assembled nanotubes from **24b** are characterized by inner diameter of 3.5 nm, outer diameter of 14 nm, and wall thickness of 5.25 nm [36]. This result means that the nanotube wall comprises three interdigitated and/or tilted bilayer membranes. The obtained inner diameter is categorized as relatively smaller size among the organic nanotubes self-assembled from synthetic amphiphiles.

2.3.1.4 Embedding of Hydrophobic Molecules

Co-assembly of glucopyranosyl amide *N*-(11-*cis*-octadecenoyl)-β-D-glucopyranosylamine **23b**, carrying a mono-unsaturated long alkyl chain and hydrophobic molecule such as 8-anilino-1-naphthalenesulfonic acid **27** (Chart 2.7), allowed for the fabrication of hybrid organic nanotube of 60 nm in inner diameters in a mixed solvent of water and organic solvents [37]. The fluorescence spectroscopy revealed that the resultant interdigitated bilayer membranes immobilized the guest molecule **27** within the hydrophobic region (Fig. 2.8). At room temperature, self-assembled nanotubes from the single component **23b** eject gradually the encapsulated molecules **27** in their lumen from their opened both ends, whereas the membrane-embedded ones **27** remain within the bilayers. Meanwhile, upon warming up the solution of the hybrid **23b/27** nanotubes to temperatures above the $T_{g\text{-}1}$ (= 59 °C), the solid bilayer membranes transform into flexible ones, thus, causing immediate release of **27** [37].

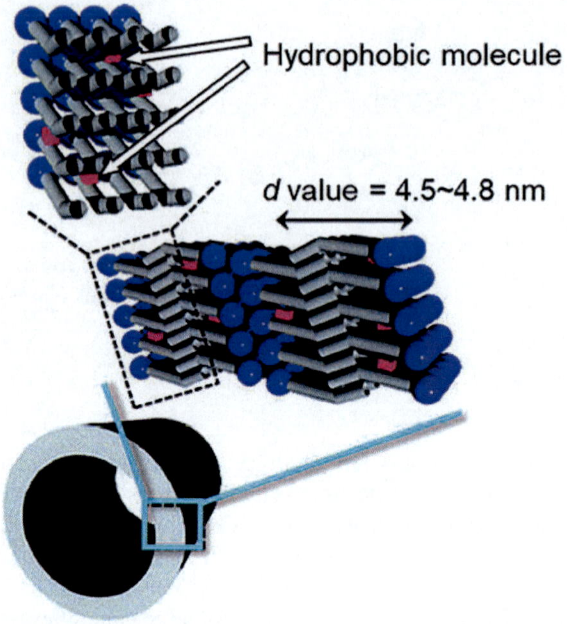

Fig. 2.8 Embedding of the hydrophobic molecule **27** (purple colored) within the bilayer membranes of the self-assembled nanotube. Reproduced with permission from Ref. [37], © The Royal Society of Chemistry 2011

2.3.1.5 Solubilization and Refolding of Protein

Morphological transformation of self-assembled nanostructures that accompany sol-to-gel thermal-phase transition was shown to exhibit smart functions. Specifically, the fluid bilayer membranes of the vesicles, self-assembled from the binary building blocks of the glucosamide lipid **23b** and its amino derivative **28** (Chart 2.7), act as good solubilizers of denatured GFP. The solid membranes of the nanotubes, converted from the vesicles by cooling the hot solutions of **23b** and **28**, displayed efficient release of the embedded GFP aggregates and the facilitation of their refolding (Fig. 2.9) [38].

2.3.1.6 Dispersion and Aggregation Control of Fullerene

SMNTs with interdigitated bilayer membranes, the inner and outer surfaces of which are covered with hydrophobic long alkyl chains, were prepared from *N*-stearoyl-β-D-glucopyranosylamine **23f** in alcohols. Interestingly, the topmost hydrophobic layer of the nanotubes, which trap fullerene (C60) molecules in the hollow cylinder, can be converted into the hydrophilic nature due to molecular rearrangement induced by thermal heating of the aqueous nanotube solutions at 85 °C [39]. As a result of this treatment, the C60-encapsulated nanotubes disperse efficiently in water. Moreover, we succeeded in covering the one end of nanotube channel with a lump of Au NPs. Irradiation of Nd-YAG laser at 532 nm was found to induce the cleavage of the

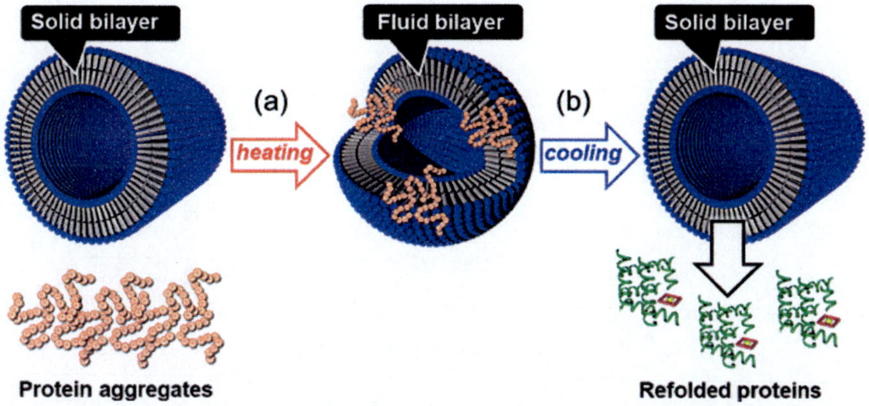

Fig. 2.9 **a** Vesicle formation from a solid bilayer-based nanotube upon heating to temperatures above the $T_{g\text{-}1}$ and solubilization of aggregates of denatured proteins. **b** Morphological transformation from vesicles to nanotubes at temperatures below $T_{g\text{-}1}$ and subsequent release of refolded proteins. Reproduced with permission from Ref. [38], © The Royal Society of Chemistry 2017

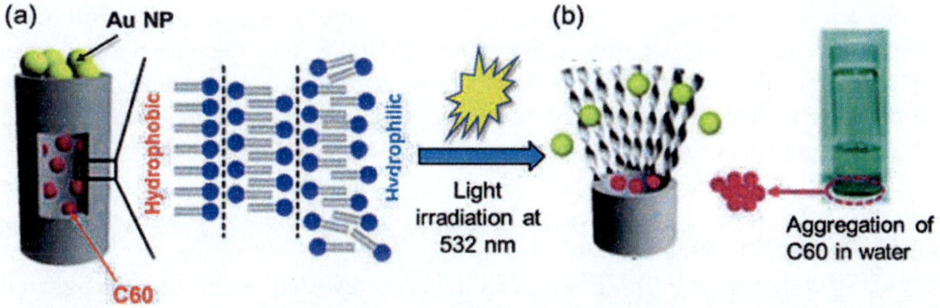

Fig. 2.10 **a** Encapsulated C60 molecules in the hydrophobic interior of an Au-NPs-capped LNT. **b** Enforced ejection of the encapsulated C60 triggered by the photothermal unfolding of the LNT upon light irradiation at 532 nm and resultant aggregated precipitate of the released C60 molecules

nanotube end into fibrous nanostructures through photothermal effect, resulting in the forced ejection of the encapsulated C60 molecules into the bulk water (Fig. 2.10) [39].

2.3.1.7 Light-Harvesting Antenna System

It is well-known that boronic acid derivatives react with diol molecules such as saccharides to result in the formation of cyclic boronic esters. Three different class of glycolipids (**29a**), (**29b** and **29c**), and (**29d** and **29e**) (Chart 2.8) having pyrene-, naphthalene-, or biphenyl-moiety, respectively, were prepared by dehydration. The

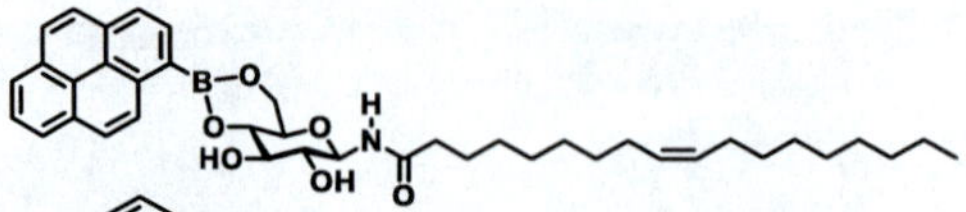

29a

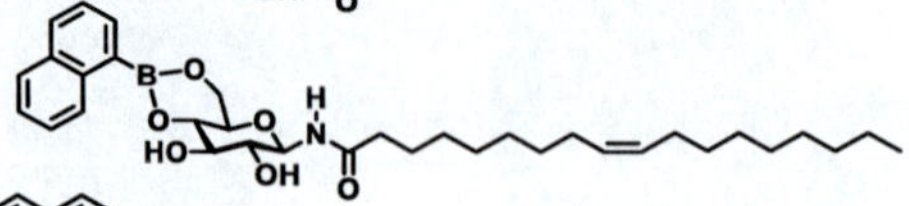

29b

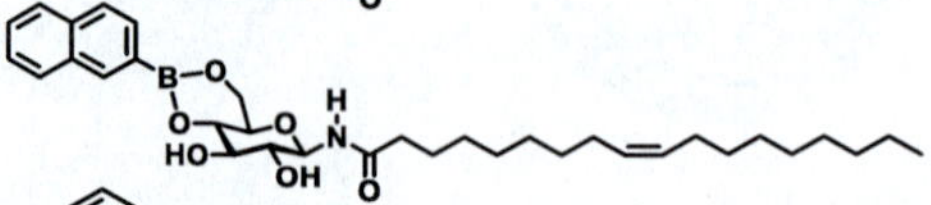

29c

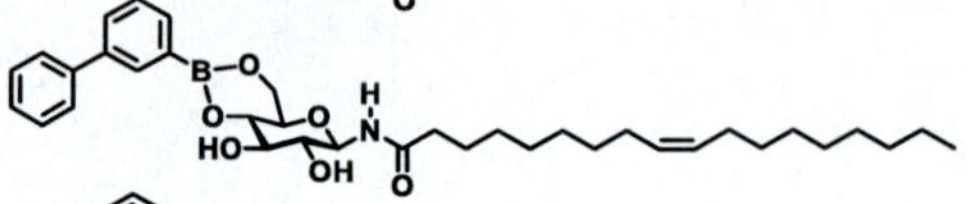

29d

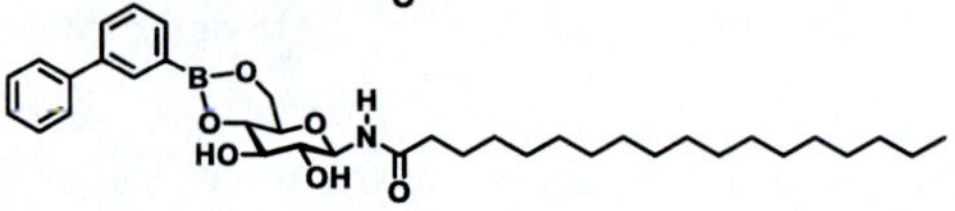

29e

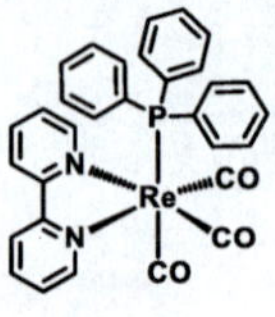

30

Chart 2.8 Chemical structures of **29**–**30**

combination of 1-naphthaleneboronic acid or 2-naphthaleneboronic acid with *N*-(9-cis-octadecenoyl)-β-D-glucopyranosylamine) **23c** [40], 1-pyreneboronic acid with the glycolipid **23c** [41], and 3-biphenylboronic acid and the glycolipid **23c** or its saturated derivative **23f** produced pyrene-bearing **29a**, naphthalene-bearing **29b** and **29c**, and biphenyl-bearing glucosamide derivatives **29d** and **29e**, respectively [42].

Self-assembly of the glycolipid **29b** having 2-naphthalene moiety in a toluene solution yields tubular structures of 15 nm in inner diameters, whereas the same solution of the 1-naphthalene-moiety-appended glycolipid **29c** gives tape-like structures 20–100 nm wide [40]. TEM and IR analyses revealed that three interdigitated bilayer membranes, stabilized by intermolecular hydrogen bonding and π–π staking among the molecules, organize the membrane wall of the nanotubes. Anthracene was, then, encapsulated in the nanotube channel and utilized as a guest fluorescent acceptor for energy transfer. High energy transfer from the donor naphthalene moieties, regularly and densely arranged in the membrane wall, to the encapsulated anthracene molecules as an acceptor was observed with 100% quantum efficiency in the supramolecular nanotube system (Fig. 2.11) [40]. Thus, the rational arrangement of energy donor and acceptor molecules by using the organic nanotube as a scaffold can provide a new model for light-harvesting antenna system.

2.3.1.8 Chiral Sensing of Amino Acids

The pyrene-appended glycolipid **29a** was shown to exhibit a chiral sensing ability toward a certain amino acid among twenty amino acids [41]. The self-assembly of this molecule in water forms vesicles (D-vesicle in Fig. 2.12) with spherical hollows of 32–68 nm in diameters, the wall of which consist of six layers of interdigitated

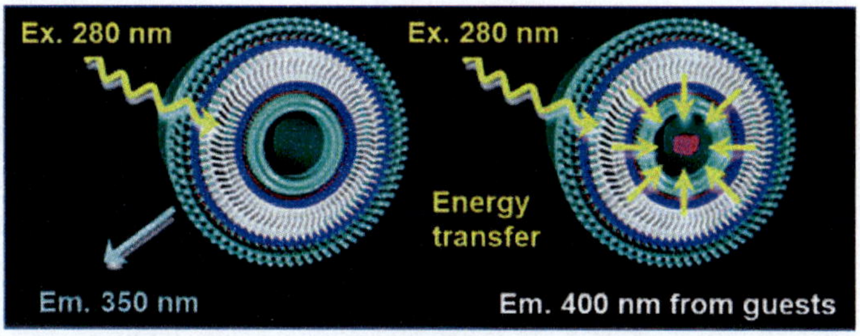

Fig. 2.11 Energy transfer from the naphthalene groups, densely organized within bilayer membranes of the nanotube wall as the host, to the encapsulated anthracene as the guest molecule in the nanochannel. Reproduced with permission from Ref. [40], © 2012 American Chemical Society

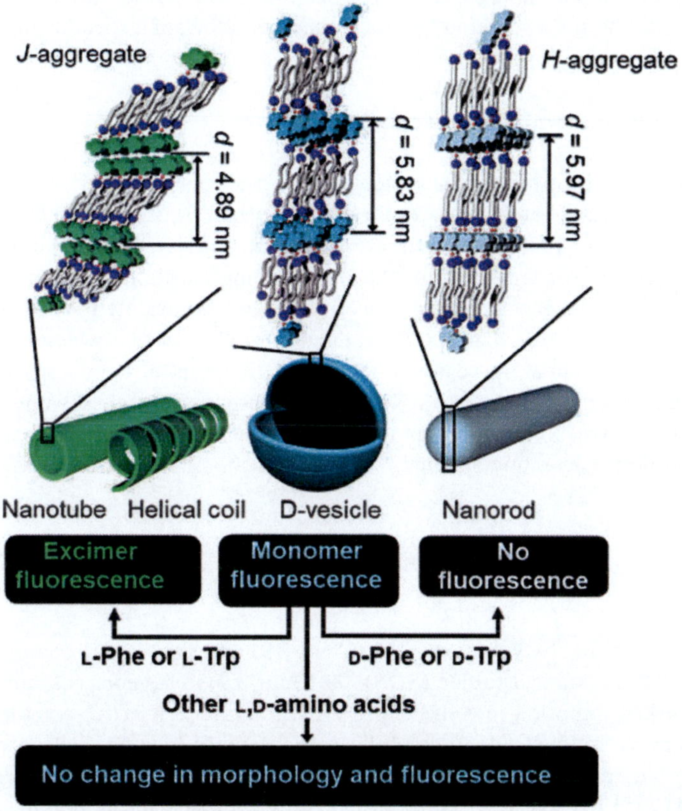

Fig. 2.12 Morphological transformations of the D-vesicle upon addition of L,D-amino acids at 45 °C. The molecular packing within the bilayer membranes of the nanotube, helical coil, nanotube, D-vesicle, and nanorod. Reproduced with permission from Ref. [41], © The Royal Society of Chemistry 2015

bilayer membranes. Addition of L-tryptophan and L-phenylalanine at 45 °C induced remarkable morphological change from the vesicular to helically coiled structures of 85–110 nm wide and a network of tubular architectures of 38 nm in inner diameters, respectively (Fig. 2.12). The sensitive response to each amino acid is visible to the naked eye, thus displaying fluorescence color changes from monomer fluorescence-based blue emission for the vesicles to excimer fluorescence-based green one for the nanotube or helical coils. Interestingly, upon addition of L-tryptophan results in the green aqueous dispersion of the helical coils, while addition of L-phenylalanine causes the hierarchical formation of green hydrogels [41]. In contrast, addition of D-tryptophan or D-phenylalanine caused no fluorescence in the aqueous dispersion of the D-vesicles (Fig. 2.12). The rearrangement of the molecular packing into J-aggregate and H-aggregate in their bilayer membranes drives the emission and quenching of fluorescent. Chiral organic nanotube and helical coil superstructures enabled, thus chiral sensing and qualitative analyses toward a specific amino acid.

2.3.1.9 Enhancement of Photocatalytic Activity

Boiled dispersions of the biphenyl-appended glycolipids **29d** and **29e** in toluene were allowed to cool slowly, thus, resulted in the formation of organogels that emit an intense fluorescence [42]. Electron microscopic, IR, and XRD studies confirmed that the organogels self-assembled from **29d** and **29e** comprise discrete organic nanotubes of 5 and 15 nm in inner diameters, respectively. The photocatalytic activity of Re(I) bipyridine complex **30** (Chart 2.8) for reduction of CO_2 to CO, when encapsulated in the nanotube channel, increases much higher than that of the complex in a free state in bulk solutions (Fig. 2.13). Notably, the activity depends sensitively on the inner diameter sizes, indicating that the nanotube with 5-nm inner diameters act as an efficient light-harvesting antenna.

2.3.2 Cyclodextrin

It is well-known that β-cyclodextrin **31** (Chart 2.9) undergoes a 2:1 complex formation with sodium dodecyl sulfate (SDS) **32** (Chart 2.9) in aqueous solutions. Huang and co-workers reported that the entirely hydrophilic 2:1 **31/32** complex undergoes an unconventional nonamphiphilic self-assembly [43, 44] in aqueous solutions to yield microtubes with outer diameter of ≈1 μm and two opened ends (Fig. 2.14) [45]. All the microtubes are existing mutually discretely, having characteristic rigidity, uniform outer diameters, and long persistence lengths of submillimeter order. The self-assembly is driven exclusively by hydrogen bonds that are relatively strong, directional interactions. In this context, this hydrophilic assembly is different from conventional self-assembly, in which the hydrophobic effect instead of hydrogen bonds dominates the behavior. The microtubes are composed of multi-walled structures of a 4-nm-thick bilayer. Freeze-fracture TEM analyses revealed

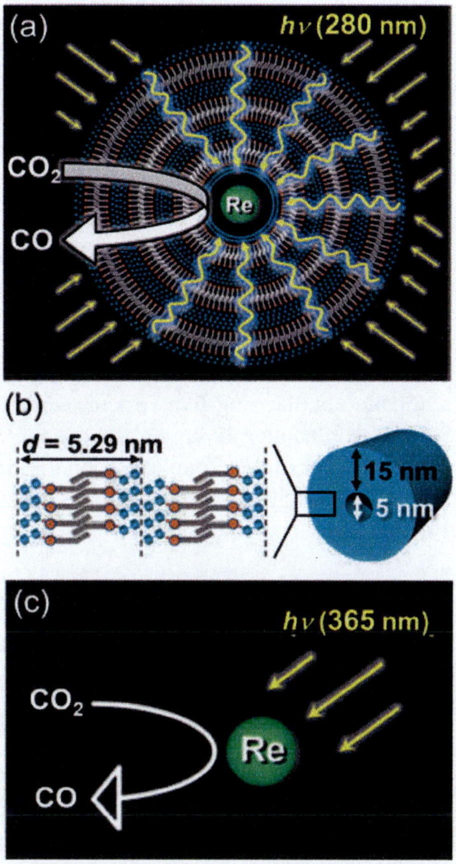

Fig. 2.13 **a** Photocatalytic reduction of CO_2 by excitation at 280 nm and the three bilayer membranes of **29d**. Effective energy transfer occurs from the biphenyl moieties, densely packed in the bilayer membranes of the nanotubes, to the encapsulated Re(I) complex **30** in the nanochannel. **b** Interdigitated bilayer membranes that, however, are depicted as two layers here. **c** Photocatalytic reduction of CO_2 by direct excitation at 365 nm of the free Re(I) complex **30** in the bulk solution. Reproduced with permission from Ref. [42], © The Royal Society of Chemistry 2017

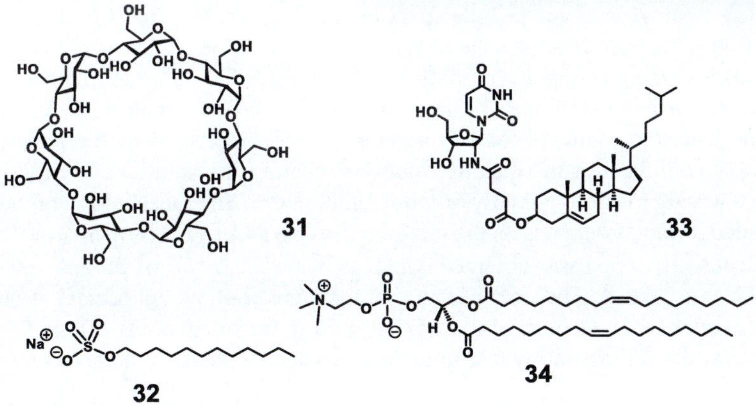

Chart 2.9 Chemical structures of **31**–**34**

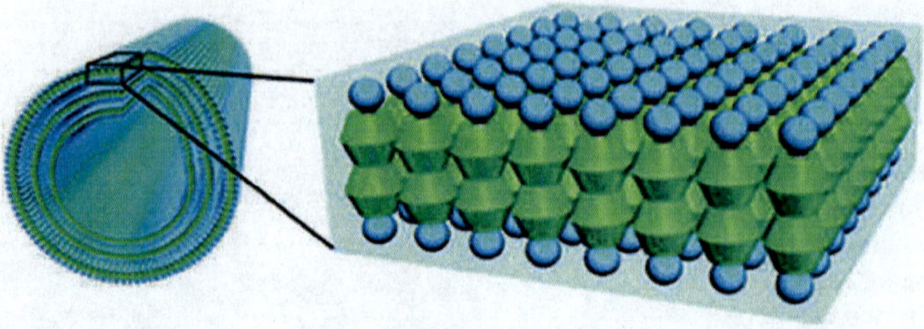

Fig. 2.14 Self-assembly of the 2:1 **31/32** complex into a multiwalled nanotube structure (conc. = 25– 6 wt%). Reproduced with permission from Ref. [45], © The Royal Society of Chemistry 2010

that a set of coaxial, equally spaced, and hollow cylindrical structures construct a notable annular-ring-like structures in the cross section of the microtubes [45]. Self-assembled morphologies of the 2:1 **31/32** complex strongly depend on the total concentrations of the complex. Upon dilution, the aggregation morphology changes from lamellae (50–25 wt%) via microtubes (25–6 wt%) to vesicles (6–4 wt%) [45, 46].

2.4 Cholesterol-Modified Nucleoside

There have been several reports on the spontaneous assembly of nanotubes and helical ribbon structures from multicomponent systems containing a natural sterol or amphiphilic nucleobase derivatives.[47, 48] Pescador, Arbuzova, and co-workers demonstrated for the first time the production of microtubes, the dimensions of which are 2–3 μm in outer diameters and 20–40 μm in length, from the binary mixture of lipophilic nucleoside **33** (Chart 2.9) (20–50 mol%) and the phospholipid 1,2-dioleoyl-*sn*-glycero-3-phosphatidylcholine (DOPC) **34** (Chart 2.9) [49]. A cholesterol moiety is connected to the 2′-position of the nucleoside 2′-deoxy-2′-aminouridine, thus, giving the lipophilic nucleoside **33**. The cholesterol moiety of **33** serve as a hydrophobic segment, whereas the uridine moiety a molecular recognition element. The procedure for the self-assembly is accomplished by the rehydration of a mixture lipid film in aqueous solutions, heating the solutions to 70 °C, and finally allowing to cool gradually to room temperature. The morphology of the self-assembled products depends on the mixing ratio of **33** and **34**. Above 50 mol% of **33** no aqueous dispersion was obtained, whereas below 20 mol% of **33** only vesicular assemblies appeared. The obtained microtubes have hollow cylindrical structures with two open ends. Control of the mixing ratio in the range of 20–50 mol% of **33** allowed for the self-assembly into thinner microtubes with outer diameters less than 1 μm.

2.5 Amphiphilic Azobenzene Derivative

The photochromic properties of azobenzene and its related derivatives have been of great interest so far. The reason why is that photoisomerization between cis and trans isomers of azobenzene moiety not only takes place as a structural change on molecular scale but also leads to a dynamic morphological change on larger length scales. Photo-responsive morphological transformation of molecular self-assemblies containing azobenzene moiety proceeds in response to light. Moreover, light serves as an external stimulus that can allow for accurate, quick, and remote photo-switcher while irradiating a target. Controlled release of a molecular guest was, thus, attempted by a morphological transformation from a tube to a cylindrical fiber with light stimulus. We prepared organic nanotubes of 20 nm in inner diameter through the self-assembly of newly designed amphiphilic azobenzene derivative Gly_1Azo **35a** (Chart 2.10) that carries *N*-terminated glycine moiety in water [50]. Two solid bilayers organize the nanotube membrane wall 8 nm thick. UV irradiation causes the trans-to-cis photoisomerization within the membrane wall, leading to a dramatic change in morphology from tubular to hollow cylindrical fibers with inner diameters of 1 nm (Fig. 2.15). Reversed cis-to-trans photoisomerization of the azo-moiety by visible-light irradiation induced no reconstruction of original nanotubes, but helical nanotapes that are unable to encapsulate any guest molecules (Fig. 2.15). By utilizing the photostimulated morphological change, we, thus, demonstrated a novel release control of carboxyfluorescein **36** (Chart 2.10) as an encapsulated guest [50]. The findings mentioned above greatly differ from the tube-to-fiber or tube-to-sphere morphological transitions caused by the change in external conditions or

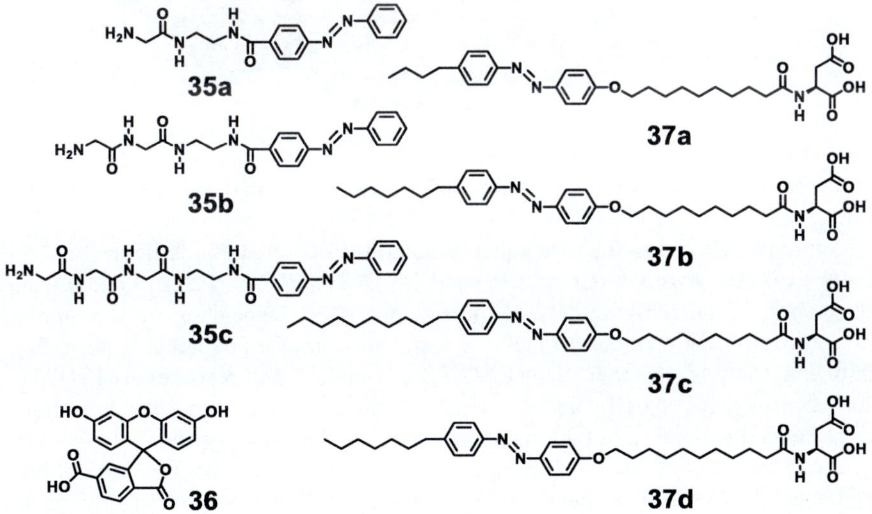

Chart 2.10 Chemical structures of **35**–**37**

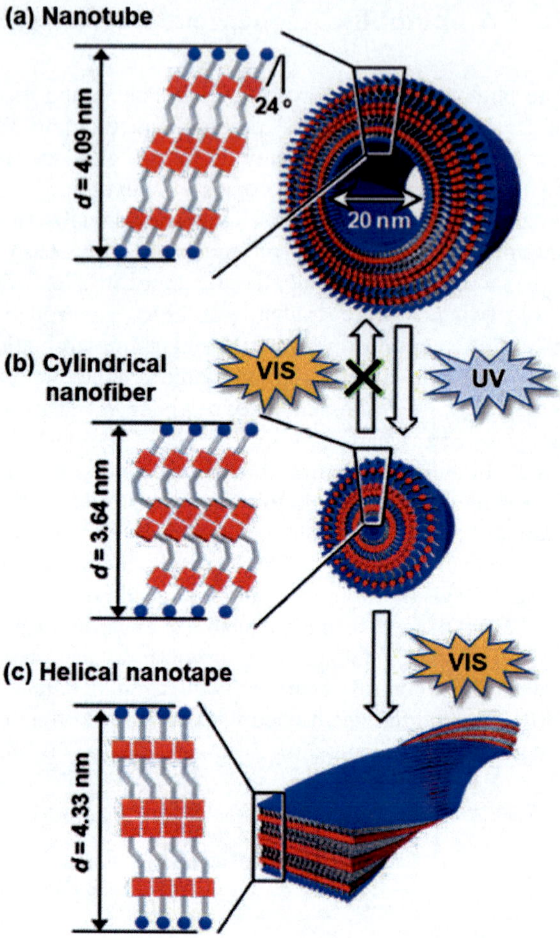

Fig. 2.15 Morphological transformations from a nanotube to a helical nanotape via cylindrical nanofiber, which are caused by photoisomerization of the azobenzene unit within the solid bilayer membranes of **35a**. Reproduced with permission from Ref. [50], © 2011 WILEY–VCH Verlag GmbH & Co. KGaA, Weinheim

surroundings, i.e., temperature, dilution, salt concentration, pH, and complex reaction [16].

Similarly, we found that the aqueous solutions of the amphiphilic azobenzene derivatives Gly_2Azo **35b** (Chart 2.10) and Gly_3Azo **35c** (Chart 2.10) yield organic nanotubes by the self-assembly at room temperature. Depending on the number of the constituting glycine residues, the resultant nanotube architectures from **35a**, **35b**, and **35c** possess inner diameters of 20, 7, and 13 nm, respectively [51]. The self-assembly, particularly under coexistence of a protein, can provide protein-encapsulated nanotubes in their nanochannels [52]. Intriguingly, the Gly_2Azo **35b**-derived nanotubes of 7 nm in inner diameters (i.d.) accelerate the protein refolding, while the Gly_3Azo **35c**-derived ones of 13 nm i.d. facilitate the protein aggregation (Fig. 2.16b, c). Meanwhile, no confinement effect was observed for the refolding behavior of the encapsulated protein in the nanotube channels with the widest i.d. of

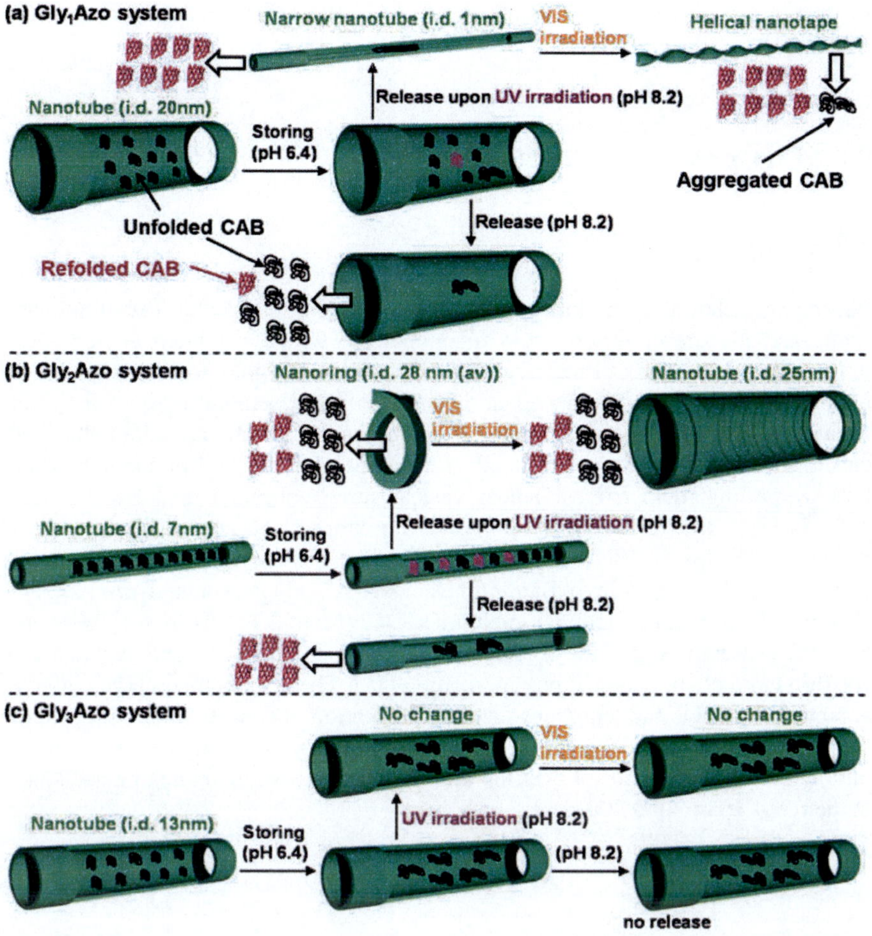

Fig. 2.16 Photoinduced morphological transformations of Gly_nAzo (n = 1, 2, and 3)-derived nanotubes and refolding of denatured CAB with and without photoirradiation scheme. Reproduced with permission from Ref. [52], © 2016 WILEY–VCH Verlag GmbH & Co. KGaA, Weinheim

20 nm from Gly_1Azo **35a**. Photo-triggered shrinkage of the inner diameters of the Gly_1Azo **35a** nanotube by UV-irradiation causes a compulsive ejection of the protein carbonic anhydrase (CAB) into the external solution, thus, enhancing the refolding efficiency furthermore (Fig. 2.16a) [52].

Barclay and co-workers reported that the self-assembled morphologies of the amphiphilic azobenzene derivatives **37a–37d** (Chart 2.10) having L-aspartic acid as a hydrophilic headgroup are helical structures-based nanotubes coexisting with helically twisted and/or coiled ribbon structures [53]. The self-assembly behavior of the amphiphiles **37a–37d** in aqueous methanolic solutions is, thus, strongly affected by the number of carbons in both the proximal (C10 or C11) and distal chains (C4, C7, or C10).

2.6 Phosphocholine Derivative

2.6.1 Phosphatidylcholine

Phosphatidylcholine is a family of phospholipids that consist of a choline headgroup and glycerophosphoric acid functionalized with a saturated and/or unsaturated fatty acid. They serve as a vital component of biological membranes. In nonbiological systems, the phosphatidylcholines self-assemble in aqueous buffers to form spherical liposomes including synthetic multilamellar and unilamellar vesicles. Tirrel and co-workers developed a new preparation method of nanotube channels by pulling them from a feeder vesicle surface through mechanical retraction [54, 55]. The vesicle is composed of a mixture of 66 mol % stearoyl-oleoyl phosphatidyl choline **38** (Chart 2.11), 33 mol % cholesterol **39** (Chart 2.11), and 1 mol % *N*-([6-(biotinyoyl)amino]hexanoyl)-1,2-dihexadecanoyl-*sn*-glycero-3-phosphoethanolamine, triethylammonium salt **40** (Chart 2.11). Thus, they were able to arrange the networks of the organic nanotube and stabilize the resultant patterns in site through a photochemical polymerization [55]. The resultant nanotubes serve as nanoscale channels and templates for the construction of nanoscale conduit patterns connecting liposome containers. Notably, only tuning the suction pressure in the micropipette enables control over the internal diameters of the nanotube accurately in the range from 20 to 200 nm.

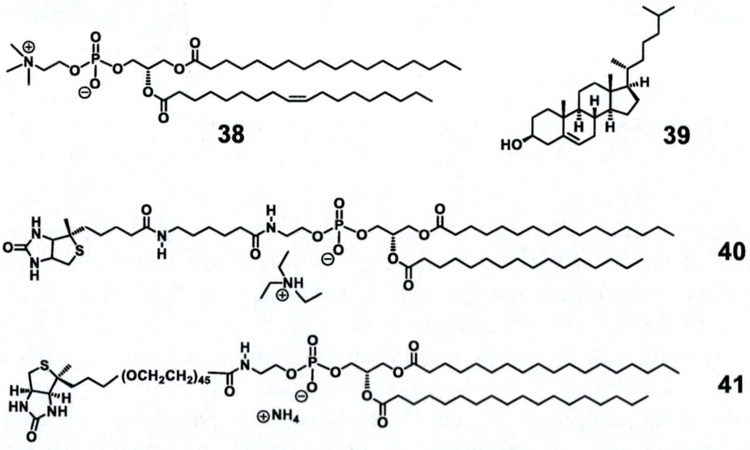

Chart 2.11 Chemical structures of **38**–**41**

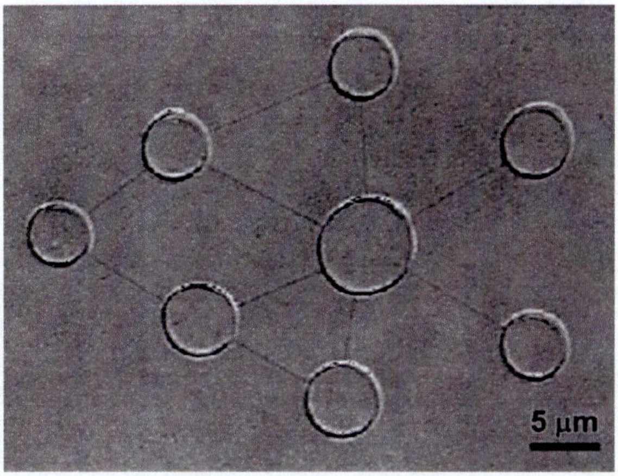

Fig. 2.17 A nanotube–vesicle network consisting of eight containers. The network was produced from a giant vesicle, in which erythrocyte membrane protein was embedded. Reproduced with permission from Ref. [60], © 2003 American Chemical Society

Subsequently to the nanotube conduits and networks by Tirrel et al. [55], Orwar and co-workers also reported an electroinjection approach that enables the fabrication of nanotubes as well as complex 2D microscopic networks [33, 56]. Organic nanotubes with diameters of 100–300 nm, thus, interconnect liposome containers self-assembled from either from egg yolk-derived L-α-phosphatidylcholine or from acetone-purified soybean lecithin. These nanotube networks provide us with prototype devices that are applicable to study chemical computations [57], intracellular transport [58], and biochemical reactions in confined nanospaces[56]. Remarkably, the research group can regulate the nanotube length, container size, the connectivity of the nanotubes and containers, and the angle formed by two nanotube lines (Fig. 2.17) [33]. The obtained nanofluidic devices should be more advanced than microfluidic devices created by top-down fabrication, leading to attractive applications [59].

Sasaki, Akiyoshi, and co-workers proposed a new preparation method of lipid nanotubes, in which a shear flow stress is applied to surface-immobilized liposomes onto the wall of a flow channel substrate [61]. Giant liposomes were self-assembled from a mixture of 1,2-dioleoyl-*sn*-glycero-3-phosphatidylcholine **34** and biotin-modified phospholipid (1,2-distearoyl-*sn*-glycero-3-phosphoethanolamine-*N*-[biotinyl(polyethyleneglycol)-2000] **41** (Chart 2.11). They successfully prepared extremely long nanotube at most several millimeters long with regulated direction. The obtained nanotube conduit can be expected as a transportation nanochannel or nanovessel reactor for small molecules and proteins.

2.6.2 *Diacetylenic Phosphocholine Derivative*

In 1984, double-chain diacetylenic phospholipids **42** [1,2-bis(tricosa-10,12-diynoyl)-*sn*-glycero-3-phosphocholine] (Chart 2.12) gave one of the first documented examples for the self-assembly of SMNTs in aqueous solutions. Yager and Schoen aimed originally at stabilization of self-assembled nanostructures such as bilayered vesicles by the polymerization of the diacetylenic groups on exposure to UV light [62]. The self-assembly of the polymerizable double-chain phospholipid resulted, however, in the formation of tubular architectures against the intended molecular design (Fig. 2.18). Introduced kink conformation of the hydrophobic tails, affected by unsaturation around the middle part of the long chains, serves a function to promote chiral molecular packing that eventually leads to the nanotube formation. This background is always fascinating large number of researchers, who have contributed to interesting findings of SMNTs from diacetylenic amphiphiles.

2.7 Diacetylenic Derivative

A family of amphiphiles with amine salts connecting to diacetylenic long single-chain was examined with regard to the self-assembly in water to produce nanotubes [63].

Chart 2.12 Chemical structures of **42**–**43**

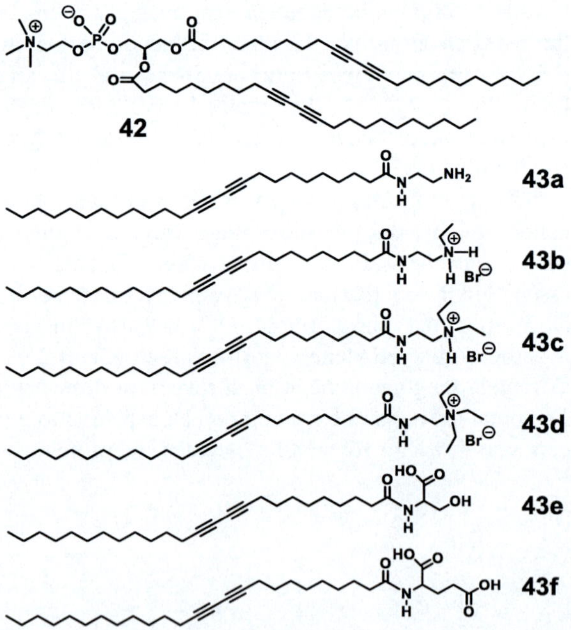

Fig. 2.18 Phase-contrast optical micrograph of unpolymerized hollow cylindrical tubules at high magnification. Reproduced with permission from Ref. [62], © 1984 Gordon and Breach, Science Publishers, Inc.

Strikingly, among the six synthesized lipids **43a–43f** (Chart 2.12) that include quaternary ammonium salt **43d**, amine salts **43b** and **43c**, and related amide derivatives (**43e** and **43f**) of diacetylenic compound, only the secondary amine salt **43b** gave uniform nanotube structures through self-assembly in high yields. The hydrophilic headgroup of the nanotube-forming amphiphiles comprises the hydrogen bromide salts of a secondary amine. Their dimensions are characterized by 89 nm outer diameters and 27 nm wall thickness corresponding to five bilayer membranes. More interestingly, nanotubular morphology emerges in spite of the absence of a chiral center in their chemical structures. This means that the nanotube self-assembly from diacetylenic amphiphiles may not always require molecular chirality. The research group proposed the initial formation of single bilayer, subsequent bending of the bilayer into a hollow cylinder, and repeating of this process five times (Fig. 2.19). It should also be noted that the vertical and dense alignment of the organic nanotubes was attained on a glass slide to give a nanotube carpet [63]. The antimicrobial activity, response of polymerized nanotubes to bacteria, and nanotube–bacteria interactions were also examined to explore biological applications. In contrast to the self-assembly of the diacetylenic amphiphiles with the headgroups of secondary amine salt derivative, no complete self-assembly into tubular structures as main components were observed for the diacetylenic amphiphile having the serine **43e** or glutamic acid headgroup **43f** [64].

Mésini and co-workers developed the self-assembly behavior of newly synthesized nonionic diacetylenic lipids **44a–44d** (Chart 2.13) having no chiral center [65]. The molecular design is simple and based on the combination of a single hydrophobic diacetylenic chain with an hydrophilic oligo(ethylene oxide) via amide bond linkage. When allowed to gradually cool from above 36 °C to room temperature, the aqueous solution of **44a** yielded nanotubes (outer diameters: 59 ± 1 nm and wall thickness: 10 ± 1 nm) or nanotapes. The obtained cryo-TEM images for the nanotube wall are consistent with a single bilayer structure, in which the hydrophobic alkyldiyne chains reside in the wall center with a tilting angle of 40°. Under the same conditions as above, the lipids **44b** and **44c** self-assemble into spherical micelles, whereas the lipid **44d** is insoluble to give dispersion. Interestingly, UV irradiation to the nanotube-containing aqueous dispersion of **44a** induced the color change into

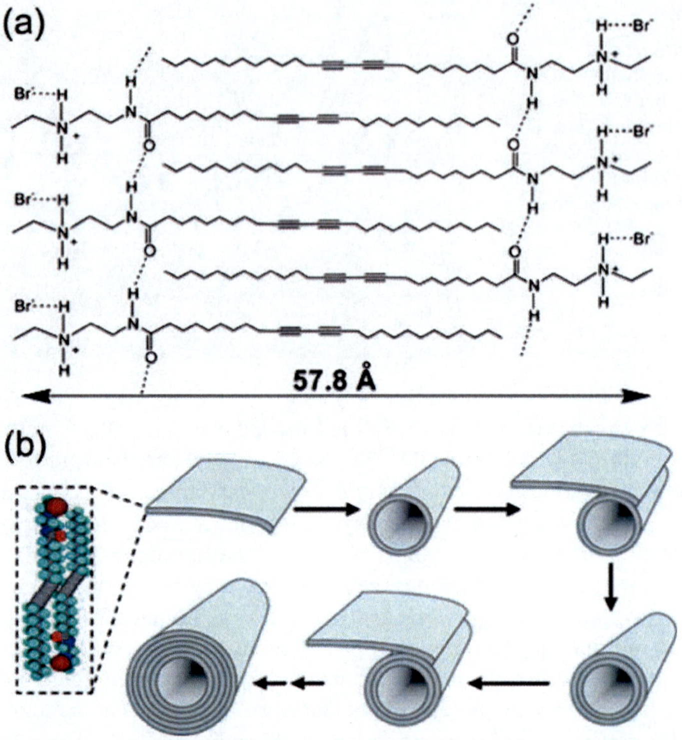

Fig. 2.19 Proposed **a** interdigitated bilayer structure that constructs a nanotube membrane wall and **b** self-assembly scheme into the nanotube structures from **43b**. Reproduced with permission from Ref. [63], © 2004 American Chemical Society)

purple, displaying the formation of loosely twisted ribbons as a major product after partial photopolymerization [65].

Kawano and Urban demonstrated that the nanotube dimensions sensitively respond to temperatures in polymeric hybrid nanotube system comprising poly(*N*-isopropylacrylamide) (PNIPAM) **45** (Chart 2.13) and the diacetylenic phospholipid **42** [66]. They thus prepared the temperature-responsive, reversibly expandable hybrid nanotubes by the polymerization of *N*-isopropylacrylamide (NIPAM) within the hydrophilic phospholipid interlayers that organize a part of membrane wall (Fig. 2.20). Morphological analyses with TEM revealed that heating the nanotubes from 25 to 40 °C causes a shrinkage of the outer diameters from 660 to 530 nm as well as the reduction in wall thickness from 142 to 67 nm. The observed dimensional transformation is attributed to molecular rearrangements due to coil-to-globule transitions of PNIPAM at temperatures below or above the lower critical solution temperature of 37 °C.

Khiar et al. newly designed diacetylene-containing glycolipids, in which tetra(ethylene glycol)-appended mannose moiety is combined with long hydrophobic

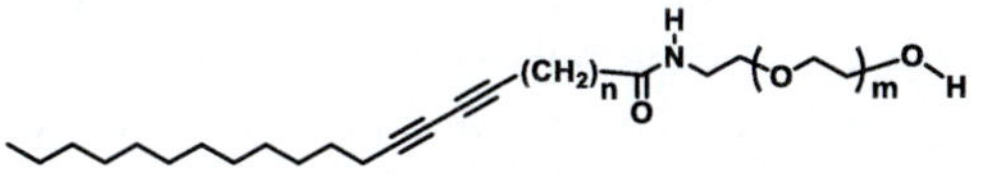

44a: n = 10, m = 5
44b: n = 8, m = 7
44c: n = 8, m = 5
44d: n = 8, m = 1

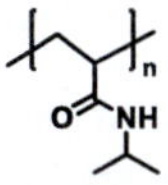

45

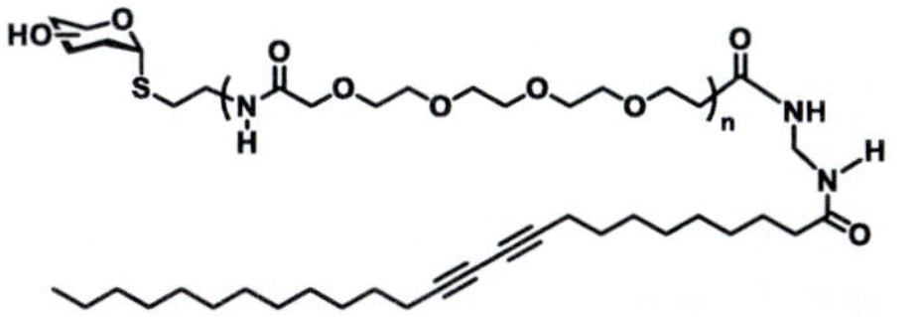

46: n = 0

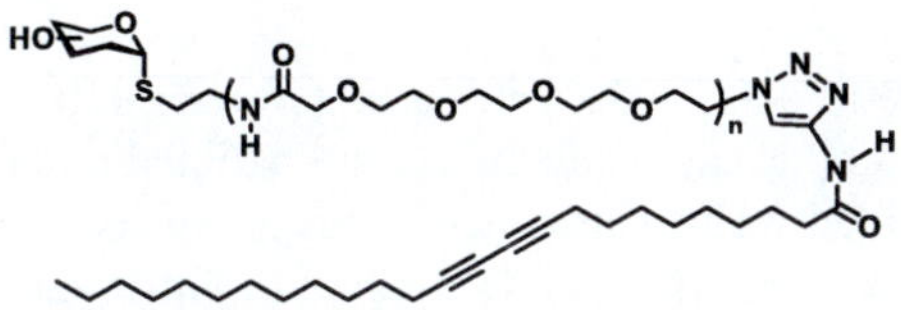

47a: n = 0
47b: n = 1
47c: n = 2

Chart 2.13 Chemical structures of **44**–**47**

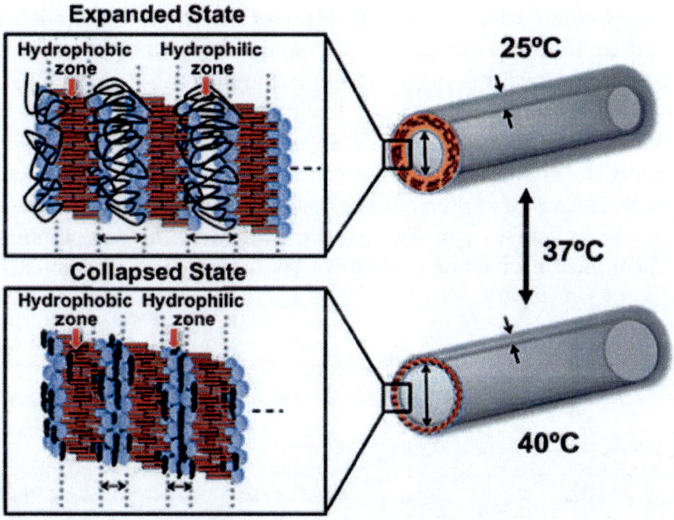

Fig. 2.20 Schematic illustration of temperature-induced expandable PNIPAM **45**–phospholipid **42** nanotubes (Reproduced with permission from Ref. [66], © 2012 American Chemical Society)

chain carrying a photopolymerizable diacetylenic functionality.[67] For the glycolipid **46** (Chart 2.13), the hydrophilic group is connected to the long hydrophobic chain via amide linkage. The glycolipids **47a**, **47b**, and **47c** (Chart 2.13) possess a triazole ring that ligates the mannose or tetra(ethylene glycol)-tethered mannose segment with the diacetylene-containing long hydrophobic chain. Remarkably, the glycolipids **47a**, **47b**, and **47c** with a triazole linkage self-assemble in water to produce nanotube structures, whereas the glycolipid **46** only gave spherical micelles through self-assembly. The research group discusses the effect of π–π stacking of the triazole rings and larger distance generated by the ring as a linkage on the nanotube formation. Additionally, the obtained nanotubes further assemble into 3D networks through their bundling at high concentrations.

2.8 Stearamide Derivative

The self-assembly experiments of the achiral amphiphile **43b**, where diacetylene group and secondary amine salt headgroups are rationally incorporated to the long alkyl chains, provided well-defined nanotubes in aqueous solutions.[63] New nanotube formation from achiral stearamide amphiphile with nondiacetylenic moiety is, however, highly challenging in molecular self-assembly. Kim and co-workers synthesized a family of amphiphilic stearamide derivatives **48a–48d** (Chart 2.14) obtained by connecting stearic acid with oligomethylenediamine via amide linkage [68]. The self-assembly behavior of **48a–48d** was shown to strongly depend on the solvent properties used, concentrations of amphiphiles, and the number of methylene segments in the hydrophilic region. For example, 1,2-dichloroethane (DCE) solutions were effective to undergo satisfactory self-assembly among a variety of solvents, e.g., DCE, chloroform, DMF, THF, ethanol, ethanol/H_2O, DMF/H_2O, and THF/H_2O. Only the amphiphile **48a** derived from ethylenediamine yielded tubular structures of 80 nm in inner diameters at low concentrations in DCE. In contrast, other stearamide derivatives **48b–48d** as well as the modified amphiphiles **49** and **50** (Chart 2.14) gave no nanotube products but a variety of assembled structures

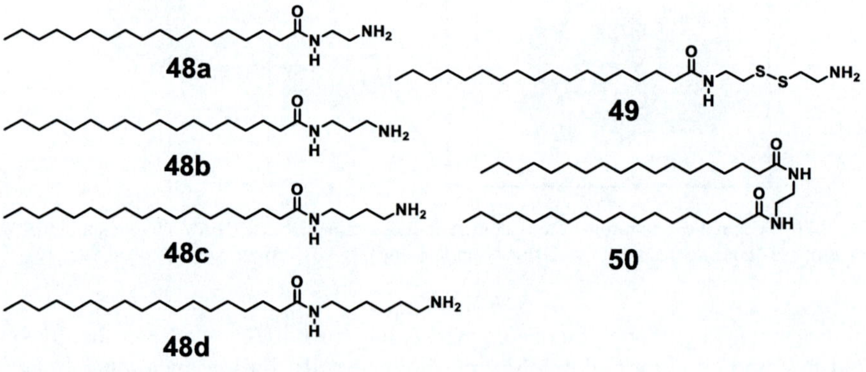

Chart 2.14 Chemical structures of **48–50**

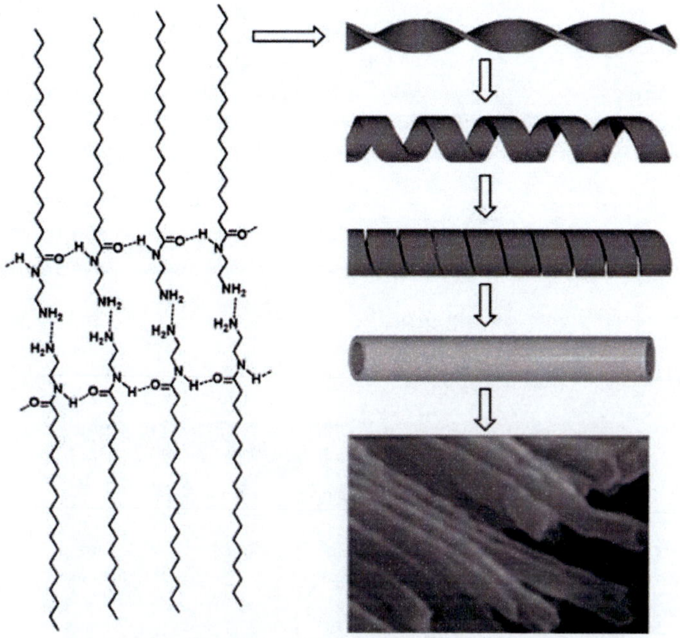

Fig. 2.21 Plausible self-assembly pathway from a bilayer of monochain stearic acid derivative **48a** in DCE. Reproduced with permission Ref. [68], © 2010 American Chemical Society)

including nanobelts, microspheres, and microclusters. With regard to the formation mechanism of nanotubes, they proposed a twisting pathway of self-assembled nanobelt, which result in the formation of multi-layered nanotubes (Fig. 2.21).

2.9 Other Amide-Group Containing Amphiphile

2.9.1 N-Amidated 4-Aminobenzoic Acid Sodium Salt

Schmidt and co-workers investigated the hydrogelation ability of eleven amphiphiles derivatized from *N*-amidated 3- and 4-aminobenzoic acids that link to long hydrophobic alkyl chains with from 3 to 13 methylene units. Among this series of amphiphilic derivatives, the 4-(octanoylamino)benzoic acid sodium salt **51** (Chart 2.15) was observed to exhibit extraordinary hydrogelation ability in aqueous solutions [69]. The hydrogel formation is thermoreversible. The stable gel state, obtained by cooling the hot alkaline aqueous solutions to room temperature, can be transformed to the sol state by heating again. Moreover, the research group succeeded in the preparation of self-supporting nanofiber mats (diameter: 65 mm) by a mold casting and succeeding process (Fig. 2.22) [69]. Notably, the hydrogel fiber mats

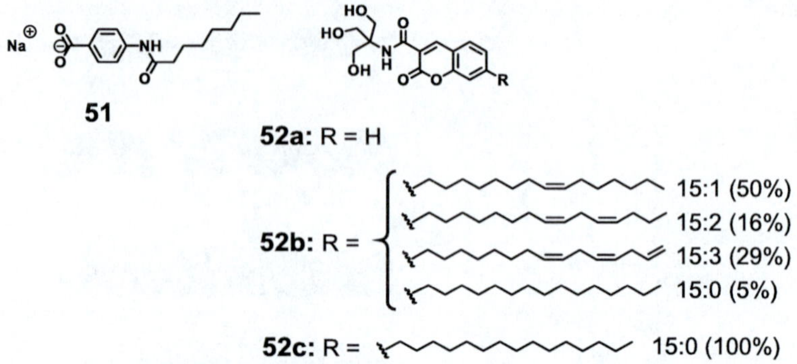

Chart 2.15 Chemical structures of **51**–**52**

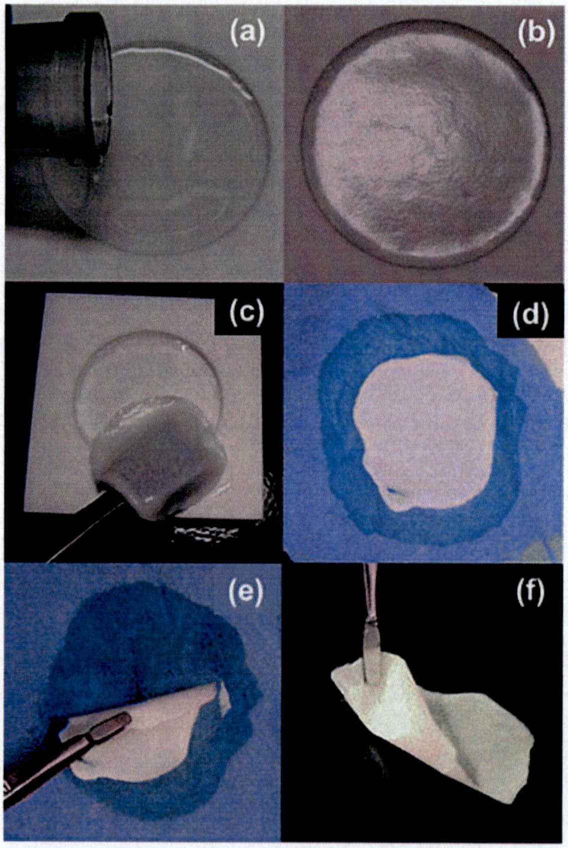

Fig. 2.22 Preparation of supramolecular nanofibrillar fiber mats from **51**: **a** mold casting; **b** gelation; **c** removal of the gel sample; **d** removal of the solvent; **e** lifting off the sample; and **f** fiber mat after oven drying. Reproduced with permission from Ref. [69], © The Royal Society of Chemistry 2011

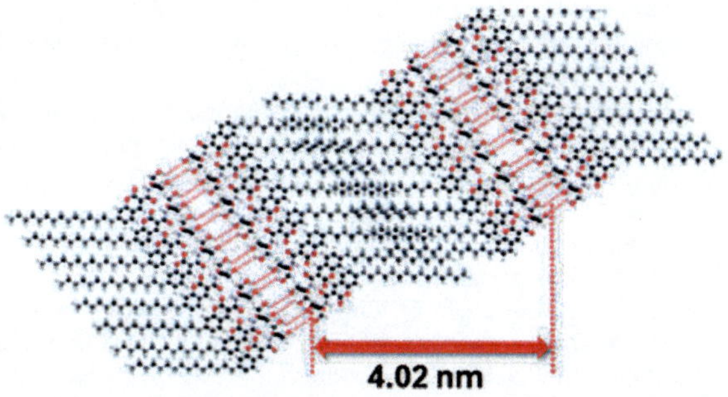

Fig. 2.23 Schematic drawing of a plausible molecular packing model of **52c** for the hydrogel. Reproduced with permission from Ref. [70], © The Royal Society of Chemistry 2015

comprise supramolecular assemblies of nanotube structures with inner and outer diameters of 22 ± 3 and 88 ± 14 nm, respectively. The wall membrane of each nanotube is composed of 9–10 layers of bilayer structures. The measured thermal and mechanical robustness of the fiber mats as well as the resistance to organic solvents suggest the potential usefulness as a filtration and template material for inorganic substances.

2.9.2 Coumarin-Tris-Based Amphiphile

Three different coumarin-appended tris amphiphiles **52a–52c** (Chart 2.15) were synthesized from cardanol derivatives that are obtainable from renewable biomaterial, cashew nut shell liquid [70]. The tris moiety serves as a biocompatible hydrophilic group, whereas the coumarin segment with long saturated and unsaturated hydrocarbon chains a hydrophobic group. As a result of hydrogelation experiments, the amphiphile **52c** showed prominent hydrogelation ability toward a mixture of DMSO and water (1:2, v/v) with a critical gelation concentration of 0.8% (Fig. 2.23) [70]. The amphiphiles **52a** and **52b**, however, exhibited no remarkable hydrogelation capacity. Lacking of a hydrophobic long chain and the presence of a kink conformation due to the unsaturated chain are responsible for the incapability of hydrogelation of **52a** and **52b**, respectively. Significantly, the fibrillar morphology that organizes the hydrogel structure is stable under neutral and basic aqueous conditions. Upon addition of acidic aqueous medium, the fibers are transformed to vesicular and nanotubular architectures that remain stable for more than 3 months [70].

References

1. Zhu X, Li Y, Duan P, Liu M (2010) Self-assembled ultralong chiral nanotubes and tuning of their chirality through the mixing of enantiomeric components. Chem Eur J 16:8034–8040. https://doi.org/10.1002/chem.201000595

2. Green MM, Garetz BA, Munoz B, Chang HP, Hoke S, Cooks RG (1995) Majority rules in the copolymerization of mirror-image isomers. J Am Chem Soc 117:4181–4182. https://doi.org/10.1021/ja00119a039
3. Smulders MMJ, Stals PJM, Mes T, Paffen TFE, Schenning APHJ, Palmans ARA, Meijer EW (2010) Probing the limits of the majority-rules principle in a dynamic supramolecular polymer. J Am Chem Soc 132:620–626. https://doi.org/10.1021/ja9080875
4. Wang X, Duan P, Liu M (2014) Self-assembly of π-conjugated gelators into emissive chiral nanotubes: emission enhancement and chiral detection. Chem Asian J 9:770–778. https://doi.org/10.1002/asia.201301518
5. Zhang L, Liu C, Jin Q, Zhu X, Liu M (2013) Pyrene-Functionalized organogel and spacer effect: from emissive nanofiber to nanotube and inversion of supramolecular chirality. Soft Matter 9:7966–7973. https://doi.org/10.1039/C3sm51204e
6. Ji L, Sang Y, Ouyang G, Yang D, Duan P, Jiang Y, Liu M (2019) Cooperative chirality and sequential energy transfer in a supramolecular light-harvesting nanotube. Angew Chem Int Ed 58:844–848. https://doi.org/10.1002/anie.201812642
7. Goto T, Okazaki Y, Ueki M, Kuwahara Y, Takafuji M, Oda R, Ihara H (2017) Induction of strong and tunable circularly polarized luminescence of nonchiral, nonmetal, low-molecular-weight fluorophores using chiral nanotemplates. Angew Chem Int Ed 56:2989–2993. https://doi.org/10.1002/anie.201612331
8. Wei X, Su X, Cao P, Liu X, Chang WR, Li M, Zhang X, Liu Z (2016) Structure of spinach photosystem II-LHCII supercomplex at 3.2 angstrom resolution. Nature 534:69–74. https://doi.org/10.1038/nature18020
9. Terech P, Jean B, Ne F (2006) Hexagonally ordered ammonium lithocholate self-assembled nanotubes with highly monodisperse sections. Adv Mater 18:1571–1574. https://doi.org/10.1002/adma.200502358
10. Terech P, Velu SKP, Pernot P, Wiegart L (2012) Salt effects in the formation of self-assembled lithocholate helical ribbons and tubes. J Phys Chem B 116:11344–11355. https://doi.org/10.1021/jp305365m
11. Anelli PL, Lattuada L, Uggeri F (1998) One-pot Mitsunobu-Staudinger preparation of 3-aminocholan-24-oic acid esters from 3-hydroxycholan-24-oic acid esters. Synth Commun 28:109–117. https://doi.org/10.1080/00397919808005079
12. Babu P, Maitra U (2005) Synthesis and in vitro cholesterol dissolution by 23-and 24-phosphonobile acids. Steroids 70:681–689. https://doi.org/10.1016/j.steroids.2005.03.008
13. Manghisi N, Leggio C, Jover A, Meijide F, Pavel NV, Tellini VHS, Tato JV, Agostino RG, Galantini L (2010) Catanionic tubules with tunable charge. Angew Chem Int Ed 49:6604–6607. https://doi.org/10.1002/anie.201000951
14. Travaglini L, D'Annibale A, Schillen K, Olsson U, Sennato S, Pavel NV, Galantini L (2012) Amino acid-bile acid based molecules: extremely narrow surfactant nanotubes formed by a phenylalanine-substituted cholic acid. Chem Commun 48:12011–12013. https://doi.org/10.1039/c2cc36030f
15. di Gregorio MC, Pavel NV, Jover A, Meijide F, Tato JV, Tellini VHS, Vargas AA, Regev O, Kasavi Y, Schillen K, Galantini L (2013) pH sensitive tubules of a bile acid derivative: a tubule opening by release of wall leaves. PCCP 15:7560–7566. https://doi.org/10.1039/c3cp00121k
16. Shimizu T (2018) Self-assembly of discrete organic nanotubes. Bull Chem Soc Jpn 91:623–668. https://doi.org/10.1246/bcsj.20170424
17. di Gregorio MC, Varenik M, Gubitosi M, Travaglini L, Pavel NV, Jover A, Meijide F, Regev O, Galantini L (2015) Multi stimuli response of a single surfactant presenting a rich self-assembly behavior. RSC Adv 5:37800–37806. https://doi.org/10.1039/c5ra01394a
18. Kameta N, Minamikawa H, Masuda M (2011) Supramolecular organic nanotubes: how to utilize the inner nanospace and the outer space. Soft Matter 7:4539–4561. https://doi.org/10.1039/c0sm01559h
19. Gubitosi M, Travaglini L, di Gregorio MC, Pavel NV, Tato JV, Sennato S, Olsson U, Schillen K, Galantini L (2015) Tailoring supramolecular nanotubes by bile salt based surfactant mixtures. Angew Chem Int Ed 54:7018–7021. https://doi.org/10.1002/anie.201500445

20. Zhang X, Zou J, Tamhane K, Kobzeff FF, Fang J (2010) Self-assembly of pH-switchable spiral tubes: supramolecular chemical springs. Small 6:217–220. https://doi.org/10.1002/smll.200901067
21. Lu Q, Kim Y, Bassim N, Collins GE (2015) Impact of confinement on proteins concentrated in lithocholic acid based organic nanotubes. J Colloid Interface Sci 454:97–104. https://doi.org/10.1016/j.jcis.2015.05.004
22. Kameta N, Minamikawa H, Someya Y, Yui H, Masuda M, Shimizu T (2010) Confinement effect of organic nanotubes toward green fluorescent protein (GFP) depending on the inner diameter size. Chem Eur J 16:4217–4223. https://doi.org/10.1002/chem.200903413
23. Lu Q, Kim Y, Bassim N, Raman N, Collins GE (2016) Catalytic activity and thermal stability of horseradish peroxidase encapsulated in self-assembled organic nanotubes. Analyst 141:2191–2198. https://doi.org/10.1039/c5an02655e
24. Liang W, He S, Fang J (2014) Self-assembly of J-aggregate nanotubes and their applications for sensing dopamine. Langmuir 30:805–811. https://doi.org/10.1021/la404022q
25. Zhang X, Bera T, Liang W, Fang J (2011) Longitudinal zipping/unzipping of self-assembled organic tubes. J Phys Chem B 115:14445–14449. https://doi.org/10.1021/Jp2064276
26. Shimizu T, Masuda M, Minamikawa H (2005) Supramolecular nanotube architectures based on amphiphilic molecules. Chem Rev 105:1401–1443. https://doi.org/10.1021/cr030072j
27. John G, Masuda M, Okada Y, Yase K, Shimizu T (2001) Nanotube formation from renewable resources via coiled nanofibers. Adv Mater 13:715–718. https://doi.org/10.1002/1521-4095(200105)13:10%3c715::AID-ADMA715%3e3.0.CO;2-Z
28. John G, Jung JH, Minamikawa H, Yoshida K, Shimizu T (2002) Morphological control of helical solid bilayers in high-axial-ratio nanostructures through binary self-assembly. Chem Eur J 8:5494–5500. https://doi.org/10.1002/1521-3765(20021202)8:23%3c5494::AID-CHEM5494%3e3.0.CO;2-P
29. Jung JH, John G, Yoshida K, Shimizu T (2002) Self-Assembling structures of long-chain phenyl glucoside influenced by the introduction of double bonds. J Am Chem Soc 124:10674–10675. https://doi.org/10.1021/ja020752o
30. Kamiya S, Minamikawa H, Jung JH, Yang B, Masuda M, Shimizu T (2005) Molecular structure of glucopyranosylamide lipid and nanotube morphology. Langmuir 21:743–750. https://doi.org/10.1021/la047765v
31. Frusawa H, Fukagawa A, Ikeda Y, Araki J, Ito K, John G, Shimizu T (2003) Aligning a single-lipid nanotube with moderate stiffness. Angew Chem Int Ed 42:72–74. https://doi.org/10.1002/anie.200390056
32. Felgner H, Frank R, Schliwa M (1996) Flexural rigidity of microtubules measured with the use of optical tweezers. J Cell Sci 109:509–516
33. Karlsson A, Karlsson R, Karlsson M, Cans A-S, Stroemberg A, Ryttsen F, Orwar O (2001) Networks of nanotubes and containers. Nature 409:150–152. https://doi.org/10.1038/35051656
34. Nogawa K, Tagawa Y, Nakajima M, Arai F, Shimizu T, Kamiya S, Fukuda T (2007) Development of novel nanopipette with a lipid nanotube as nanochannel. J Robot Mechatr 19:528–534. https://doi.org/10.20965/jrm.2007.p0528
35. Giulieri F, Guillod F, Greiner J, Krafft M-P (1996) Anionic Glucophospholipids-a new family of tubule-forming amphiphiles. Chem Eur J 2:1335–1339. https://doi.org/10.1002/chem.19960021022
36. Imae T, Funayama K, Krafft MP, Giulieri F, Tada T, Matsumoto T (1999) Small-angle scattering and electron microscopy investigation of nanotubules made from a perfluoroalkylated glucophospholipid. J Colloid Interface Sci 212:330–337. https://doi.org/10.1006/jcis.1999.6094
37. Kameta N, Asakawa M, Masuda M, Shimizu T (2011) Self-Assembled organic nanotubes embedding hydrophobic molecules within solid bilayer membranes. Soft Matter 7:85–90. https://doi.org/10.1039/C0sm00375a
38. Kameta N, Matsuzawa T, Yaoi K, Fukuda J, Masuda M (2017) Glycolipid-based nanostructures with thermal-phase transition behavior functioning as solubilizers and refolding accelerators for protein aggregates. Soft Matter 13:3084–3090. https://doi.org/10.1039/c7sm00310b

39. Ishikawa K, Kameta N, Aoyagi M, Asakawa M, Shimizu T (2013) Soft nanotubes with a hydrophobic channel hybridized with au nanoparticles: photothermal dispersion/aggregation control of C60 in Water. Adv Funct Mater 23:1677–1683. https://doi.org/10.1002/adfm.201202160
40. Kameta N, Ishikawa K, Masuda M, Asakawa M, Shimizu T (2012) Soft nanotubes acting as a light-harvesting antenna system. Chem Mater 24:209–214. https://doi.org/10.1021/Cm2030526
41. Kameta N, Masuda M, Shimizu T (2015) Qualitative/chiral sensing of amino acids by naked-eye fluorescence change based on morphological transformation and hierarchizing in supramolecular assemblies of pyrene-conjugated glycolipids. Chem Commun 51:11104–11107. https://doi.org/10.1039/c5cc03843j
42. Kameta N, Aoyagi M, Asakawa M (2017) Enhancement of the photocatalytic activity of Rhenium(I) complexes by encapsulation in light-harvesting soft nanotubes. Chem Commun 53:10116–10119. https://doi.org/10.1039/c7cc05337a
43. Liu T, Diemann E, Li H, Dress AWM, Muller A (2003) Self-assembly in aqueous solution of wheel-shaped Mo-154 oxide clusters into vesicles. Nature 426:59–62. https://doi.org/10.1038/nature02036
44. Wu L, Lal J, Simon KA, Burton EA, Luk YY (2009) Nonamphiphilic assembly in water: polymorphic nature, thread structure, and thermodynamic incompatibility. J Am Chem Soc 131:7430–7443. https://doi.org/10.1021/ja9015149
45. Jiang L, Peng Y, Yan Y, Deng M, Wang Y, Huang J (2010) Annular Ring" microtubes formed by SDS@2 beta-CD complexes in aqueous solution. Soft Matter 6:1731–1736. https://doi.org/10.1039/b920608f
46. Jiang L, Peng Y, Yan Y, Huang J (2011) Aqueous self-assembly of SDS@2 beta-CD complexes: lamellae and vesicles. Soft Matter 7:1726–1731. https://doi.org/10.1039/c0sm00917b
47. Khaykovich B, Hossain C, McManus JJ, Lomakin A, Moncton DE, Benedek GB (2007) Structure of cholesterol helical ribbons and self-assembling biological springs. Proc Natl Acad Sci USA 104:9656–9660. https://doi.org/10.1073/pnas.0702967104
48. Aime C, Tamoto R, Satoh T, Grelard A, Dufourc EJ, Buffeteau T, Ihara H, Oda R (2009) Nucleotide-promoted morphogenesis in amphiphile assemblies: kinetic control of micrometric helix formation. Langmuir 25:8489–8496. https://doi.org/10.1021/la8043297
49. Pescador P, Brodersen N, Scheidt HA, Loew M, Holland G, Bannert N, Liebscher J, Herrmann A, Huster D, Arbuzova A (2010) Microtubes self-assembled from a cholesterol-modified nucleoside. Chem Commun 46:5358–5360. https://doi.org/10.1039/c0cc00562b
50. Kameta N, Tanaka A, Akiyama H, Minamikawa H, Masuda M, Shimizu T (2011) Photoresponsive soft nanotubes for controlled guest release. Chem Eur J 17:5251–5255. https://doi.org/10.1002/chem.201100179
51. Kameta N, Masuda M, Shimizu T (2015) Photoinduced morphological transformations of soft nanotubes. Chem Eur J 21:8832–8839. https://doi.org/10.1002/chem.201500430
52. Kameta N, Akiyama H, Masuda M, Shimizu T (2016) Effect of photoinduced size changes on protein refolding and transport abilities of soft nanotubes. Chem Eur J 22:7198–7205. https://doi.org/10.1002/chem.201504613
53. Barclay TG, Constantopoulos K, Zhang W, Fujiki M, Matisons JG (2012) Chiral Self-assembly of designed amphiphiles: optimization for nanotube formation. Langmuir 28:14172–14179. https://doi.org/10.1021/la3030606
54. Waugh RE, Hochmuth RM (1987) Mechanical equilibrium of thick, hollow, liquid membrane cylinders. Biophys J 1987:391–400. https://doi.org/10.1016/S0006-3495(87)83227-7
55. Evans E, Bowman H, Leung A, Needham D, Tirrell D (1996) Biomembrane templates for nanoscale conduits and networks. Science 273:933–935. https://doi.org/10.1126/science.273.5277.933
56. Karlsson M, Nolkrantz K, Davidson MJ, Stroemberg A, Ryttsen F, Akerman B, Orwar O (2000) Electroinjection of colloid particles and biopolymers into single unilamellar liposomes and cells for bioanalytical applications. Anal Chem 72:5857–5862. https://doi.org/10.1021/ac0003246

57. Laplante JP, Pemberton M, Hjelmfelt A, Ross J (1995) Experiments on pattern-recognition by chemical-kinetics. J Phys Chem 99:10063–10065. https://doi.org/10.1021/j100025a001
58. Sciaky N, Presley J, Smith C, Zaal KJM, Cole N, Moreira JE, Terasaki M, Siggia E, Lippincott-Schwartz J (1997) Golgi tubule traffic and the effects of brefeldin a visualized in living cells. J Cell Biol 139:1137–1155. https://doi.org/10.1083/jcb.139.5.1137
59. Lobovkina T, Jesorka A, Onfelt B, Lagerwall J, Dommersnes P, Orwar O (2011) Soft-matter nanotubes. In: Hayden O, Nielsch K (eds) Molecular- and nano-tubes, pp 75–125. https://doi.org/10.1007/978-1-4419-9443-1_4
60. Davidson M, Karlsson M, Sinclair J, Sott K, Orwar O (2003) Nanotube-vesicle networks with functionalized membranes and interiors. J Am Chem Soc 125:374–378. https://doi.org/10.1021/ja027699o
61. Sekine Y, Abe K, Shimizu A, Sasaki Y, Sawada S, Akiyoshi K (2012) Shear flow-induced nanotubulation of surface-immobilized liposomes. RSC Adv 2:2682–2684. https://doi.org/10.1039/C2ra00629d
62. Yager P, Schoen PE (1984) Formation of tubules by polymerizable surfactant. Mol Cryst Liq Cryst 106:371–381. https://doi.org/10.1080/00268948408071454
63. Lee SB, Koepsel R, Stolz DB, Warriner HE, Russell AJ (2004) Self-assembly of biocidal nanotubes from a single-chain diacetylene amine salt. J Am Chem Soc 126:13400–13405. https://doi.org/10.1021/ja048463i
64. Barclay T, Constantopoulos K, Matisons J (2011) Self-assembled lipid nanotubes by rational design. J Mater Res 26:322–335. https://doi.org/10.1557/Jmr.2010.3
65. Perino A, Schmutz M, Meunier S, Mésini PJ, Wagner A (2011) Self-assembled nanotubes and helical tapes from diacetylene nonionic amphiphiles. Structural studies before and after polymerization. Langmuir 27:12149–12155. https://doi.org/10.1021/la202162q
66. Kawano S, Urban MW (2012) Expandable temperature-responsive polymeric nanotubes. ACS Macro Lett 1:232–235. https://doi.org/10.1021/mz2000303
67. Assali M, Cid JJ, Fernandez I, Khiar N (2013) Supramolecular diversity through click chemistry: switching from nanomicelles to 1D-nanotubes and tridimensional hydrogels. Chem Mater 25:4250–4261. https://doi.org/10.1021/cm4022613
68. Zhang L, Li H, Ha CS, Suh H, Kim I (2010) Fabrication of nanotubules and microspheres from the self-assembly of amphiphilic monochain stearic acid derivatives. Langmuir 26:17890–17895. https://doi.org/10.1021/la103480p
69. Bernet A, Behr M, Schmidt HW (2011) Supramolecular nanotube-based fiber mats by self-assembly of a tailored amphiphilic low molecular weight hydrogelator. Soft Matter 7:1058–1065. https://doi.org/10.1039/C0sm00456a
70. Lalitha K, Prasad YS, Maheswari CU, Sridharan V, John G, Nagarajan S (2015) Stimuli responsive hydrogels derived from a renewable resource: synthesis, self-assembly in water and application in drug delivery. J Mater Chem B 3:5560–5568. https://doi.org/10.1039/c5tb00864f

Chapter 3
Bolaamphiphile-Based Nanotubes

3.1 Glucose–Amine and Glucose–Carboxylic Acid Bolaamphiphiles

Chemically modified organic nanotubes with rationally selected and different inner and outer functional groups are of significantly interest to perform their intrinsic ability in encapsulation, detection, stabilization, diffusion, and ejection of nanometer scaled guest substances, e.g., proteins, DNA, and viruses [1]. A single-crystal structural analysis of unsymmetrical bolaamphiphile **1** (Chart 3.1) with a 1-galactosamide and carboxylic acid headgroups at each end confirmed that the obtained crystal from methanol solution form unsymmetrical MLMs characteristic of a head-to-tail interface [2, 3]. Exclusive preparation of organic nanotubes, stabilized by unsymmetrical MLMs with the head-to-tail molecular stacks in aqueous environment, remained challenging at that time. With the aim of the self-assembly into monolayer-stacked SMNTs, we first prepared simple 1-glucosamide-based bolaamphiphiles **2(n)** (n = 12, 14, 16, 18, and 20) (Chart 3.1) having hydrophobic long alkyl chains and a hydrophilic carboxylic acid headgroup at one end [4]. The self-assembly of **2(n)** in water forms nanotube structures with 30–43 nm outer diameters, which co-exist with microtubes with relatively larger 118–190 nm outer diameters (Fig. 3.1). The membrane wall of the nanotubes comprises two to three layers of unsymmetrical MLMs. Notably, the bolaamphiphilic molecules pack in a parallel manner, thus forming a head-to-tail interface. In contrast to this homogeneous molecular packing in the nanotubes, the microtubes were shown to form three distinct molecular packing, i.e., symmetrical MLMs, unsymmetrical MLMs stacked in head-to tail, and unsymmetrical MLMs stacked in head-to-head manners (Fig. 3.2). FT-IR measurement for the nano and microtubes confirmed that the long oligomethylene spacers take an all-trans conformation, suggesting the absence of a gauche-derived bent structure of the spacer chain.

Similar molecular design strategy allowed us to synthesize unsymmetrical bolaamphiphiles **3 (n)** (n = 12, 14, 16, 17, 18, and 20) (Chart 3.1) [5, 6]. The characteristic molecular structure is that 1-N-glucosamide headgroup is connected

T. Shimizu, *Smart Soft-Matter Nanotubes*, Nanostructure Science and Technology,
https://doi.org/10.1007/978-981-16-2685-2_3

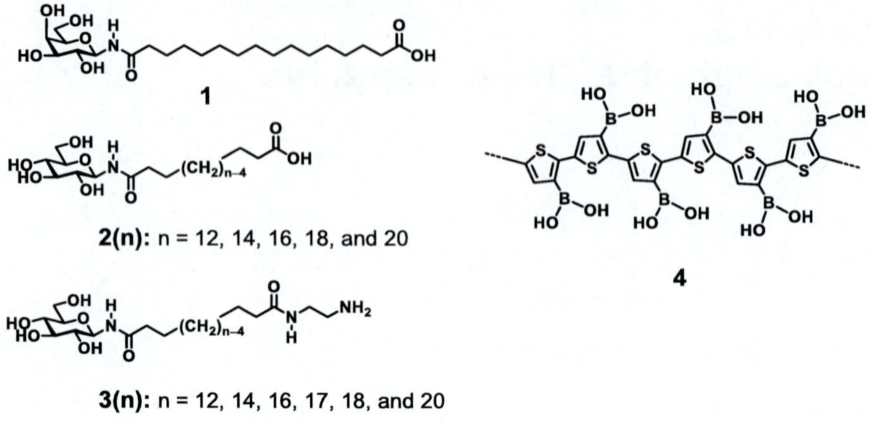

Chart 3.1 Chemical structures of **1–4**

to another amine headgroup through different length of long alkyl chains via two separate amide linkages. Self-assembly was conducted in water at pH 10 by using a solvent-evaporated film from the DMF solution as a seed. Slow cooling the hot aqueous solution of the film, prepared by the long-chain bolaamphiphiles **3 (n)** (n = 18 and 20), to room temperature produced mainly tubular structures of 70–80 nm inner and 100–110 nm in outer diameters. Meanwhile, the short-chain derivatives **3 (n)** ($n = 12$, 14, 16, and 17) self-assembled to yield nanotapes 80–250 nm wide.

IR spectroscopy, XRD, and differential scanning calorimetry (DSC) corroborated the strong dependence of the behavior on the length of the spacer chain length. A series of bolaamphiphiles **3 (n)** can be, thus, categorized into two groups, i.e., S: short chain bolaamphiphiles ($n = 12$, 14, 16, and 17) and L: long-chain ones ($n = 18$ and 20) [6]. The nanotapes, prepared by the self-assembly of the S-group, showed double peaks in the $\delta(CH_2)$ IR band, suggesting the presence of a symmetrical MLM with the subcell structure of the orthorhombic ($O_{\perp}$) type. On the other hand, the single peaks in the $\delta(CH_2)$ IR band, obtained for the self-assembled nanotubes from the L-group, prove that the tubes form the triclinic ($T_{//}$) subcell structure to result in an unsymmetrical MLM. The resultant *polymorph* of the nanotapes and nanotubes correspond reasonably with those of the solid film employed as a seed. The small angle region of the XRD profiles for the nanotapes and nanotubes gives a single sharp peak corresponding to the stacking periodicity (d) of each MLM. The d values of the nanotapes are longer than the extended molecular length (L). The nanotapes consist of, thus, symmetrical MLMs. Meanwhile, the nanotubes form unsymmetrical MLMs due to the shorter d values than the L values.

By using variable-temperature (VT)-IR and VT-XRD, we investigated various packing feature of the S- and L-group molecules **3(n)** within the MLMs when the molecular assemblies were heated and self-assembled in water [6]. The L-group molecules (n = 18 and 20) exhibit crystal polymorphism (Cr-1 and Cr-2) while forming an unsymmetrical MLM (Fig. 3.3). Upon further heating, isotropic phase

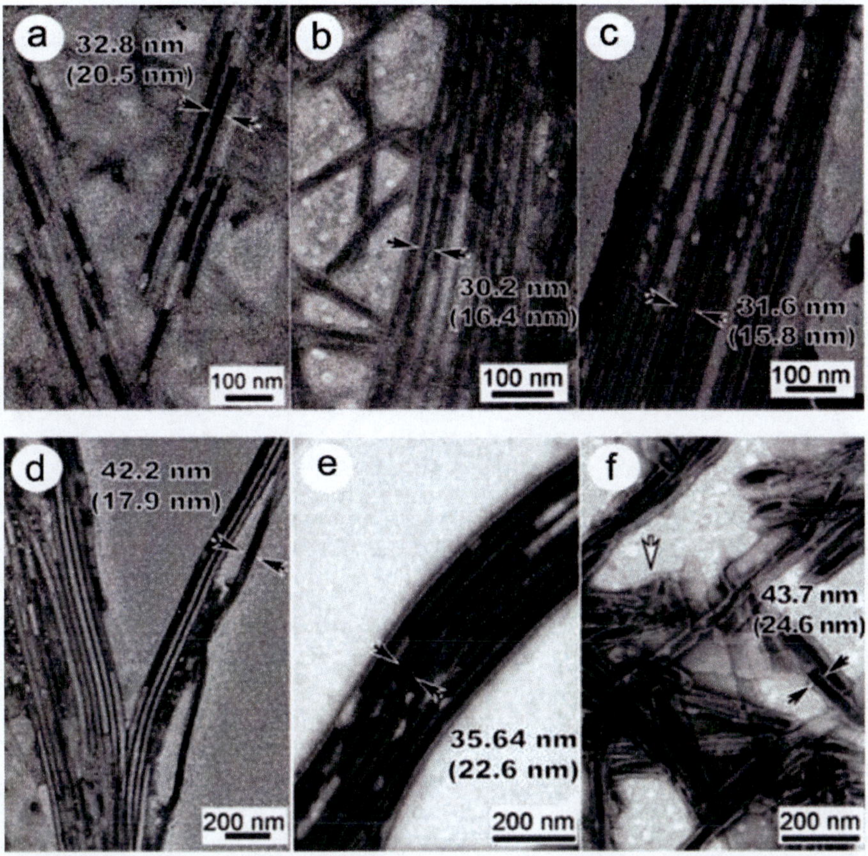

Fig. 3.1 TEM pictures of the nanotubes self-assembled from **a 2 (12)**, **b 2 (14)**, **c 2 (16)**, **d 2 (18)**, **e 2 (20)**, and **f** the sodium salt of **2 (18)**. The numerical values, indicated by arrows, represent the outer and inner diameters (in parentheses) of the nanotubes (Reproduced with permission from Ref. [4], © 2004 American Chemical Society)

was observed to follow smectic liquid-crystalline phases. It is noteworthy that the symmetrical MLM packing is maintained even after the gradual cooling of the smectic mesophase. The molecular packing of the S-group molecules ($n = 12$, 14, 16, and 17) adopts a symmetrical MLM, exhibiting only a single crystalline phase (Cr-1) upon heating (Fig. 3.3). The symmetrical MLM of the S-group molecules keeps staying through the self-assembly of the seed film into the nanotapes in water. The seed films with unsymmetrical MLMs of the L-group molecules change likewise into the nanotube structures with unsymmetrical MLMs.

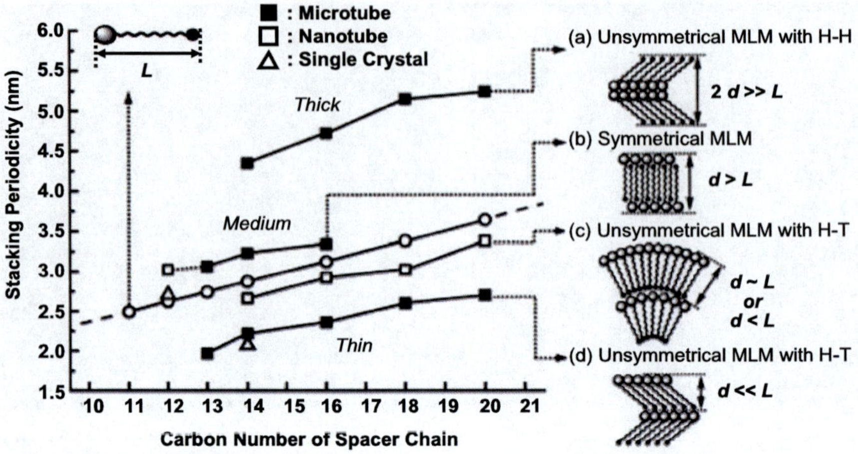

Fig. 3.2 MLM stacking periodicity (*filled square box* and *open square box*) calculated for the micro and nanotubes of **2** (***n***) from XRD data, which is plotted against the carbon number of the spacer chain. For reference, the calculated length (*L*) (*open circle*) for each extended molecule of **2** (***n***) was also shown. The same calculations were also conducted for a symmetrical MLM of **2** **(12)** (Δ at $n = 12$) and an unsymmetrical MLM of **1** having head-to-tail interface (Δ at $n = 14$) (Reproduced with permission from Ref. [4], © 2004 American Chemical Society)

3.1.1 *Control of Inner Diameter*

By employing the aforementioned unsymmetrical bolaamphiphiles **2(n)**, our research group demonstrated a practical control over the inner diameters of the self-assembled nanotubes [4]. These molecules carry a glucose headgroup at one end and a carboxyl headgroup that is relatively smaller than the glucose group at the other. The both headgroups are bridged by a series of oligomethylene chains with an even number of carbons (C12, C14, C16, C18, and C20). We assume here that the nanotubular structures are composed of unsymmetrical MLMs of wedge-shaped bolaamphiphiles. Moreover, we postulate that the larger glucose headgroups occupy the external surface areas of the MLM (Fig. 3.4). The innermost diameter ($D_{\text{calc-in}}$) can, thus, be defined by Eq. (3.1), where a_s, a_l, and L are the cross-sectional areas of the small headgroup, those of the large headgroup, and the molecular length, respectively.

$$D_{\text{calc-in}} = 2a_sL/(a_l - a_s) \quad (3.1)$$

If the a_s and a_l values have fixed ones, the $D_{\text{calc-in}}$ value is proportional to L, i.e., proportional to the length of the bridging oligomethylene chains. Varying the carbon number of the oligomethylene bridge from $n = 14$ to $n = 20$ allows for the control over the inner diameters in the range of 17.7–22.2 nm in steps of approximately 1.5 nm per two carbons.

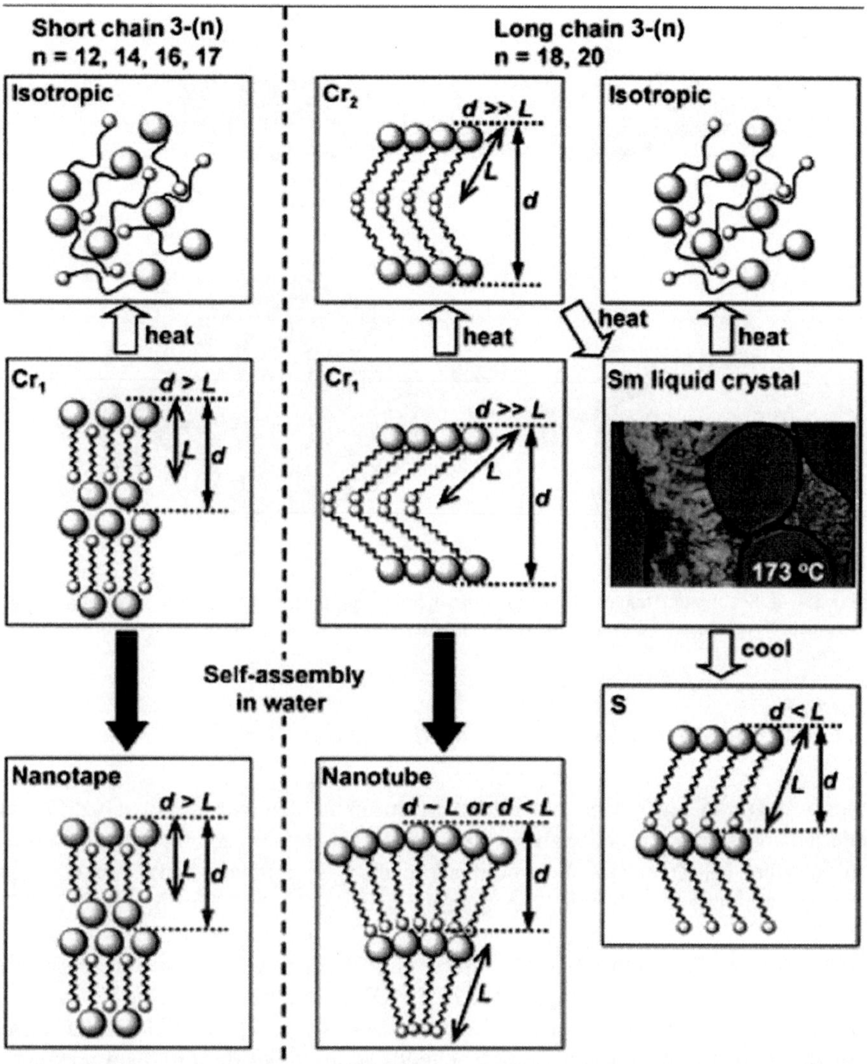

Fig. 3.3 Variation of the molecular packing of **3 (n)** induced by thermal phase transition and self-assembly in water. (right middle) Polarized light micrograph of a fanlike texture of the mesophase of **3 (20)** observed at 173 °C on cooling from the isotropic phase (Reproduced with permission from Ref. [6], © 2007 American Chemical Society)

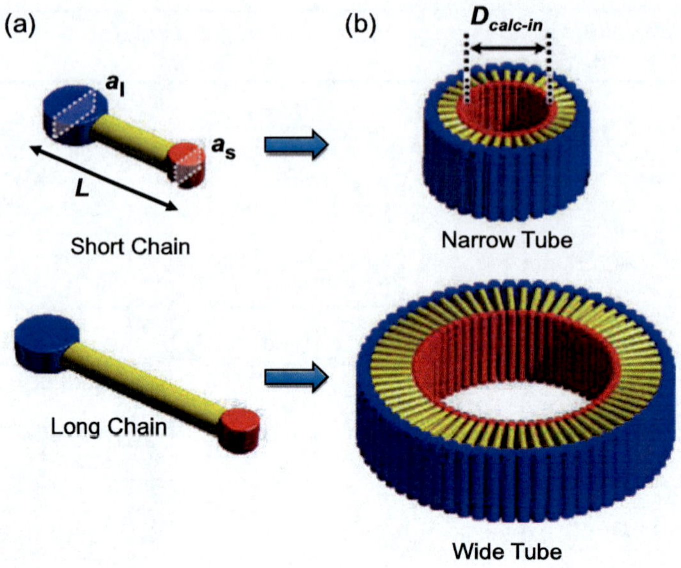

Fig. 3.4 **a** Schematic image of unsymmetrical bolaamphiphiles **2(n)** with a short and a long oligomethylene chain and **b** self-assembled tubular architectures from the short- and long-chain bolaamphiphiles

3.1.2 Polymorph *Control Using a Polymer Template*

Mono- and multi-layered unsymmetrical organic nanotubes with completely different internal and external functionalities take parallel molecular packing with a head-to-tail interface [1]. To realize such molecular packing system with **3(n)**, we attempted to employ synthetic poly(thiopheneboronic acid) (PTB, **4**) (Chart 3.1) ($M_n = 710$ and $M_w/M_n = 1.83$) as a template [7]. The boronic acid side chains of the thiophene polymer **4** can bind with the glucosamide diol groups of **3(n)** through reversible boronate ester formation. Evaporation of the DMF solution containing the powder of **3(n)** and 0.5 or 1.0 equivalent of the polymer **4** in vacuo produced lamellar films through dehydration reaction. Boiled aqueous solutions of the film as a precursor for self-assembly were slowly cooled to room temperature. The hydration process and the subsequent separation from the polymer template by centrifugation process resulted in the formation of targeting self-assembled nanostructures. XRD and FT-IR spectroscopy disclosed that the starting Film-A of **3(n)** complexed with **4** [**3(n)/4** = 1/1, molar ratio] takes symmetrical MLMs. Meanwhile, the MLMs of the starting Film-B [**3(n)/4** = 2/1, molar ratio] take unsymmetrical MLMs [7]. The dissociation reaction of the PTB polymer **4** through the hydrolysis of the boronate esters yielded self-assembled nanotape and nanotube structures from the Film-A and Film-B, respectively, (Fig. 3.5). Significantly, the nanotapes and nanotubes as final products memorize the initial antiparallel and parallel molecular packing of the starting films,

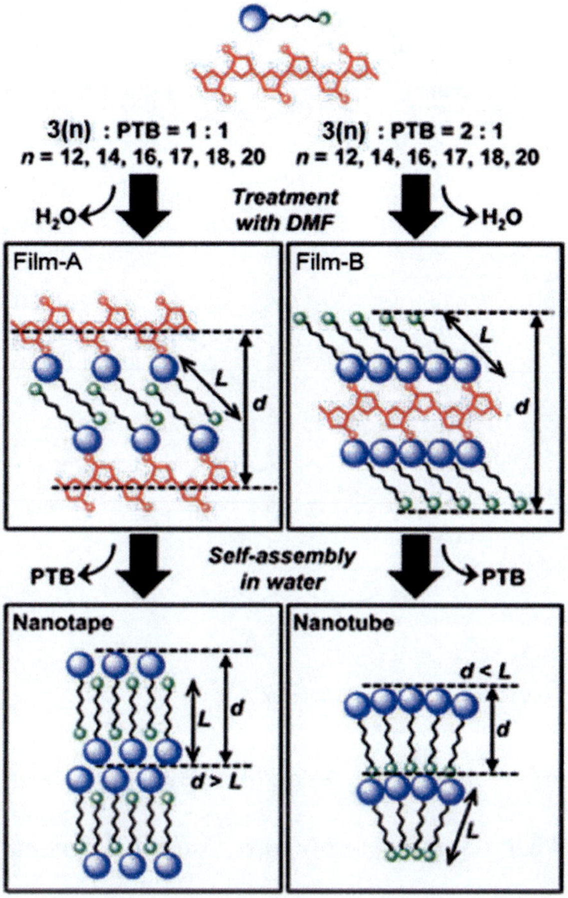

Fig. 3.5 Molecular packing of the starting Film-A and Film-B and the resultant self-assemblies of **3(n)** after dissociation of PTB **4** from the films (Reproduced with permission from Ref. [7], © 2013 American Chemical Society)

respectively. The present template approach is reminiscent of the self-organization of TMV by using an RNA chain as a template. This methodology allowed for the conservation of the unsymmetrical *polymorph*, resulting in the control of the head-to-tail *polytype*. Furthermore, this technique can avoid the spacer length dependence of the molecular packing that occurred upon self-assembly of the single component **3-(n)** without the template [6].

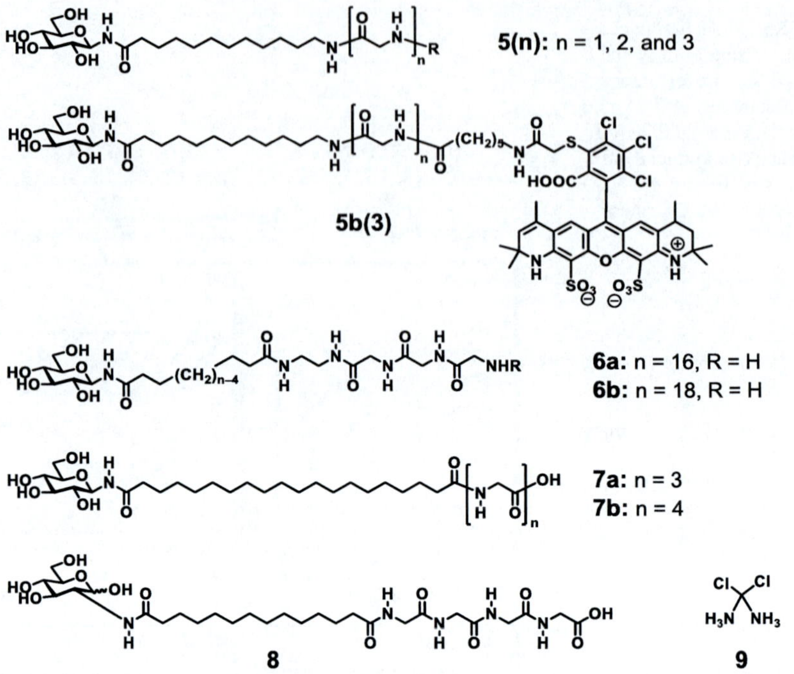

Chart 3.2 Chemical structures of **5–9**

3.2 Glucose–Oligoglycine Bolaamphiphile

3.2.1 Self-assembly into Nanotube Structures

Directional and multiple hydroxy groups on glucopyranose rings can be involved with diverse cooperative and complementary hydrogen-bond networks. Meanwhile, the carbonyl and amino groups of the diglycine residues are deeply associated with the formation of polyglycine II-type hydrogen-bond networks (Fig. 3.6) [8, 9]. Aiming at the preparation of SMNTs with unsymmetrical MLMs, we adopted this combination of glucopyranosyl and diglycine moieties as terminal headgroups of bolaamphiphiles. The newly synthesized wedge-shaped molecules **5(n)** (n = 1, 2, and 3) (Chart 3.2) comprise 1-β-*N*-glucosamide and oligoglycine (i.e., mono-, di-, and triglycine residues) terminals connected by a long hydrophobic chain through two amide linkages [10]. In the self-assembly in water, the bolaamphiphile **5(3)** formed very thin organic nanotubes of 7–9 nm in internal diameters and 3–4 nm in wall thickness, whereas the derivatives **5(1)** and **5(2)** produced helical nanofibers of 10–25 nm wide. Increasing the number of glycine segment resulted in the stabilization of unsymmetrical MLM in a single monolayered nanotube. The clear appearance of both the CH skeletal vibration band at 1026 cm^{-1} and the CH deformation band at

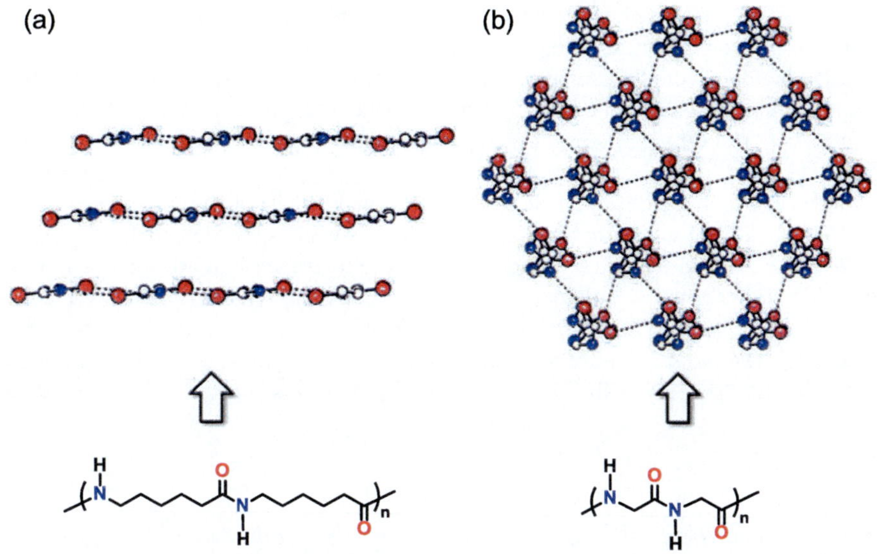

Fig. 3.6 Representative crystal structures of aliphatic polyamide; **a** β-sheet structure of nylon 6 and **b** polyglycine II-type structure of nylon 2 (polyglycine)

1420 cm^{-1} strongly indicates the formation of polyglycine II-type hydrogen bonds [8] among the triglycine residues (Fig. 3.7). This feature is quite distinct from the FT-IR spectra of the self-assembled nanofibers from **5(1)** and **5(2)**. As a result of the exclusive *polymorph* with unsymmetrical packing, the obtained nanotubes provide completely different inner and outer functionalities of monolayer-based wall [10].

Similarly, the aforementioned molecular design allowed us to synthesize some other glucose–oligoglycine bolaamphiphiles having amino terminal **6a** and **6b** [11,

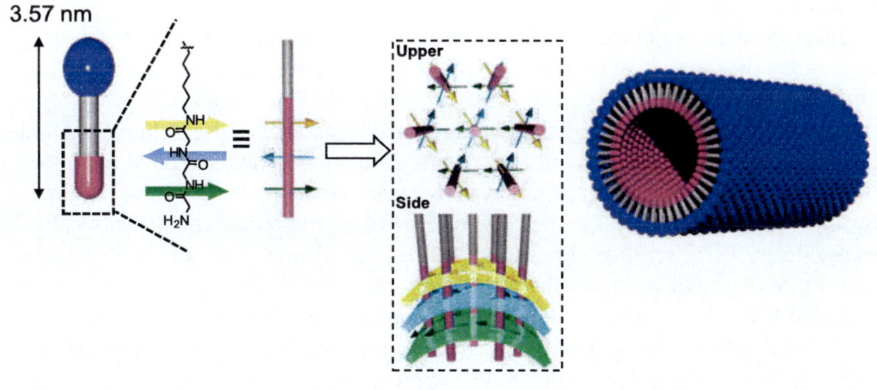

Fig. 3.7 Schematic illustration for the nanotube based on the single monolayer lined by pseudo polyglycine II-type hydrogen-bond networks among triglycine residues in **5(3)**

12] (Chart 3.2) or those having carboxyl terminal **7a** [12], **7b** [13], and **8** (Chart 3.2) [14]. Discrete organic nanotubes with inner diameters of less than 10 nm were shown to self-assemble from the amine-terminated glucose bolaamphiphiles of **6a** [11] and **6b** [12], and the carboxylic acid-terminated bolaamphiphiles of **7a** [12], **7b** [13], and **8** [14]. The inner surfaces of the obtained nanotubes from the bolaamphiphiles (**6a** or **6b**) and the carboxylic acid-terminated derivatives (**7a**, **7b**, or **8**) are covered with the amino and carboxyl headgroups, respectively. The outstanding appearance of the CH skeletal vibration band at 1026 cm^{-1} as well as CH deformation band at 1420 cm^{-1} also evidences the polyglycine II-type hydrogen-bond formation among the triglycine segments. Notably, we can synthesize the 2-*N*-glucosamide derivative **8** in only three steps [14], whereas more than seven steps are needed for the other analogues. Additionally, only reprecipitation process can make the bolaamphiphile **8** purify, and the yield of the resultant nanotube from **8** is considerably higher than others.

3.2.2 Nanotube with Metallodrug-Coordinated Inner Surfaces

The self-assembly of the acidic form of the bolaamphiphile **8** gives nanotube structures with outer and inner diameters of 14.3 nm and 7.0 nm, respectively, by acidification of the alkaline solution of **8** to pH 4 [14]. Interestingly, *endo*-complexation of the terminal carboxylate headgroup of **8** with cisplatin [cis-dichlorodiamineplatinum (II), CDDP] **9** (Chart 3.2) results in the formation of a CDDP-encapsulated organic nanotube (Fig. 3.8) [14]. Cisplatin is commonly used as a chemotherapeutic platinum coordination drug, exhibiting anticancer activity. A number of spherical and fibrous nanostructures have been prepared by substitution of the chloride ligands of **9** with carboxylate functional groups [15, 16]. The outer and inner surfaces of the self-assembled nanotubes were, however, differently covered with hydroxyl groups and CDDP, respectively.

Initially, self-assembled nanofibers from the sodium salt of **8** in water are transformed to the tubular assemblies after the addition of cisplatin **9**. Notably, the *endo*-complexation of the carboxyl group with CDDP takes place exclusively in the nanotube interior surface through complexation-assisted self-assembly. A single cisplatin molecule chelates with two carboxylate anions through metal complexation. The presence of polyglycine II-type hydrogen-bond networks can be confirmed by the occurrence of the CH deformation band of the tetraglycine moiety. Each molecule packs in an unsymmetrical fashion with tilting angle of 32.8° to organize a monolayer-based nanotube wall. Metal-complexed organic nanotubes can be prepared from, for example, tube-forming peptidic amphiphiles [17, 18]. However, the internal and external surfaces of the metal-complexed nanotubes are covered with identical functionality containing metal cations [19].

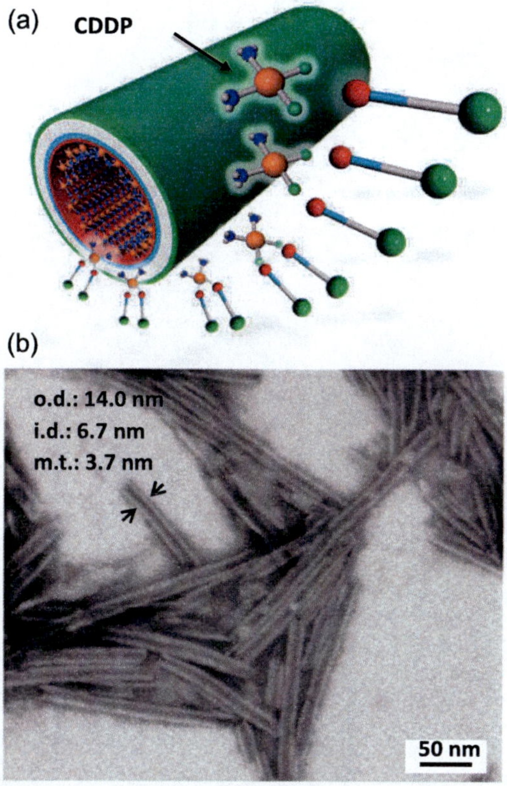

Fig. 3.8 **a** *Endo*-complexation-triggered self-assembly of **8** into a nanotube. **b** Scanning transmission electron microscopic (STEM) image of the CDDP-complexed SMNTs, negatively stained with 2% phosphotungstate (o.d., i.d., and m.t. means outer diameter, inner diameter, and membrane thickness of the nanotube, respectively) (Reproduced with permission from Ref. [1], © 2016 American Chemical Society)

3.2.3 Tailored Fabrication of Unique-Shaped Polydopamine Nanofibers

Chemistry and potent applications of polydopamine (PDA) **10** (Chart 3.3) (shown as one of the plausible polymer structures), inspired by the strong adhesion of mussels to a variety of solid surfaces, have intrigued many materials scientists in terms of surface modifications and coatings [20]. The separate formation of PDA nanofibers, nanocoils, and nanotapes, however, still remains fairly difficult. We utilized two organic nanotubes with different structures and components to template the polymerization of dopamine **11** (Chart 3.3). As a result of oxidative templated polymerization, linear and helically coiled PDA nanofibers were formed separately (Fig. 3.9). For example, the glucose–tetraglycine bolaamphiphile **7b** undergoes the self-assembly in water to give nanotube structures, the inner surface of which is covered with carboxylate anions [13]. The generated hollow cylindrical nanospace with a diameter of 8 nm serves as a template to fabricate linear PDA nanofibers. In aqueous solution at pH 8.6, electrostatic interactions allow for the preferential adsorption of

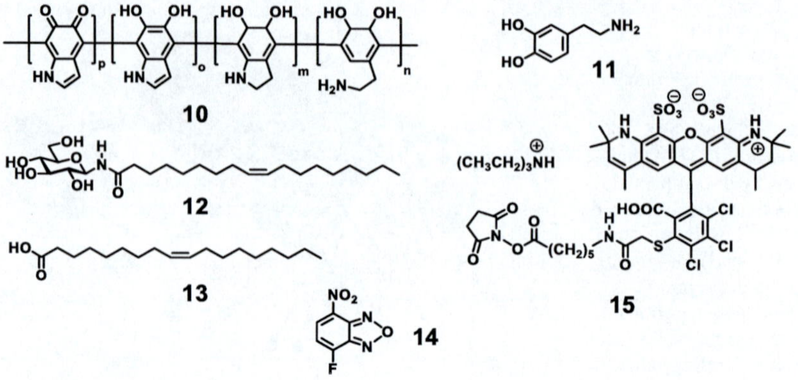

Chart 3.3 Chemical structures of **10–15**

Fig. 3.9 Production of **a** dumbbell-like PDA–nanotube hybrids and **b** helically coiled PDA nanofibers by utilizing the self-assembled monolayered and binary co-assembled SMNTs, respectively, as a polymerization template (Reproduced with permission from Ref. [1], © 2016 American Chemical Society)

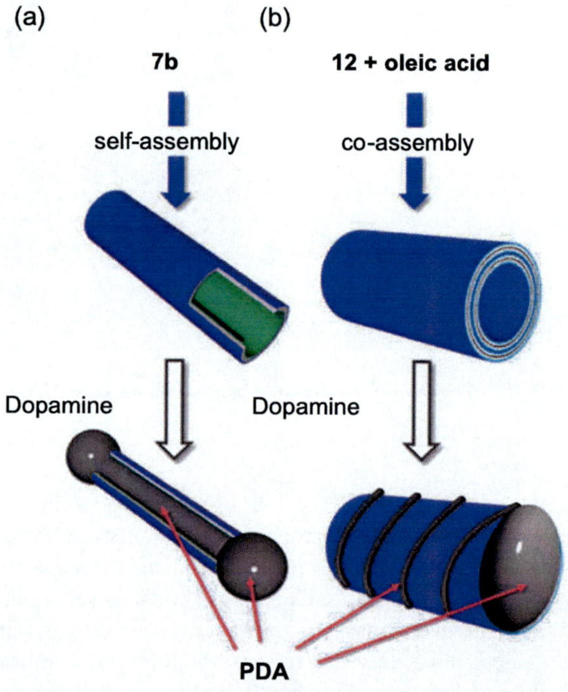

dopamine **11** onto the anionic inner surface of the nanotube from **7b**. STEM observation presented high-contrast image for the resultant PDA–nanotube hybrids with dumbbell- and match-like morphology (Fig. 3.9a). Interestingly, PDA nanoparticles of 30–40 nm that capped the nanotube ends were also observed. These findings mean that the polymerization of dopamine **11** takes place not only in the nanotube channels but also around both open ends of the nanotubes (PDA/nanotube = 0.6:1 w/w).

Meanwhile, binary co-assembly of the unsaturated glucosamide amphiphile **12** (Chart 3.3) and oleic acid **13** (Chart 3.3) in methanol produced bilayer-based nanotubes with 70 nm average inner diameter and 190 nm outer diameter. Polymerization of dopamine **11** proceeded on the outer surface to yield helically coiled nanofibers of PDA (Fig. 3.9b). Of particular interest is that higher photothermal conversion effect as well as higher light absorbance were observed for the helically coiled PDA as compared to the linear PDA nanofibers [21]. More specifically, laser irradiation at a concentration of 0.8 mg/mL induced a remarkable temperature increase (ΔT) of 52.3 °C in the coiled PDA fibers after 10 min. By contrast, the linear PDA nanofibers and spherical PDA nanoparticles displayed only moderate temperature increases of 19.7 and 23.4 °C, respectively.

3.2.4 Transportation of Proteins in a Confined Nanochannel

A part of the amino groups located on the inner surface of the self-assembled nanotubes (i.d.: 80 nm) from **3(18)** underwent the reaction with 4-fluoro-7-nitrobenzofurazan (NBD-F) **14** (Chart 3.3) as a fluorescent donor in an aqueous phase. The resultant nanotubes can emit fluorescence, proving a typical example of the effective post-modification of the inner surfaces [22]. We were also able to functionalize the interior of the nanotubes of **3(18)** or **6a** with Alexa Fluoro 546 (**15**: Alexa Fluor 546 carboxylic acid succinimidyl ester) (Chart 3.3) as a fluorescent acceptor of GFP through succinimidyl ester linkage [23]. The resultant nanotubes from **6a** possess 10 nm inner diameter, whereas the nanotube from **3(18)** gives different inner diameters, 20 and 80 nm under different pH conditions. With the use of the three different nanotubes, we attempted to visualize the transport feature of confined proteins in the nanochannels but also to evaluate the diffusion constants of the proteins. Figure 3.10 shows a fluorescence microscopic image of a single nanotube of **3(18)**, the interior of which was partially modified with **14**. A linear fluorescent image is emitted from nitrobenzofurazan moiety that is covalently connected to the nanotube inner surfaces. Figure 3.10b, c also represent time-lapse fluorescence micrographs upon addition of the QSY7 (a fluorescent acceptor dye)-immobilized ferritin to aqueous nanotube solution [22]. Quenching phenomenon starts from the two open ends due to FRET mechanism from NBD to QSY7. The quenching edge gradually moves along the long axis to the nanotube center. It took 3.5 s to complete the quenching (Fig. 3.10d).

When we employed QSY7-immobilized Au NPs of 1.4 nm wide as a reference nanostructure, quenching was completed within 0.84 s. This time is considerably shorter than that required for ferritin [22]. To perform further systematic and quantitative studies, we investigated the diffusion behavior of GFP in the Alexa-immobilized nanotube channels with three different i.d. (10, 20, and 80 nm) [23]. This FRET-based analysis also enables the satisfactory visualization of GFP diffusion in the nanotube channel of **3(18)** or **6a**.

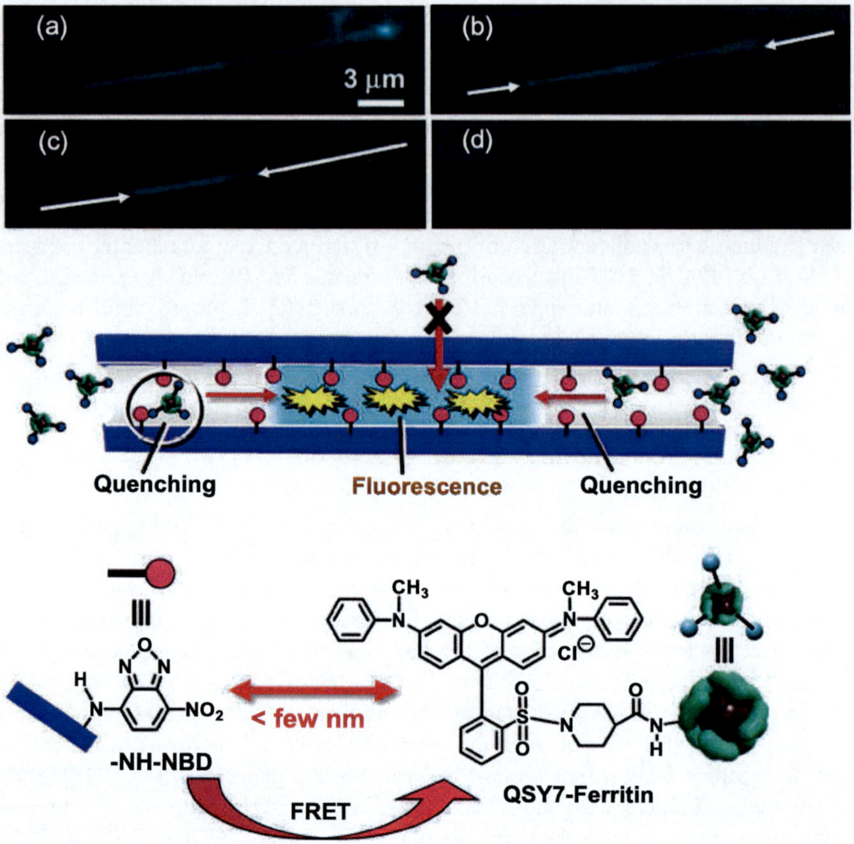

Fig. 3.10 Schematic images of FRET that take place in the nanotube channel and (**a–d**) time-lapse fluorescence microscopic images for the NBD-modified SMNT self-assembled from **3(18)** upon addition of QSY7-immobilized ferritin. The time elapsed is **a** 0, **b** 1.0, **c** 2.0, and **d** 3.5 s (Reproduced with permission from Ref. [24], © 2008 John Wiley and Sons)

Table 3.1 arranges the obtained diffusion constants (*D*) of three guest substances (GFP, ferritin, and latex beads) under various conditions. The *D* value of 0.7×10^{-11} $m^2\ s^{-1}$ of ferritin is one-fifth that in bulk water (3.4×10^{-11} $m^2\ s^{-1}$). The *D* values obtained for GFP remarkably decreased with the decrease in the nanotube inner diameter. The smaller *D* values should be due to the electrostatic interaction between the inner surfaces and guest molecules, the confined geometry of the nanochannel, and the viscosity enhancement in the nanochannels. Relatively rapid transportation was observed for the guest molecules such as ferritin. Meanwhile, the diffusion of dye molecules in silica gel pores is reported to decrease with the decrease in the pore size from 30 to 15, 6, and 3 nm [25]. Our experimental findings about the diffusion behavior of the proteins coincide well with such results.

Table 3.1 Evaluated diffusion constants (*D*) of various guest substances in bulk water and SMNT channels

Nanochannel					
Amphiphile		Inner diameter (nm)	Guest (size, nm)	*D* (10^{-11} m^2 s^{-1})	References
	Bulk		GFP (3–4)	12.0	[23]
3(18)		80	GFP (3–4)	2.6	[23]
3(18)		20	GFP (3–4)	0.14	[23]
6a		10	GFP (3–4)	0.036	[23]
	Bulk		QSY7-ferritin (15)	3.4	[22]
3(18)		80	QSY7-ferritin (15)	0.7	[22]
	Bulk		Latex beads (30)	2.0	[26]
Soybean lecithin		100	Latex beads (30)	0.9	[26]

Higgins and Ito's group disclosed the diffusion mechanism and behavior of anionic sulforhodamine B (SRB) dye **16** (Chart 3.4) in the nanochannel of the self-assembled nanotube from the glucose–triglycine bolaamphiphile **6b** [27]. Imaging fluorescence correlation spectroscopy [28] made them conclude that a desorption-mediated Fickian diffusion mechanism [29] can describe the mass transport behavior of the SRB molecule in the nanochannel. Namely, the SRB molecules frequently adsorb onto the nanotube internal surfaces with diffusing by Brownian-like motion. Coulombic interactions between the ammonium cations of the nanotube inner surfaces and anionic SRB molecule compel the SRB motion rate to be low. The same technique has recently elucidated three critical factors for Nile Red **17** (Chart 3.4) and hydroxylated Nile Red **18** (Chart 3.4) to transport in the nanotube channel [30]. Those aspects are associated with the location effect of molecules on the mobility, differential distribution of dipolar molecules depending on the environmental polarity, and finally dielectric properties of various regions in the nanochannels.

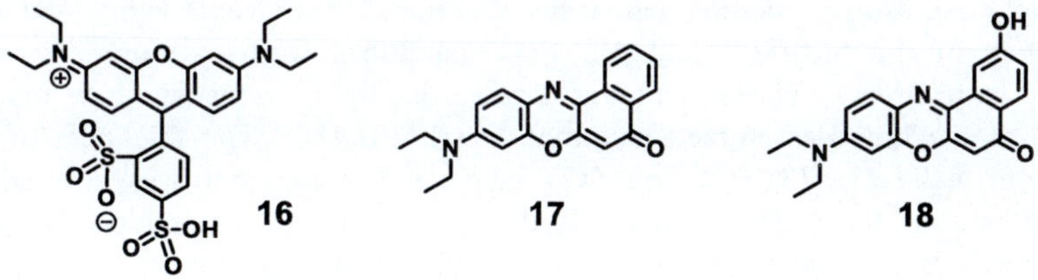

Chart 3.4 Chemical structures of **16–18**

3.2.5 Binary Co-assembly

In a self-assembly process, appropriate combination of *polymorph* control with binary co-assembly technique enables the exclusive modification of a functional moiety onto the inner or outer surfaces of assembled organic nanotubes. Favorably, the adjustment of the mixing ratio of nanotube-forming amphiphiles as a mother component and pre-functionalized amphiphiles as a doped component leads to the regulation of the functionalization degree over a wide range. For example, the binary co-assembly was performed to fabricate the supramolecular nanotubes that can sense the capture and release of GFP [31]. In this case, only the inner surface of the unsymmetrical nanotubes from **5(3)** with the inner diameter of 8 nm was partially covered with the large fluorescent recognition probe Alexa **15** toward the targeting GFP. Figure 3.11 illustrates that multiple hydrogen bonds are formed between the mother bolaamphiphile **5(3)** and a small amount of doped bolaamphiphilic analogue **5b(3)** (Chart 3.2) labeled with Alexa. This circumstance should enforce the adoption of parallel molecular packing within the Alexa-immobilized nanotube wall of a single monolayer. A single sharp peak assignable to the $\gamma(CH_2)$ rocking vibration appeared at 719 cm^{-1}. This indicates that the lateral packing of the oligomethylene chain takes a triclinic ($T_{//}$) subcell structure. The Alexa moiety is, thus, exclusively located on the inner surface of the nanotube. Binary co-assembly technique also allows for partial hydrophobization on the nanotube interior surfaces with the benzyloxycarbonyl (Cbz) or Boc group [13, 32]. It is worthy of note that the hydrophobicity of the interior surfaces is accurately controllable. This self-assembly strategy resulted in the realization of slow drug release [13] and structural optimization of organic nanotubes toward artificial chaperone [32].

3.2.6 Chirality Induction to Entrapped Achiral Polythiophenes

Binary co-assembly technique using the combination of **5(3)/5(2)** (1:1, mol/mol) or **5(3)/5(1)** (1:1, mol/mol) produced a mixture of nanofibers and nanotubes. The co-assembly of **5(3)/5(2)** (2:1, mol/mol) or **5(3)/5(1)** (2:1, mol/mol), however, exclusively yielded nanotube structures [33]. Remarkably, the inner diameters of the **5(3)/5(2)** and **5(3)/5(1)** nanotubes were 4–5 nm and of 2–3 nm, respectively. These diameter sizes are slightly smaller than that (7–9 nm) of the homo-assembled nanotube from **5(3)** [10]. The thickness of both nanotube walls of **5(3)/5(2)** and **5(3)/5(1)** are 2–4 nm that was also equivalent to that of the **5(3)** nanotube. Both types of the co-assembled nanotubes are, therefore, composed of a single monolayer membrane of a mixture of **5(3)** and **5(2)**, or a mixture of **5(3)** and **5(1)**. We found that three different nanotubes with different inner diameters can regulate the conformation and aggregation state of entrapped achiral poly(thiophene boronic acid) (M_n = 24,300)

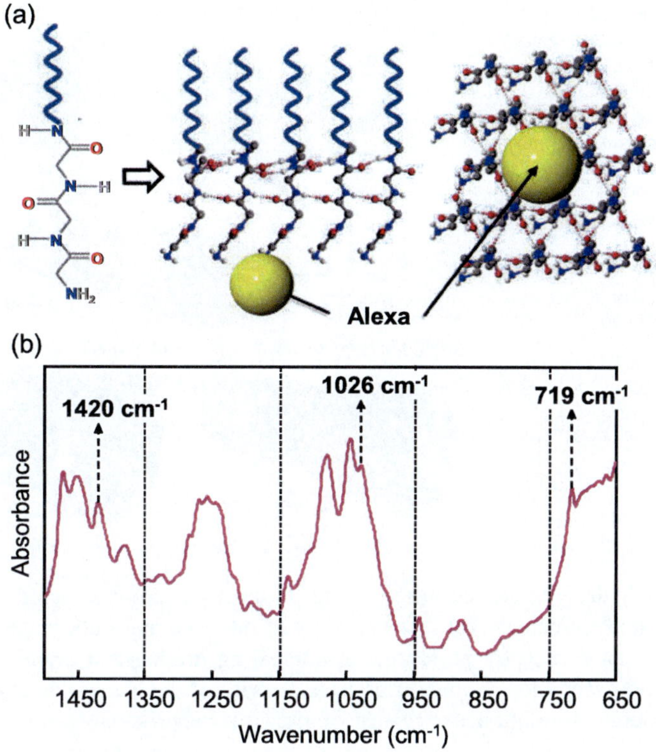

Fig. 3.11 **a** Formation of polyglycine II-type hydrogen-bond network between the mother bolaamphiphile **5(3)** and Alexa-appended derivative **5b(3)**. (Left) Building block **5(3)**, (middle) side view, and (right) view from the nanotube interior. **b** IR spectrum of lyophilized Alexa-modified SMNT produced by the binary co-assembly of **5(3)** with **5b(3)** (Reproduced with permission from Ref. [1], © 2016 American Chemical Society)

4. Indeed, depending on the sizes of the inner diameter (7–9, 4–5, and 2–3 nm), the entrapped achiral polymer **4** into each nanotube takes aggregated, random coil, and planar and 1D conformation with an apparently longer conjugation length, respectively, (Fig. 3.12) [33]. Simultaneously, the polythiophene chains with the planar conformation undergo chirality induction, exhibiting a recognition capability for the chirality of D- and L-sugars.

3.2.7 Duplex Formation of Short Nucleotides

Confined nanospaces such as supramolecular nanocavity [34] and mesoporous silica pore [35] are known to support the duplex formation of short nucleotides in spite of energetically unfavorable event. No reports have, however, described the duplex

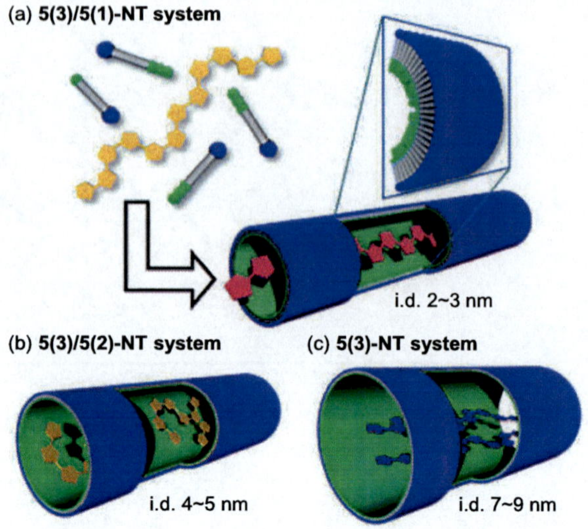

Fig. 3.12 Schematic images of 1D **a** extension, **b** random coil conformation, and **c** aggregation of the poly(thiopheneboronic acid) **4** chain by encapsulation into the nanochannels of **5(3)/5(1)-NT**, **5(3)/5(2)-NT**, and **5(3)-NT**, respectively, (Reproduced with permission from Ref. [33], © The Royal Society of Chemistry 2016)

formation or hydrogen-bonded base pairing of short nucleotides including mono-, di-, and trinucleotides as well as their release into external bulk water. Kameta and Akiyama have recently presented an interesting finding that photoresponsive shrinkable self-assembled nanotubes assisted a series of the nucleotide entrapment, subsequent duplex formation, and release of the duplexed product while utilizing the fields of interior and around the open ends of the nanotubes [36]. They synthesized the glucose–triglycine bolaamphiphiles **19a**, **19b**, and **19c** (Chart 3.5), bearing a 2-nitrobenzyl group as a photoresponsive moiety. The self-assembly of **19a** in water produces monolayer-based nanotubes with an inner diameter of 18 nm, whereas that of **19b** and **19c** only gave tape- and sheet-like nanostructures, respectively. Photochemical dissociation of the 2-nitrobenzyl group in **19a** with a high-pressure Hg lamp triggered a shrinkage in the nanotube inner diameter from 18 to less than 2 nm without changing the wall thickness. Concurrently, the dinucleotides were compelled to release from the nanochannel to the external aqueous media (Fig. 3.13). Importantly, the formation of the dinucleotide duplexes concurrently accelerates in the confined nanochannels of less than 2 nm wide. Confined water environment featured by lower polarity as well as the shrinkable size effect contributes to the stabilization of the hydrogen-bonded base pairing.

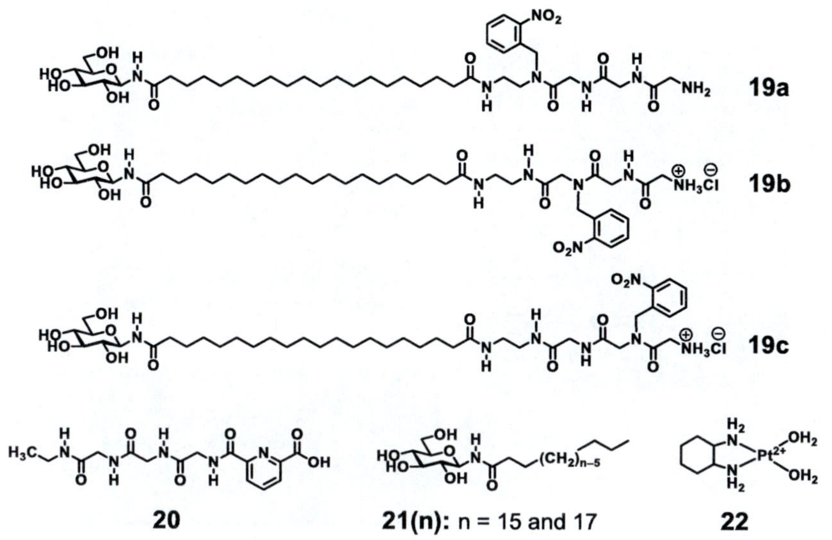

Chart 3.5 Chemical structures of **19–22**

3.2.8 Two-Step Self-assembly for Interior Surface Modification

Two-step self-assembly process with three different components enables the selective localization of the ligand for an anticancer platinum complex. We chose the nanotube-forming unsymmetrical bolaamphiphile (**6b** or **7a**), the triglycine derivative (**20**) (Chart 3.5) carrying a pyridine carboxylate segment as the ligand, and the nanotube-forming glucosamide amphiphile **21(n)** (*n* = 15 and 17) (Chart 3.5). As a final result of the two-step self-assembly, nanotube or nanotape structures were formed (Fig. 3.14) [12]. The obtained nanotubes and nanotapes are composed of an unsymmetrical MLM, the inner surface or one side of which is partially modified with the ligand for the anticancer complex **22** (Chart 3.5).

Homo-self-assembly of the single component **6b** or **7a** in water produces nanotubular structures with an inner diameter of 8 nm and a wall thickness of 4 nm independently [12]. Binary co-assembly of **6b/20** and that of **7a/20** yielded nanotubes and nanotapes, respectively. The obtained $T_{g\text{-}l}$ (= 51 and 64 °C) for the hybridized **6b/20** nanotube and **7a/20** nanotape, respectively, was found to be much lower than that of the single components ($T_{g\text{-}l}$ > 115 °C). The lower $T_{g\text{-}l}$ responsible for the lower thermal stability of the nanotubes and nanotapes is ascribable to the lack of both a glucose headgroup and a long alkyl chain. Importantly, the addition of the glycolipid **21(n)** (*n* = 15 and 17) as the second step process was able to make up for the present unfavorable state. Indeed, the observed $T_{g\text{-}l}$ above 115 °C for the **6b/20/21(17)** nanotube or **7a/20/21(15)** nanotape coincides with the fact that the

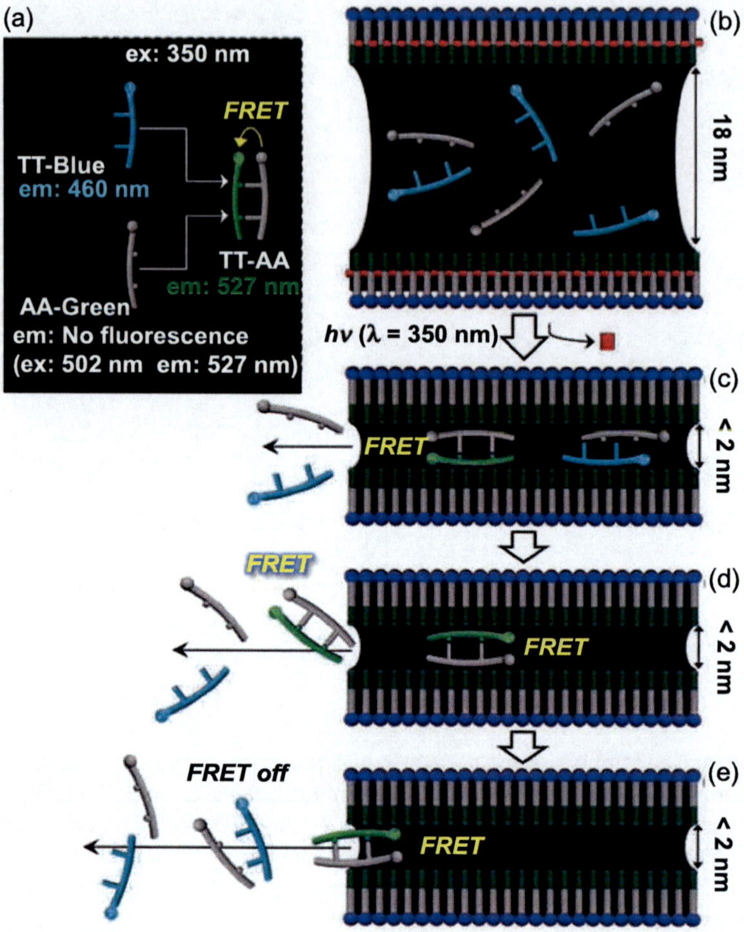

Fig. 3.13 **a** Fluorescence property of each dinucleotide and the duplexed dinucleotide (ex: excitation wavelength; em: emission wavelength). **b** Duplex formation of the dinucleotides in the nanotube channel. **c–e** Shrinkage of the nanotube channel upon photoirradiation results in squeezing out of the duplexed dinucleotides and subsequent dissociation of the dinucleotides in the bulk solution (Reproduced with permission from Ref. [36], © 2018 WILEY-VCH Verlag GmbH & Co. KGaA, Weinheim)

third component glycolipid **21(n)** fills well the void spaces within the monolayered molecular packing of the nanotube and nanotape structures (Fig. 3.14).

We, next, discuss the reason why similar two-step assembly results in the different self-assembled products, nanotubes, and nanotapes. The observed two peaks at 1420 and 1026 cm^{-1} for the **6b/20/21(17)** nanotube strongly indicate the stabilization through the formation of the polyglycine II-type hydrogen bond [12]. The bolaamphiphile **6b** and the triglycine-appended ligand **20** should pack in parallel fashion

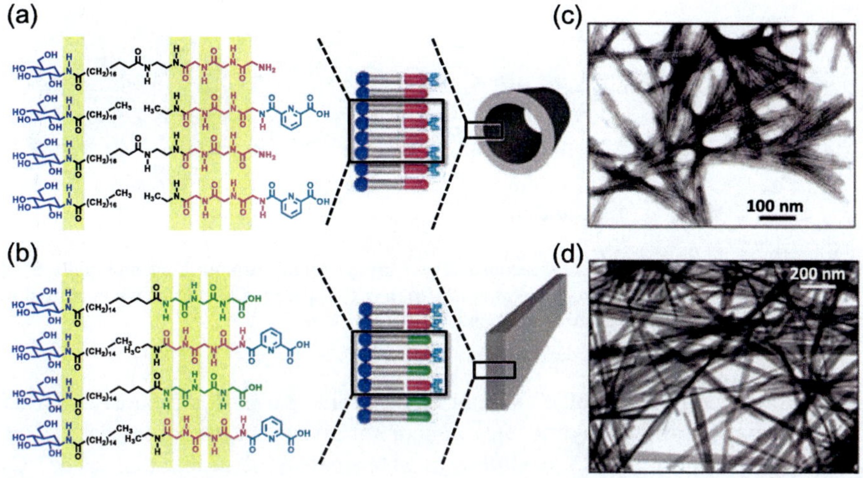

Fig. 3.14 Plausible molecular packing within the MLMs of the self-assembled **a** nanotubes from the combination of **6b/20/21(17)** and **b** nanotapes from the combination of **7a/20/21(15)**. TEM images of the obtained **c** nanotubes and **d** nanotapes (Reproduced with permission from Ref. [1], © 2016 American Chemical Society)

while sharing the common triglycine residues. Conversely, the **7a/20/21(15)** nanotape displayed no outstanding peaks in the identical IR band regions. Additionally, the observed frequency (1619 cm^{-1}) of the C = O stretching vibrational band for the nanotape suggests the formation of polyglycine I-type hydrogen bonding (like a typical parallel β-sheet structure). Thus, the bolaamphiphile **7a** and the ligand **20** should pack in an antiparallel fashion. As a result of the binary self-assembly, the polyglycine II-type hydrogen bonds between **6b** and **20** drive the exclusive formation of the nanotubes. By contrast, antiparallel β-sheet-like hydrogen bonds between **7a** and **20** result in the self-assembly of the nanotape morphology.

3.2.9 Two-Step Self-assembly for Exterior Surface Modification

The two-step self-assembly procedure also permits self-assembled nanotubes to install a desired functionality onto the exterior surface exclusively. For instance, a recognition site for concanavalin A (Con A) can be successfully located onto the outer surface of the self-assembled nanotube from the bolaamphiphile **5(3)** (Fig. 3.15) [37]. First, the binary co-assembly of the nanotube-forming bolaamphiphile **5(3)** with the triglycine derivative **23** (Fig. 3.15) yielded nanotube structures with similar dimensions to those of the homo-self-assembled nanotubes from the single component **5(3)** [10, 37]. A deficiency of the glucose headgroup and long hydrocarbon chain

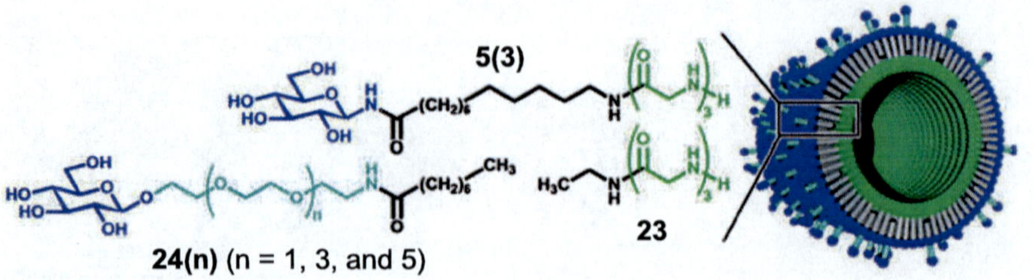

Fig. 3.15 Monolayer-based nanotube consisting of the nanotube-forming bolaamphiphile **5(3)**, triglycine derivative **23**, and glucose derivatives **24(n)** ($n = 1, 3$, and 5) (Reproduced with permission from Ref. [37], © The Royal Society of Chemistry 2015)

in the molecular structure of **23** should induce void spaces in a monolayer-based molecular packing of the hybrid **5(3)/23** nanotubes. In fact, the observed $T_{g\text{-}l}$ (= 52 °C) of the hybrid **5(3)/23** nanotubes in water is lower than that ($T_{g\text{-}l}$ = 67 °C) of the single component of **5(3)**. The glucose derivatives **24(n)** (n = 1, 3, and 5) (Fig. 3.15) were, however, added to the aqueous dispersion of the hybrid **5(3)/23** nanotubes. Heating the dispersion resulted in the increase in the $T_{g\text{-}l}$ value to 65 –67 °C. This finding clearly indicates that the glucose derivatives **24(n)** bearing different ethylene glycol (EG) chains well compensate the void spaces within the MLM. FT-IR studies revealed that the observed two CH deformation (1420 cm^{-1}) and skeletal (1026 cm^{-1}) vibrational bands are compatible with the formation of a polyglycine II-type hydrogen-bond network among the triglycine segments. All the findings strongly suggest that the glucose moiety of **24(n)** (n = 1, 3, and 5) was located at a distance from the outer surface of the ternary hybrid nanotubes via the EG spacer chain.

We examined the effect of the spacer length of ethylene glycol on the detection of Con A as a targeting lectin [37]. The addition of Con A causes the appearance of the aqueous dispersions containing the hybrid nanotubes of **5(3)/23/24(n)** (n = 1, 3, and 5). The findings sensitively depend on the difference in the spacer length (n = 1, 3, and 5) (Fig. 3.16). Upon addition of Con A, the nanotube of **5(3)/23/24(1)** resulted in the formation of precipitate with the decrease in the nanotube dispersibility (Fig. 3.16b). The nanotube of **5(3)/23/24(3)** exhibited the two-step sensitiveness composed of the formation of liquid crystal and subsequent hydrogel, which depends on the concentration of Con A (Fig. 3.16a, b). Finally, the nanotube of **5(3)/23/24(5)** responded to a relatively high concentration of Con A, resulting in the hydrogel formation. Favorably, the binding behavior of the **5(3)/23/24(3)**- and **5(3)/23/24(5)**-based nanotubes can be easily detected by the appearance of the dispersion as well as the given color with using naked eyes (Fig. 3.16b) [37]. However, authentic self-assembled nanotube from **5(3)** alone, which has the same dimensions and structural parameters, is unable to detect Con A (Fig. 3.16b). This is because the external surfaces are directly covered with glucose headgroups.

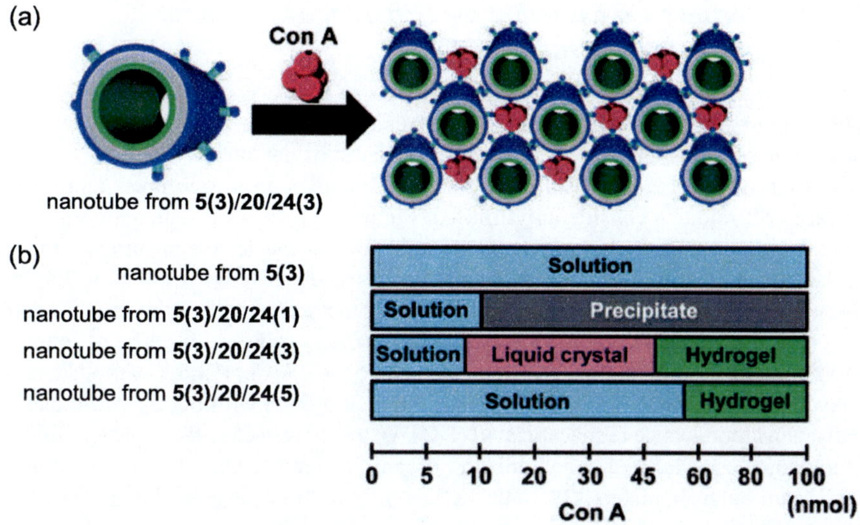

Fig. 3.16 **a** Side-by-side alignment of the self-assembled nanotubes from the combination of **5(3)/20/24(3)**, which are complexed with Con A. **b** Responsiveness of the nanotube dispersions from **5(3)** and **5(3)/20/24(n)** ($n = 1, 3$, and 5) toward different concentrations of Con A (Reproduced with permission from Ref. [37], © The Royal Society of Chemistry 2015)

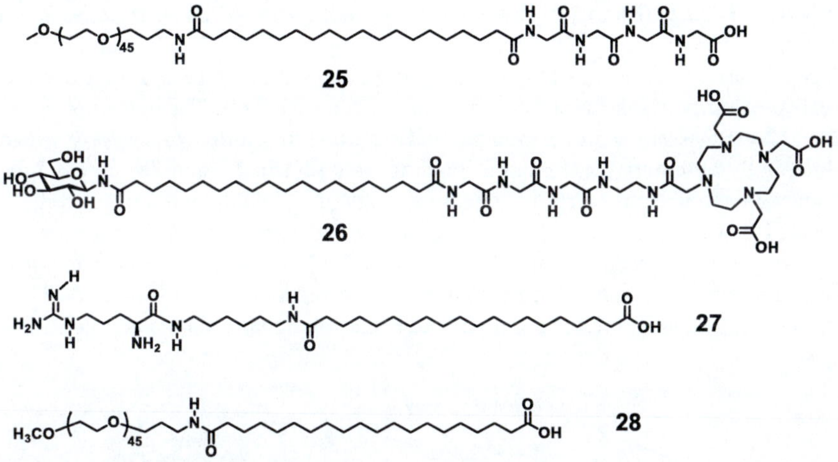

Chart 3.6 Chemical structures of **25–28**

3.2.10 Ternary Co-assembly for Different Modification of Inner and Outer Surfaces

The aforementioned co-assembly strategy also benefits the incorporation of dual functionalities onto the inner and/or outer surfaces of nanotubes. However, simultaneous and separate install of different functionalities onto each inner and outer surface still remains considerably limited. Figure 3.17 shows a schematic illustration, in which a PEG chain and metal cation ligand are localized on each outer surface covered with hydroxyl group and inner surface covered with amino group, respectively. Aiming at the preparation of organic nanotubes with desired functionalities at desired sites, we adopted a ternary co-assembly process [38]. The nanotube-forming bolaamphiphile **7b**, PEG-appended analogue **25** (Chart 3.6) (10 mol%), and unsymmetrical bolaamphiphile **26** (Chart 3.6) (10 mol%) with both glucose and a tetraazacyclododecane tetraacetic acid (DOTA) group were chosen for the assembly. We, then, successfully obtained uniform organic nanotubes with 7–9 nm inner and 16–18 nm outer diameters [38]. The XRD and DSC studies showed that 10 mol% of the PEG amphiphile **25** is incorporatable into the resulting nanotubes. Notably, the ternary co-assembled nanotubes with the DOTA ligand were shown to improve the dispersibility of the nanotube drastically in PBS (pH 7.4). This result directly indicated that the PEG chain is actually anchored onto the nanotube outer surface as required.

The ternary co-assembly of the mother amphiphile **2(18)** with other two bolaamphiphiles **27** and **28** (Chart 3.6) also enables the modification of the outer surfaces with both PEG chains and cationic arginine residues (Fig. 3.18) [39]. The bolaamphiphiles **27** and **28** serve to accelerate DNA binding to the nanotube surface and yet to enhance the dispersibility of the nanotube–DNA complex, respectively. The bolaamphiphiles **27** and **28** self-assemble into nanotapes and spherical structures, respectively. Meanwhile, both the binary [**2(18)** and **27**] and ternary [**2(18)**, **27**, and **28**] co-assembly produced nanotube structures in aqueous dispersions. Zeta potential measurements revealed that the surface charges of the binary and ternary co-assemblies exhibit monodisperse positive peaks at +37.7 and +21.3 mV, respectively. The zeta potential of the homo-assemblies from **2(18)** displayed a negative peak at −25.8 mV. These findings suggest that both the arginine- and PEG-functionalized bolaamphiphiles **27**

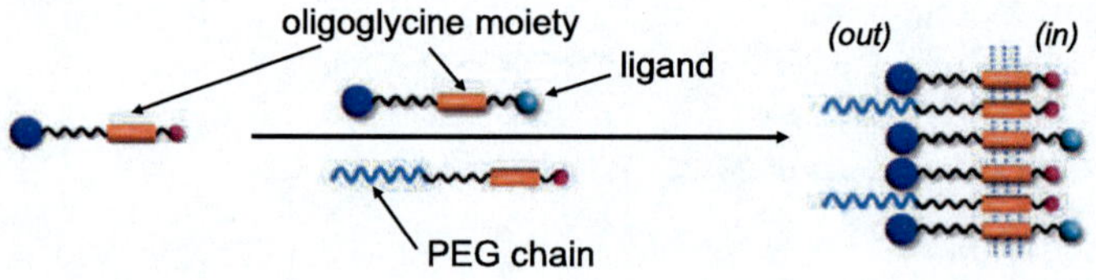

Fig. 3.17 Selective formation of unsymmetrical MLMs with different inner and outer surfaces from unsymmetrical bolaamphiphiles through ternary co-assembly

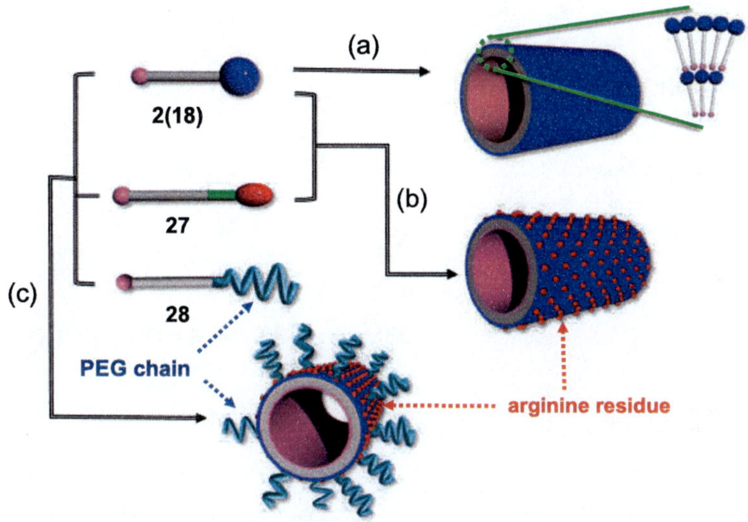

Fig. 3.18 Formation of exterior-functionalized nanotubes through **a** self-assembly of a mother bolaamphiphile **2(18)**, **b** binary co-assembly of the mother **2(18)** and arginine-appended bolaamphiphiles **27**, and **c** ternary co-assembly of the mother **2(18)**, arginine-appended **27**, and PEG-attended bolaamphiphiles **28**

and **28**, respectively, are indeed anchored onto the nanotube outer surface. Moreover, the PEG chain layers mildly shield the surface charges on the outer surface [40]. The appearance of a single δ(CH_2) scissoring band at 1471 cm^{-1} is assignable to a triclinic ($T_{//}$) subcell structure, indicating a parallel molecular packing of the ternary component amphiphiles in the PEG-arginine nanotubes [39].

3.2.11 Thermally Controllable Extraction and Separation of Peptides

PEG chains play a significant role in medical and biological applications to display advantageous functions such as antigenicity, low toxicity, high water solubility, and thermal responsiveness [41, 42]. The topological structures as well as the molecular weights of the PEG chains have a profound effect on their physical and chemical properties. For instance, the performance of the PEG chains is strongly influenced by their confinement circumstance in nanopores [43] and nanochannels [44]. A two-step co-assembly with ternary components [**29(16)/30(16)/31** or **29(16)/21(17)/32**] (Chart 3.7) yielded two different monolayer-based nanotubes (WNT-exPEG and WNT-inPEG) (Fig. 3.19). The WNT-exPEG is featured by the dense modification with PEG chains on the outer surface, whereas the WNT-inPEG involves the dense

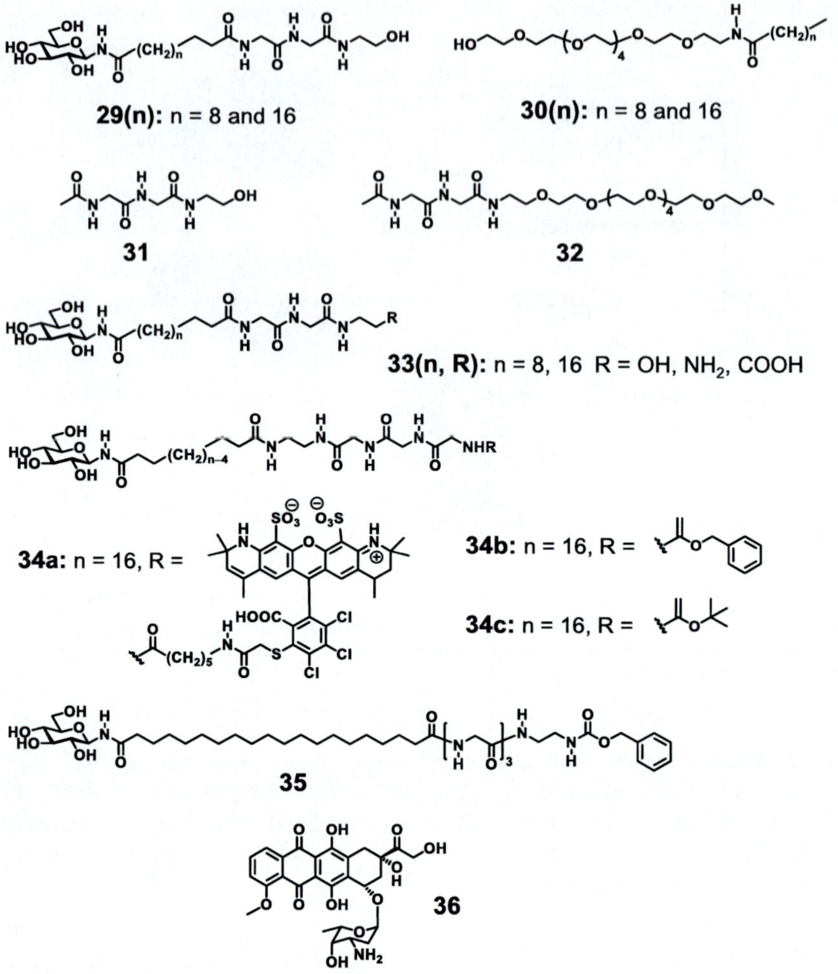

Chart 3.7 Chemical structures of **29–36**

functionalization with PEG chains on the inner surface. Importantly, the PEGylated stearamide **30(16)** alone, diglycine derivative **31** alone, glycolipid **21(17)** alone, and PEGylated diglycine derivative **32** alone self-assemble to form nanofibers of 100–500 nm wide, micelles, bilayered nanotubes, and micelles, respectively. Kameta and co-workers demonstrated that the WNT-inPEG can efficiently separate hydrophobic aliphatic amino acids (i.e., Ala, Val, Met, and Ile) over those carrying an aromatic side chain (i.e., Phe and Trp) [45]. Dehydration of the interior PEG chains upon heating changes the nanotube interior into hydrophobic environment, thus encapsulating hydrophobic amino acids from surrounding bulk solutions. Meanwhile, rehydration of the PEG chains upon cooling causes the release of the encapsulated amino acids

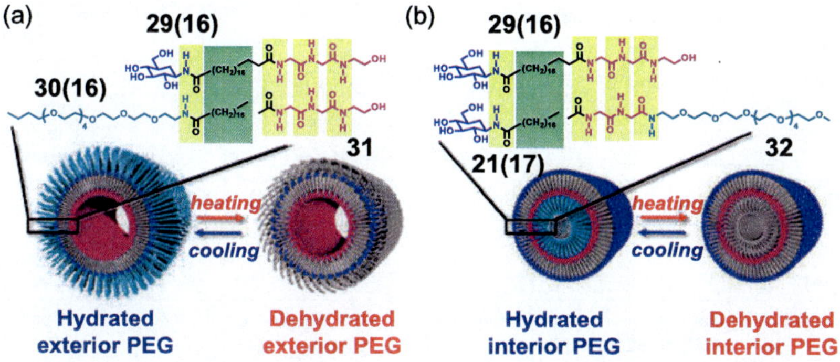

Fig. 3.19 Monolayer-based nanotubes: **a** Nanotubes with PEG chains on the outer surface, self-assembled from the ternary component of **29(16)/30(16)/31**. **b** Nanotubes with PEG chains on the inner surface, self-assembled from the ternary component of **29(16)/21(17)/32**. The lower illustration shows the thermal dehydration and rehydration behavior of each nanotube (Reproduced with permission from Ref. [45], © 2017 American Chemical Society)

into outer bulk solutions. This thermoresponsive dehydration and rehydration of the PEG chain-modified nanotube in the interior surface can also separate efficiently three peptides differing in a single amino acid in the amino acid sequences [45].

3.2.12 Chiral Separation by Thermoresponsive PEG-Coated Nanotubes

In the same way, we prepared ternary co-assembled nanotubes, WNT-outPEG and WNT-inPEG, from the combination of **29(16)/30(16)/31** and **29(16)/21(17)/32**, respectively. We also synthesized three different D-glucose-based bolaamphiphiles [**33(8, R)**: R = OH, NH_2, COOH] (Chart 3.7) having a hydroxyl, amino, or carboxyl group at another end. The enantiomers of **33(8, OH)** and **30(8)**, i.e., **33(8, OH)*** and **30(8)*** that contain an L-glucose unit were also prepared to perform the ternary co-assembly. We eventually constructed the five nanotubes with an internal diameter of 4 nm, i.e., inPEG-NT, inPEG-NT*, inPEG-(+)-NT, inPEG-(−)-NT, and outPEG-NT. The inner surfaces of the inPEG-NT and inPEG-NT* covered with a dense layer of the short PEG chains serve to detect the chirality of amino acids (i.e., Ser, Thr, Cys, and Tyr) (Table 3.2) [46]. The inner diameters of the hybrid nanotubes and the location of the PEG layers (inner or outer surfaces) have a strong influence on the selectivity and efficiencies of the chiral separation. Various hydrophobic groups were incorporated to the nanotube inner surface through the amide linkage between the amino groups of **33(8, NH_2)** and several hydrophobic acid chlorides. Significantly, the incorporation of branched alkyl chains enables the discrimination of chiral peptides including enkephalin and three enkephalin analogues, which contain hydrophobic amino acids [46].

Table 3.2 Components of the ternary co-assembled nanotubes and their chiral separation properties for amino acids

Nanotube	Components	PEG	I.D.[a] (nm)	Charge	T_D^b, T_R^c (°C)	Isolatable amino acids (> 20%[d])
inPEG-NT	**33(8, OH)/30(8)/32**	Inner	4	Nonionic	41, 44	D-Ser, D-Thr, D-Cys, D-Tyr
inPEG-NT*	**33(8, OH)*[e]/30(8)*[f]/32**	Inner	4	Nonionic	40, 46	L-Ser, L-Thr, L-Cys, L-Tyr
inPEG-(+)-NT	**33(8, NH_2)/30(8)/32**	Inner	4	Cationic	49, 35	D-Asp, D-Glu
inPEG-(−)-NT	**33(8, COOH)/30(8)/32**	Inner	4	Anionic	47, 37	D-Lys, D-Arg, D-His
outPEG-NT	**33(8, OH)/30(8)/31**	Outer	4	Nonionic	56, 42	None

[a]Inner diameter
[b]Dehydration temperature of the poly(ethylene glycol) (PEG) chains
[c]Rehydration temperature of the PEG chains
[d]20% < 100%-back-extraction ratio %
[e]Enantiomer of **33(8, OH)** containing a D-glucose unit
[f]Enantiomer of **30(8)** containing a D-glucose unit

3.2.13 Protein Stabilization

Practical use of enzymatic proteins as biological catalyst has grown rapidly in a variety of industry. External conditions, however, such as organic phase environment, high temperatures, and existence of denaturants, make the activity of the proteins which extremely reduce due to the transformation of their 3D conformations into unfavorable state. Entrapment or immobilization of the proteins onto and into the solid support such as mesoporous silicates serves to keep the activity even under unfavorable conditions [47].

Self-assembled organic nanotubes from **6a** can also encapsulate several proteins into their lumens with rationally designed inner surfaces as well as inner diameter sizes. Intriguingly, the notable size effect of internal cross sections on the protein stabilization was manifested in the encapsulated GFP in the nanochannel of the **6a** nanotubes [23]. Free GFP and the encapsulated GFP in relatively larger channels (i.d.: 80 nm and 20 nm) of the self-assembled nanotubes (**80-AlexaNT** and **20-AlexaNT**, respectively) from **3(18)** [22] drastically denatured upon increase in temperatures (Fig. 3.20). Conversely, the entrapped GFP in the small channel (i.d.: 10 nm) of the self-assembled nanotube (**10-AlexaNT**) from **6a** [11] almost maintains its intrinsic

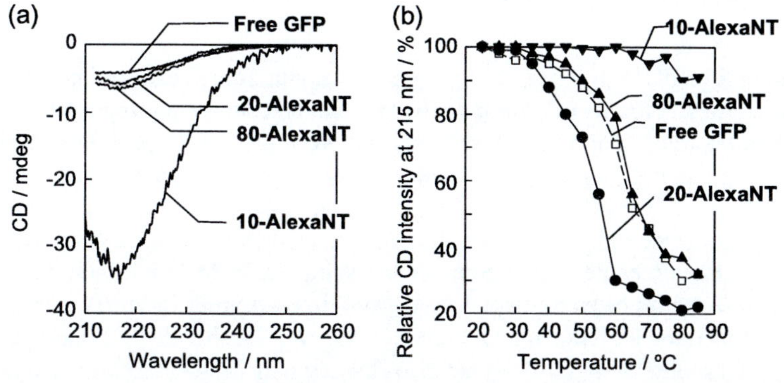

Fig. 3.20 **a** CD spectra of the encapsulated GFP in various nanochannels of the AlexaNT with different inner diameters and free GFP after incubation at 80 °C for 24 h. **b** Thermal stability of the encapsulated GFP encapsulated in various nanochannels of the AlexaNT with different inner diameters and free GFP at pH 6.8 (Reproduced with permission from Ref. [23], © 2010 WILEY-VCH Verlag GmbH & Co. KGaA, Weinheim)

conformations, displaying no thermal denaturation (Fig. 3.20). Moreover, 90% of the confined GFP took native state even in the presence of high urea concentration (6 mol/l) that is enough to induce the protein denaturation. The nanochannel with inner diameters of 10 nm should confine the protein in a tightly restricted fashion, thus resulting in the protection against thermal and chemical denaturation.

The same nanotubes from **6a** are also able to confine oxygen storage hemoprotein, myoglobin (Mb) of 3–4 nm in dimensions in their lumens. Even at high concentrations of urea as a denaturant, the entrapped Mb was observed to exhibit oxygen-binding activity [11]. Half lifetime of the active Mb in the auto-oxidation reaction from oxy-Mb to met-Mb depends on the surrounding media. Entrapped oxy-Mb displayed much longer half lifetime ($\tau = 14$ h) as compared to that of free oxy-Mb ($\tau = 7.0$ h). A decline in the nucleophilic attack by water molecules on the heme is strongly involved in the preservation of the enzymatic activity of the entrapped oxy-Mb. This is because the solvent polarity [$E_T(30) = 50$ kcal/mol] of the confined water in the nanotube hollow cylinders is relatively lower than that (63 kcal/mol) of bulk water [48].

Furusho and co-workers took advantage of the ability for nanotube channels to stabilize proteins. They improved TEM CT analysis and explored a novel sample immobilization technique for 3D TEM analyses. The organic nanotube-encapsulating technique enabled 3D imaging of a cage-shaped protein with high resolutions [49, 50].

3.2.14 Protein Refolding

Rationally folded 3D structures of proteins play a vital role in expressing their functional activities. A family of proteins called molecular chaperone supports the correct folding and unfolding of protein structures and prevention of the protein aggregation in living organism [51]. Sequential multiple processes comprising entrap of a denatured protein into the affinity site, compulsive change of the protein conformation, and subsequent ejection of the protein is associated with the chaperon systems. Multifarious organic molecules and their assemblies, inorganic, and their hybrid nanomaterials have been reported to display chaperone-like activities in vitro environments [52]. In connection with the tubular architectures, polysaccharide hydrogels [53] and mesoporous zeolite materials [54] are also characteristic of well-defined nanospaces and nanopores that can selectively encapsulate denatured proteins.

How organic nanotubes function as artificial chaperones that can refold the conformation of chemically denatured proteins? Dimensionally controlled organic nanotubes modified with rational inner and outer functional groups can perform entrapment, storage, stabilization, transport, and ejection activities cooperatively [1, 52]. We found that under controlled pH conditions the self-assembled nanotubes from **6a** aggregate hierarchically to form nanotube hydrogels [11, 32]. The nanotube membrane wall consists of 3 nm thick monolayer. Characteristic polyglycine II-type hydrogen-bond network forces the constituent molecules to take a parallel packing. A typical experimental procedure for the refolding of denatured GFP is shown in Fig. 3.21. The bolaamphiphile **6a** and GFP were blended in water at pH 5 (Fig. 3.21a). The addition of sodium hydroxide neutralizes the aqueous solutions, resulting in the hydrogelation (Fig. 3.21b). Washing the produced hydrogel with water can eliminate unencapsulated GFP in the nanochannels (Fig. 3.21c). The initial refolding stage in the tube channels proceeds in this fashion (Fig. 3.21d). The final step for refolding takes place during the subsequent recovery process in buffer solutions at pH 7.8 (Fig. 3.21e). To investigate the effect of the i.d. sizes and the internal hydrophobicity of the nanotubes, we prepared three types of nanotube hydrogels and compared the total refolding ratios with that obtained by a dilution method (Table 3.3) [32]. The nanotube prepared by co-assembly of **6a** and **34b** (Chart 3.7) increased the total refolding ratio to 84% toward CAB (3–4 nm) because of partial incorporation of the Cbz hydrophobic group. The nanotubes (20 nm i.d.) based on **3(18)** showed no chaperone ability for either GFP or CAB (0%), but relatively larger refolding ratio (62%) for citrate synthase with dimensions of 7.5 × 6.0 × 9.0 nm. Suitable size fitting between the protein and channel sensitively promotes the chaperone ability by the nanotubes.

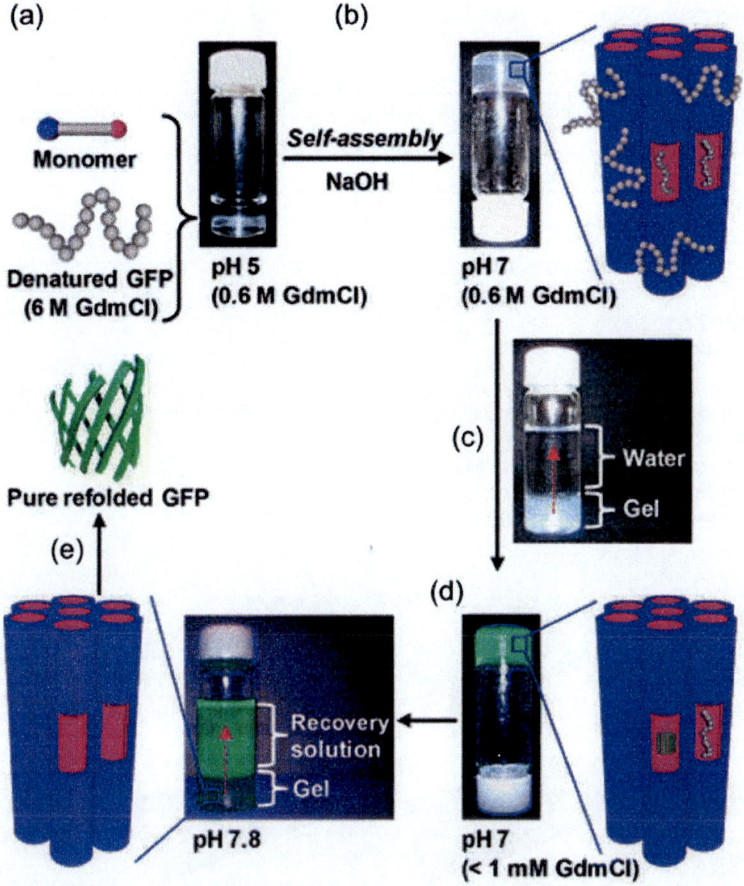

Fig. 3.21 Refolding procedure of denatured GFP in the nanotube hydrogel (reproduced with permission from Ref. [32], © 2012 American Chemical Society)

3.2.15 Loading of Anticancer Drugs

Novel nanotubes, which act as cylindrical drug nanocarriers, were prepared by co-assembly of the tube-forming glucose–triglycine bolaamphiphile **7b** and its derivative with hydrophobic terminal group **35** (Chart 3.7) [13]. The Cbz group was employed to control the release behavior of doxorubicin (DOX) **36** (Chart 3.7) and hybridized with the anionic inner surface. The self-assembled nanotubes can entrap DOX **36** efficiently, forming ion complexes between the anionic COO^- and cationic DOX **36** moieties on the nanotube inner surfaces. By adjusting the molar ratio of **7b/35** to be 6:5, we demonstrated the outstanding decrease in the release amount of DOX from 60% to less than 10% after 48 h (Fig. 3.22).

Table 3.3 Refolding ratios of denatured proteins in the nanotube hydrogel and dilution method systems (Reproduced with permission from Ref. [32])

	GFP		CAB	
Media for refolding	Refolding I (%)	Refolding I + II (%)	Refolding I (%)	Refolding I + II (%)
Alexa-hydrogel[a]	35	71 (max.)		
H-hydrogel[b]			38	53 (max.)
LZ-hydrogel[c]			44	84 (max.)
LBoc-hydrogel[d]			46	70 (max.)
20-H-hydrogel[e]	0	0	0	0
bulk (dilution method)	14		16	

[a]Alexa-hydrogel: composed of **6a** and **34a** (Chart 3.7)
[b]H-hydrogel: composed of **6a**
[c]LZ-hydrogel: composed of **6a** and **34b**
[d]LBoc-hydrogel: **6a** and **34c** (Chart 3.7)
[e]20-H-hydrogel: composed of **3(18)**

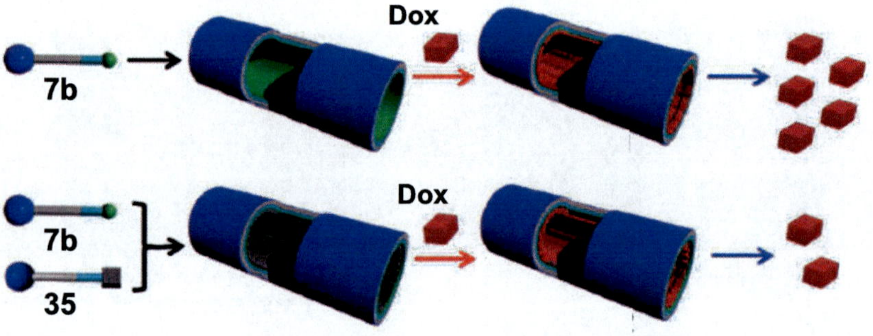

Fig. 3.22 Construction of unsymmetrical nanotubes used as DOX nanocarriers, by the self- and co-assembly of glycolipids **7b** and **7b/35**, respectively, (Reproduced with permission from Ref. [13], © 2012 WILEY-VCH Verlag GmbH & Co. KGaA, Weinheim)

As mentioned in **3.2.2**, we prepared cisplatin-complexed SMNTs on the interior surface by the coordination of platinum to the two carboxyl groups of **8** [14]. The platinum-based anticancer drug was thus incorporated in the nanotube interior with high efficiency (220 mg per 1 g nanotube). The drug-carried nanotubes displayed a prominently lengthy release of CDDP under a physiological condition.

To produce sophisticated nanotubes that can bind an anticancer Pt drug more strongly in the interior surface, we explored a two-step self-assembly procedure by employing ternary components, i.e., two types of combination of **7a/20/21(15)** and **6b/20/21(17)** [12]. The former combination gave single monolayered nanotape structures, while the latter produced nanotubes with single monolayers as well as the pyridine carboxylate ligand located at the inner surface. Both the nanotapes and nanotubes can form stable complexes with the metallodrug in Milli-Q water [12]. In terms of slow and sustainable release in PBS media, the nanotubes having cylindrical nanochannel inside outperform the nanotapes.

3.3 Glucose–3-Hydroxy-Propionyl Bolaamphiphile

3.3.1 Enzymatic Channel Reactor

Immobilization and concurrent catalytic performance of enzymes in organic nanotube channels were attempted to construct enzymatic channel bioreactor. We designed a new unsymmetrical bolaamphiphile **37** (Chart 3.8) that can undergo photoresponsive structural isomerization [55]. The molecule has both glucose moiety as a relatively larger headgroup and 3-hydroxy-propionyl group as a smaller headgroup that links to the both ends of a hydrophobic spacer chain having azobenzene unit as a photoresponsive segment. The self-assembly of the bolaamphiphile **37** in water to yield nanotube structures with inner diameter of 10–12 nm and wall thickness of 3–4 nm under the pH conditions in the range of 3–10. The molecules pack in a parallel fashion to form unsymmetrical monolayer membrane of the nanotube wall.

Capillary action was found to enable the entrapment of HRP into the nanotube channel from **37** [55]. Significantly, pyridyl group-modified magnetic nanoparticles of iron oxide (II, III) can cover only both ends of the nanotube, leading to the prevention of the HRP release from the nanochannel. UV-light irradiation causes the trans-to-cis isomerization of the azobenzene unit, resulting in the morphological transformation from nanotubes to helically coiled ribbons (Fig. 3.23). As a result of

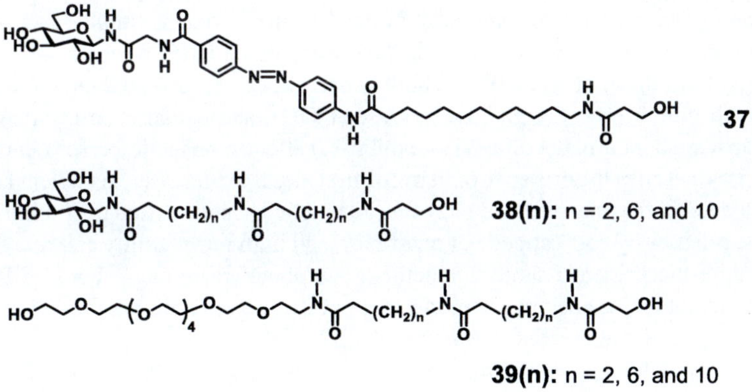

Chart 3.8 Chemical structures of **37–39**

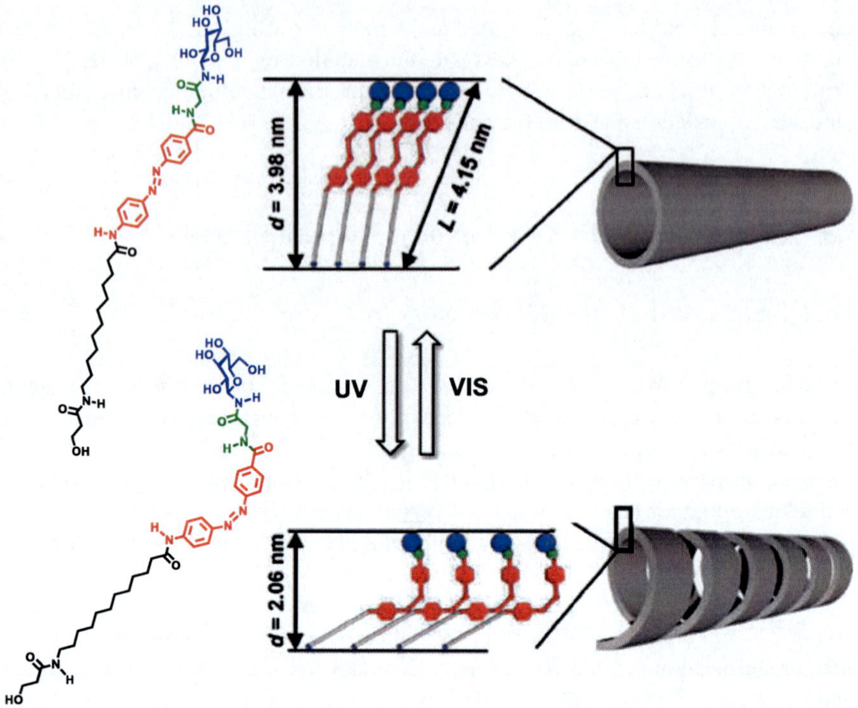

Fig. 3.23 Photoisomerization of the unsymmetrical bolaamphiphile **37** and reversible photo-induced morphological transformation of a nanotube into a nanocoil. Both nanotubes comprise a monolayer membrane with parallel molecular packing

this morphological change as a trigger, the nanotube bioreactor encapsulating HRP initiates the catalytic reaction to oxidize *o*-phenylenediamine [55]. The newly created narrow gaps in the coil structures serve as size-exclusion passages, through which substrates penetrate from an external solution. The measured maximum reaction rate (*V*max) and catalytic constant (k_{cat}) of the encapsulated HRP in the nanocoil were comparable to those of free HRP in bulk water. Other encapsulated enzymes, such as Cytochrome c (Cyt C), glucose oxidase (GOD), and catalase, in the magnetic nanoparticle-capped nanocoilds also exhibit excellent enzymatic performance and operate well even in the presence of harmful enzymes like proteases. Excellent kinetic performance of the encapsulated enzymes, magnetic properties, and the inhibition of enzyme releases by end-capped nanoparticles, and high recyclability and reusability can feature the present nanotube reactors. The obtained findings should differentiate the nanotube bioreactor from other immobilization system with metal-organic frameworks [56] and mesoporous materials [57].

3.3.2 Liquid Crystal as a Template for Construction of Surfactant-Free Gold Nanorods

As components for binary co-assembly, the glucose-based bolaamphiphile **38(n)** ($n = 2$, 6, and 10) (Chart 3.8) terminated with 3-hydroxy-propionamide group and PEG derivative **39(n)** ($n = 2$, 6, and 10) (Chart 3.8) with the same terminal fragment containing 3-hydroxy-propionamide group was synthesized [58]. Homo-assembly of **39(2)**, **39(6)**, and **39(10)** in water gives micelle, nanofiber, and sheet-like structures, respectively. Binary co-assembly of **38(n)** and **39(n)** with the same **n** values in water yielded only monolayer-based nanotube architectures. The glucose bolaamphiphiles and PEG-derivatized molecules take an unsymmetrical packing stabilized by both hydrophobic interactions between the two sets of oligomethylene chains and intermolecular hydrogen bonds between the amide groups. The larger glucose headgroups occupy the outer surface that is partially covered with PEG chains, whereas the smaller hydroxy headgroups locate at the interior surface. The internal diameter of the obtained nanotubes increases from 18 nm to 23 and 31 nm with increase in the **n** number from 2 to 6 and 10, respectively. Under diluted conditions (< 0.2 wt%), the aqueous dispersions appear clear to suggest random orientation of the nanotubes. Under relatively concentrated conditions (0.2–5 wt%), however, the dispersions display bluish color ascribable to Rayleigh scattering. Birefringence observation with a cross-polarizer shows the formation of lyotropic liquid crystals, indicating side-by-side arrangement of the nanotubes [58].

Kameta and Shiroishi demonstrated that the nanogrooves generated by the side-by-side alignment of the PEG-coated nanotubes from **38(n)** and **39(n)** can serve as a template to produce surfactant-free gold nanorods (Au NRs) (Fig. 3.24) [58]. In the presence of thiol-crown ether compounds, photothermal dehydration of the external PEG chains upon near-infrared irradiation was shown to liberate the generated Au NRs into external bulk solution. The diameters of the thiol-functionalized Au NRs can be regulated by the nanotube outer diameters. Furthermore, the Au NRs modified with thiol-crown ether compounds can selectively recognize alkaline metal cations and chiral amines. The use of various organic nanotubes as a template has ever been known to afford abundant 1D nanostructures, e.g., 1D arrays of metal NPs [59–61], diverse helical structures [62], and concentric tubular hybrids [63]. The present results, thus, give the first example that the grooves, created by the side-by-side alignment of nanotube, serve to endow the generated Au NRs with chiral activity.

3.4 Amino Acid-Based Bolaform Amphiphile

3.4.1 Controlled Polymerization of Imine

As already discussed in **3.1.1**, the innermost diameter ($D_{\text{calc-in}}$) can be defined as the equation: $D_{\text{calc-in}} = 2a_sL/(a_l - a_s)$, where a_s, a_l, and L are the cross-sectional areas of

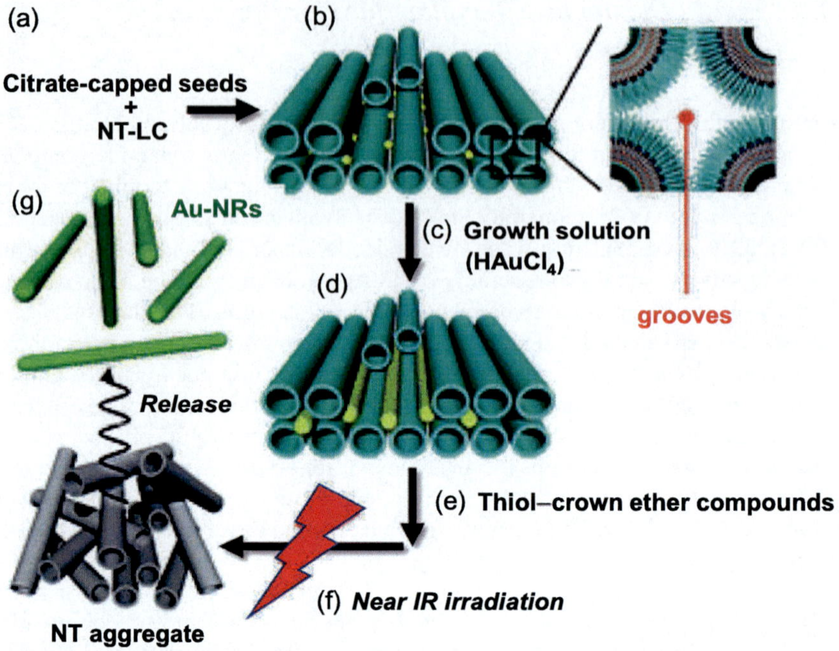

Fig. 3.24 Production of Au NRs using the lyotropic liquid crystal of nanotubes as a template (Reproduced with permission from Ref. [58], © The Royal Society of Chemistry 2018)

the small headgroup, those of the large headgroup, and the molecular length, respectively [4]. Namely, if the a_s and a_l values are constant, the resultant inner diameters are proportional to the spacer length. However, we need to postulate here that the nanotube structures comprise unsymmetrical MLMs of wedge-shaped bolaamphiphiles. Then, if the a_s and L values keep constant, the inner diameters should also depend on the a_l value, i.e., the cross-sectional areas of the relatively larger headgroup. Based on this guiding principle of molecular design, we aimed at the fabrication of self-assembled nanotubes with wide-ranging internal diameters.

For example, we newly designed amino acid-based bolaform amphiphiles **40(Gly, AA)** (AA = Gly, Ala, Val, Leu, Ile, Phe, and Trp) (Chart 3.9) [64]. To be exact, the amphiphiles **40(Gly, AA)** are not bolaamphiphiles. A carboxyl-terminated amino acid (Gly, Ala, Val, Leu, Ile, Phe, or Trp) as a large headgroup and *N*-propyonylglycine as a small headgroup are connected to both ends of 12-aminododecanoic acid. The glycine-based amphiphile **40(Gly, Gly)** self-assemble in water to yield only nanofibers with no hollow cylinders. Notably, the self-assembly of the other bolaform amphiphiles **40(Gly, AA)** (AA = Ala, Val, Leu, Ile, Phe, and Trp) gave nanotube structures with inner diameter of 29 nm, 16 nm, 12 nm, 9 nm, 4 nm, and 1 nm, respectively, [64]. The increase in the bulkiness of the large headgroup

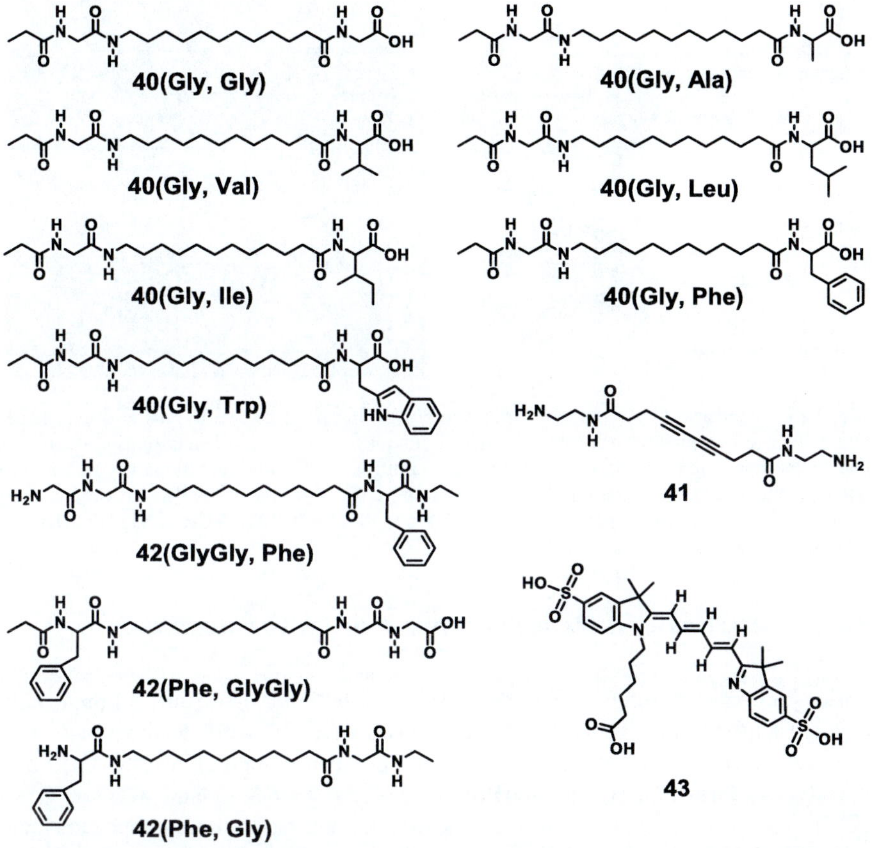

Chart 3.9 Chemical structures of **40–43**

was shown to decrease the nanotube inner diameters. This finding is well compatible with the equation mentioned above.

The obtained six nanotubes with different inner diameters were observed to encapsulate benzaldehyde and diamine derivative **41** (Chart 3.9) with a diacetylene moiety as imine monomers into the hydrophobic channels and thereby accelerate the imine formation remarkably [64]. This results mean that the obtained nanotubes serve as supramolecular channel reactors to yield regulated polymerization products with different morphologies. Depending on the nanotube inner diameters of 4, 9, 12, 16, and 29 nm, polymerization upon UV irradiation resulted in the formation of twisted nanofibers, nanocoils, nanotapes of 10 nm wide, nanotapes of 12 nm wide, and nanoparticles, respectively, (Fig. 3.25). Although a variety of SMNTs have functioned to template the fabrication of unique inorganic and metal nanostructures [65], the polymerization behavior of encapsulated monomers in a confined manner by the nanotube channels has been reported very little [21, 66].

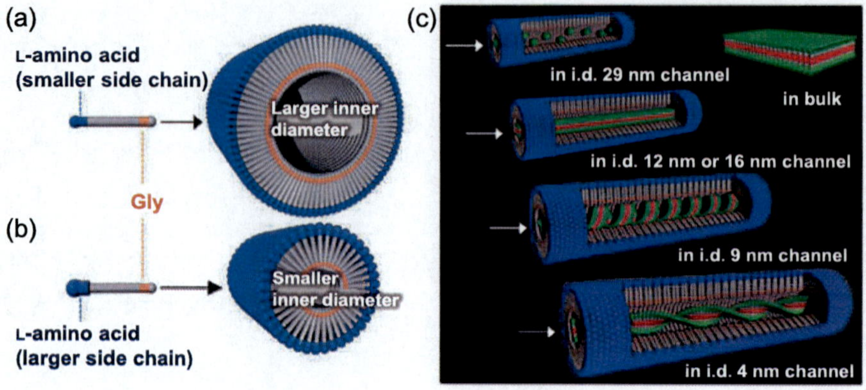

Fig. 3.25 Schematic illustrations of the nanotubes having **a** larger and **b** smaller inner diameters, which self-assembled from **40(Gly, AA)** containing L-amino acid headgroups with smaller and larger side chains, respectively. **c** Illustrated images for the conformations of the obtained nanostructures by polymerization of imine monomer in the nanotube channels with various inner diameters (Reproduced with permission from Ref. [64], © 2019 WILEY-VCH Verlag GmbH & Co. KGaA, Weinheim)

3.4.2 Interlink of a Heterogeneous Pair of Nanochannels

Direct joining of nanotubes with identical outer and inner diameters but oppositely charged surfaces still remains extremely challenging [67–69]. Kameta and Ding synthesized four different amino acid-based bolaform amphiphiles[**40(Gly, Phe)**, **42(GlyGly, Phe)** (Chart 3.9), **42(Phe, GlyGly)** (Chart 3.9), and **42(Phe, Gly)**] (Chart 3.9) [70]. The amphiphiles have a spacer chain of 12-aminododecanoic acid that is flanked by a L-phenylalanine derivative as a large headgroup and a glycine derivative as a small one. All the amphiphiles were shown to self-assemble in water to form only nanotubular structures. Notably, the dimensions of the nanotubes from the four different amphiphiles are almost same; 22 nm outer diameter, 16 nm inner diameter, and 3 nm wall thickness. Moreover, all the molecules pack in parallel fashion to form unsymmetrical monolayer membrane wall. These results clearly show that the headgroup of the glycine derivative occupy the inner surface and that of the L-phenylalanine derivative the outer surface. Thus, four different types of monolayer-based nanotubes, i.e., the nanotubes featured by cationic external surface [designated as (+)-out-NT], anionic external one [(−)-out-NT], cationic internal one [(+)-in-NT], and anionic internal one [(−)-in-NT], have been successfully prepared independently through self-assembly [70].

Mixing of the (+)-out-NT and (−)-out-NT in transparent aqueous dispersion changed into turbid phase due to aggregation of the two nanotubes [70]. Meanwhile, mixing of the (+)-out-NT and (−)-in-NT, or (−)-out-NT and (+)-in-NT remained clear dispersion, resulting in nonoccurrence of remarkable interactions. However, mixing of the (+)-in-NT and (−)-in-NT resulted in the formation of hydrogel comprising several micrometer long heterogeneous nanotubes generated by end-to-end joining of the two nanotubes (Fig. 3.26). Electrostatic interaction between the open nanotube ends allowed for the highly efficient direct joining, in which the (+)-in-NT interlinked to (−)-in-NT alternatively. To corroborate the realization of alternating acidic and basic nanochannel, we performed the experiment on the transport of a fluorophore in the nanochannel. As a result, the entrapped, pH-responsive fluorophore CypHer5 **43** (Chart 3.9) fluoresced only during transporting within the acidic nanochannels of the (+)-in-NT. No remarkable fluorescence was observed within the basic nanochannels of the (−)-in-NT. We were, thus, able to regulate acid–base reactions of the fluorophore during transport through alternating heterogeneous nanochannels [70].

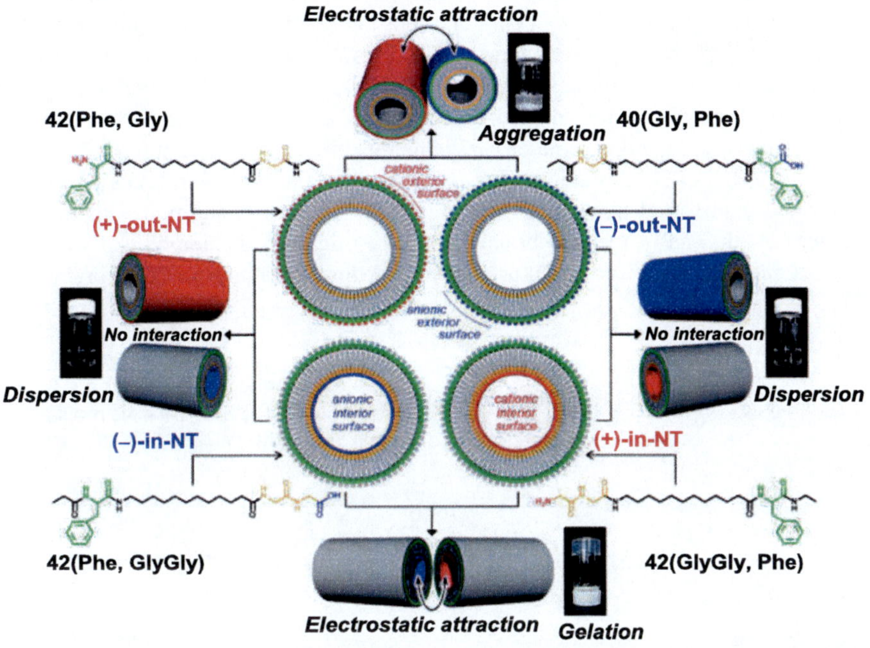

Fig. 3.26 Schematic illustration of single monolayer-based nanotubes self-assembled from the different amino acid-based bolaform amphiphiles and photographs that display the mixing results of clear aqueous dispersions of different pairs of nanotube types (Reproduced with permission from Ref. [70], © The Chemical Society of Japan)

3.5 Amino Acid Bolaamphiphile

3.5.1 *Diglycine and Triglycine Bolaamphiphiles*

We designed and synthesized a new family of diglycine- (**44a**–**44h**) (Chart 3.10) and triglycine-based bolaamphiphiles (**45a** and **45b**) (Chart 3.10) with individual dicarboxylic headgroup at each end. Among these bolaamphiphiles, the four different molecular building blocks, **44a**, **44c**, **44e**, and **45b**, self-assemble in water to give vesicle-encapsulated microtubes [71, 72]. Phase-contrast optical and laser scanning microscopic measurement confirmed the generation of well-defined microtube structures with a uniform diameter (about 1–2 μm) and closed ends (Fig. 3.27a–d). All microtubular assemblies were shown to encapsulate a number of spherical vesicles in a confined aqueous hollow cylinder. Microtubes self-assemble only when the spacer length of the oligomethylene chain takes an even number of carbons (C6, C8, and C10) and yet the headgroup comprises diglycine and triglycine residues [72]. Directional formation of acid–anion dimers as well as loose hydrogen-bond networks between the diglycine or triglycine residues stabilize the microtube structures (Fig. 3.27e). Elaborated AFM measurement for the microtubes from **44e** confirmed a distorted hexagonal packing of the peptide headgroups on the tube surfaces. Additionally, the membrane wall of the microtube is hierarchically formed in the order of bent molecule, distorted layer, columnar domain, and membrane wall [74].

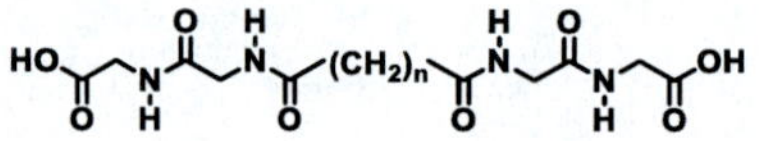

44a: n = 6 **44b:** n = 7 **44c:** n = 8
44d: n = 9 **44e:** n = 10 **44f:** n = 11
44g: n = 12 **44h:** n = 14

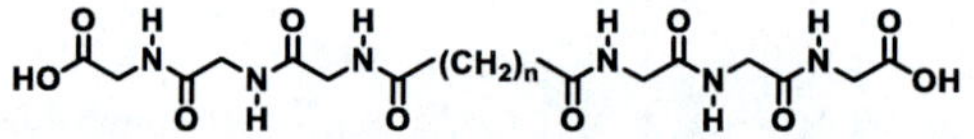

45a: n = 6
45b: n = 10

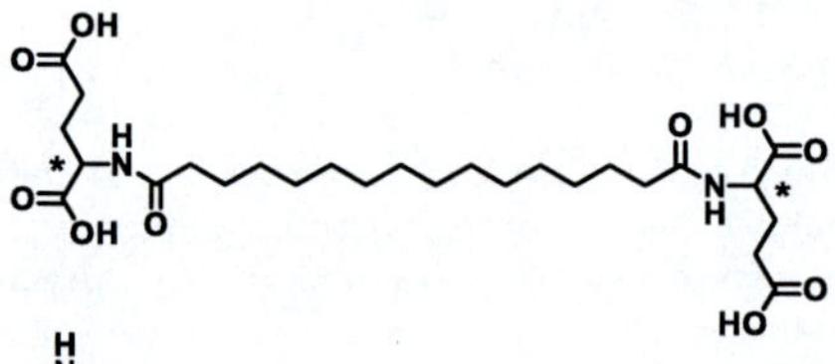

46L and 46D

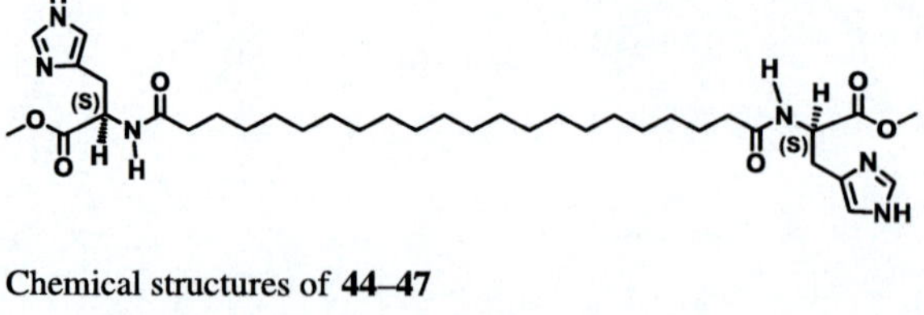

47

Chart 3.10 Chemical structures of **44**–**47**

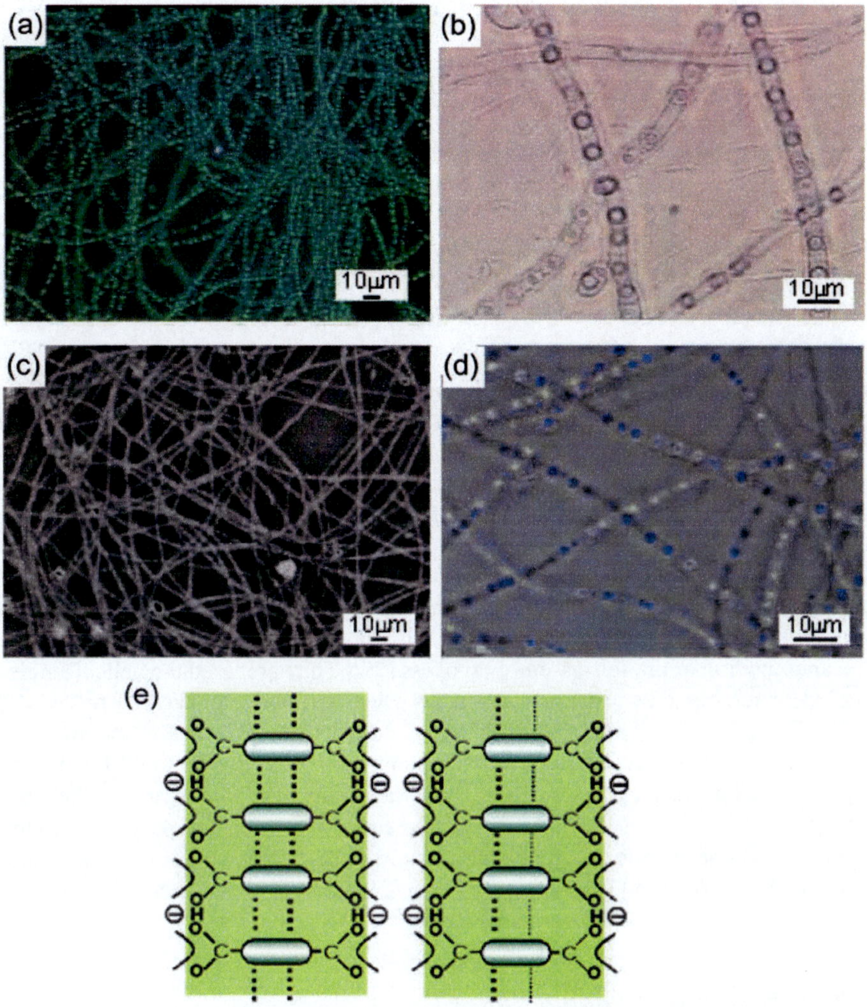

Fig. 3.27 **a**, **b** Phase-contrast, **c**, **d** dark field optical micrographs for the vesicle-encapsulated microtubes self-assembled from **44e**, and **e** intralayer acid–anion interaction that drives the microtube formation (Reproduced with permission from Ref. [73], © 2005 American Chemical Society)

3.5.2 Glutamic Acid Bolaamphiphile

Liu and co-workers have carried out comprehensive research on the hierarchical self-assembly of a family of glutamic acid-based bolaamphiphiles with different hydrophobic spacers. As a result of the self-assembly under various conditions, the bolaamphiphiles were shown to yield nanoribbons and nanofibers except for helical

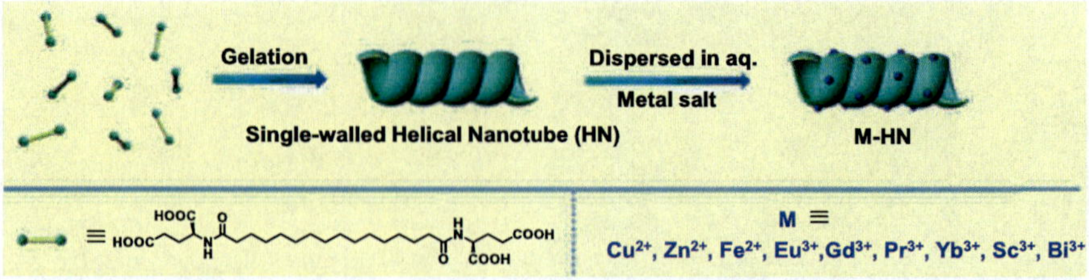

Fig. 3.28 Preparation of the catalyst based on metal-coordinated helical nanotube (M-HN) through self-assembly of the glutamic acid-based bolaamphiphiles **46L** and **46D** (Reproduced with permission from Ref. [79], © 2016 American Chemical Society)

nanotubes [75, 76]. For example, the glutamic acid-based bolaamphiphiles, *N*,*N*-hexadecanedioyl-di-L-glutamic acid (**46L**) (Chart 3.10), and its enantiomer (**46D**) (Chart 3.10) undergo the self-assembly in water to form hydrogels comprising right-handed helical nanotubes and left-handed ones, respectively, [77]. Interestingly, the addition of Cu^{2+} to the hydrogel system alters the present monolayer-based nanotube to multi-layer-based nanotubes that can exhibit excellent chiral catalytic activity for an asymmetric Diels–Alder cycloaddition [78]. To improve the defect that one can utilize only the coordination existing on the outermost layers, the research group developed metal-coordinating helical nanotubes with a single wall through the optimization of the self-assembly process [79]. They prepared the helical single-walled nanotubes that can bind to various metal cations, e.g., Fe^{2+}, Zn^{2+}, Cu^{2+}, Gd^{3+}, Eu^{3+}, Pr^{3+}, Yb^{3+}, Sc^{3+}, and Bi^{3+} (Fig. 3.28). Among them, Bi^{3+}-coordinated helical nanotube was found to display high enantioselectivity (up to 97% ee) in an aqueous medium as a catalytic substrate for the asymmetric Mukaiyama aldol reaction (Fig. 3.28). Moreover, Cu^{2+}-coordinated single-walled nanotube also exhibited highly excellent enantioselectivity (at most 91% ee) for asymmetric Diels–Alder reaction within 60 min [79].

3.5.3 Histidine Bolaamphiphile

Liu and co-workers newly designed and synthesized the bolaamphiphile, *N*,*N*-eicosanedioyl-di-L-histidine methyl esters **47** (Chart 3.10), in which two L-histidine methyl ester headgroups are connected to a long alkylene chain [80]. They found the formation of hydrogels comprising single-walled nanotubes and nanofibers of single molecular thick (~3 nm) upon addition of concentrated H^+ and Cu^{2+} ions, respectively (Fig. 3.29). Protonation or metal cation coordination allows the imidazole moiety to be hydrophilic, resulting in the hydrogelation through hierarchical self-assembly. Among meal cations that can coordinate, e.g., Fe^{3+}, Fe^{2+}, Zn^{2+}, Ni^{2+}, Co^{2+}, Ag^+, and Cu^{2+}, only Cu^{2+} binding can commit the hydrogel formation. Surprisingly, hydrogelation with the histidine ester terminated bolaamphiphile **47** takes place even in the presence of concentrated hydrochloric or sulfuric acid (approximately 8 N) [80].

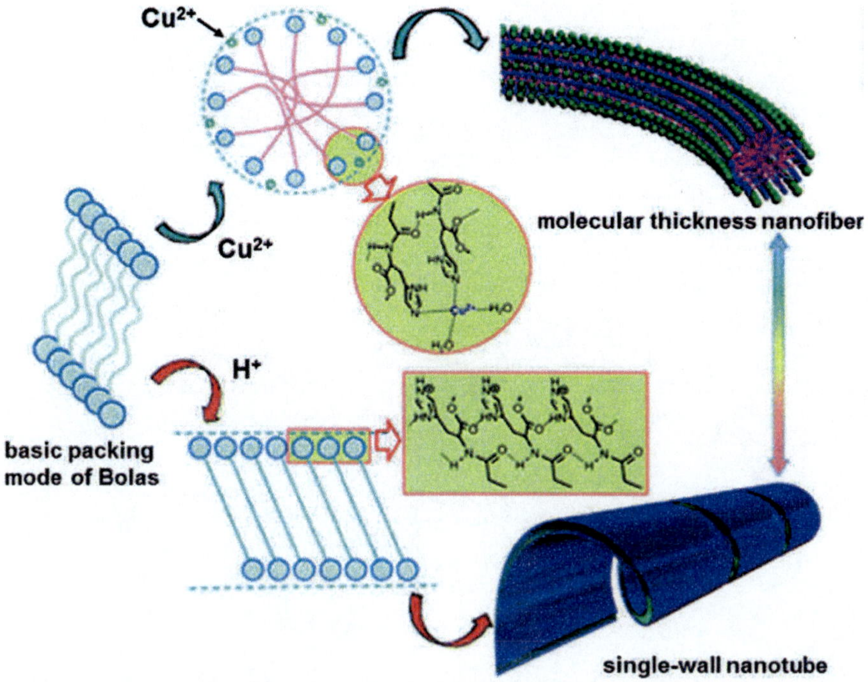

Fig. 3.29 Schematic representation of two major self-assembled morphologies from L-histidine ester-based bolaamphiphile **47** upon complexation with Cu^{2+} or H^+ (Reproduced with permission from Ref. [80], © The Royal Society of Chemistry 2013)

The obtained 3D network of single-walled nanotubes and tough molecular packing contribute to the stabilization of the hydrogel that exhibited excellent tolerance to a concentrated and strong acid environment.

3.6 Cinnamic Acid Bolaamphiphile

Incorporation of interfacial approach into the self-assembly sometimes enables the realization of supramolecular chirality from achiral molecules. For example, the achiral, fused-ring pyrazine derivative **48** (Chart 3.11), hierarchically assembles upon compression at the air/water interface, producing chiral nanotube structures with rolled up thin membranes [81]. Wang, Liu, and co-workers attempted the fabrication of tubular objects by combining Langmuir–Blodgett technique with a family of bolaamphiphiles **49a**–**49c** (Chart 3.11) having cinnamoyl derivatives as headgroups [82]. The phenolic hydroxyl-terminated bolaamphiphile **49c** resulted in the formation of nanotubes with supramolecular chirality, whereas other analogues **49a** and **49b** assembled to give only thin films. Notably, photodimerization of the cinnamoyl

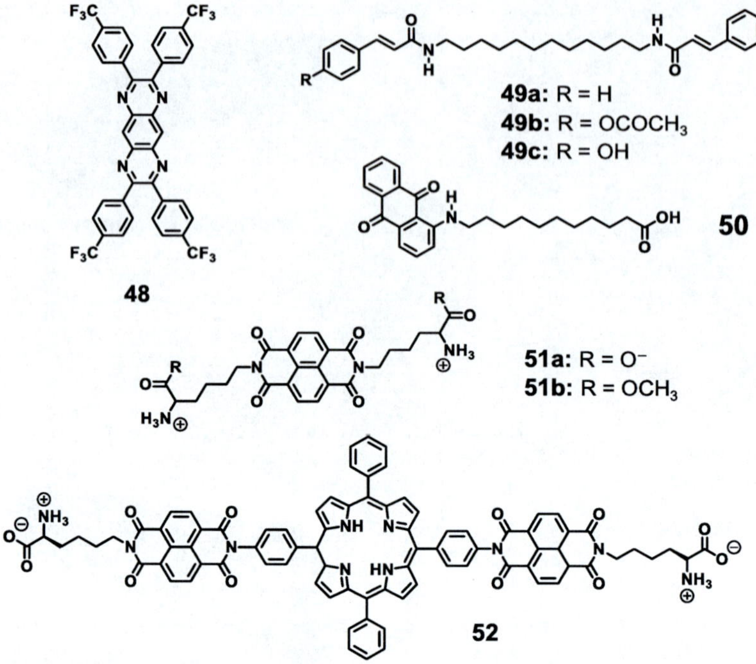

Chart 3.11 Chemical structures of **48–52**

moiety underwent in the nanotube assemblies upon UV irradiation, inducing the stabilization of the nanotube through photochemical sewing. This process affects no remarkable change in the tubular structures as well as supramolecular chirality. More interestingly, the obtained chiral nanotube was shown to provide an achiral porphyrin derivative with the chirality, which assembled on the nanotube surface (Fig. 3.30).

3.7 Anthraquinone–Carboxylic Acid Bolaamphiphile

Unsal and Aydogan synthesized anthraquinone headgroup-containing bolaamphiphile **50** (Chart 3.11) that has carboxyl terminal spanned by the C10 long alkyl chain [83]. The anthraquinone and carboxylic acid segments were designed in expectation of their redox-active and pH sensitive functionalities, respectively. The ethanolamine salt of **50** yields nanotube structures with helical markings on the surfaces through the self-assembly in aqueous solutions. The nanotube dimensions are characterized by the outer diameter of 110–190 nm and wall thicknesses of 21.81 $\pm$ 4 nm, indicating the membrane walls with multiple bilayer structures. The results of zeta potential and cryo-TEM measurement are compatible with a symmetrical MLM within each layer [84]. Interestingly, reversible morphological transformation

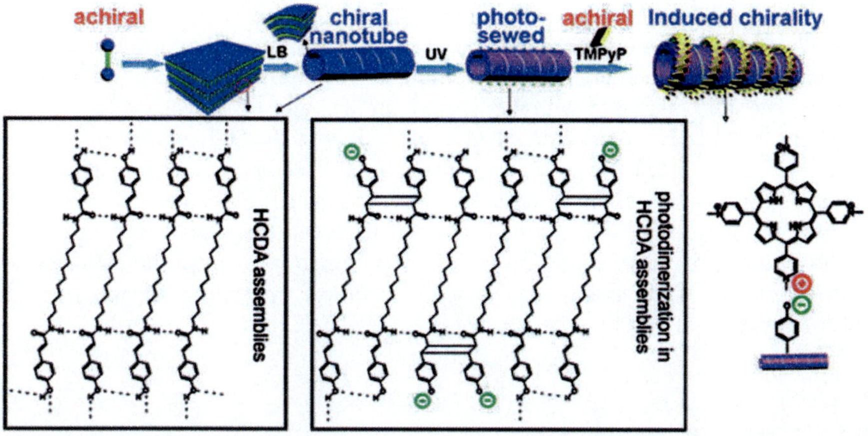

Fig. 3.30 Possible formation mechanisms of self-assembled chiral nanotubes from cinnamoyl-terminated bolaamphiphile **49c**, subsequent photo-sewing of the chiral nanotube and eventual chirality induction of the achiral porphyrin assembled on the nanotube (Reproduced with permission from Ref. [82], © The Owner Societies 2011)

between nanotube and ribbon structures realizes upon chemical reduction and re-oxidation of the nanotubes (Fig. 3.31) [84]. At the same time, electrically conductive and insulate state can be tuned reversibly by the redox reaction. The reversibility of the pH-triggered morphological change between nanotubes and nanosheets was also observable in the range of pH 3–12.

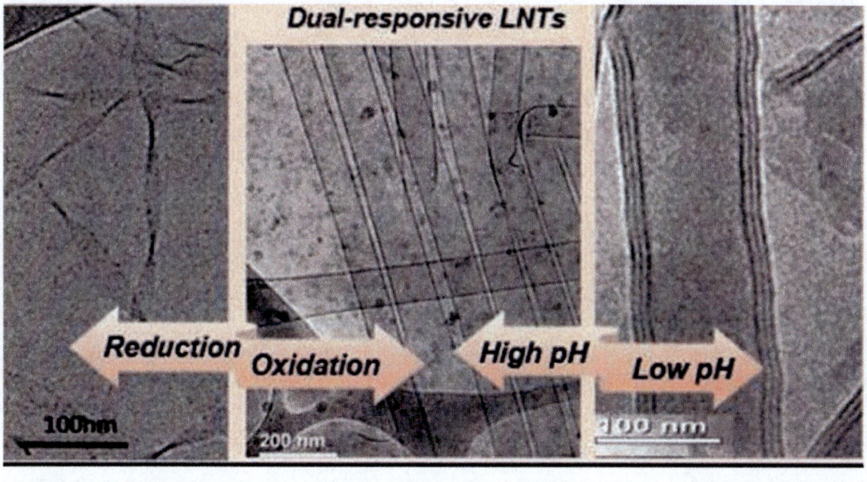

Fig. 3.31 Cryo-TEM images of dual-responsive lipid nanotubes whose self-assembled morphology can be reversibly regulated by both redox and pH effects (Reproduced with permission from Ref. [84], © 2016 American Chemical Society)

Aydogan and co-workers also investigated the release behavior of DOX-conjugated nanotubes of **50** in different formulations through two different types of normal dialysis membrane and colonic cell membrane. The loading efficiency is proportionally increased with the increase in pH value of aqueous solution, thus reaching 96% at pH 9 at an equivalent DOX weight of the **50**-based nanotube used [85]. DOX-conjugated nanotubes were also immobilized in four different gel formulation using chitosan, alginate, hydroxypropyl methylcellulose, and polycarbophil. Among the gels employed, the chitosan gel displayed the most excellent release ability. As compared with the release of DOX itself (80%) from the chitosan gel formulation, the DOX-conjugated nanotubes showed controlled release behavior (67%) at pH 7.4 after 24 h.

3.8 NDI Bolaamphiphile

Carbon nanotubes, probably the most elaborated π-electronic tubular architectures, exhibit unique tubular morphology and exceptional properties. Such nanomaterials have inspired us to realize 1D molecular assembly of π-conjugated building blocks into well-defined fibers and tubes [86]. Aqueous self-assembly of discrete organic nanotubes from n-type molecular building blocks still remains a laborious challenge. Parquette and co-workers developed the self-assembly of the bolaamphiphiles **51a** and **51b** (Chart 3.11) into n-type nanotubes in water [87]. The 1,4,5,8-naphthalenetetracarboxylic acid diimide (NDI)-based bolaamphiphiles **51a** and **51b** carry two lysine headgroups that are conjugated with NDI as a hydrophobic spacer. Intriguingly, self-assembled monolayer-based nanoring structures stack in 1D manner to result in the tubular objects that further hierarchically form hydrogels (Fig. 3.32).

In nature, the delocalization of exited state and energy transfer takes place with high efficiency over multiple molecules. Based on experimentally observed restraints on interatomic distances by SSNMR spectroscopy, XRD, and TEM, Herbert, Parquette, Jaroniec, and co-workers proposed a structural model for the semiconductor bolaamphiphile **51a**-based nanotubes [88]. The resultant model is consistent with the 2D-crystal-like structures, in which the NDI segments self-assemble to stack a monolayer ring comprising approximately 50 NDI–lysine molecules, thus forming nanotube structures. Remarkably, important π–π stacking interactions occur within the ring boundaries and along the long nanotube axes. By calculating first-order exciton couplings based on an incoherent hopping model and time-dependent DFT, they revealed how excitation energy migrates through the self-assembled nanotubes.

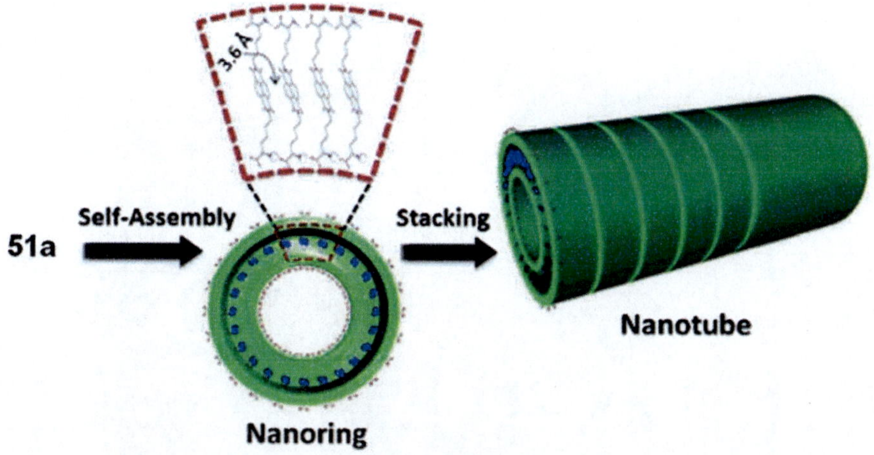

Fig. 3.32 Self-assembly of **51a** into nanorings, which stack to give nanotubes (Reproduced with permission from Ref. [87], © 2010 WILEY-VCH Verlag GmbH & Co. KGaA, Weinheim)

3.9 NDI–Lysine/Tetraphenylporphyrin/NDI–Lysine Bolaamphiphile

Self-assembly of a molecular building block with D/A dyads into nanotubular structures still remains difficult problem for the optimization of charge separation and transport [89]. Modarelli, Parquette, and co-workers prepared the bolaamphiphile **52** (Chart *3.11*) with D/A dyads, in which two NDI–lysine segments as acceptors are covalently linked to the both ends of a central tetraphenylporphyrin chromophore as a donor [90]. The bolaamphiphile **52** self-assembles in 10% MeOH/H_2O to mainly form monolayer-based nanotube structures with an outer diameter of 13.6 nm and wall thickness of 4.6 nm. AFM observation disclosed that the nanotubes are formed by the stacking of self-assembled nanorings on top of one another in columnar fashion (Fig. 3.33). Both the NDI chromophore and porphyrin moiety are associated with strong J-type π–π stacking interactions within self-assembled monolayer-based nanorings. Steady-state, time-resolved fluorescence, and femtosecond transient absorption spectroscopy corroborated that the time constants for electron transfer and charge recombination depend on the character and aggregated forms of the self-assemblies.

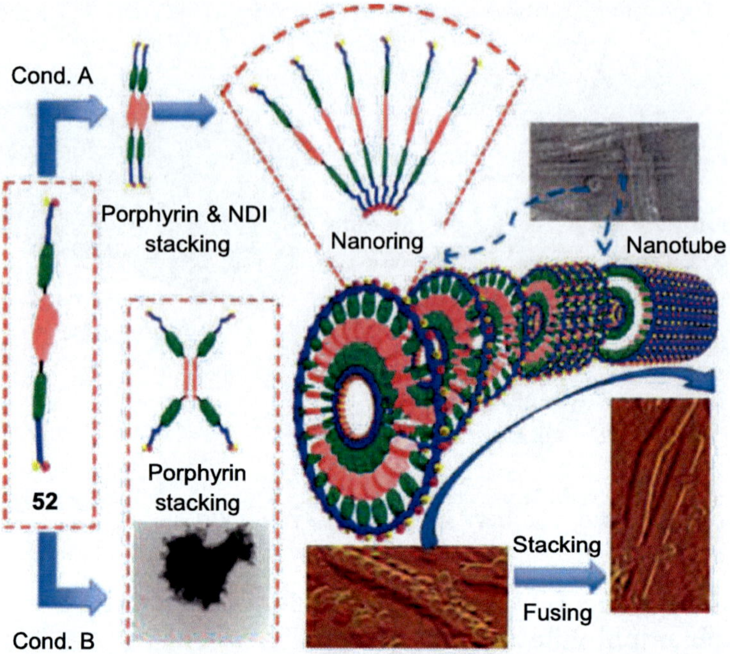

Fig. 3.33 Schematic illustration of the self-assembly of bolaamphiphile **52**. In 10% MeOH/H_2O (condition A), the bolaamphiphile **52** self-assembles to form monolayer nanorings, which further stack into nanotubes. In pure MeOH, or at pH 1 or 11 in 10% MeOH/H_2O (condition B), porphyrin-driven nonspecific aggregation takes place. (Top right) TEM images that show obtained nanotubes and nanorings and (bottom left) nonspecific aggregates (10% MeOH/H_2O, pH 1). (Bottom right) a portion of the AFM images that show the supramolecular stacking of nanorings into nanotubes (Reproduced with permission from Ref. [90], © 2011 American Chemical Society)

References

1. Shimizu T, Kameta N, Ding W, Masuda M (2016) Supramolecular self-assembly into biofunctional soft nanotubes: from bilayers to monolayers. Langmuir 32:12242–12264. https://doi.org/10.1021/acs.langmuir.6b01632
2. Masuda M, Shimizu T (2001) Multilayer structure of unsymmetrical monolayer lipid membrane with head-to-tail interface. Chem Commun 2001:2442–2443. https://doi.org/10.1039/B106581P
3. Masuda M, Yoza K, Shimizu T (2005) Polymorphism of monolayer lipid membrane structures made from unsymmetrical bolaamphiphiles. Carbohydr Res 340:2502–2509. https://doi.org/10.1016/j.carres.2005.08.005
4. Masuda M, Shimizu T (2004) Lipid nanotubes and microtubes: experimental evidence for unsymmetrical monolayer membrane formation from unsymmetrical bolaamphiphiles. Langmuir 20:5969–5977. https://doi.org/10.1021/la049085y
5. Kameta N, Masuda M, Minamikawa H, Goutev NV, Rim JA, Jung JH, Shimizu T (2005) Selective construction of supramolecular nanotube hosts with cationic inner surfaces. Adv Mater 17:2732–2736. https://doi.org/10.1002/adma.200501092

6. Kameta N, Masuda M, Minamikawa H, Shimizu T (2007) Self-assembly and thermal phase transition behavior of unsymmetrical bolaamphiphiles having glucose- and amino-hydrophilic headgroups. Langmuir 23:4634–4641. https://doi.org/10.1021/La063542o
7. Kameta N, Ishikawa K, Masuda M, Shimizu T (2013) Control of self-assembled morphology and molecular packing of asymmetric glycolipids by association/dissociation with poly(thiopheneboronic acid). Langmuir 29:13291–13298. https://doi.org/10.1021/La4028018
8. Shimizu T, Kogiso M, Masuda M (1997) Noncovalent formation of polyglycine II-type structure by hexagonal self-assembly of linear polymolecular chains. J Am Chem Soc 119:6209–6210. https://doi.org/10.1021/ja970844r
9. Kogiso M, Masuda M, Shimizu T (1998) Supramolecular polyglycine II-type structure of glycylglycine bolaamphiphile. Supramol Chem 9:183–189. https://doi.org/10.1080/10610279808034985
10. Kameta N, Mizuno G, Masuda M, Minamikawa H, Kogiso M, Shimizu T (2007) Molecular monolayer nanotubes having 7–9 nm inner diameters covered with different inner and outer surfaces. Chem Lett 36:896–897. https://doi.org/10.1246/Cl.2007.896
11. Kameta N, Yoshida K, Masuda M, Shimizu T (2009) Supramolecular nanotube hydrogels: remarkable resistance effect of confined proteins to denaturants. Chem Mater 21:5892–5898. https://doi.org/10.1021/Cm903108h
12. Kameta N, Lee SJ, Masuda M, Shimizu T (2013) Biologically responsive, sustainable release from metallo-drug coordinated 1d nanostructures. J Mater Chem B 1:276–283. https://doi.org/10.1039/C2tb00101b
13. Ding W, Kameta N, Minamikawa H, Wada M, Shimizu T, Masuda M (2012) Hybrid organic nanotubes with dual functionalities localized on cylindrical nanochannels control the release of doxorubicin. Adv Healthcare Mater 1:699–706. https://doi.org/10.1002/adhm.201200133
14. Ding W, Wada M, Minamikawa H, Kameta N, Masuda M, Shimizu T (2012) Cisplatin-encapsulated organic nanotubes by *endo*-complexation in the hollow cylinder. Chem Commun 48:8625–8627. https://doi.org/10.1039/C2cc33970f
15. Cabral H, Miyata K, Osada K, Kataoka K (2018) Block copolymer micelles in nanomedicine applications. Chem Rev 118:6844–6892. https://doi.org/10.1021/acs.chemrev.8b00199
16. Kim JK, Anderson J, Jun HW, Repka MA, Jo S (2009) Self-assembling peptide amphiphile-based nanofiber gel for bioresponsive cisplatin delivery. Mol Pharm 6:978–985. https://doi.org/10.1021/mp900009n
17. Kogiso M, Zhou Y, Shimizu T (2007) Instant preparation of self-assembled metal-complexed lipid nanotubes that act as templates to produce metal oxide nanotubes. Adv Mater 19:242–246. https://doi.org/10.1002/adma.200601117
18. Mukai M, Aoyagi M, Minamikawa H, Asakawa M, Shimizu T, Kogiso M (2011) A simple *N*-Acyl-L-amino acid constructed metal-complexed organic nanotube having an inner diameter below 10 nm. Chem Lett 40:218–220. https://doi.org/10.1246/Cl.2011.218
19. Shimizu T, Minamikawa H, Kogiso M, Aoyagi M, Kameta N, Ding W, Masuda M (2014) Self-organized nanotube materials and their application in bioengineering. Polym J 46:831–858. https://doi.org/10.1038/Pj.2014.72
20. Liu Y, Ai K, Lu L (2014) Polydopamine and its derivative materials: synthesis and promising applications in energy, environmental, and biomedical fields. Chem Rev 114:5057–5115. https://doi.org/10.1021/cr400407a
21. Ding W, Chechetka SA, Masuda M, Shimizu T, Aoyagi M, Minamikawa H, Miyako E (2016) Lipid nanotube tailored fabrication of uniquely shaped polydopamine nanofibers as photothermal converters. Chem Eur J 22:4345–4350. https://doi.org/10.1002/chem.201504958
22. Kameta N, Masuda M, Minamikawa H, Mishima Y, Yamashita I, Shimizu T (2007) Functionalizable organic nanochannels based on lipid nanotubes: encapsulation and nanofluidic behavior of biomacromolecules. Chem Mater 19:3553–3560. https://doi.org/10.1021/Cm070626p
23. Kameta N, Minamikawa H, Someya Y, Yui H, Masuda M, Shimizu T (2010) Confinement effect of organic nanotubes toward green fluorescent protein (GFP) depending on the inner diameter size. Chem Eur J 16:4217–4223. https://doi.org/10.1002/chem.200903413

24. Shimizu T (2008) Self-assembled organic nanotubes: toward attoliter chemistry. J Polym Sci, Part A: Polym Chem 46:2601–2611. https://doi.org/10.1002/Pola.22652
25. Sekine T, Nakatani K (2004) Nanometer pore size dependence of intraparticle diffusion in silica gel. Chem Lett 33:600–601. https://doi.org/10.1246/Cl.2004.600
26. Karlsson R, Karlsson M, Karlsson A, Cans A-S, Bergenholtz J, Akerman B, Ewing AG, Voinova M, Orwar O (2002) Moving-wall-driven flows in nanofluidic systems. Langmuir 18:4186–4190. https://doi.org/10.1021/la025533v
27. Xu H, Nagasaka S, Kameta N, Masuda M, Ito T, Higgins DA (2016) Imaging fluorescence correlation spectroscopy studies of dye diffusion in self-assembled organic nanotubes. PCCP 18:16766–16774. https://doi.org/10.1039/c6cp03069f
28. Kisley L, Brunetti R, Tauzin LJ, Shuang B, Yi XY, Kirkeminde AW, Higgins DA, Weiss S, Landes CF (2015) Characterization of porous materials by fluorescence correlation spectroscopy super-resolution optical fluctuation imaging. ACS Nano 9:9158–9166. https://doi.org/10.1021/acsnano.5b03430
29. Fu Y, Kao WJ (2010) Drug release kinetics and transport mechanisms of non-degradable and degradable polymeric delivery systems. Expert Opin Drug Deliv 7:429–444. https://doi.org/10.1517/17425241003602259
30. Xu H, Nagasaka S, Kameta N, Masuda M, Ito T, Higgins DA (2017) Spectroscopic imaging studies of nanoscale polarity and mass transport phenomena in self-assembled organic nanotubes. PCCP 19:20040–20048. https://doi.org/10.1039/c7cp03672h
31. Kameta N, Masuda M, Mizuno G, Morii N, Shimizu T (2008) Supramolecular nanotube *endo* sensing for a guest protein. Small 4:561–565. https://doi.org/10.1002/smll.200700710
32. Kameta N, Masuda M, Shimizu T (2012) Soft nanotube hydrogels functioning as artificial chaperones. ACS Nano 6:5249–5258. https://doi.org/10.1021/Nn301041y
33. Kameta N, Masuda M, Shimizu T (2016) Soft nanotubes acting as confinement effecters and chirality inducers for achiral polythiophenes. Chem Commun 52:1346–1349. https://doi.org/10.1039/c5cc08035e
34. Sawada T, Yoshizawa M, Sato S, Fujita M (2009) Minimal nucleotide duplex formation in water through enclathration in self-assembled hosts. Nat Chem 1:53–56. https://doi.org/10.1038/Nchem.100
35. Arafune H, Yamaguchi A, Namekawa M, Sato Y, Itoh T, Yoshida R, Teramae N (2014) Trinucleotide duplex formation inside a confined nanospace under supercooled conditions. Nat Commun 5:5151. https://doi.org/10.1038/ncomms6151
36. Kameta N, Akiyama H (2018) Shrinkable nanotubes for duplex formation of short nucleotides. Small 14:1801967. https://doi.org/10.1002/smll.201801967
37. Kameta N, Masuda M, Shimizu T (2015) Two-step naked-eye detection of lectin by hierarchical organization of soft nanotubes into liquid crystal and gel phases. Chem Commun 51:6816–6819. https://doi.org/10.1039/C5cc01464f
38. Ding W, Minamikawa H, Kameta N, Shimizu T, Masuda M (2014) Effects of PEGylation on the physicochemical properties and in vivo distribution of organic nanotubes. Int J Nanomed 9:5811–5823. https://doi.org/10.2147/Ijn.S75604
39. Ding W, Wada M, Kameta N, Minamikawa H, Shimizu T, Masuda M (2011) Functionalized organic nanotubes as tubular nonviral gene transfer vector. J Controlled Release 156:70–75. https://doi.org/10.1016/j.jconrel.2011.07.007
40. Meyer O, Kirpotin D, Hong KL, Sternberg B, Park JW, Woodle MC, Papahadjopoulos D (1998) Cationic liposomes coated with polyethylene glycol as carriers for oligonucleotides. J Biol Chem 273:15621–15627. https://doi.org/10.1074/jbc.273.25.15621
41. Vermonden T, Censi R, Hennink WE (2012) Hydrogels for protein delivery. Chem Rev 112:2853–2888. https://doi.org/10.1021/cr200157d
42. Peng L, Chang L, Liu X, Lin J, Liu H, Han B, Wang S (2017) Antibacterial property of a polyethylene glycol-grafted dental material. ACS Appl Mater Interf 9:17688–17692. https://doi.org/10.1021/acsami.7b05284
43. Uemura T, Yanai N, Watanabe S, Tanaka H, Numaguchi R, Miyahara MT, Ohta Y, Nagaoka M, Kitagawa S (2010) Unveiling thermal transitions of polymers in subnanometre pores. Nat Commun 1:83. https://doi.org/10.1038/ncomms1091

44. Kameta N, Matsuzawa T, Yaoi K, Masuda M (2016) Short polyethylene glycol chains densely bound to soft nanotube channels for inhibition of protein aggregation. RSC Adv 6:36744–36750. https://doi.org/10.1039/c6ra06793j
45. Kameta N, Ding W, Dong J (2017) Soft nanotubes derivatized with short PEG chains for thermally controllable extraction and separation of peptides. ACS Omega 2:6143–6150. https://doi.org/10.1021/acsomega.7b00838
46. Kameta N, Dong J, Yui H (2018) Thermoresponsive PEG-coated nanotubes as chiral selectors of amino acids and peptides. Small 14:1800030. https://doi.org/10.1002/smll.201800030
47. Hudson S, Cooney J, Magner E (2008) Proteins in mesoporous silicates. Angew Chem Int Ed 47:8582–8594. https://doi.org/10.1002/anie.200705238
48. Yui H, Guo Y, Koyama K, Sawada T, John G, Yang B, Masuda M, Shimizu T (2005) Local environment and property of water inside the hollow cylinder of a lipid nanotube. Langmuir 21:721–727. https://doi.org/10.1021/La040109a
49. Furusho H, Mishima Y, Kameta N, Masuda M, Yamashita I, Shimizu T (2008) Lipid nanotube encapsulating method for two- and three-dimensional transmission electron microscopy analyses of cage-shaped proteins. Jpn J Appl Phys 47:394–399. https://doi.org/10.1143/Jjap.47.394
50. Furusho H, Mishima Y, Kameta N, Yamane M, Masuda M, Asakawa M, Yamashita I, Mori H, Takaoka A, Shimizu T (2009) Lipid nanotube encapsulating method in low-energy scanning transmission electron microscopy analyses. Jpn J Appl Phys 48: https://doi.org/10.1143/Jjap.48.097001
51. Saibil HR, Fenton WA, Clare DK, Horwich AL (2013) Structure and allostery of the chaperonin GroEL. J Mol Biol 425:1476–1487. https://doi.org/10.1016/j.jmb.2012.11.028
52. Shimizu T, Ding W, Kameta N (2020) Soft-matter nanotubes: a platform for diverse functions and applications. Chem Rev 120:2347–2407. https://doi.org/10.1021/acs.chemrev.9b00509
53. Kawasaki R, Sasaki Y, Katagiri K, Mukai S, Sawada S, Akiyoshi K (2016) Magnetically guided protein transduction by hybrid nanogel chaperones with iron oxide nanoparticles. Angew Chem Int Ed 55:11377–11381. https://doi.org/10.1002/anie.201602577
54. Nara TY, Togashi H, Sekikawa C, Kawakami M, Yaginuma N, Sakaguchi K, Mizukami F, Tsunoda T (2009) Use of zeolite to refold a disulfide-bonded protein. Colloids Surf B-Biointerf 68:68–73. https://doi.org/10.1016/j.colsurfb.2008.09.012
55. Kameta N, Manaka Y, Akiyama H, Shimizu T (2018) Bioreactors based on enzymes encapsulated in photoresponsive transformable nanotubes and nanocoils end-capped with magnetic nanoparticles. Adv Biosys 2:1700214. https://doi.org/10.1002/adbi.201700214
56. Lian XZ, Fang Y, Joseph E, Wang Q, Li JL, Banerjee S, Lollar C, Wang X, Zhou HC (2017) Enzyme-MOF (metal-organic framework) composites. Chem Soc Rev 46:3386–3401. https://doi.org/10.1039/c7cs00058h
57. Magner E (2013) Immobilisation of enzymes on mesoporous silicate materials. Chem Soc Rev 42:6213–6222. https://doi.org/10.1039/c2cs35450k
58. Kameta N, Shiroishi H (2018) PEG-nanotube liquid crystals as templates for construction of surfactant-free gold nanorods. Chem Commun 54:4665–4668. https://doi.org/10.1039/c8cc02013b
59. Yang B, Kamiya S, Shimizu Y, Koshizaki N, Shimizu T (2004) Glycolipid nanotube hollow cylinders as substrates: fabrication of one-dimensional metallic-organic nanocomposites and metal nanowires. Chem Mater 16:2826–2831. https://doi.org/10.1021/cm049695j
60. Yang B, Kamiya S, Yoshida K, Shimizu T (2004) Confined organization of au nanocrystals in glycolipid nanotube hollow cylinders. Chem Commun 500–501. https://doi.org/10.1039/B313100A
61. Eisele DM, Berlepsch HV, Bottcher C, Stevenson KJ, Bout DAV, Kirstein S, Rabe JP (2010) Photoinitiated growth of sub-7 nm silver nanowires within a chemically active organic nanotubular template. J Am Chem Soc 132:2104–2105. https://doi.org/10.1021/Ja907373h
62. Liu M, Zhang L, Wang T (2015) Supramolecular chirality in self-assembled systems. Chem Rev 115:7304–7397. https://doi.org/10.1021/cr500671p

63. Ji QM, Kamiya S, Jung JH, Shimizu T (2005) Self-assembly of glycolipids on silica nanotube templates yielding hybrid nanotubes with concentric organic and inorganic layers. J Mater Chem 15:743–748. https://doi.org/10.1039/B412725k
64. Kameta N, Ding W (2019) Supramolecular nanotube reactors for production of imine polymers with controlled conformation, size, and chirality. Small 15:1900682. https://doi.org/10.1002/smll.201900682
65. Zhou Y, Shimizu T (2008) Lipid nanotubes: a unique template to create diverse one-dimensional nanostructures. Chem Mater 20:625–633. https://doi.org/10.1021/Cm701999m
66. Sada K, Takeuchi M, Fujita N, Numata M, Shinkai S (2007) Post-polymerization of preorganized assemblies for creating shape-controlled functional materials. Chem Soc Rev 36:415–435. https://doi.org/10.1039/b603555h
67. Zhang W, Jin W, Fukushima T, Saeki A, Seki S, Aida T (2011) Supramolecular linear heterojunction composed of graphite-like semiconducting nanotubular segments. Science 334:340–343. https://doi.org/10.1126/science.1210369
68. Hattori T, Itagaki T, Uji H, Kimura S (2018) Temperature-induced phase separation in molecular assembly of nanotubes comprising amphiphilic polypeptoid with poly(*N*-Ethyl Glycine) in water by a hydrophilic-region-driven-type mechanism. J Phys Chem B 122:7178–7184. https://doi.org/10.1021/acs.jpcb.8b03419
69. Itagaki T, Kurauchi S, Uebayashi T, Uji H, Kimura S (2018) Phase-separated molecular assembly of a nanotube composed of amphiphilic polypeptides having a helical hydrophobic block. ACS Omega 3:7158–7164. https://doi.org/10.1021/acsomega.8b01073
70. Kameta N, Ding W (2019) Direct joining of a heterogeneous pair of supramolecular nanotubes and reaction control of a guest compound by transportation in the nanochannels. Bull Chem Soc Jpn 92:1053–1059. https://doi.org/10.1246/bcsj.20190046
71. Shimizu T, Kogiso M, Masuda M (1996) Vesicle assembly in microtubes. Nature 383:487–488. https://doi.org/10.1038/383487b0
72. Kogiso M, Ohnishi S, Yase K, Masuda M, Shimizu T (1998) Dicarboxylic oligopeptide bolaamphiphiles: proton-triggered self-assembly of microtubes with loose solid surfaces. Langmuir 14:4978–4986. https://doi.org/10.1021/la9802419
73. Shimizu T, Masuda M, Minamikawa H (2005) Supramolecular nanotube architectures based on amphiphilic molecules. Chem Rev 105:1401–1443. https://doi.org/10.1021/cr030072j
74. Shimizu T, Ohnishi S, Kogiso M (1998) Cross section molecular imaging of supramolecular microtube using contact atomic force microscopy. Angew Chem Int Ed 37:3260–3262. https://doi.org/10.1002/(SICI)1521-3773(19981217)37:23%3c3260:AID-ANIE3260%3e3.0.CO;2-4
75. Wang T, Jiang J, Liu Y, Li Z, Liu M (2010) Hierarchical self-assembly of bolaamphiphiles with a hybrid spacer and L-glutamic acid headgroup: pH- and surface-triggered hydrogels, vesicles, nanofibers, and nanotubes. Langmuir 26:18694–18700. https://doi.org/10.1021/la103435t
76. Zhang L, Wang T, Shen Z, Liu M (2016) Chiral nanoarchitectonics: towards the design, self-assembly, and function of nanoscale chiral twists and helices. Adv Mater 28:1044–1059. https://doi.org/10.1002/adma.201502590
77. Jiang J, Wang T, Liu M (2010) Creating chirality in the inner walls of silica nanotubes through a hydrogel template: chiral transcription and chiroptical switch. Chem Commun 46:7178–7180. https://doi.org/10.1039/c0cc00891e
78. Jin Q, Zhang L, Cao H, Wang T, Zhu X, Jiang J, Liu M (2011) Self-assembly of copper(II) Ion-mediated nanotube and its supramolecular chiral catalytic behavior. Langmuir 27:13847–13853. https://doi.org/10.1021/La203110z
79. Jiang J, Meng Y, Zhang L, Liu M (2016) Self-assembled single-walled metal-helical nanotube (M-HN): creation of efficient supramolecular catalysts for asymmetric reaction. J Am Chem Soc 138:15629–15635. https://doi.org/10.1021/jacs.6b08808
80. Liu Y, Wang T, Li Z, Liu M (2013) Copper(II) ion selective and strong acid-tolerable hydrogels formed by an L-histidine ester terminated bolaamphiphile: from single molecular thick nanofibers to single-wall nanotubes. Chem Commun 49:4767–4769. https://doi.org/10.1039/c3cc41786g

81. Yao P, Wang H, Chen P, Zhan X, Kuang X, Zhu D, Liu M (2009) Hierarchical assembly of an achiral π-conjugated molecule into a chiral nanotube through the air/water interface. Langmuir 25:6633–6636. https://doi.org/10.1021/la901435s
82. Liu X, Wang T, Liu M (2011) Interfacial assembly of cinnamoyl-terminated bolaamphiphiles through the air/water interface: headgroup-dependent assembly, supramolecular nanotube and photochemical sewing. PCCP 13:16520–16529. https://doi.org/10.1039/C1cp21561b
83. Unsal H, Aydogan N (2013) Formation of chiral nanotubes by the novel anthraquinone containing-achiral molecule. J Colloid Interface Sci 394:301–311. https://doi.org/10.1016/j.jcis.2012.12.023
84. Unsal H, Schmidt J, Talmon Y, Yildirim LT, Aydogan N (2016) Dual-responsive lipid nanotubes: two-way morphology control by pH and redox effects. Langmuir 32:5324–5332. https://doi.org/10.1021/acs.langmuir.6b00350
85. Ilbasmis-Tamer S, Unsal H, Tugcu-Demiroz F, Kalaycioglu GD, Degim IT, Aydogan N (2016) Stimuli-responsive lipid nanotubes in gel formulations for the delivery of doxorubicin. Colloid Surf B Biointerfaces 143:406–414. https://doi.org/10.1016/j.colsurfb.2016.03.070
86. Yamamoto T, Fukushima T, Aida T (2008) Self-assembled nanotubes and nanocoils from pi-conjugated building blocks. Adv Polym Sci 220:1–27. https://doi.org/10.1007/12_2008_171
87. Shao H, Seifert J, Romano NC, Gao M, Helmus JJ, Jaroniec CP, Modarelli DA, Parquette JR (2010) Amphiphilic self-assembly of an n-type nanotube. Angew Chem Int Ed 49:7688–7691. https://doi.org/10.1002/anie.201003415
88. Gao M, Paul S, Schwieters CD, You ZQ, Shao H, Herbert JM, Parquette JR, Jaroniec CP (2015) A structural model for a self-assembled nanotube provides insight into its exciton dynamics. J Phys Chem C 119:13948–13956. https://doi.org/10.1021/acs.jpcc.5b03398
89. Tu S, Kim SH, Joseph J, Modarelli DA, Parquette JR (2013) Proton-coupled self-assembly of a porphyrin-naphthalenediimide dyad. Chem Phys Chem 14:1609–1617. https://doi.org/10.1002/cphc.201300023
90. Tu S, Kim SH, Joseph J, Modarelli DA, Parquette JR (2011) Self-assembly of a donor-acceptor nanotube. A strategy to create bicontinuous arrays. J Am Chem Soc 133:19125–19130. https://doi.org/10.1021/ja205868b

Chapter 4
Di-phenylalanine-Based Nanotubes

4.1 Di-phenylalanine

Proteins are not only chemically and structurally diverse macromolecules but also functional polypeptides that comprise a great deal of covalently linked amino acid residues. The suitable combinations and sequences of the amino acids result in the construction of rational 3D structures, thereby exhibiting a variety of vital functions including mechanical, structural, enzymatic, chaperone, immune, storage, and transport actions. Notably, dipeptides represented by di-L-phenylalanine (H_2N–L–Phe–L–Phe–COOH, FF) **1** (Chart 4.1), although the structure is the shortest and simple as a peptide, can play a talented role in the self-assembly into nanostructures. This FF sequence is known as the recognition motif (–F19–F20–) of Alzheimer's β-amyloid polypeptide (Fig. 4.1a). For example, Gazit's research group has carried out pioneering work on self-assembled tubular structures from the FF dipeptide [1]. Self-assembly of FF by diluting the 1,1,1,3,3,3-hexafluoro-2-propanol (HFIP) solution with water was shown to give tough and discrete nanotube structures with high axial ratios (Fig. 4.1b) [2]. Statistical distribution of the outer diameters is in the range of 50–150 nm for 50% of nanotubes measured. In turn, Shelnutt and co-workers presented alternative approach for the nanotube production with the enantiomer of **1** (H_2N–D–Phe–D–Phe–COOH) using Nanopure water (purified by using a Barnstead/Thrmolyne Nanopure laboratory water system) as the self-assembly medium in place of the HFIP [3]. The selected D-phenylalanine residues have a high resistance to decomposition by enzymes such as Proteinase K. The dimensions of the obtained nanotubes are relatively larger (diameters: 100 nm to 2 μm, lengths: exceed 100 μm) as compared those produced in mixtures of HFIP and water.

In 2001, Görbitz reported an intriguing finding on a single crystal structure of self-assembled needles from FF in water [5]. In the ordered 3D structure, the hydrophobic dipeptide molecules form a hydrophilic channel shaped by helically arranged peptide backbone chains in head-to-tail manner, which were stabilized by the hydrogen bonding of $-NH_3^+\cdots{}^-OOC-$ as well as π–π stacking between the aromatic residues.

T. Shimizu, *Smart Soft-Matter Nanotubes*, Nanostructure Science and Technology,
https://doi.org/10.1007/978-981-16-2685-2_4

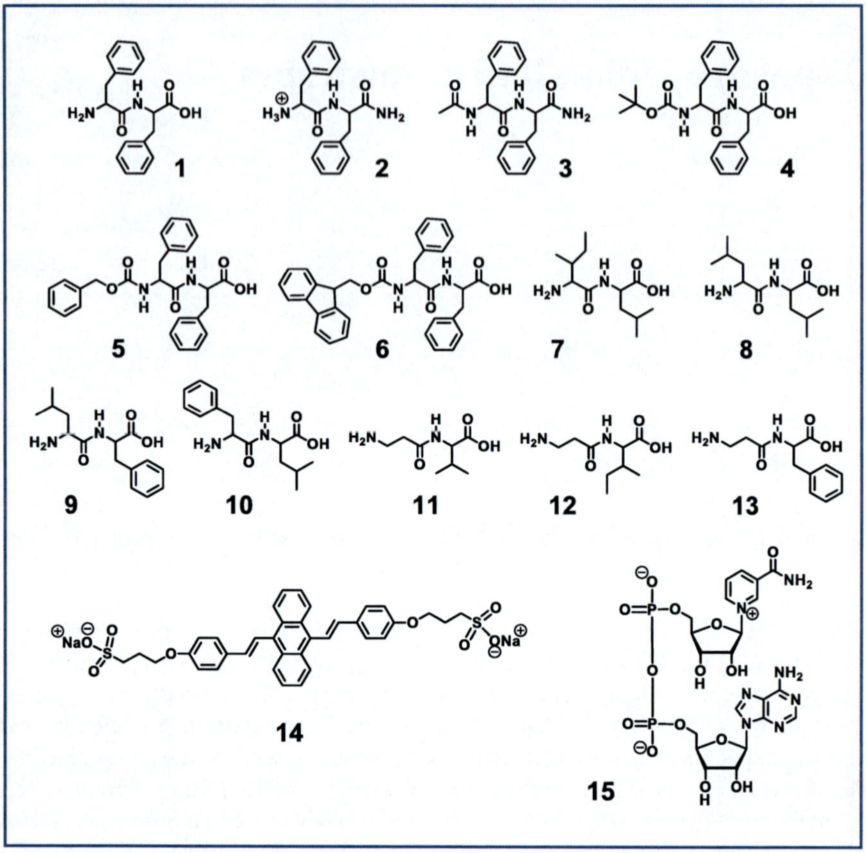

Chart 4.1 Chemical structures of **1–15**

Five years later, he produced stiff nanotubes by himself by using the same self-assembly procedure as that reported by Reches and Gazit [2]. As a result of his scrutinized measurement, he corroborated that the X-ray powder diffraction pattern for the present nanotube is almost consistent with the simulated one from the single crystal structure [4]. Fig. 4.2 represents a possible packing model for the FF nanotube. A cross-sectional area of the FF nanotube with an outer diameter of 110 nm and inner diameter of 50 nm is represented. It is unknown whether the inner surface of the nanotube is featured by a mixed system of hydrophobic and hydrophilic nature or completely hydrophobic one. Meanwhile, the respective nanochannel with 1 nm van der Waals' diameter presents hydrophilic character that can encapsulate guest substances such as hydrated metal cations. Taken together, the aromatic side chains form an elaborate and sophisticated 3D arrangement, thus resulting in the formation of the nanotube wall. Moreover, the aromatic groups assist the generation of the

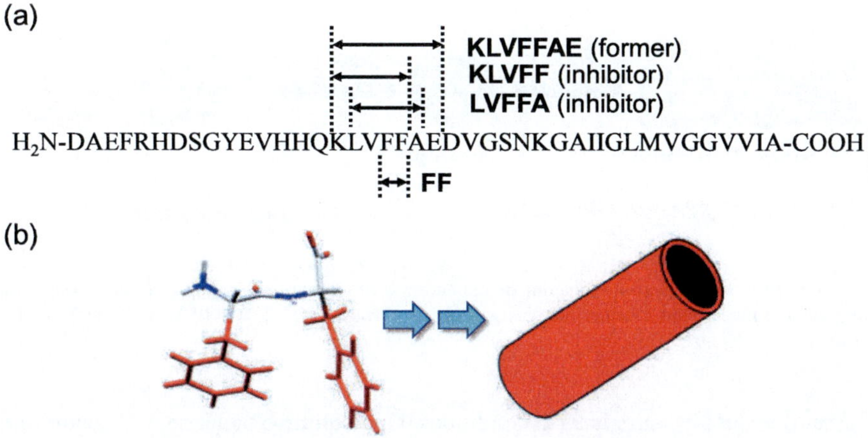

Fig. 4.1 **a** Recognition motif of Alzheimer's β-amyloid polypeptide and several fragments that form amyloid fibrils or inhibit their formation. **b** A single-crystal structure of FF peptide and illustration of the self-assembled nanotube (Reproduced with permission from Ref. [4], © The Royal Society of Chemistry 2006)

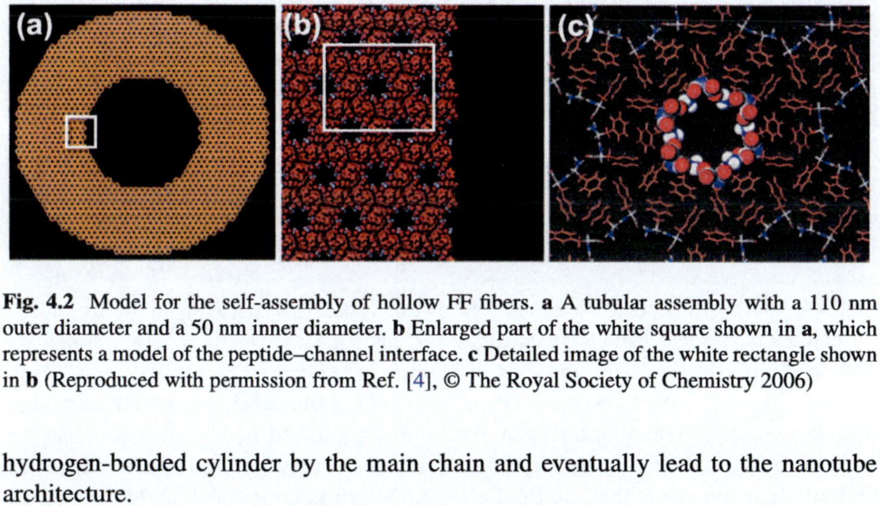

Fig. 4.2 Model for the self-assembly of hollow FF fibers. **a** A tubular assembly with a 110 nm outer diameter and a 50 nm inner diameter. **b** Enlarged part of the white square shown in **a**, which represents a model of the peptide–channel interface. **c** Detailed image of the white rectangle shown in **b** (Reproduced with permission from Ref. [4], © The Royal Society of Chemistry 2006)

hydrogen-bonded cylinder by the main chain and eventually lead to the nanotube architecture.

4.2 Termini-Modified FF Dipeptide

Aiming at efficient binding to a negatively charged oligonucleotide, Li and coworkers developed positively charged dipeptide nanotubes [6]. The designed dipeptide is cationic FF derivative (H_3N^+–L–Phe–L–Phe–$CONH_2$, **2**) (Chart 4.1) that self-assembles in a mixture of HFIP and water to yield nanotubes (diameters: ca.

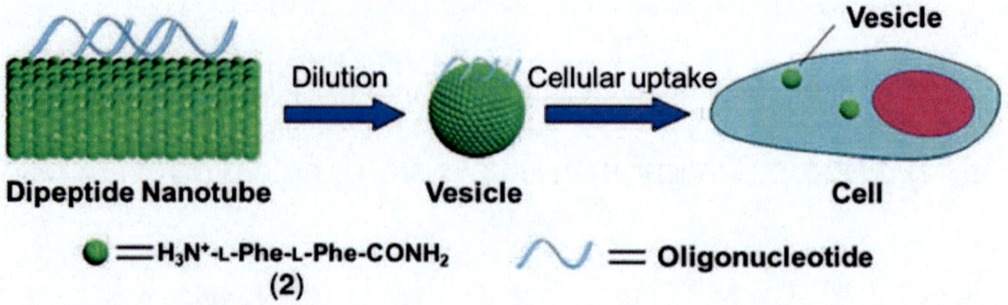

Fig. 4.3 Proposed transition behavior of the cationic FF dipeptide **2** nanotubes into vesicles for oligonucleotide delivery (Reproduced with permission from Ref. [6], © 2007 WILEY-VCH Verlag GmbH & Co. KGaA, Weinheim)

200 nm) under physiological pH conditions. Upon dilution under neutral conditions, the obtained nanotubes further transform their morphologies into spherical vesicles. Strong binding of anionic single-stranded DNA to the vesicles was shown to enable the intracellular delivery with HeLa cells (Fig. 4.3).

Reches and Gazit newly synthesized noncharged FF derivative **3** (Chart 4.1) having acetylated *N*-terminal as well as amidated *C*-terminal [7]. The modified dipeptide **3** forms well-defined nanotubes by self-assembly in a similar manner as **1**. Conversely, the FF analogues **4**, **5**, and **6** (Chart 4.1) with bulkier aromatic protecting groups than that of **3** resulted in the formation of no nanotube structures but nanofibers. These findings suggest that aromatic interactions play a more significant role in the nanotube self-assembly than electrostatic interactions.

4.3 Related Dipeptide

Görbitz has performed a systematic study on molecular packing of hydrophobic dipeptides in single crystal X-ray structures [5, 8, 9]. The investigated dipeptides are composed of arbitrary combinations of alanine (Ala), valine (Val), isoleucine (Ile), leucine (Leu), and phenylalanine (Phe). He divided the obtained crystal structures into two major classes that differ in the dimensionality of hydrogen-bond chain as well as in hydrophobic or hydrophilic nature of the nanopore channels (Fig. 4.4). Of particular interest is that the Ile–Leu **7**, Leu–Leu **8**, Leu–Phe **9**, and Phe–Leu **10** (Chart 4.1) except for Phe–Phe **1** dipeptides form 2D hydrogen-bond sheets and then fold them into tubular architectures. The resulting hydrophilic inner surfaces face cylindrical clusters of water molecules. Dipeptide nanotubes with several hundred nm in diameters provide the tube walls comprising interesting nanoporous crystalline materials.

Guha and Banerjee developed pH, proteolytically, and thermally stable robust nanotubes self-assembled from the dipeptides β-Ala–L-Val **11**, β-Ala–L-Ile **12**, and β-Ala–L-Phe **13** (Chart 4.1), in which β-alanine and α-amino acid residues locate as

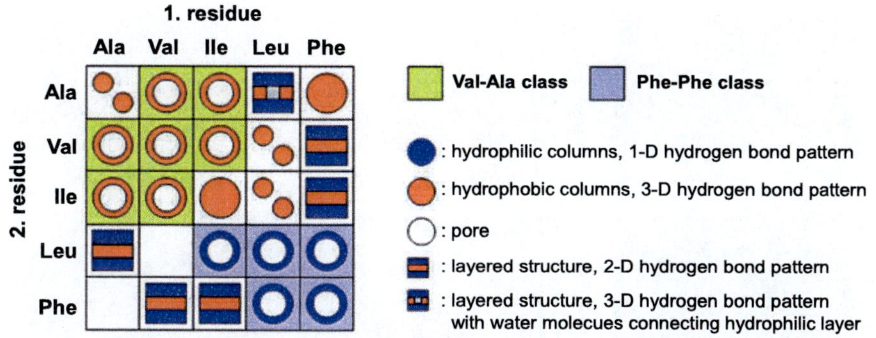

Fig. 4.4 Summary for the crystal packing arrangements obtained by various combinations of two hydrophobic amino acids. The matrix represents pattern for separation between hydrophobic and hydrophilic moieties and the presence of pores (Reproduced with permission from Ref. [9], © 2007 WILEY-VCH Verlag GmbH & Co. KGaA, Weinheim)

the *N*- and *C*-terminus, respectively. In solid states, supramolecular double helical structures are formed from all the three dipeptides **11–13**, whereas nanotube architectures appear in solution state via slow evaporation of the aqueous solution and two-day drying in vacuo at 30 °C. The dimensions of the generated nanotubes from **11**, **12**, and **13** are characterized by the average outer diameters of 37, 36, and 35 nm and inner diameter of 7, 6, and 5 nm, respectively. Notably, those nanotube structures exhibited high robustness that are stable against heating to 80 °C and a broad range of pH 2–10. Additionally, the nanotubes have a resistance to proteolytic degradation by Proteinase K.

4.4 Potent Applications of FF-Based Nanotubes

Among a great deal of organic nanotubes reported so far, self-assembled nanotubes from the FF dipeptide have displayed excellent abilities in multifarious applications. The applications of the FF-based nanotubes cover chemical, physical, and medical fields. For example, sensing of water molecules, ferro- and piezoelectrics, ultracapacitor, second harmonic generation (SHG), and detection of cancer cell should be included. Details will be discussed next. Recent outstanding progress of the FF self-assembly and diverse applications achieved so far was overviewed elsewhere [1].

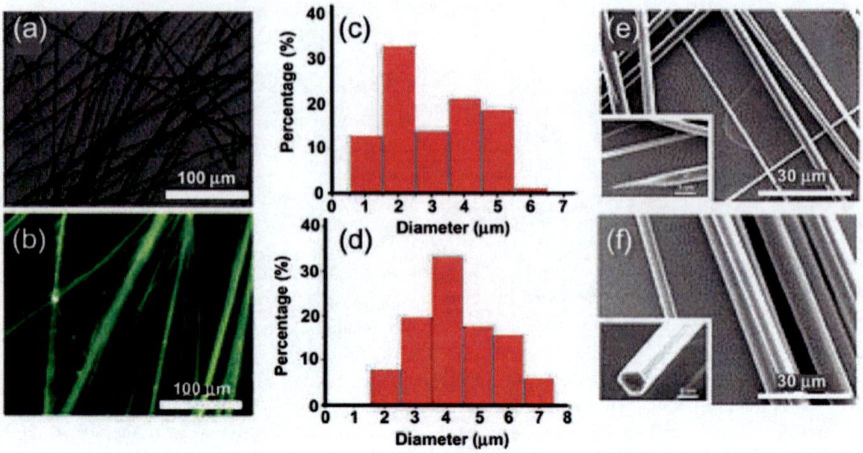

Fig. 4.5 Fluorescence optical micrographs of FF dipeptide self-assemblies in **a** water and **b** BSPSA solution. (**c** and **d**) Statistical size distribution of diameters for obtained microtubes in **a** and **b**. **e**, **f** Magnified SEM images for the microtubes in **a** and **b** (Reproduced with permission from Ref. [11], © The Royal Society of Chemistry 2013)

4.4.1 Microtubes with "Turn-on" Fluorescence

Aggregation-induced emission (AIE) is a phenomenon, in which a certain molecule emits strong luminescence through the molecular aggregation in solutions [10]. The AIE-active molecule is, in contrast, nonfluorescent when dispersed as a monomer in solutions. Ouyang and co-workers demonstrated the self-assembly of microtubes that can exhibit "turn-on" fluorescence [11]. When the FF peptide undergoes the self-assembly in aqueous solutions of 9,10–bis[4–(3–sulfonatopropoxyl)–styryl] anthracene (BSPSA) sodium salt **14** (Chart 4.1) as an AIE-active molecule, the generated microtubes emit bright yellow green fluorescence (Fig. 4.5). The microtubes self-assembled in the aqueous solution of **14** have hexagonal hollow cylinders with flat tips, the outer diameters of which are in the range of 2–8 μm. The self-assembly of the FF peptide in water in the absence of **14** gave no marked fluorescence. The BSPSA **14** molecules aggregate in an encapsulated environment of the microtube channel, acting as a fluorescent label [11]. The fluorescent intensity of the microtubes increases with the increasing the concentrations of **14** in the aqueous solutions.

4.4.2 Recognition and Sensing

The interaction between the FF dipeptide and water molecules plays a critical role in the nanochannel formation in a crystalline solid [9]. Wu, Chu, and co-workers reported that the photoluminescence (PL) spectra measured for the FF-based

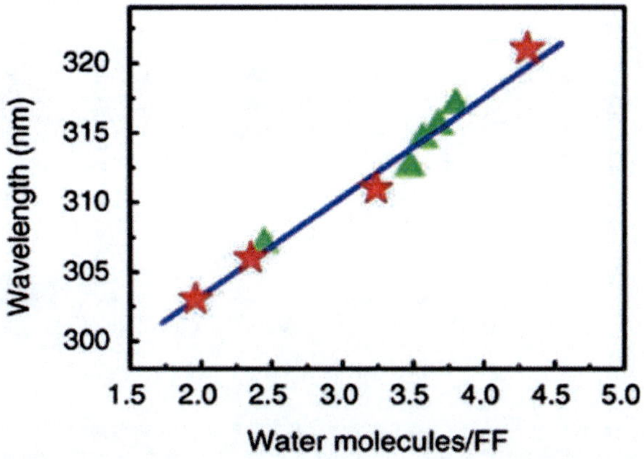

Fig. 4.6 Linear dependence of the UV PL peak position on the average number of water molecules per FF molecule. This demonstrates that the peak position can be employed as an indicator to identify the modification of water content in the dipeptide nanotube samples (Reproduced with permission from Ref. [12], © 2011 WILEY–VCH Verlag GmbH & Co. KGaA, Weinheim)

nanotubes gave a prominent peak between 300 and 320 nm, which is ascribable to deficiency of water molecules in the nanotube channel [12]. Tuning of the FF concentration and different vapor pressures of water allowed for the self-assembly of FF nanotube with a hexagonal shape. The peak position of the UV PL is thus found to be linearly proportional to the water content (Fig. 4.6). This finding means that the PL spectra of the FF nanotubes can be employed to evaluate the number of water molecules bound to the FF dipeptides. Raman spectra of the FF-based nanotubes also revealed that the bonded water to the FF molecules undergo remarkable peak separation of the benzene segment at 1034 cm^{-1} into doublets [13]. Simultaneously, the separation becomes smaller with the decrease in water content. The FF-based nanotubes, thus, served as sensing probes to determine the number of water molecules that interact with the FF dipeptide in the nanochannel pores.

Precise acquisition of thermal information in living cells is of vital importance to monitor the cell activities. Although temperature sensing platform that monitors temperature-dependent PL intensity has developed, there still remains many problems to be solved [14, 15]. Wu, Chu, and co-workers examined the capability of FF-based nanotubes to sense the temperature on a micro and nanometer scale. The FF-based nanotubes was shown to display an excellent function as a precise temperature sensor [16]. For instance, monitoring of the temperature dependence of the PL lifetime and intensity for the FF-based nanotubes enables the evaluation of the absolute temperature in situ with high accuracy of 1 °C and detection of local temperature change, respectively (Fig. 4.7). All the obtained results demonstrated that the FF nanotubes should serve as precise and micro and nanoscale thermometer with good sensitivity and spatial resolution in microbiological environment.

Phenolic compounds are familiar organic substances in our daily life and become often hazardous, causing a notable environmental risk. Gazit, Rishpon, and co-workers explored amperometric biosensors with Au electrode, the surface of which were modified with dense arrays of FF nanotubes [17]. Electrochemical studies using the nanotube-modified biosensors corroborated that they displayed up to 17-fold higher sensitivity for the detection of phenol as compared to CNT-coated or uncoated

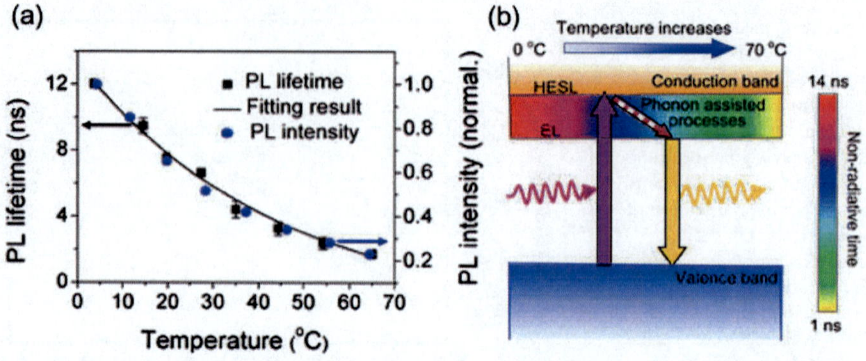

Fig. 4.7 **a** Plots of the average PL intensity (circle) and PL lifetime (square) versus temperature. The curved line shows the fitted lifetime as a function of temperature. **b** Schematic illustration of the qualitative mechanism that explains the temperature-dependent PL lifetime (Reproduced with permission from Ref. [16], © 2013 American Chemical Society)

electrode. For example, chronoamperometry experiments showed that one-layer deposition of the FF nanotube at a concentration of 2 mg/mL gave the highest sensitivity at different phenol concentrations (100, 500, and 1000 nM) (Fig. 4.8a). Additionally, the FF-based nanotubes were subjected to the integration with MWCNTs. Novel electrodes modified with the FF nanotube/CNT composites were thus fabricated. This combination aims at synergistic effect by the excellent biocompatibility of the FF peptide and electroconductivity of CNTs. Indeed, this hybrid type of electrode can detect lower concentration of phenol (50 nM) (Fig. 4.8b) [17]. Similar fabrication

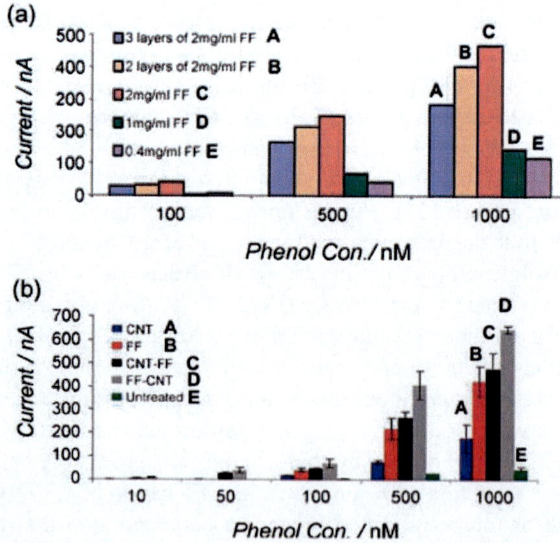

Fig. 4.8 **a** Amperometric response of a screen-printed electrode modified with FF peptide nanotubes under various conditions to 100, 500, and 1000 nM phenol. **b** Amperometric response of CNT- and FF-peptide-nanotube-modified screen-printed electrodes and untreated electrodes to different phenol concentrations (Reproduced with permission from Ref. [17], © 2010 WILEY-VCH Verlag GmbH & Co. KGaA, Weinheim)

of composite electrodes with the FF nanotubes and MWCNTs resulted in the exploration of a new biosensor to monitor nicotinamide adenine dinucleotide (NADH) **15** (Chart 4.1) [18]. FT-IR measurement suggested that both hydrogen-bonding and π–π stacking interactions make the self-assembled FF dipeptide nanosheet wrap the exterior of the MWCNT. The FF dipeptides and CNTs contribute to the reduction of the overpotential and current of NADH oxidation, respectively.

4.4.3 Vertical Alignment of FF Nanotubes

Gazit group developed a new self-assembly approach of FF nanotubes to fabricate large arrays of the peptide nanotube on a glass substrate (Fig. 4.9) [19]. The advantage of this technique is that the density and length of the arrayed nanotubes are tunable precisely by carefully regulating the supply of the FF monomer from the gas phase. The formation of uniquely functional surface that can exhibit high hydrophobicity was prepared by finely controlled self-assembly of the FF-based nanotubes. The FF nanotube arrays with vertical alignment are also applicable to blue luminescent material, [20] microfluidic device, self-cleaning surface, and high surface area electrodes for energy storage [1, 19].

The FF peptide is known to also self-assemble to give nanowire structures except for nanotubes under different conditions. Park and co-workers developed a solid-phase self-assembly of the FF peptide by exposing the assembly to pentafluoroaniline vapor at high temperature [21]. This process satisfactorily produced vertically aligned

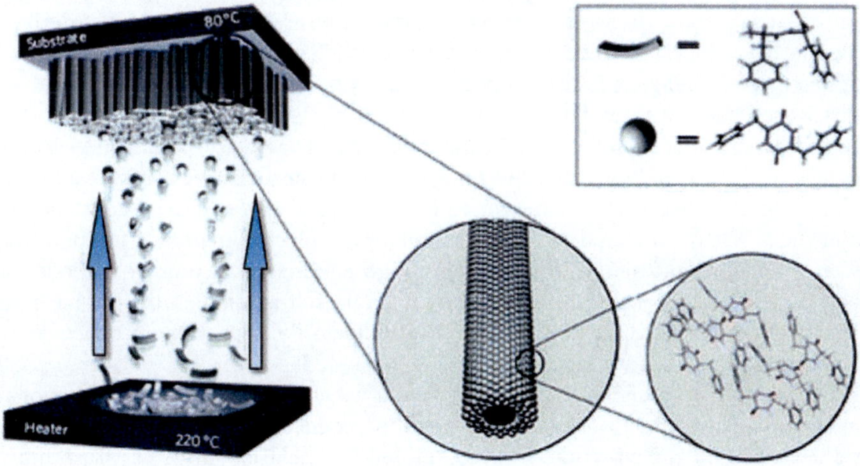

Fig. 4.9 Schematic illustration of the physical vapor deposition (PVD) technique. During evaporation, the FF peptide, which is heated to 220 °C, took a cyclic structure and then self-assembled on a substrate to yield ordered, vertically aligned nanotubes (Reproduced with permission from Ref. [1], © The Royal Society of Chemistry 2014)

FF nanowires. Moreover, capillary force caused by a solvent evaporation enables the fabrication of hierarchical reorganization of the peptide nanowires, which gives a hybrid of nano and microstructures. If we ignore the lack of the hollow cylinder, the fabricated hill-and-valley-like structures can present a superhydrophobic surface.

4.4.4 Ferroelectrics and Piezoelectrics

Dielectric materials become polarized when an external electric field is applied. Among the dielectric materials, ferroelectric materials can produce a spontaneous electric polarization capable of switching by an applied external electric field. Likewise, piezoelectric materials can generate an electrical potential in response to externally applied mechanical stress. Biological source-derived ferroelectric crystals of nanometers scale are attractive as green nanodevices because of their environmentally benign natures. FF dipeptide-based nanostructures including wire, tube, and forest morphologies, particularly, can be highly expected as bioinspired nanomaterials in various applications. In the course of the measurements for hysteresis loop using the FF-based nanotubes and microtubes, Wu and co-workers observed the appearance of a saturated hysteresis loop after irradiation of xenon lamp [22]. The obtained polarization–electric field (P–E) loop evidently contains a concave region, exactly representing the ferroelectric behavior (Fig. 4.10). The interaction between the spontaneous polarization field and applied external field is associated with the light-induced ferroelectricity in the self-assembled FF-based tubes.

Inorganic piezoelectric materials are generally harmful. Self-assembled organic nanomaterials that show intense piezoelectricity, therefore, are gaining a lot of attention because of their bioinspired bottom-up production as well as bioresource-derived materials. The measured piezoelectric coefficient (~20 pm/V at most) for bovine and human scleral collagens is relatively small and yet, not always satisfactory values [23]. Rosenman and co-workers observed remarkably large shear piezoelectricity for the self-assembled nanotubes from FF dipeptide [24]. The electric polarization occurs along the direction of their axes, thus representing the piezoelectric coefficient (leastwise 60 pm/V, shear response measured for nanotubes with approximately 200 nm diameter). When compared with conventional piezoelectric materials $LiNbO_3$, the obtained values correspond to a strikingly high coefficient. A prototype model of the piezoelectric resonator constructed by a single-self-assembled microtube from FF dipeptide exhibited the performance of satisfactorily high-resonance frequency (~2.7 MHz) and a quality factor of 114 (Fig. 4.11) [25].

Fabrication of scalable piezoelectric energy harvesters having unidirectional polarization still remains challenging for the production of practical piezoelectric devices. Lee and co-workers have explored the meniscus-driven self-assembly technique that can fabricate massive-scale aligned arrays of FF-based nanotubes (Fig. 4.12a, b) [26]. When 42 N of force was applied, the resultant FF-based piezoelectric devices were able to display the electric performance of current (37.4 nA),

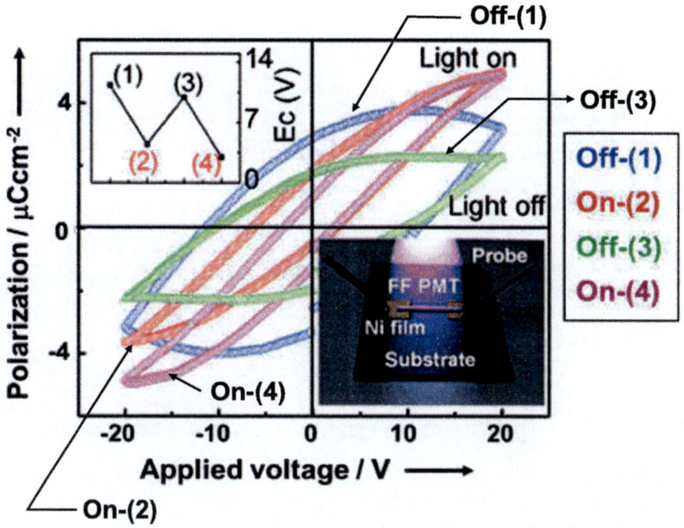

Fig. 4.10 Four P–E curves obtained from the FF peptide microtube during the off/on switching of the light. The inset shows the light-induced coercive field variations with sequent light-off [(1) and (3)] and light-on [(2) and (4)] switching. (bottom right) Schematic image of the set-up for acquisition of the P–E curves (Reproduced with permission from Ref. [22], © 2013 WILEY-VCH Verlag GmbH & Co. KGaA, Weinheim)

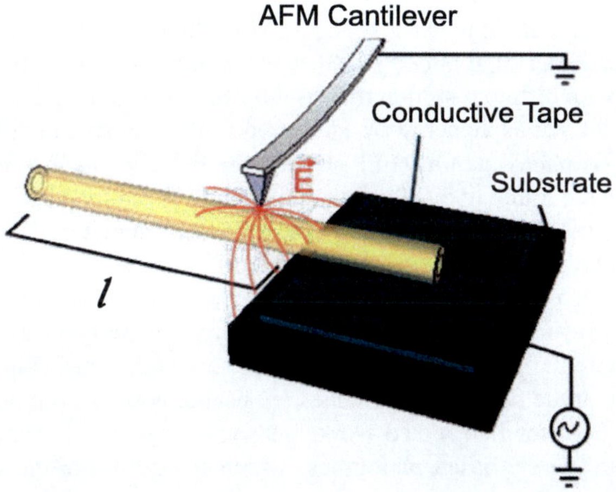

Fig. 4.11 Schematic image of the FF microtube arrangement on the rigid substrate (Reproduced with permission from Ref. [25], © 2013 American Institute of Physics)

voltage (2.8 V), and power (8.2 nW) (Fig. 4.12c). Thus, they demonstrated that, as a proof of concept, the newly fabricated devices can power multiple LC display panels.

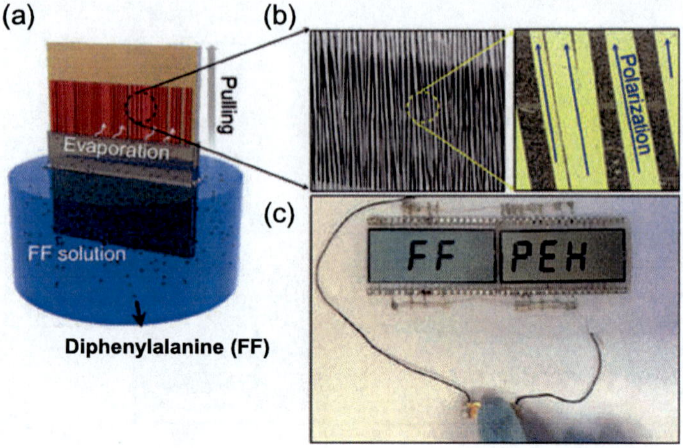

Fig. 4.12 **a** Meniscus-driven dip-coating process to fabricate aligned FF nanotubes with unidirectional piezoelectric polarization and **b** SEM images of the FF nanostructures. **c** Photograph of the LCD driven by pressing the FF peptide-based piezoelectric energy harvester with a finger (Reproduced with permission from Ref. [26], © 2018 American Chemical Society)

4.4.5 Nonlinear Optical Effect

The SHG , which is a representative nonlinear optical phenomenon, is strongly associated with both the symmetry of crystalline materials and electronic characteristics of dielectric materials. The SHG is, then, known to emerge only in crystals that lack a center of symmetry [20]. Various kinds of self-assembled peptide nanostructures with different symmetries and morphologies, i.e., nanotubes, nanofibers, nanobelts, and nanospheres, were subjected to the measurement of the SHG phenomena [27]. Tri-phenylalanine (FFF)-based nanobelts and nanospheres, and FF-based nanotubes were found to display an outstanding SHG effect. The FF and FFF-based nanostructures can be potential nonlinear optical biomaterials as novel frequency converters and photonic devices. When two different light frequencies ($\lambda = 800$ nm and 400 nm as a fundamental and double SHG frequency, respectively) directly illuminate vertical arrays of FF nanotubes, the light flux transmitted through the nanotube hollow and circular brilliance around the outer nanotube shell (Fig. 4.13). This finding indicates that the FF-based nanotubes are usable both as nonlinear and linear waveguides.

Rosenman and co-workers also found prominent and heat-sensitive phase change in the FF-based nanotubes, which have a profound effect on SHG intensity, the piezoelectric properties, and macroscopic wettability [28]. Heat treatment of vertically aligned FF nanotubes around 160 °C causes intramolecular cyclization of the FF molecule from linear conformation to ring structure of diketopiperazine family. This drastic phase change result in the space group transformation from a noncentrosymmetric hexagonal to a centrosymmetric orthorhombic (Fig. 4.14).

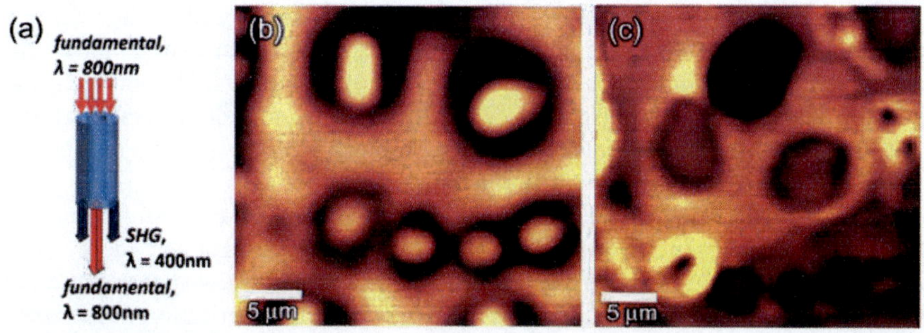

Fig. 4.13 **a** Schematic image of the fundamental and SHG wave (fundamental: $\lambda = 800$ nm, SHG: $\lambda = 400$ nm), **b** linear waveguiding of the FF-based peptide nanotubes, and **c** nonlinear waveguiding of the FF nanotubes (Reproduced with permission from Ref. [27], © 2013 WILEY-VCH Verlag GmbH & Co. KGaA, Weinheim)

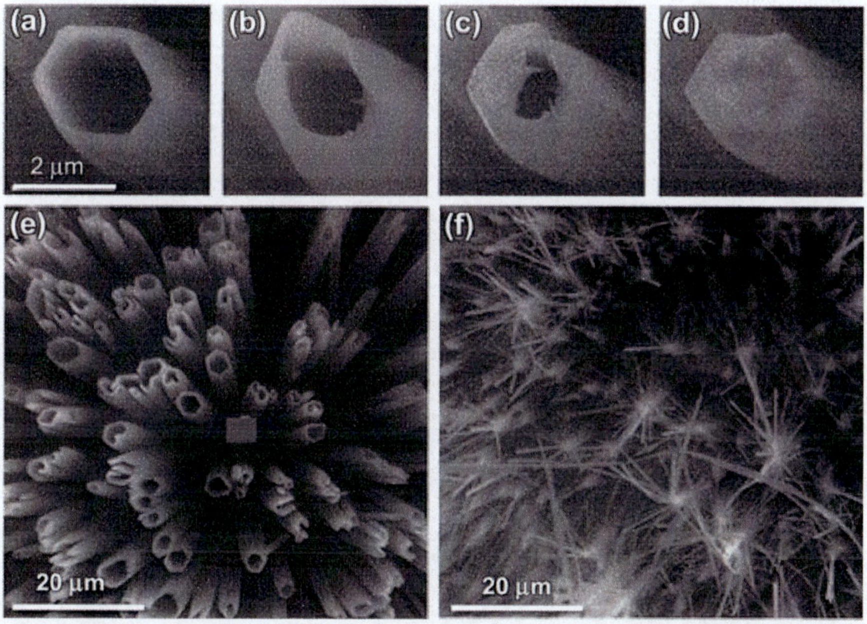

Fig. 4.14 **a–d** Environmental SEM images of an FF-based nanotube tip while heating with the electron beam at $t = 0$, 10, 15, and 20 s, respectively. The scale bar in **a** is valid to **a–d**. **e**, **f** Environmental SEM image of vertically aligned FF-based nanotubes prior to the heating and after heating, respectively (Reproduced with permission from Ref. [28], © 2011 American Chemical Society)

Newly developed approach of solvent thermal annealing allowed for the assembly of hexagonal crystalline microtubes. During this hierarchical and ordered self-assembly process, FF dipeptide molecules first assemble into nanotubes in hexagonal packing manner and subsequently the nanotube seeds get together to form hexagonal microtubes comprising ordered nanotube arrays (Fig. 4.15). The resultant microtubes was shown to perform active optical waveguiding [29].

Nonlinear and linear optical waveguiding actions were studied using ultrashort FF-based nanotubes of 40 μm long with two different alignment modes, i.e., vertical and horizontal nanotube configurations [30]. Rapid solvent evaporation of the HFIP solution containing FF dipeptide allowed for the fabrication of the vertically aligned FF nanotubes that can display the function of a diffractive multifocal lens (Fig. 4.16). Meanwhile, dehydration of the aqueous solution containing the FF dipeptide successfully gave horizontally aligned FF nanotubes. Simulation and experimental results confirmed that the horizontally aligned FF nanotubes displayed eminent optical confinement effect along the nanotube axis. They also demonstrated an outstanding SHG directionality and power conversions efficiency of SHG ~ 10^{-5} for the horizontal nanotubes.

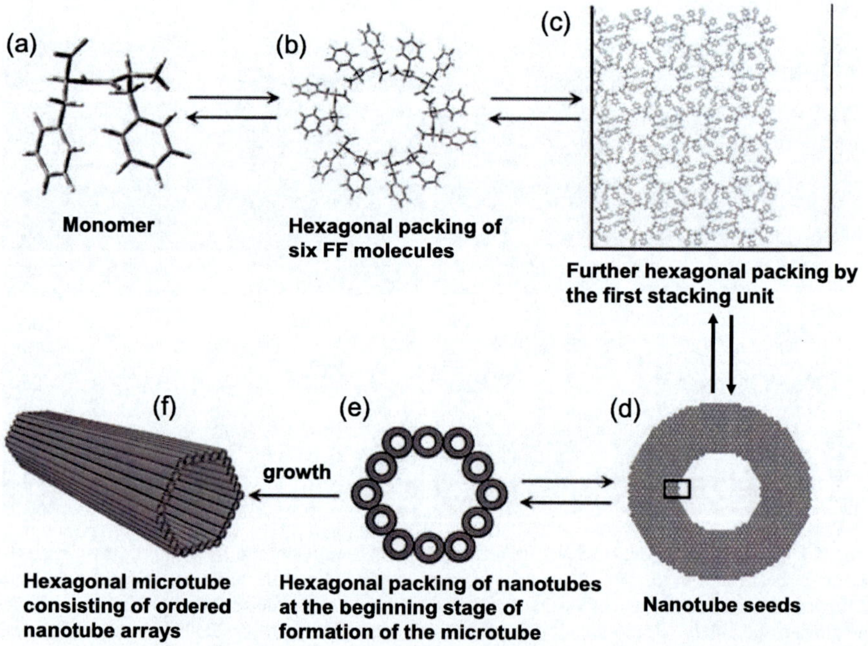

Fig. 4.15 Formation of hexagonal peptide microtubes. **a** FF molecule as a monomer, **b–d** self-assembly of the FF molecules into nanotube seeds in a hexagonal packing pattern, (**e** and **f**) further spontaneous self-assembly into hexagonal microtubes. Panel **c** represents an enlarged view of the area indicated with a black rectangle in **d** (Reproduced with permission from Ref. [29], © 2011 WILEY–VCH Verlag GmbH & Co. KGaA, Weinheim)

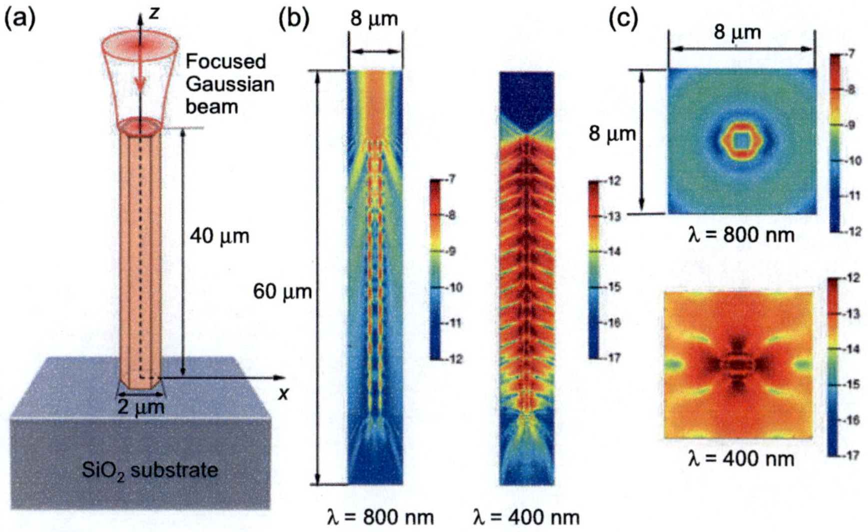

Fig. 4.16 **a** Simulation of linear and nonlinear waveguiding effect in vertically aligned FF nanotube with hexagonal cross section. **b** 2D distributions of E^2 in $y = 0$ plane for fundamental wavelength ($\lambda = 800$ nm) and second harmonic wavelength ($\lambda = 400$ nm) in log scale and **c** the cross sections of the hexagonal FF-based nanotube, which were taken in $z = 20$ μm (Reproduced with permission from Ref. [30], © 2015 Acta Materialia Inc.)

4.4.6 Quantum Confinement

When the dimensions of a semiconductor as a nanomaterial are substantially small, e.g., 10 nm or less, the electronic and optical properties of the corresponding materials show large deviation from those of bulky materials having a large size. Quantum confinement (QC) is closely associated with the profound changes that can be observed in the semiconductors. Decreasing the nanomaterial size increases the bandgap conversely. Thus, QC transforms the continuous energy bands of a solid nanomaterial into discrete energy levels. The optical absorption of an FF dipeptide monomer presents a narrow peak at 257 nm in aqueous solution. Meanwhile, the absorption spectrum of aligned FF nanotubes gave two discriminated steps that appeared in the regions of 245–264 and 300–370 nm [31]. The observed step-like absorption strongly suggests the association of 2D quantum wells. Moreover, strong PL appeared in the UV (305 nm) and blue (400–500 nm) spectra of exciton origin [31].

Similar QC phenomena is also observable in hierarchically assembled hydrogels from the FF dipeptide derivative **6**, the *N*-terminus of which is protected by *N*-fluorenylmethoxycarbonyl (Fmoc) group [32]. The FF derivative **6** form either self-assembled hydrogel in water or aggregate in the solution that is unsuitable for the self-assembly. Figure 4.17 represents the PL excitation spectra of the self-assembled hydrogel and aggregates from **6**. The hydrogel gives a narrow excitation at 310 nm

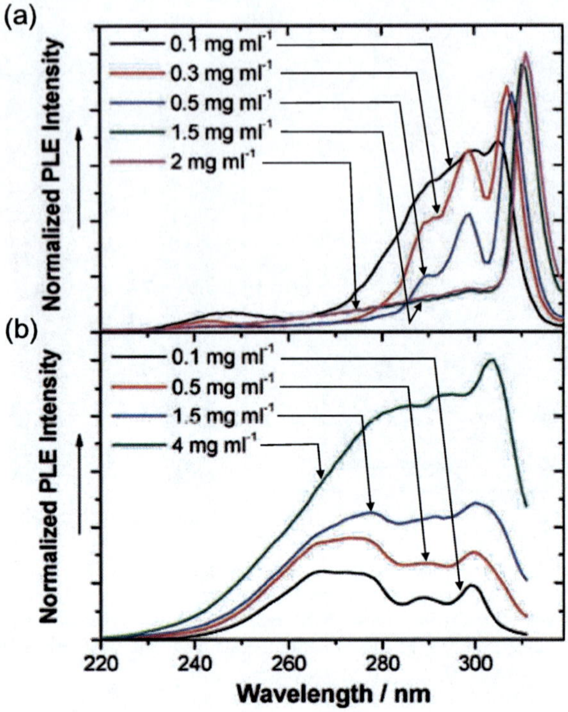

Fig. 4.17 Formation of the ultranarrow PL excitation peak. Excitation spectrum of the **a** hydrogels and **b** aggregates of Fmoc-FF **6** at several concentrations. The observed emission wavelength is 325 nm (Reproduced with permission from Ref. [32], © 2010 WILEY-VCH Verlag GmbH & Co. KGaA, Weinheim)

after the concentration increase and reaches 2 mg/mL, whereas the aggregate only complex and wide spectrum.

4.4.7 Light Harvesting

Natural photosynthesis is strongly involved with sophisticated integration of two large protein complexes, i.e., photosystem I and II, which are organized from the protein-embedded metal catalyst and light-harvesting antenna, i.e., chlorophyll a and b [33]. Mimicking such a natural photosynthetic system, Park and co-workers demonstrated the self-assembly of light-harvesting FF-based nanotubes that are hybridized with tetra(*p*-hydroxyphenyl) porphyrin **16** (Chart 4.2) and Pt NPs (Fig. 4.18) [34]. The porphyrin **16** serves as a molecular model of light harvesting. The Pt NPs were employed for the efficient separation and transfer of the exited electrons from the porphyrin **16** to the electron mediator (M) **17** (Chart 4.2). Electrostatic attraction and hydrogen bonding enabled the formation of J-aggregate of **16** through the

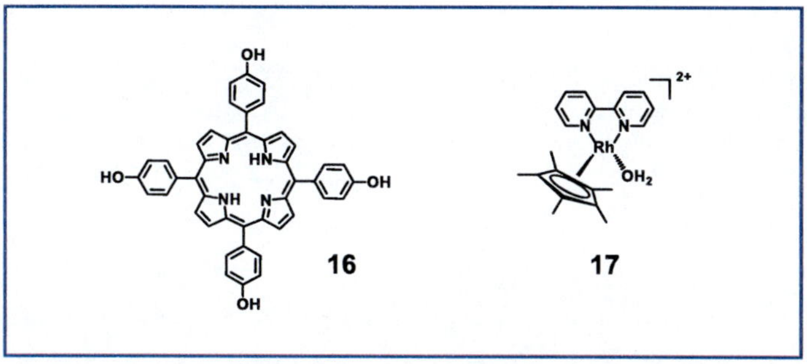

Chart 4.2 Chemical structures of **16** and **17**

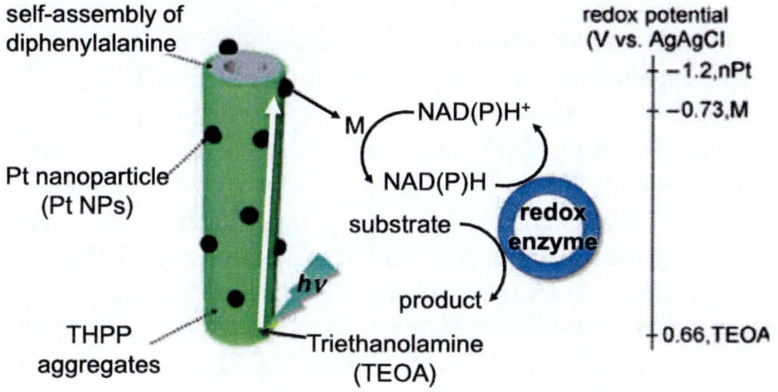

Fig. 4.18 Biomimetic photosynthesis by light-harvesting FF-based nanotubes. White and black arrows represent photoinduced electron excitation and electron transfer, respectively (Reproduced with permission from Ref. [34], © 2012 WILEY-VCH Verlag GmbH & Co. KGaA, Weinheim)

self-assembly. The Pt NPs were immobilized on the nanotube surfaces via a self-metallization process under visible light. The photocatalytic hybrid nanotube materials, thus, demonstrated efficient regeneration of visible-light-induced NADH **15** in cooperation with redox enzymatic synthesis of L-glutamate [34].

4.4.8 Electrode and Supercapacitor

Electrical double-layer supercapacitors (SCs) are modern devices with a high capacity. In terms of electric capacitance, the SCs rank between conventional capacitor and rechargeable battery. Unlike batteries, the SCs need to operate at rapid charge/discharge cycling rates over a great deal of cycles. The energy density of the SCs is much lower than those of batteries. Fabrication of SC electrodes with large surface area is of particular interest in the field of energy devices. PVD methodology [19], developed by Rosenman, Gazit, and co-workers, enabled the coating

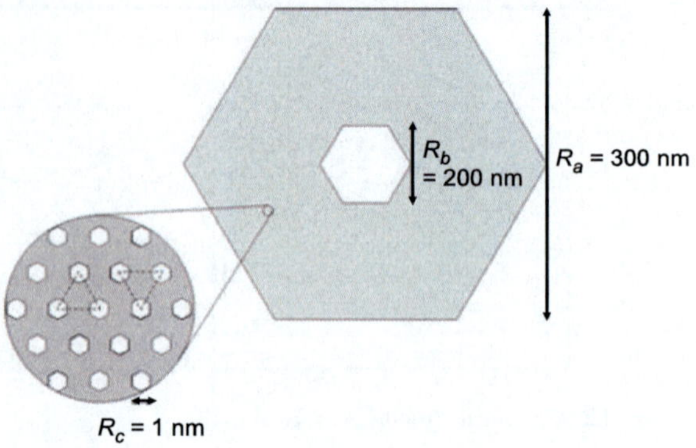

Fig. 4.19 Schematic illustration of the arrangement of hydrophilic channels within the FF-based nanotube (Reproduced with permission from Ref. [35], © Springer Science + Business Media, LLC 2010)

of FF nanotubes on carbon electrode. The nanotubes are normally oriented and distributed with regulated spatial arrangement on the electrode. This FF nanotube-modified electrodes increase the electrical double-layer capacitance by 50 times compared to the inherent carbon electrode [35, 36]. The hollow channels of the FF nanotubes with open tops are strongly involved with the increase in the electrode capacitance. The existence of hydrophilic nanochannels with 1 nm diameters promotes the transportation and accumulation of electrolyte ions (Fig. 4.19).

Golodnitsky and co-workers investigated the influence of magnetic FF nanotubes as fillers on the ion transport of polymer electrolytes comprising concentrated $LiCF_3SO_3$ (LiTf) and polyethylene oxide (PEO) [37]. The magnetic nanotubes were obtained by modifying the surface with ferrofluids containing magnetic NP. The presence of only 1% (w/w) magnetic nanotubes in the LiTf:$(PEO)_3$ polymer electrolyte was found to markedly enhance the total ionic conductivity that reaches over 100 times larger values than those of pristine polymer electrolytes. Interestingly, they found no outstanding variation in the conductivity even after the storage of the solid electrolyte for one month at 70 °C.

4.4.9 Mechanical Reinforcement

The apparent persistence length of over 10 μm for a single piece of FF nanotube strongly suggests an excellent stiffness of the nanotubes. Indeed, AFM analysis based on indentation type experiments disclosed the outstanding stiffness to be approximately 19 GPa [38]. In order not only to avoid the difficulty in the indentation technique in the AFM operation but also to obtain the shear modulus, Tendler employed the bending beam model for AFM imaging [39]. The estimated E-value and shear modulus for the FF nanotube, when hanged over the silicon grid hole, are 27 ± 4 GPa and 0.21 ± 0.03 GPa, respectively, at room temperature (Fig. 4.20). This

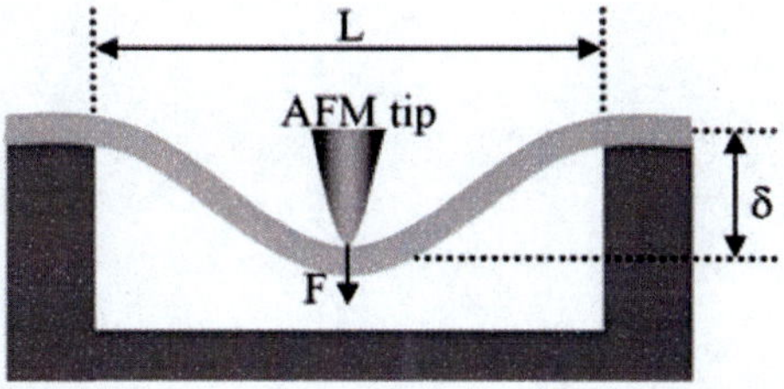

Fig. 4.20 Schematic illustration of the measuring device, where the FF-based fibrils are lying over the holes of the silicon grid. This set-up enables the production of a suspended beam configuration at the nanoscale. The AFM is employed to apply a loading force to the fibril and to directly measure the resulting deflection (Reproduced with permission from Ref. [39], © 2007 American Chemical Society)

distinguished stiffness can maintain under relative humidity (0–70%) and temperature conditions (from room temperature to 100 °C). Both typical analyses represent the reason for their existence as the stiffest self-assembled organic nanotubes. The E-values of 3.3 ± 0.4 GPa evaluated for insulin amyloid fibrils means that the protein fibers are one of the most robust natural fibers originated from β-sheet-rich 3D structures [40]. The convincing reason for the approximately ten times higher rigidity of the FF nanotubes is, however, not fully elucidated. In response to this question, the first principle calculation demonstrated that the unexpectedly high stiffness results from the distinct array structures of the FF nanotube backbones, which are constructed by the interlocking "zipper-like" structures of aromatic segments (Fig. 4.21) [41].

The mechanical stiffness of self-assembled organic nanomaterials is intrinsically low due to their noncovalent bonding feature. This background sometimes disadvantages the organic nanotubes in terms of their practical use to potent applications. Various types of self-assembled nanostructures from FF dipeptide, however, have demonstrated eminent thermal, chemical, and mechanical functions as compared to any other noncovalently fabricated nanomaterials [1]. Gazit's research group attempted to employ FF-based nanotubes for the reinforcement of the epoxy polymer resins [42]. As a result of the incorporation of the nanotube fillers, the resultant polymer composites were shown to enhance the shear and peel strengths by approximately 70 and 450%, respectively (Fig. 4.22).

4.4.10 Detection of Cancer Cell and Neurotoxin

A graphene electrode was modified with FF-based nanotubes/folic acid (FA) conjugates to detect Hela cells (Fig. 4.23) [43]. Amide coupling reaction between the amine groups of the FF nanotubes and the carbonyl groups of FA enabled the conjugation of FA onto the nanotube surfaces. Folate receptors (FRs) are generally overexpressed on the cancer cell surface. The high affinity of FA to FRs are thus realized on the

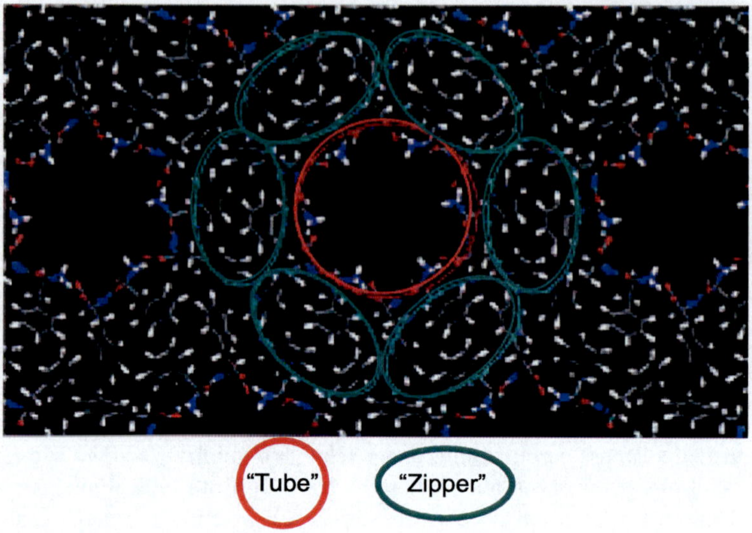

Fig. 4.21 Schematic partition of the FF-based molecular solid into repeating building blocks that composed of a backbone-based-"tube" surrounded by six "zipper" units consisting of two FF peptides each (Reproduced with permission from Ref. [1], © The Royal Society of Chemistry 2014)

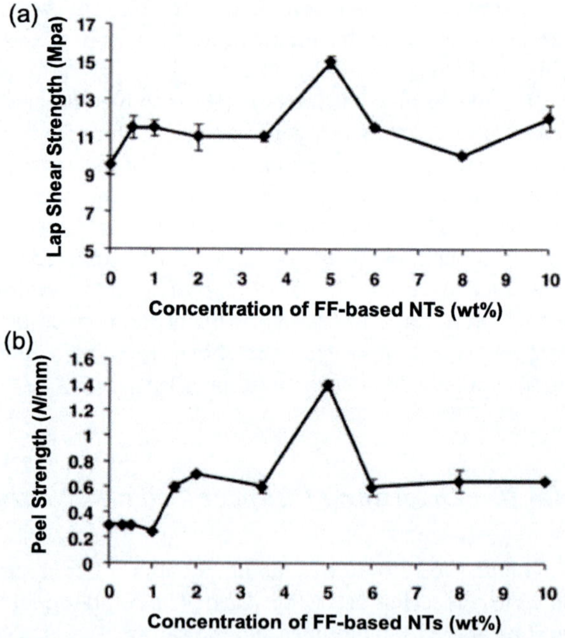

Fig. 4.22 **a** Change in the lap shear of epoxy nanocomposites and **b** the change in T-peel strength with increasing the concentration of FF-based nanotubes. An optimal concentration of 5 wt% nanotubes was obtained for shear and peel systems (Reproduced with permission from Ref. [42], © 2011 WILEY-VCH Verlag GmbH & Co. KGaA, Weinheim)

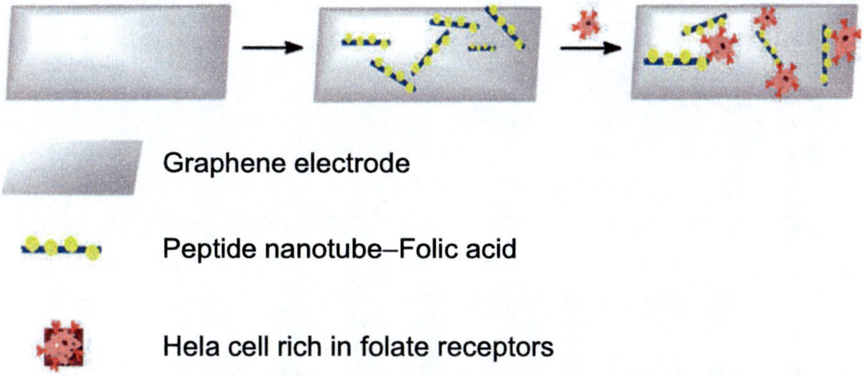

Fig. 4.23 Preparation of a peptide nanotube–FAmodified graphene electrode and the interaction between FA and FRs on HeLa cells (Reproduced with permission from Ref. [43], © The Royal Society of Chemistry 2013)

modified electrode with the nanotube/FA conjugates, accelerating the electrochemical detection of the cancer cell. As a result, the modified graphene electrode was shown to detect Hela cells selectively with a detection limit of 250 HeLa cells per mL. This value of the detection limit is lower than previously reported values measured using other electrochemical sensors with different kinds of interfaces [44–46].

A great deal of organophosphates is strong neurotoxin that can depress the activity of acetylcholinesterase in nerve cells. The sensitive and efficient detection of the organophosphates is thus of great importance for health safety and security. Park and co-workers reported that photoluminescent FF-based nanotubes conjugated with lanthanide complexes [47] rapidly and selectively respond to the existence of an organophosphates. For example, paraoxon **18** (Chart 4.3) is detectable selectively by quick quenching of PL among the organophosphates, i.e., phosmet **19** (Chart 4.3), malathion **20** (Chart 4.3), dichlovos **21** (Chart 4.3), diethylchlorophosphate **22** (Chart 4.3), and paraoxon **18** analyzed (Fig. 4.24) [48]. The FF nanotubes play an critical role to serve as stable host matrices for lanthanide complexes, suppressing the dissociation of the complexes.

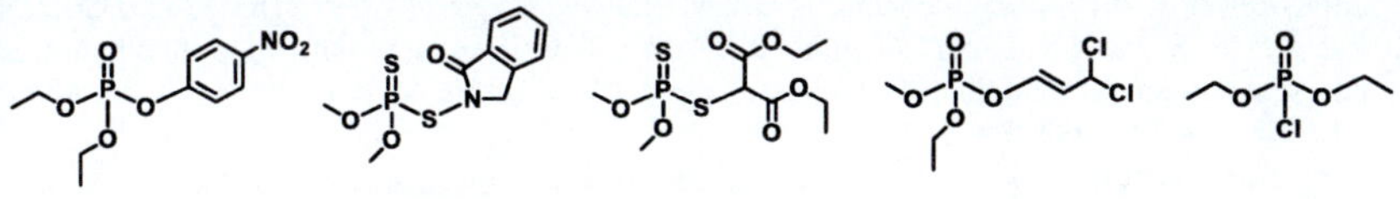

18 19 20 21 22

Chart 4.3 Chemical structures of **18–22**

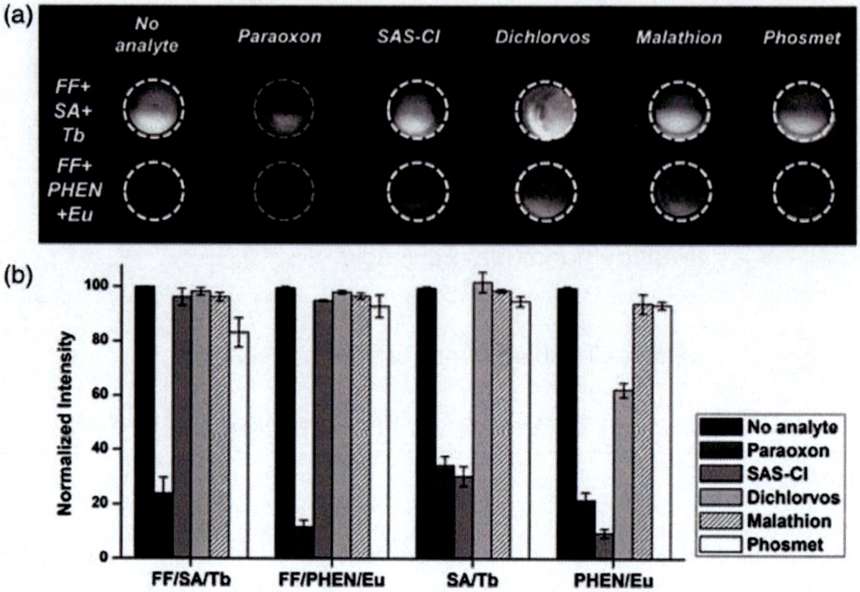

Fig. 4.24 **a** Fluorescence photograph of photoluminescent FF-based nanotubes under UV excitation at 312 nm, demonstrating the selective detection of paraoxon **18** by photoluminescent FF nanotubes. **b** Normalized PL intensity of salicylic acid (SA)/Tb and 1,10-phenanthroline/Eu upon exposure to various organophosphates with or without incorporation into FF nanotubes (Reproduced with permission from Ref. [48], © 2011 WILEY-VCH Verlag GmbH & Co. KGaA, Weinheim)

References

1. Adler-Abramovich L, Gazit E (2014) The physical properties of supramolecular peptide assemblies: from building block association to technological applications. Chem Soc Rev 43:6881–6893. https://doi.org/10.1039/c4cs00164h
2. Reches M, Gazit E (2003) Casting metal nanowires within discrete self-assembled peptide nanotubes. Science 300:625–627. https://doi.org/10.1126/science.1082387
3. Song Y, Challa SR, Medforth CJ, Qiu Y, Watt RK, Pena D, Miller JE, van Swol F, Shelnutt JA (2004) Synthesis of peptide-nanotube platinum-nanoparticle composites. Chem Commun 1044–1045. https://doi.org/10.1039/B402126f
4. Görbitz CH (2006) The structure of nanotubes formed by diphenylalanine, the core recognition motif of Alzheimer's beta-amyloid polypeptide. Chem Commun 2332–2334. https://doi.org/10.1039/b603080g
5. Görbitz CH (2001) Nanotube formation by hydrophobic dipeptides. Chem Eur J 7:5153–5159. https://doi.org/10.1002/1521-3765(20011203)7:23%3c5153:Aid-Chem5153%3e3.0.Co;2-N
6. Yan X, He Q, Wang K, Duan L, Cui Y, Li J (2007) Transition of cationic dipeptide nanotubes into vesicles and oligonucleotide delivery. Angew Chem Int Ed 46:2431–2434. https://doi.org/10.1002/anie.200603387
7. Reches M, Gazit E (2005) Self-assembly of peptide nanotubes and amyloid-like structures by charged-termini-capped diphenylalanine peptide analogues. Isr J Chem 45:363–371. https://doi.org/10.1560/5mc0-V3dx-Ke0b-Yf3j
8. Görbitz CH (2003) Nanotubes from hydrophobic dipeptides: pore size regulation through side chain substitution. New J Chem 27:1789–1793. https://doi.org/10.1039/b305984g

9. Görbitz CH (2007) Microporous organic materials from hydrophobic dipeptides. Chem Eur J 13:1022–1031. https://doi.org/10.1002/chem.200601427
10. Luo JD, Xie ZL, Lam JWY, Cheng L, Chen HY, Qiu CF, Kwok HS, Zhan XW, Liu YQ, Zhu DB, Tang BZ (2001) Aggregation-induced emission of 1-methyl-1,2,3,4,5-pentaphenylsilole. Chem Commun 1740–1741. https://doi.org/10.1039/b105159h
11. Na N, Mu X, Liu Q, Wen J, Wang F, Ouyang J (2013) Self-assembly of diphenylalanine peptides into microtubes with "turn on" fluorescence using an aggregation-induced emission molecule. Chem Commun 49:10076–10078. https://doi.org/10.1039/c3cc45320k
12. Wang M, Xiong S, Wu X, Chu PK (2011) Effects of water molecules on photoluminescence from hierarchical peptide nanotubes and water probing capability. Small 7:2801–2807. https://doi.org/10.1002/smll.201100353
13. Wu X, Xiong S, Wang M, Shen J, Chu PK (2012) Water-sensitive high-frequency molecular vibrations in self-assembled diphenylalanine nanotubes. J Phys Chem C 116:9793–9799. https://doi.org/10.1021/jp212087h
14. Brites CDS, Lima PP, Silva NJO, Millan A, Amaral VS, Palacio F, Carlos LD (2012) Thermometry at the nanoscale. Nanoscale 4:4799–4829. https://doi.org/10.1039/c2nr30663h
15. Jaque D, Vetrone F (2012) Luminescence nanothermometry. Nanoscale 4:4301–4326. https://doi.org/10.1039/c2nr30764b
16. Gan Z, Wu X, Zhang J, Zhu X, Chu PK (2013) In situ thermal imaging and absolute temperature monitoring by luminescent diphenylalanine nanotubes. Biomacromol 14:2112–2116. https://doi.org/10.1021/bm400562c
17. Adler-Abramovich L, Badihi-Mossberg M, Gazit E, Rishpon J (2010) Characterization of peptide-nanostructure-modified electrodes and their application for ultrasensitive environmental monitoring. Small 6:825–831. https://doi.org/10.1002/smll.200902186
18. Yuan J, Chen J, Wu X, Fang K, Niu L (2011) A NADH biosensor based on diphenylalanine peptide/carbon nanotube nanocomposite. J Electroanal Chem 656:120–124. https://doi.org/10.1016/j.jelechem.2010.12.018
19. Adler-Abramovich L, Aronov D, Beker P, Yevnin M, Stempler S, Buzhansky L, Rosenman G, Gazit E (2009) Self-assembled arrays of peptide nanotubes by vapour deposition. Nat Nanotechnol 4:849–854. https://doi.org/10.1038/nnano.2009.298
20. Handelman A, Beker P, Amdursky N, Rosenman G (2012) Physics and engineering of peptide supramolecular nanostructures. PCCP 14:6391–6408. https://doi.org/10.1039/c2cp40157f
21. Lee JS, Ryu J, Park CB (2009) Bio-inspired fabrication of superhydrophobic surfaces through peptide self-assembly. Soft Matter 5:2717–2720. https://doi.org/10.1039/b906783c
22. Gan Z, Wu X, Zhu X, Shen J (2013) Light-induced ferroelectricity in bioinspired self-assembled diphenylalanine nanotubes/microtubes. Angew Chem Int Ed 52:2055–2059. https://doi.org/10.1002/anie.201207992
23. Ghosh S, Mei BZ, Lubkin V, Scheinbeim JI, Newman BA, Kramer P, Bennett G, Feit N (1998) Piezoelectric response of scleral collagen. J Biomed Mater Res 39:453–457. https://doi.org/10.1002/(Sici)1097-4636(19980305)39:3%3c453:Aid-Jbm15%3e3.0.Co;2-B
24. Kholkin A, Amdursky N, Bdikin I, Gazit E, Rosenman G (2010) Strong piezoelectricity in bioinspired peptide nanotubes. ACS Nano 4:610–614. https://doi.org/10.1021/nn901327v
25. Bosne ED, Heredia A, Kopyl S, Karpinsky DV, Pinto AG, Kholkin AL (2013) Piezoelectric resonators based on self-assembled diphenylalanine microtubes. Appl Phys Lett 102: https://doi.org/10.1063/1.4793417
26. Lee JH, Heo K, Schulz-Schonbagen K, Lee JH, Desai MS, Jin HE, Lee SW (2018) Diphenylalanine peptide nanotube energy harvesters. ACS Nano 12:8138–8144. https://doi.org/10.1021/acsnano.8b03118
27. Handelman A, Lavrov S, Kudryavtsev A, Khatchatouriants A, Rosenberg Y, Mishina E, Rosenman G (2013) Nonlinear optical bioinspired peptide nanostructures. Adv Opt Mater 1:875–884. https://doi.org/10.1002/adom.201300282
28. Amdursky N, Beker P, Koren I, Bank-Srour B, Mishina E, Semin S, Rasing T, Rosenberg Y, Barkay Z, Gazit E, Rosenman G (2011) Structural transition in peptide nanotubes. Biomacromol 12:1349–1354. https://doi.org/10.1021/bm200117w

29. Yan X, Li J, Mohwald H (2011) Self-assembly of hexagonal peptide microtubes and their optical waveguiding. Adv Mater 23:2796–2801. https://doi.org/10.1002/adma.201100353
30. Handelman A, Apter B, Turko N, Rosenman G (2016) Linear and nonlinear optical waveguiding in bio-inspired peptide nanotubes. Acta Biomater 30:72–77. https://doi.org/10.1016/j.actbio.2015.11.004
31. Amdursky N, Molotskii M, Aronov D, Adler-Abramovich L, Gazit E, Rosenman G (2009) Blue luminescence based on quantum confinement at peptide nanotubes. Nano Lett 9:3111–3115. https://doi.org/10.1021/nl9008265
32. Amdursky N, Gazit E, Rosenman G (2010) Quantum confinement in self-assembled bioinspired peptide hydrogels. Adv Mater 22:2311–2315. https://doi.org/10.1002/adma.200904034
33. Wasielewski MR (2009) Self-assembly strategies for integrating light harvesting and charge separation in artificial photosynthetic systems. Acc Chem Res 42:1910–1921. https://doi.org/10.1021/ar9001735
34. Kim JH, Lee M, Lee JS, Park CB (2012) Self-assembled light-harvesting peptide nanotubes for mimicking natural photosynthesis. Angew Chem Int Ed 51:517–520. https://doi.org/10.1002/anie.201103244
35. Beker P, Koren I, Amdursky N, Gazit E, Rosenman G (2010) Bioinspired peptide nanotubes as supercapacitor electrodes. J Mater Sci 45:6374–6378. https://doi.org/10.1007/s10853-010-4624-z
36. Beker P, Rosenman G (2010) Bioinspired nanostructural peptide materials for supercapacitor electrodes. J Mater Res 25:1661–1666. https://doi.org/10.1557/Jmr.2010.0213
37. Goldshtein K, Golodnitsky D, Peled E, Adler-Abramovich L, Gazit E, Khatun S, Stallworth P, Greenbaum S (2012) Effect of peptide nanotube filler on structural and ion-transport properties of solid polymer electrolytes. Solid State Ionics 220:39–46. https://doi.org/10.1016/j.ssi.2012.05.028
38. Kol N, Adler-Abramovich L, Barlam D, Shneck RZ, Gazit E, Rousso I (2005) Self-assembled peptide nanotubes are uniquely rigid bioinspired supramolecular structures. Nano Lett 5:1343–1346. https://doi.org/10.1021/nl0505896
39. Niu L, Chen X, Allen S, Tendler SJB (2007) Using the bending beam model to estimate the elasticity of diphenylalanine nanotubes. Langmuir 23:7443–7446. https://doi.org/10.1021/la7010106
40. Knowles TPJ, Buehler MJ (2011) Nanomechanics of functional and pathological amyloid materials. Nat Nanotechnol 6:469–479. https://doi.org/10.1038/Nnano.2011.102
41. Azuri I, Adler-Abramovich L, Gazit E, Hod O, Kronik L (2014) Why are diphenylalanine-based peptide nanostructures so rigid? insights from first principles calculations. J Am Chem Soc 136:963–969. https://doi.org/10.1021/ja408713x
42. Even N, Adler-Abramovich L, Buzhansky L, Dodiuk H, Gazit E (2011) Improvement of the mechanical properties of epoxy by peptide nanotube fillers. Small 7:1007–1011. https://doi.org/10.1002/smll.201001940
43. Castillo JJ, Svendsen WE, Rozlosnik N, Escobar P, Martineza F, Castillo-Leon J (2013) Detection of cancer cells using a peptide nanotube-folic acid modified graphene electrode. Analyst 138:1026–1031. https://doi.org/10.1039/c2an36121c
44. Wang H, Wang T, Ye YX, Zhang YX, Yang PH, Cai HH, Cai JY (2012) Construction of an electrochemical cytosensor based on polyaniline nanofiber/gold nanoparticle interface and application to detection of cancer cells. Chin J Anal Chem 40:184–190. https://doi.org/10.1016/S1872-2040(11)60527-6
45. Zheng TT, Zhang R, Zou L, Zhu JJ (2012) A label-free cytosensor for the enhanced electrochemical detection of cancer cells using polydopamine-coated carbon nanotubes. Analyst 137:1316–1318. https://doi.org/10.1039/c2an16023d
46. Qian Z, Bai HJ, Wang GL, Xu JJ, Chen HY (2010) A photoelectrochemical sensor based on CdS-polyamidoamine nano-composite film for cell capture and detection. Biosens Bioelectron 25:2045–2050. https://doi.org/10.1016/j.bios.2010.01.036
47. Ryu J, Lim SY, Park CB (2009) Photoluminescent peptide nanotubles. Adv Mater 21:1577–1581. https://doi.org/10.1002/adma.200802700
48. Kim JH, Ryu J, Park CB (2011) Selective detection of neurotoxin by photoluminescent peptide nanotubes. Small 7:718–722. https://doi.org/10.1002/smll.201002107

Chapter 5
Peptide-Based Nanotubes

5.1 Amphiphilic Peptide and Peptide Amphiphile

Amphiphilic peptides, i.e., the peptides that possess lipophilic as well as hydrophilic properties, can be divided into two classes in terms of the structural elements (Fig. 5.1). The first class is associated with genuine peptide molecules, in which rationally designed sequences of naturally occurring, hydrophobic and hydrophilic amino acid residues can furnish the amphiphilicity. These amphiphilic peptides are termed "APs" hereafter. Meanwhile, peptide amphiphiles, in which hydrophobic long chains are covalently connected to a hydrophilic peptide sequence, are termed "PAs".

5.2 Linear Amphiphilic Peptide

Excellent reviews by Hamley discuss on the self-assembly behavior of APs and PAs, focusing on the relationship between the secondary structure and resultant self-assembled structures [1–3]. Alzheimer's disease is considered to progress, while being strongly involved with the fibrillization of the amyloid β peptide through aggregation. On these backgrounds, Hamley and co-workers reported the self-assembly of peptide nanotubes from the peptide NH_2–Ala–Ala–Lys–Leu–Val–Phe–Phe–COOH (AAKLVFF) **1** (Chart 5.1) in methanol [4]. The amino acid sequence of **1** corresponds to an extended fragment of the amyloid β peptide Aβ (16–20), i.e., KLVFF that self-assembles into amyloid-like fibrils in water [see Fig. 4.1a in Chap. 4]. Aromatic π–π stacking interaction between the constituent FF segments firstly drives the self-assembly of **1** into bilayered β-sheets. Subsequently, the self-assembled β-sheets result in the formation of helical ribbons, thus, rolling-up entirely into nanotube structures (Fig. 5.2).

Intriguingly, the modified amyloid βpeptide **1** forms fibrillar and nanotubular structures by self-assembly in water and methanol, respectively. SSNMR technique disclosed slight differences in the molecular arrangement of **1** within the fibrils

T. Shimizu, *Smart Soft-Matter Nanotubes*, Nanostructure Science and Technology,
https://doi.org/10.1007/978-981-16-2685-2_5

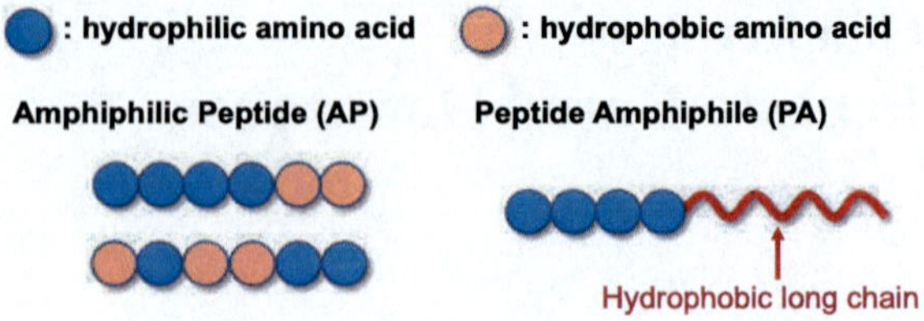

Fig. 5.1 Schematic image of amphiphilic peptide (AP) and peptide amphiphile (PA)

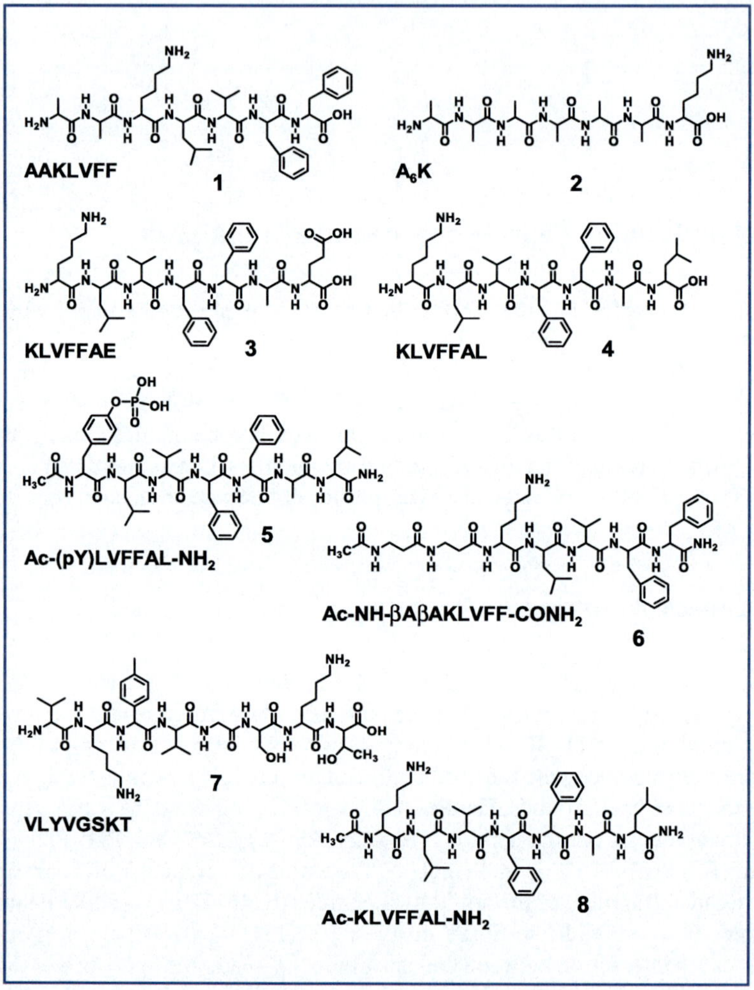

Chart 5.1 Chemical structures of **1–8**

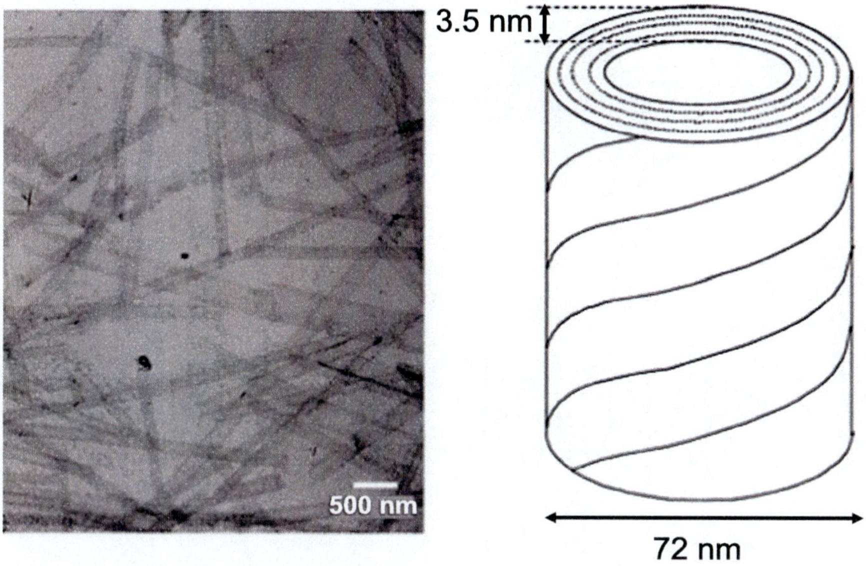

Fig. 5.2 High-resolution TEM image of mature nanotubes self-assembled from the peptide **1** and a proposed model for the nanotube structure that is based on helical wrapping of a bilayer β-sheet into a tube (Reproduced with permission from Ref. [4], © 2008 American Chemical Society)

and nanotubes [5]. The fibrils are stabilized by multiple β-sheet bilayers, whereas the nanotube structures are based on offset monolayers or partially interdigitated bilayers (Fig. 5.3). In contrast to the bilayered feature for the nanotube of **1**, the self-assembled nanotube walls from the heptapeptide A_6K **2** (Chart 5.1) form a single wall structure comprising a monolayer less than 1 nm thick [6]. In situ flow alignment XRD experiment elucidated that antiparallel dimers of **2** assemble to form β-sheet ribbons that eventually roll up helically to organize the tube wall. Subsequently to this measurement, combined technique of FT-IR with SSNMR spectroscopy successfully gave quantitative information on the molecular architecture of the peptide nanotube from **2**, i.e., the tilt angle of the hydrogen-bonding axis and molecular orientation within the nanotube wall [7]. The structural model that satisfies both the experimental NMR spectra and simulation analysis indicated that the NH bond tilts at the angels in the range of 65–70° relative to the long nanotube axis.

The Aβ (16–22) peptide fragment KLVFFAE **3** (Chart 5.1), the nucleation core of amyloid β in Alzheimer's disease, and the leucine substitution Aβ (16–22) derivative KLVFFAL **4** (Chart 5.1) form bilayer membrane-based nanotube structures with diameters of 54 ± 3 nm and 32 ± 5 nm, respectively (Fig. 5.4) [8–10]. Stacking of two amyloid β-sheets through an end-to-end interaction of the peptide termini results in the formation of the bilayer structure 4 nm thick. The peptide bilayers morphologically resemble well-known lipid bilayers [11], but considerably differ from those in structural aspects. The β-sheets arrange the peptide molecules in an

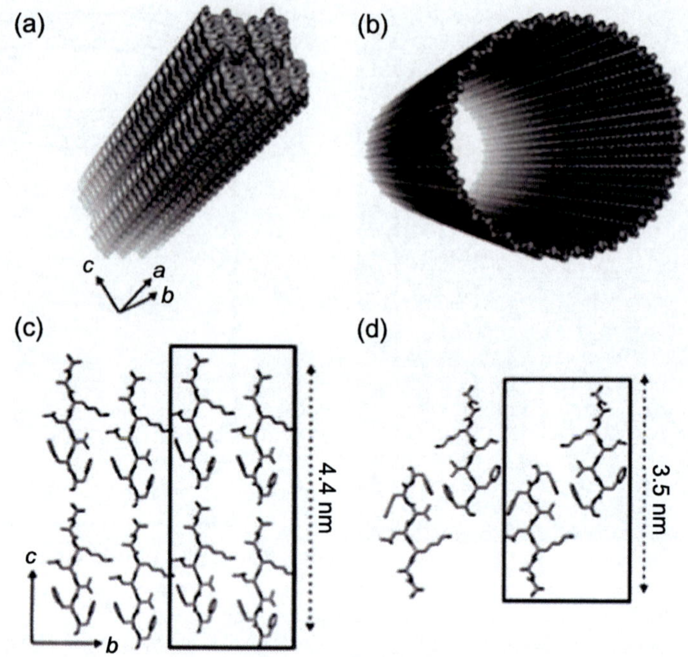

Fig. 5.3 Models of AAKLVFF **1** assemblies. Space-filling model of **a** fibrils and **b** nanotubes. **c** One possible peptide arrangement viewed along the fibril axis *a* in (**a**). **d** A slice through a single peptide layer (perpendicular to the hydrogen-bonding axis) illustrating the interdigitated bilayer arrangement. The rectangles show putative unit cells (Reproduced with permission from Ref. [5], © The Royal Society of Chemistry 2012)

eccentric, antiparallel manner. In this framework, the *N*-terminal residues are situated outside the hydrogen-bonding array.

Hyperphosphorylation of Tau proteins detaches them from the involved microtubules, thus, inducing a protein aggregation. This event subsequently leads to the formation of the neurofibrillary tangles that can be seen in Alzheimer's patients. Aiming at the construction of nanostructures as a surrogate of the neurofibrillary tangle, Lynn and co-workers fabricated a functional peptide nanotube [10]. The design is based on the preparation of the surface covered with a patterned phosphotyrosine residue. The phosphotyrosine peptide fragment Ac-(pY)LVFFAL-NH_2 **5** (Chart 5.1) self-assembles, actually, in aqueous acetonitrile solution (CH_3CN:water = 4:6) to yield nanotubes with diameters of 32 ± 3 nm (Fig. 5.5). The resultant nanotubes display intense binding ability to histone H1, acting as a surrogate of the neuronal tangle.

Several ambient conditions and parameters, e.g., temperature, pH, ionic strength, hydrogen bonding, electrostatic interactions, aromatic interactions, hydrophobic interactions, and molecular chirality, are strongly involved with variety of polymorphic states in the amyloid fibrillation process [3]. A greater understanding of the

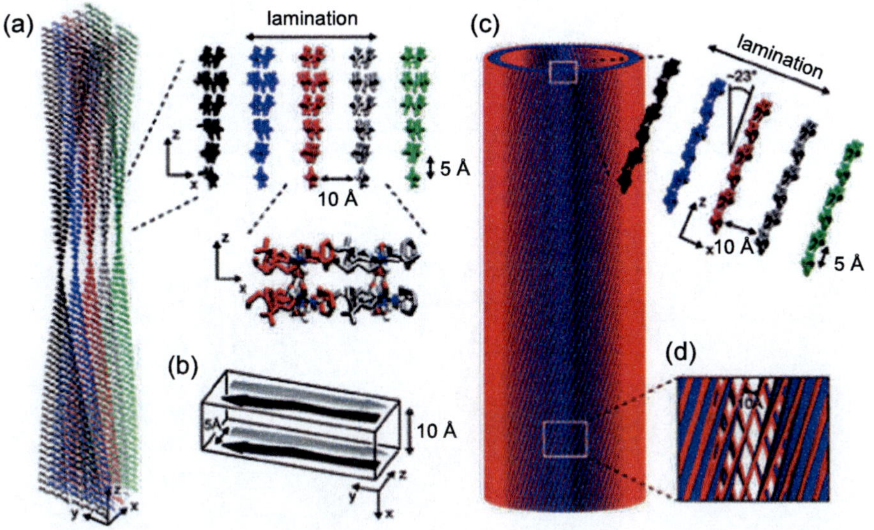

Fig. 5.4 **a** Aβ (16–22) amyloid fiber cross-β architecture, in which five β-sheets are arranged with the peptides perpendicular to the fiber *z*-axis. **b** Minimal unit cell, with β-strands depicted as arrows, consisting of two β-sheets each with two H-bonded β-strands to describe macromolecular assemblies. **c** Aβ (16–22) pH 2 nanotube, in which a total of 130 β-sheets laminate and coil up into ribbons to form nanotubes. **d** Expansion showing β-sheet tilt of the front and back nanotube walls (Reproduced with permission from Ref. [8], © 2008 American Chemical Society)

polymorphic changes should lead to the exploration of novel strategies to block the fibrillation pathway. The heptapeptide derivative from the Aβ (16–20), CH_3CONH-βAβAKLVFF-$CONH_2$ **6** (Chart 5.1) exhibited time-dependent polymorphic transitions of twisted ribbons, helical ribbons, and nanotubular structures over a prolonged period of time (10 h ~ 28 days) (Fig. 5.6) [12].

The octapeptide fragment VLYVGSKT **7** (Chart 5.1) of α-synuclein, a different Parkinson's disease-associated component from Aβ, self-assembles in water to form helical nanotubes at high concentrations (Fig. 5.7) [13]. XRD analyses for the amphiphilic bilayers revealed the diffraction patterns at 4.7–4.8 and 9.8 Å, suggesting the presence of cross-β structures. The proposed model is consistent with the molecular packing, in which the long axis of the molecule **7** is aligned perpendicular to that of the resultant nanotube.

The bolaamphiphile with different headgroups at both ends, when packing in a parallel alignment with a head-to-tail interface, can yield monolayered nanotube architectures having different types of inner and outer functionalities [11]. Unsymmetrical nanotube architectures with opposite charges on the nanotube inner (lysine-rich) and outer surfaces (phosphotyrosine-rich) were constructed through co-assembly of two different types of peptide fragments [14]. The employed peptides are the phosphotyrosine peptide fragment Ac-(pY)LVFFAL-NH_2 **5** and Ac-KLVFFAL-NH_2 **8** (Chart 5.1). Both linear APs **5** and **8** possess different *N*-terminal residues, i.e.,

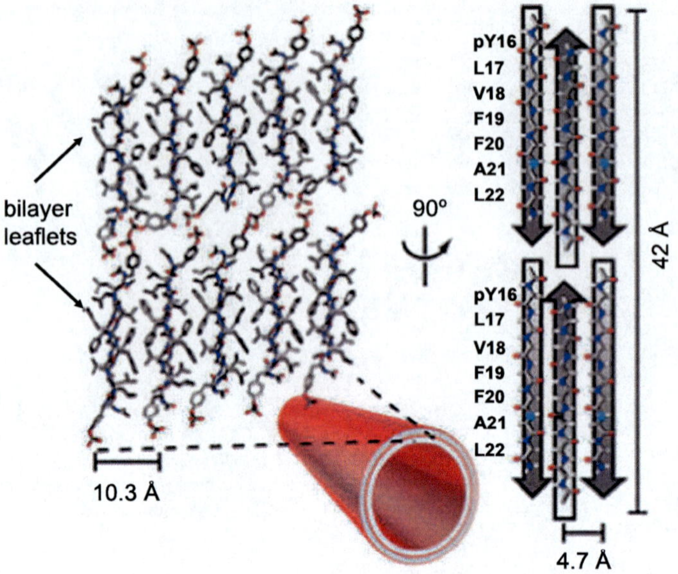

Fig. 5.5 Model for self-assembled bilayer-based nanotubes from Ac-(pY)LVFFAL-NH_2 **5** with each leaflet composed of antiparallel out-of-register β-sheets (Reproduced with permission from Ref. [10], © 2014 American Chemical Society)

negatively charged phosphotyrosine and positively charged lysine, respectively. The co-assembly procedure results in stacking of antiparallel β-sheets that organize the leaflet of the peptide bilayer (Fig. 5.8). Individually self-assembled homo-nanotubes from the peptide **5** or **8** serve as a seed that functions as a template for the growth of oppositely charged nanotube. Unsymmetrical structure control over these cross-β assemblies across the bilayer membranes present a new opportunity of molecular design for highly ordered mesoscale architectures.

5.3 Drug Amphiphile

The term "drug amphiphile (DA)" was coined by Cui et al., meaning an amphiphilic anticancer drug that has the capability of self-delivery without any drug carriers [15]. They focused on camptothecin (CPT) that can be utilized in variety of cancer chemotherapy. Although CPT exhibit outstanding anticancer activities, it has a significant drawback of insolubility in aqueous media [16]. First, the research group joined the anticancer drug CPT to a β-sheet-forming nonapeptide Ac-CGVQIVYKK, derived from the Tau protein, with a reducible disulfanyl butyrate S–S linkage. Thus, they developed the new drug amphiphiles **9a–9c** (Chart 5.2). The produced morphologies of the amphiphiles **9a–9c** through the self-assembly in water markedly depend

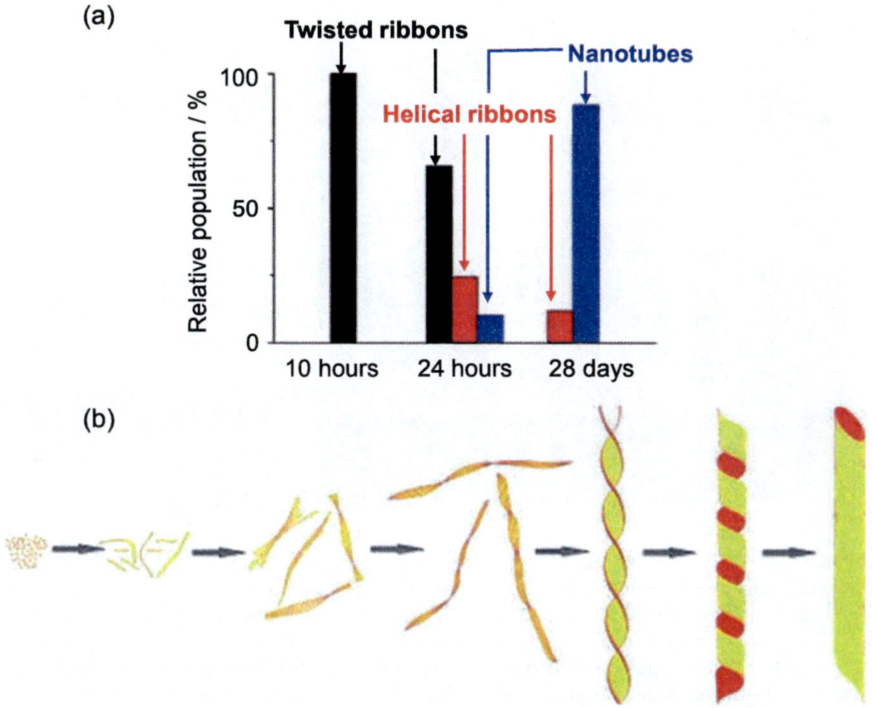

Fig. 5.6 **a** Relative populations of twisted ribbons, helical ribbons, and nanotubes at 10, 24 h, and 28 days of incubation and **b** time-dependent polymorphic changes in the fibrillation of **6** (Reproduced with permission from Ref. [12], © 2011 WILEY–VCH Verlag GmbH & Co. KGaA, Weinheim)

on the number of CPT segments (Fig. 5.9). The resultant drug loading rates of **9a**, **9b**, and **9c** were 23%, 3%, and 38%, respectively. Above all, the drug amphiphile **9c** bearing four CPT segments was shown to self-assemble into nanotube structures with 4 nm wall thickness and 9.5 ± 1 nm outer diameters. Moreover, cancer-relevant reductant, glutathione **10** (Chart 5.2), can cleave the S–S linkage, thus, releasing the bioactive CPT moiety.

5.4 Peptide–Dendron Hybrid

Rationally designed β-sheet assembly of peptide building blocks can lead to fascinating peptide-derived 1D nanomaterials with well-defined morphologies, e.g., nanofibers, nanoribbons, and nanotubes. By utilizing their sophisticated molecular packing as a scaffold, one can position some structural or electrical functionalities

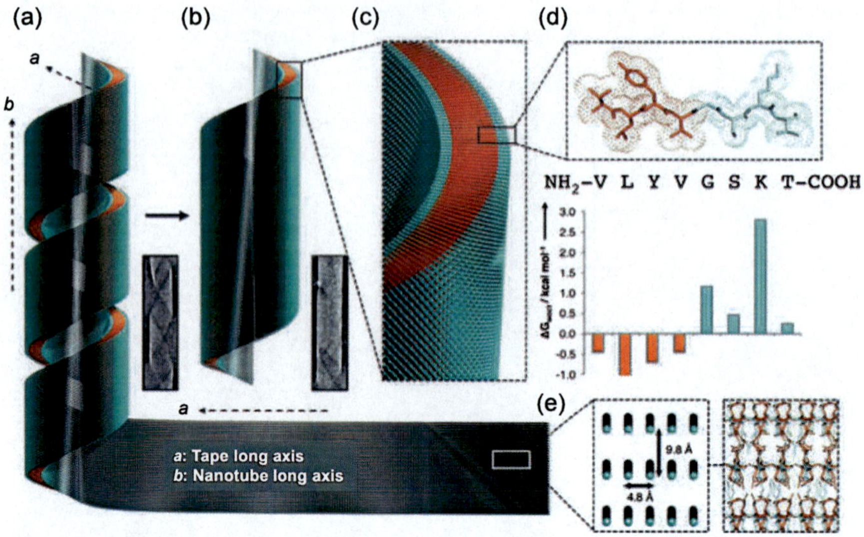

Fig. 5.7 Proposed model of the arrangement of the octapeptide fragment VLYVGSKT **7** of α-synuclein, which self-assembles into a nanotubular morphology. **a** Helical tapes form and then close into **b** mature tubes. **c** The peptides are arranged out-of-plane with respect to the nanotube wall that creates a bilayer stabilized by **d** the amphiphilic nature of the peptide **7**. **e** The orientation of the peptide **7** strands is represented in the context of the tape then leading to the nanotubes (Reproduced with permission from Ref. [13], © 2013 The Authors, Published by WILEY–VCH Verlag GmbH & Co. KGaA, Weinheim)

onto their inner or outer surfaces [17]. Shao and Parquette newly designed dendron-bearing peptides, in which the dendrons are attached to an α-helical and alanine-rich peptide backbone at the i and $i + 6$, and i and $i + 10$ positions. They, thus, demonstrated that in water, hydrophobic interaction between the dendrons of the conjugates induces a notable conformational change from the α-helix to β-sheet structure. Interestingly, the peptide–dendron hybrid **11** (Chart 5.3) containing the sequence such as Ac-AAA^{d}AKAAAAKAAA^{d}AYA-NH$_2$ (A^d: dendron-substituted alanine) experiences a morphological transition from an amyloid-like fibril to uniform nanotubes of 6 nm outer diameters (Fig. 5.10) [18]. These water-insoluble fibril and water-soluble tubular structures interconvert by the modulation of salt concentration or pH value.

5.5 Dilysine Peptide

Based on the designed sequence having alternate polar and hydrophobic amino acid residues, dilysine–NDI conjugates **12a–12c** (Chart 5.4) were synthesized by the functionalization with NDI at the ε-amino group of one lysine residue [19]. Of three dipeptides, the peptide conjugates **12a** and **12c** formed β-sheet assemblies

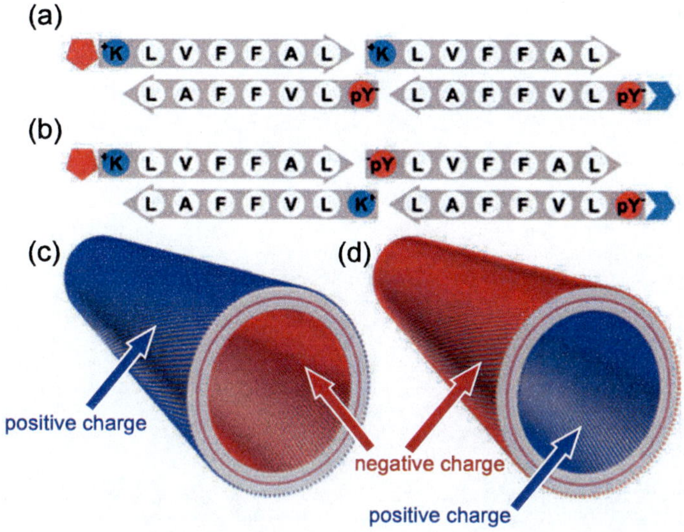

Fig. 5.8 Models for co-assembled peptide nanotubes from the peptides **5** and **8**. Two β-sheet arrangements are possible to produce **a** heterogeneous and **b** homogeneous leaflets. Each arrangement will create segregated charged surfaces (**c**) and (**d**) (Reproduced with permission from Ref. [14], © 2016 American Chemical Society)

featuring helical fibers and twisted ribbons in aqueous solutions. Similar design strategy for β-sheet assembly as described above allowed for the joining of anti-cancer CPT segment at the same position of dilysine residues. The synthesized CPT-conjugated dilysine peptides **13a** and **13b** (Chart 5.4) self-assembled in PBS to yield nanotube structures of 80–120 nm in outer diameters (Fig. 5.11) [20]. The obtained nanotubes exhibited high loading ratios of CPT (approximately 47%); furthermore, the nanoassemblies contributed to the enhanced protection of CPT from hydrolytic deactivation. The NDI-appended proline–lysine dipeptide **14** (Chart 5.4) at the ε-position of lysine also formed catalytic nanotubes through similar β-sheet assembly as those mentioned above [21]. The proline residue plays a critical role in creating hydrophobic microenvironment in the lumen, thus, catalyzing the aldol condensation efficiently in water.

5.6 Amphiphilic Block Peptide

The β-sheets, one of the most well-known secondary structures of proteins, are stabilized by intermolecular hydrogen bonding between two or more polypeptide strands. Molecular self-assemblies, if benefit from the β-sheet structures, can build up relatively robust framework [3]. Conversely, the α-helix structures arisen from intrastrand hydrogen bonding, can form higher ordered helix bundles by utilizing

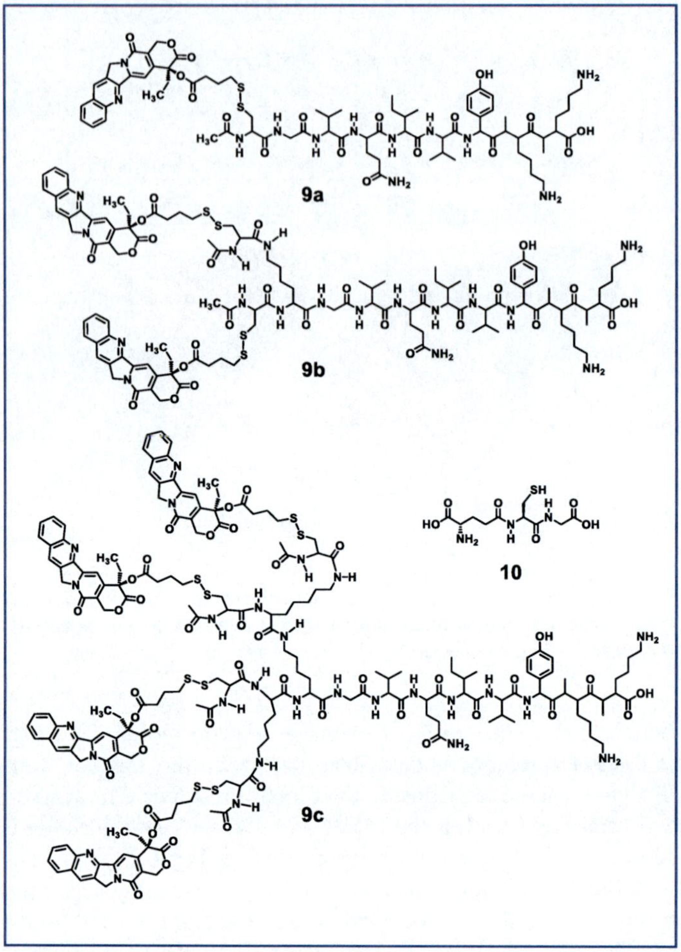

Chart 5.2 Chemical structures of **9–10**

interlocking interface between hydrophobic amino acid residues [22]. Kimura and co-workers have disclosed the self-assembly behavior of amphiphilic polypeptides that can be classified into the APs. Notably, the APs bear helix-forming peptide as a hydrophobic building block. They investigated the nanotube self-assembly of the nonionic peptides **15a** and **15b** (Chart 5.5) that consist of both dodecapeptide (Leu–Aib)$_6$ as a hydrophobic helical segment and poly(*N*-methyl glycine) [so-called poly(Sar)] as a hydrophilic nonionic segment (Fig. 5.12a) [23]. Under the conditions of aqueous media and room temperature, the amphiphilic block peptide poly(Sar)$_{27}$-(L-Leu–Aib)$_6$ **15b** initially formed characteristic curved sheet structures by self-assembly (Fig. 5.12b). Heating the solution of **15b** caused the rolling-up of the sheets, resulting in the self-assembly of nanotube structures of approximately 60 nm

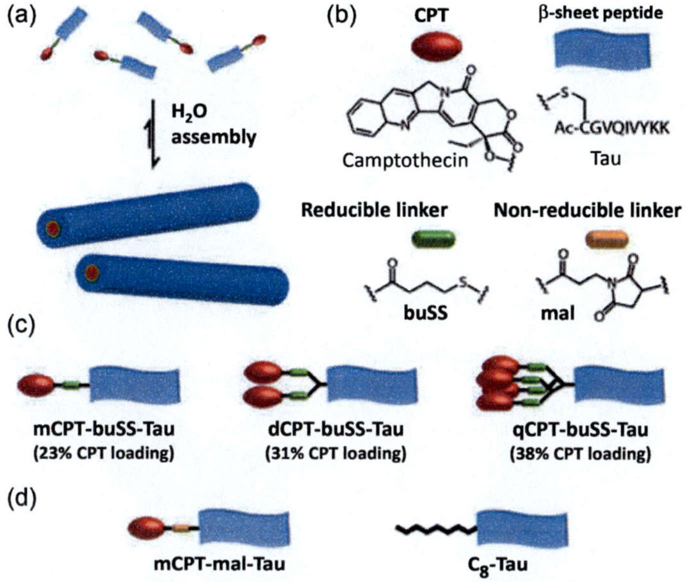

Fig. 5.9 Schematic representation and related molecular structures of DAs and control molecules. **a** Self-assembly of DAs into tubular nanostructures containing the same drug segment as the DA. **b** Three important components that constitute the DAs: the hydrophobic drug (CPT), Tau-β-sheet-forming peptide, biodegradable linker (buSS), and nonreducible linker (mal). **c** Synthesized CPT-appended DAs with quantitative CPT loadings of 23, 31, and 38%. **d** Two synthesized molecules as controls (Reproduced with permission from Ref. [15], © 2013 American Chemical Society)

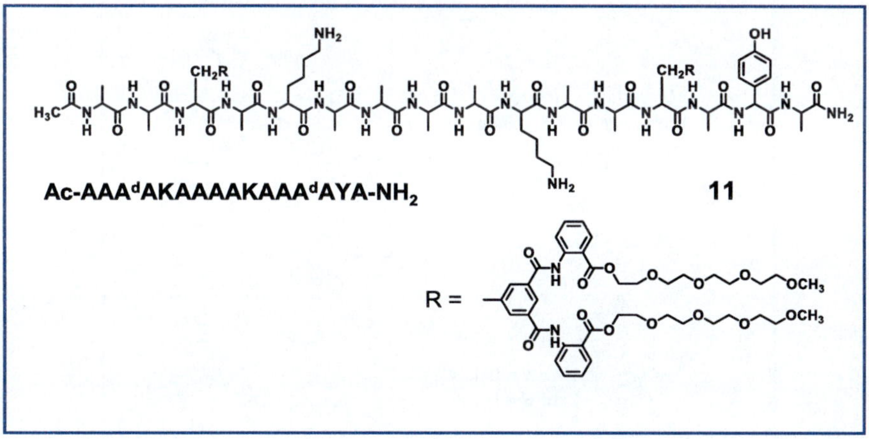

Chart 5.3 Chemical structure of **11**

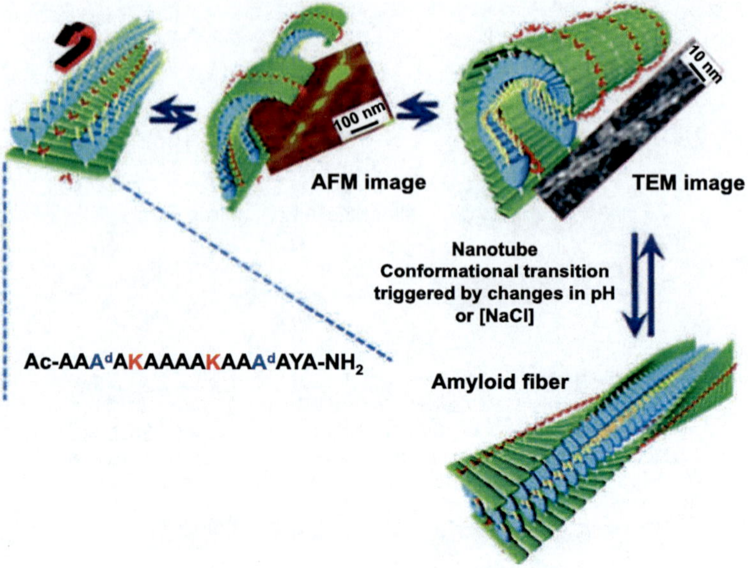

Fig. 5.10 Schematic illustration of nanotube and amyloid fiber self-assembled from the peptide–dendron hybrid **11**. AFM image shows superstructural undulations indicative of rolled-up tape as a nanotube precursor. TEM image shows a high-resolution micrograph of a single nanotube (Reproduced with permission from Ref. [18], © 2009 WILEY–VCH Verlag GmbH & Co. KGaA, Weinheim)

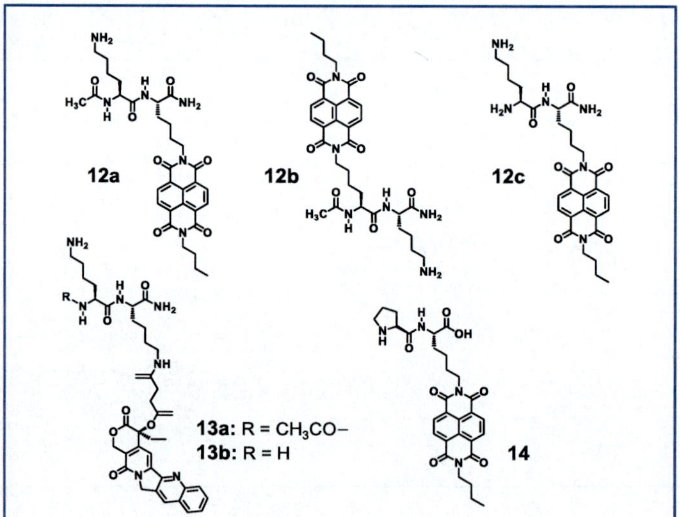

Chart 5.4 Chemical structures of **12–14**

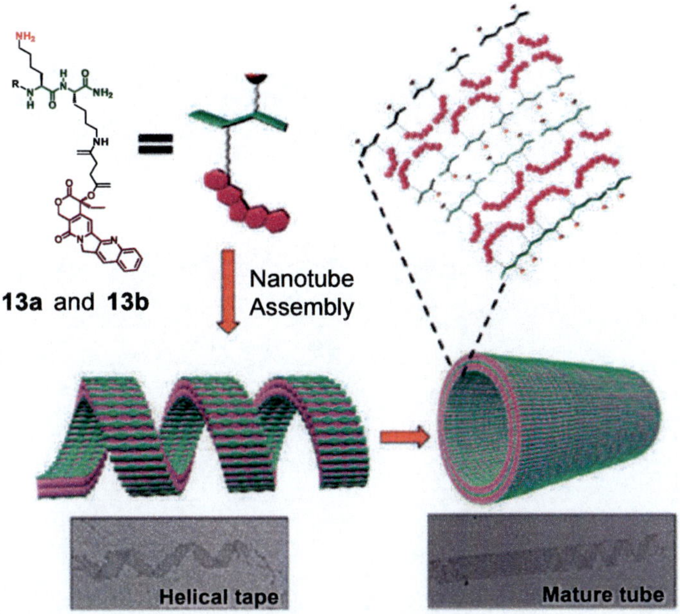

Fig. 5.11 Self-assembly of the CPT–dipeptides **13a** and **13b** into nanotubes (Reproduced with permission from Ref. [20], © 2015 WILEY–VCH Verlag GmbH & Co. KGaA, Weinheim)

in diameter and approximately 200 nm long (Fig. 5.12c). TEM observation revealed twofold, threefold, and fourfold increase in the nanotube length partly. Presumably, intrinsic nanotubes held together through the uniaxial connection between their open ends. In contrast to conventional pathway into tubular architectures via helical ribbons [24], this nanotube formation intermediates curved sheets and subsequent the rolling-up process. The observed uniaxial elongation by the interfacial joining, thus, gives a rare exception.

Triskelion AB-type amphiphilic block peptides $[\text{poly(Sar)}_{26}]_2$–His–(L–Leu–Aib)$_6$ **16** (Chart 5.5) as well as two different types of A_3B-type $[\text{poly(Sar)}_{23}]_3$–His–(L–Leu–Aib)$_6$ **17a** and $[\text{poly(Sar)}_{26}]_3$–His–(L–Leu–Aib)$_6$ **17b** (Chart 5.5) were synthesized on the basis of similar design strategy as that mentioned above [25, 26]. The peptide **16** carries two hydrophilic poly(Sar) chains and hydrophobic helical dodecapeptide as A and B blocks, respectively. Similarly, the peptides **17a** and **17b** possess three poly(Sar) chains and the dodecapeptide. One or two histidine (His) residues were included at the position between the hydrophilic and hydrophobic block, serving as a pH-sensitive segment. As a result of the self-assembly in aqueous solutions under different pH conditions, the triskelion A_2B-type peptide **16** produced nanotubes derived from loosely and tightly rolled-up sheets at pH 5.0 and 7.4, respectively [25]. Strongly depending on the protonation degrees of the two His residues, the A_3B-type peptide **17b** resulted in the formation of twisted ribbon, helical ribbon, and nanotube structures under different pH conditions (pH = 3.0, 5.0, and 7.4,

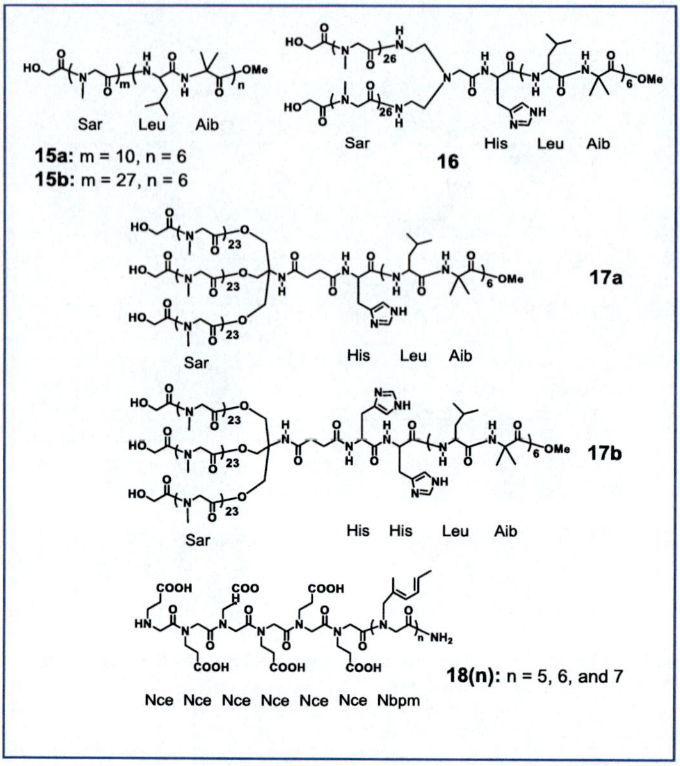

Chart 5.5 Chemical structures of **15–18**

respectively), whereas the peptide **17a** gave helical ribbons at pH 3.0 (Fig. 5.13) [26].

5.7 Amphiphilic Peptoid Oligomer

Chen and co-workers have recently synthesized the amphiphilic peptoid (poly-*N*-substituted glycine) oligomers with a defined sequence of **18(n)** ($n = 5$, 6, and 7) (Chart 5.5), where hydrophilic hexa[*N*-(2-carboxyethyl)glycine] (Nce) is coupled with different hydrophobic *N*-[(4-bromophenyl)methyl]glycine (Nbpm) oligomer (pentamer, hexamer, and heptamer) via amide linkage [27]. They found that the amphiphilic peptoids **18(n)** ($n = 5$, 6, and 7) underwent the self-assembly in a mixture of water and acetonitrile (50/50, v/v, pH = 2.5–3) to yield a new family of highly stiff, dynamic, and designable nanotube structures (Fig. 5.14a). In the self-assembly, spherical amorphous particles first form nanosheet structures that subsequently roll-up and close the sheets into the nanotubes (Fig. 5.14b). Intriguingly, these nanotubes exhibit

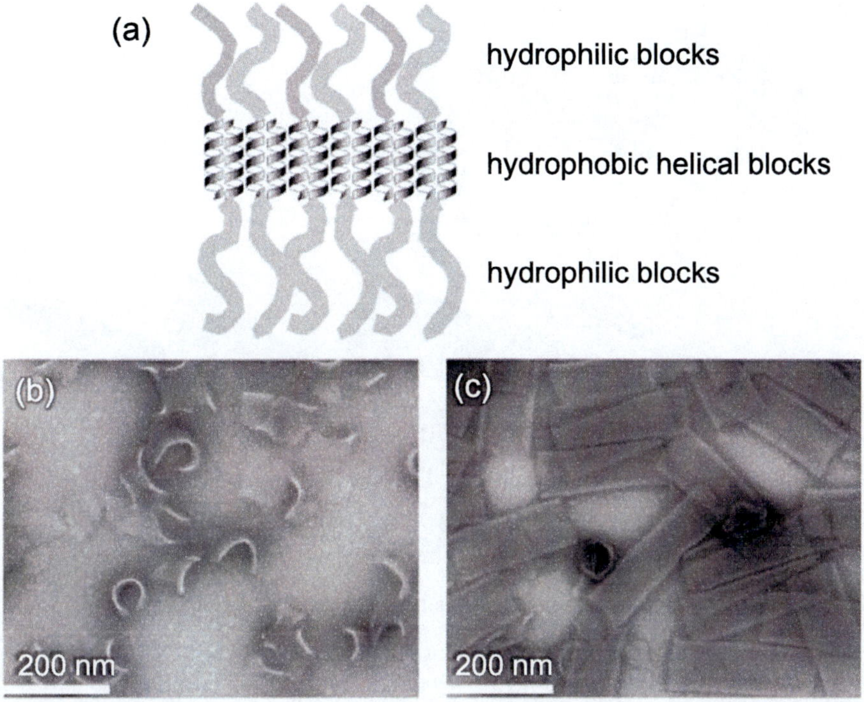

Fig. 5.12 **a** Schematic image of the "polymer brush" that covers the hydrophobic core of the helix bundles in the peptide assembly. TEM images of curved sheet assembly of the polypeptide **15b**, **b** before heating and **c** after heating at 90 °C for 10 min (Reproduced with permission from Ref. [23], © 2008 WILEY–VCH Verlag GmbH & Co. KGaA, Weinheim)

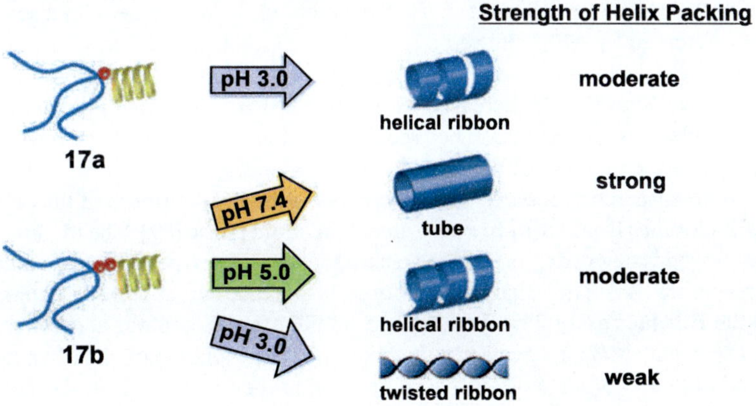

Fig. 5.13 pH-Responsive self-assembly from the A_3B-type peptides **17a** and **17b**. The obtained morphology ranges from a twisted ribbon, a helical ribbon, to a nanotube (Reproduced with permission from Ref. [26], © 2014 American Chemical Society)

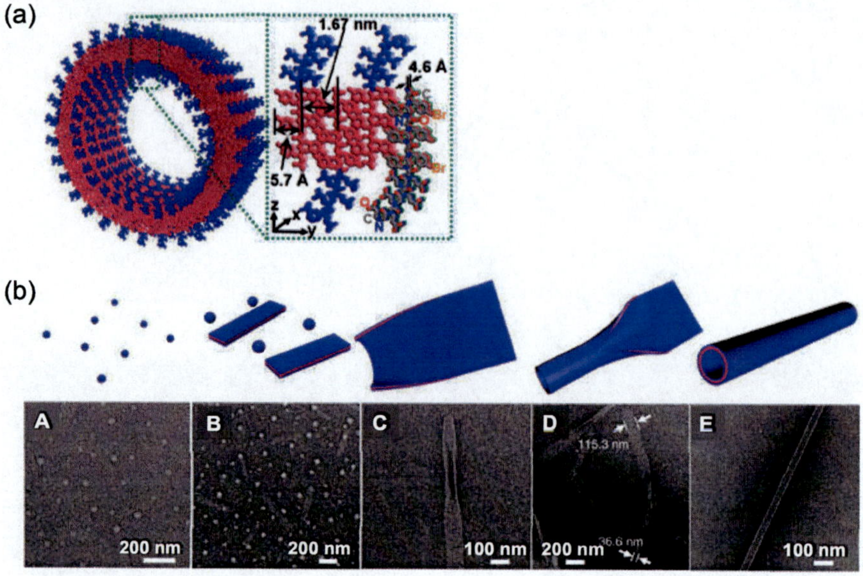

Fig. 5.14 **a** A proposed model that represents the molecular packing of **18(6)** within peptoid nanotubes. **b** Time-dependent TEM images and schematic illustration that show the self-assembly pathway of the peptoid nanotubes from **18(6)**; A Nanospheres, B a mixture of nanospheres and nanoribbons after 0.5 h crystallization, C one nanoribbon with partially rolled up edges after 24 h crystallization, D partially converted nanotubes after 48 h crystallization, and E a peptoid nanotube formed after 72 h crystallization (Reproduced with permission from Ref. [27], © The Author(s) 2018)

a pH-sensitive structural dynamics through the reversible expansion–contraction motion in the range of pH 3.6–8.0 [27]. The height of the nanotube with a sequence of **18(6)** observed by AFM showed a gradual reduction from 24.3 to 13.1 nm when the solution pH decreased from 8.0 to 3.6. However, the nanotube height returned to the original value as the pH increased from 3.6 to 8.0. Responding to the increase in pH values, electrostatic repulsion between deprotonated *N*-(2-carboxyethyl)glycine groups causes the increase of the tube diameter.

The same research group evaluated the nanomechanical stiffness of the obtained peptoid nanotubes from **18(n)** ($n = 5$, 6, and 7) in liquid phase [27]. The methodology they employed is newly explored AFM technique to be called peak force quantitative nanomechanics [28]. They analyzed the force vs distance curves carefully, thus evaluating the E-values as 13.24, 17.65, and 15.88 GPa for the peptoid nanotubes from **18(5)**, **18(6)**, and **18(7)**, respectively. It should be noted that the obtained values are comparable to that of a single FF-based nanotube reported so far [29]. Meanwhile, they have chemically incorporated a Arg–Gly–Asp–Gly (RGDG) peptide into the *N*-RGDG terminus of **18(6)** [27]. Notably, the RGDG-containing peptoid derivative also self-assembles in the nanotube structures that were utilized to decorate a glass slide. The A549 cancer cells thus adhere to the RGDG-coated nanotubes on the glass slide [27]. No significant adhesion of the cancer cells to the slide was observed for the glass slides without the RGDG-coated nanotubes.

5.8 Linear Peptide Amphiphile

5.8.1 Diglycine- and Triglycine-Based Nanotubes

Dipeptide-based nanotubes should be the most hopeful nanomaterials from the viewpoint of low-cost raw materials, less numbers of synthetic steps, relatively high yields, and facile production methodology for them. Among numerous combinations of amino acid sequences in dipeptide, diglycine, and di-phenylalanine (see Chap. 4) residues are worthy of note to play a crucial role in constructing tubular structures in liquid media. Three-dimensional polyglycine II-type hydrogen-bond networks function among oligo-glycine residues to contribute to the stabilization of the tubular architectures from diglycine-based bolaamphiphile **19** (Chart 5.6) (Fig. 5.15). The diglycine sequence is, therefore, one of highly practical molecular building blocks that form the nanotubes with high probability.

We synthesized carboxyl-terminated dipeptide amphiphiles [**20(n)** and **21(n)**] (Chart 5.6) containing diglycine or triglycine residues, respectively, as well as amine-terminated amphiphiles **22(n)** (Chart 5.6) containing diglycine residue and thus investigated the self-assembly behavior in water [31]. The PAs **20(n)** comprising diglycine and fatty acid can yield organic nanotubes stabilized by polyglycine II-type hydrogen-bond networks. The addition of acetic acid to aqueous solutions of **20(n)** (*n* = 12, 14, 15, and 16) allowed for the prompt production of the nanotube

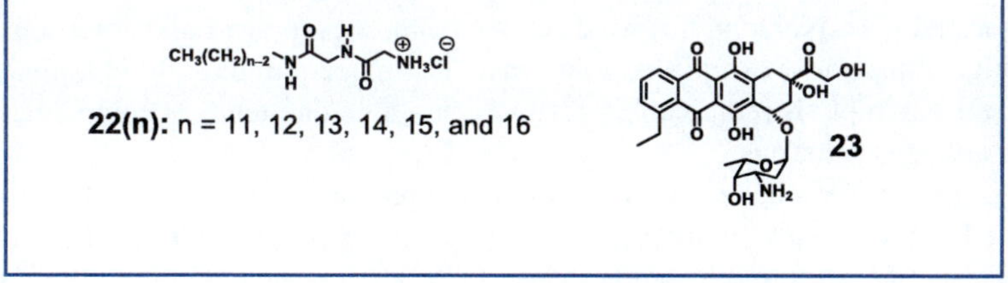

Chart 5.6 Chemical structures of **19–23**

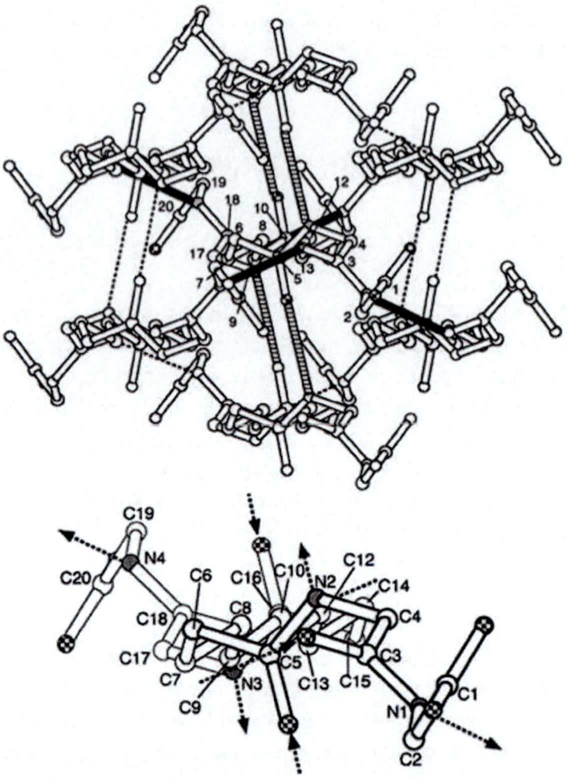

Fig. 5.15 A hexagonal lattice of the polymolecular chains in **19**, which is stabilized by three-dimensional polyglycine II-type hydrogen-bonded networks. The nonplanar (out of the carbonyl plane) hydrogen bonds are shown by bold lines, while the linear hydrogen bonds dashed lines (Reproduced with permission from Ref. [30], © 1997 American Chemical Society)

structures with diameters of 71–82 nm. The triglycine PAs **21(13)** and **21(14)** self-assemble into the nanotube structures with outer diameters of ca. 151 and 129 nm, respectively. In case of this nanotube, the inner and outer surfaces are covered with carboxyl groups [type (a) in Fig. 5.16].

Meanwhile, when a hot alcoholic solution of **20(n)** ($n = 12$ and 14) is rapidly evaporated, dry nanotubes are obtainable as solid residues (Fig. 5.17) [31]. For example, when the *n*-butanolic solutions of **20(n)** are dried, the nanotube with carboxylic acid on their surfaces [type (a)] can be produced. In contrast, the nanotubes with alkyl chains on their surfaces [type (b) in Fig. 5.16] self-assemble in methanol or ethanol to be formed [31]. Notably, the yields of the nanotubes in alcoholic solutions are 5 times higher than those in aqueous solutions. The time required for the completement of the self-assembly is remarkable short, i.e., less than one-tenth as compared to the case in aqueous solutions.

Amino-group-terminated PAs **22(n)** were prepared by condensation of *N*-protected diglycine and a long-chain alkylamine. Neutralization of the aqueous solutions of **22(n)** through vapor diffusion of dilute triethylamine successfully produced nanotube structures, the inner and outer surfaces of which are covered with amino groups [31]. These nanotube materials with three different characteristic

Fig. 5.16 Different types of self-assembled organic nanotubes with identical inner and outer surfaces that consist of solid bilayer membranes. Functional groups of **a** –COOH, –OH or –NH_2 for inner and outer surfaces; **b** –CH_3; **c** –COO^-M^+ (M means metal); **d** 2-naphthyl; and **e** –NH_2 and the nanotube is associated with photoisomerization, undergoing a morphological change

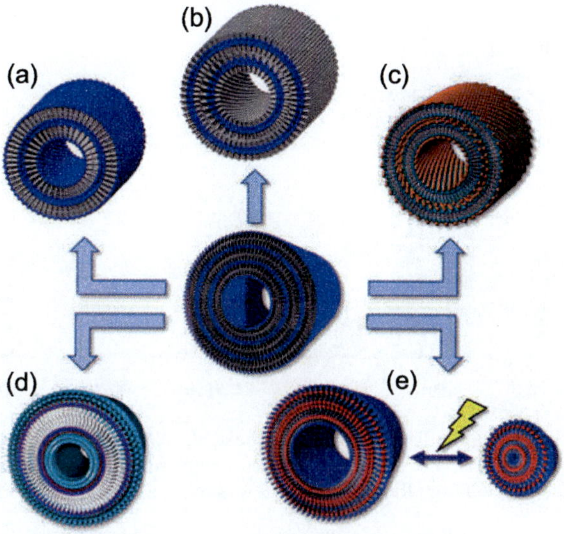

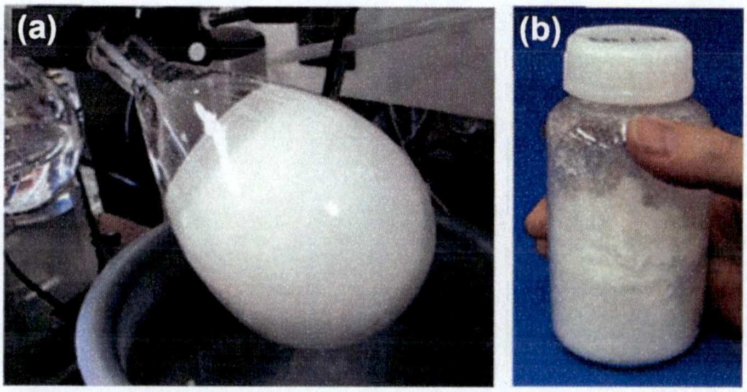

Fig. 5.17 **a** Powdered residue of dried self-assembled organic nanotubes in a 2-l round-bottom flask after evaporation of the methanol solution of **20(12)**. **b** Photograph of the dry nanotube (14 g) in a 250-ml bottle. The obtained nanotubes keep stable in air even after a few years (Reproduced with permission from [32], © 2014 The Society of Polymer Science, Japan)

surfaces covered with carboxyl, methyl, or amino groups are applicable as functional adsorbents [31].

Self-assembled nanotubes from **20(13)** and **22(16)** with carboxyl and amino group on their surfaces, respectively, were first applied for high-axial-ratio nanocarriers toward the anticancer DOX **23** (Chart 5.6) [33]. When compared with the cytotoxicity ($IC_{50} = 8.6$ μg/mL) of common liposomes toward C26 and KB cells, the **20(13)**-based nanotubes exhibited much lower cytotoxicity ($IC_{50} = 200$ and 171.1 μg/mL). This finding indicates that the present nanotubes are safe as a drug carrier. The nanotubes having carboxylic functionalities can accommodate 480 μg of DOX per

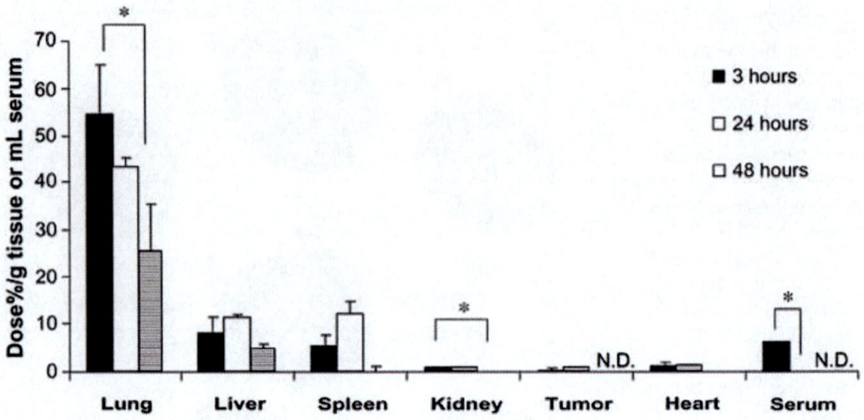

Fig. 5.18 Biodistribution of Gd-chelated, self-assembled organic nanotube from **20(13)** at 3, 24, and 48 h after a single intravenous injection into mice bearing C26 tumors (Reproduced with permission from Ref. [34], © 2013 Maitani et al., publishers and licensee Dove Medical Press Ltd.)

1 mg of the nanotube under acidic aqueous conditions. Meanwhile, the nanotubes with amino groups have little loading ability. The cells can take up the DOX-loaded nanotubes that can release DOX rapidly into the cells. In addition to the aforementioned characteristics, the **20(13)**-based nanotubes accumulate in the lungs of mice, suggesting the potential capability as a lung-targeting drug carrier (Fig. 5.18) [34].

5.8.2 *Metal-Complexed Diglycine-Based Nanotube*

Dense coating of the external or interior surface of organic nanotubes by a metal cation is of particular interest to explore novel catalysts and sensors. Few reports have, however, described direct complexation of a metal cation with the nanotube surfaces [35–37]. Mere mixing of the aqueous solutions of diglycine- **20(n)** or triglycine-based PAs **21(14)** with various metal cations, e.g., Mn^{2+}, Fe^{3+}, Co^{2+}, Ni^{2+} Cu^{2+}, Zn^{2+}, Ag^{+} and La^{3+}, was shown to produces unique metal-complexed nanotubes [type (c) in Fig. 5.16] [38]. The complexation of the PAs with metal cations including Mn^{2+}, Fe^{3+} and Cu^{2+} that favor octahedral coordination tend to construct well-defined nanotube structures under mild conditions. Similarly, Pt^{2+}- [39], Ni^{2+}- [40], Cu^{2+}-, and Au^{3+}-complexed nanotube structures [41] were shown to self-assemble from other amphiphiles. Surprisingly, only mixing an aqueous solution of a metal salt with an alcoholic dispersion of the amphiphile **20(11)** or **20(13)** enabled the fabrication of Mg^{2+}-, Co^{2+}-, Ni^{2+}-, Cu^{2+}-, Zn^{2+}-, In^{3+}-, and Gd^{3+}-complexed nanotubes in yields of 80–240 g within a few hours using a 1 L volume of solvents (Fig. 5.19 and Table 5.1) [42]. Notably, the calcination of the resultant Cu^{2+} or Mn^{2+}-complexed nanotubes resulted in the formation of CuO or Mn_2O_3 nanotubes [38].

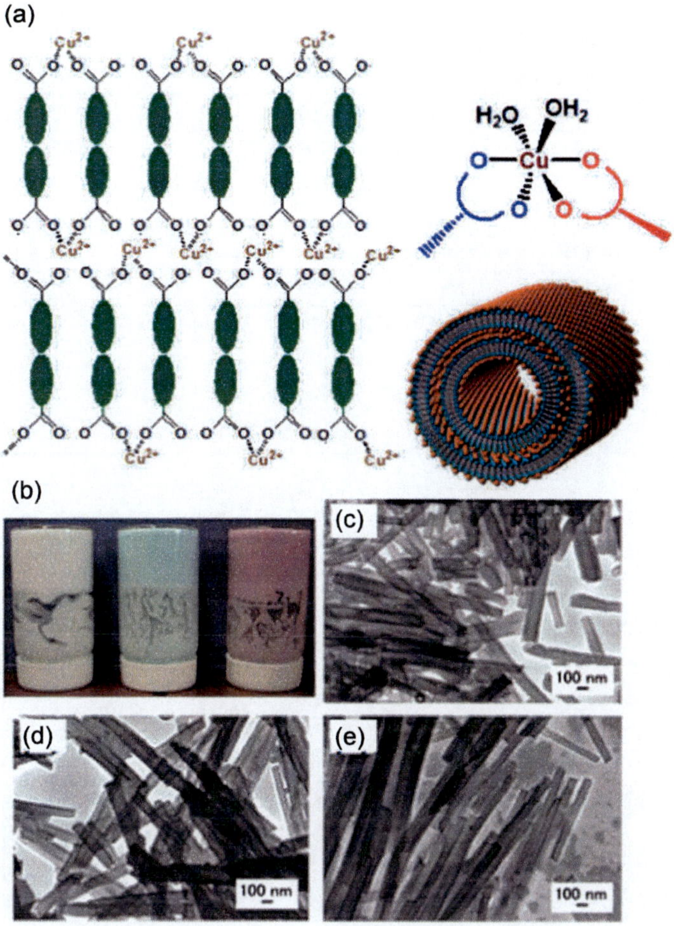

Fig. 5.19 **a** A proposed structure of the Cu^{2+} complex and a schematic image of the molecular packing within the Cu^{2+}-complexed, self-assembled organic nanotube from **20(13)**. **b** Appearance of cream-like dispersions of metal-complexed nanotubes derived from **20(13)** (left, Zn^{2+}-; middle, Cu^{2+}-; and right, Co^{2+}-complex). STEM images of **c** Zn^{2+}-, **d** Cu^{2+}- and **e** Co^{2+}-complexed nanotubes (Reproduced with permission from Ref. [32], © 2014 The Society of Polymer Science, Japan)

5.8.3 *Functional Linear Peptide Amphiphile*

Thermo-reversible morphological transition between nanotubes coexisting with helical ribbons and twisted tapes were observed for the PA, C16-KKFFVLK **24** (Chart 5.7) [43]. The PA comprise both the hydrophilic KKFFVLK peptide segment and hydrophobic hexadecanoyl group, which are linked by amide group. The self-assembly of **24** in water produced monolayer-based nanotubes together with helical

Table 5.1 Mixing conditions of a metal salt with the amphiphile **20(13)** or **20(11)** and the diameters for the obtained metal-complexed organic nanotubes

Metal salt	Amphiphile	Solvent	Alkali	Diameter (nm)[a]
$MgCl_2$	**20(13)**	MeOH: H_2O = 10:1	–	70 ± 10
$CoCl_2$	**20(13)**	EtOH: H_2O = 4:1	–	85 ± 10
$NiCl_2$	**20(11)**	EtOH: H_2O = 1:4	1 eq. NEt_3	40 ± 10
$Cu(OAc)_2$	**20(13)**	EtOH: H_2O = 1:4	1 eq. NEt_3	100 ± 20
$Zn(OAc)_2$	**20(13)**	MeOH: H_2O = 4:1	–	85 ± 15
$InCl_3$	**20(13)**	MeOH: H_2O = 10:1	–	90 ± 25
$GdCl_3$	**20(13)**	MeOH: H_2O = 4:1	1 eq. NEt_3	80 ± 10

Abbreviation NEt_3, triethylamine
[a]Average outer diameters and their s.d.

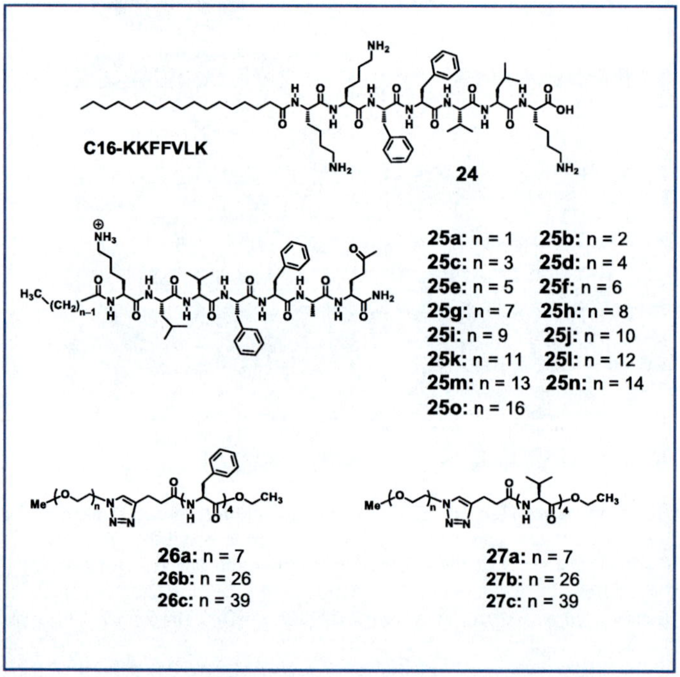

Chart 5.7 Chemical structures of **24–27**

ribbons at room temperature. Heating treatment of the aqueous solution to 55 °C induced the transformation from the nanotubes and helical ribbons into twisted tapes (Fig. 5.20). This unwinding and reversed winding processes of the monolayer sheet alternate upon heating and cooling, respectively.

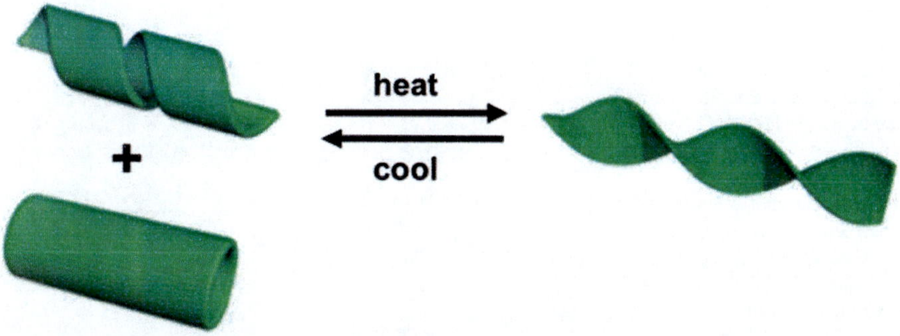

Fig. 5.20 Schematic illustration of the thermo-reversible morphological change for the self-assemblies from C16-KKFFVLK **24** (Reproduced with permission from Ref. [43], © The Royal Society of Chemistry 2013)

Newly designed PAs **25a–25o** (Chart 5.7), in which different fatty acids with from C2 through C16 alkyl chains are linked to the Alzheimer's disease-related, amyloid forming Aβ (16–22) peptide via amide group, were synthesized [44]. The length effect of the long alkyl chains on the self-assembly was then investigated in aqueous solutions. As a result of the self-assembly under acidic conditions, the C2- (**25b**), C3- (**25c**), and C4-appended (**25d**) PA derivatives formed hollow cylindrical tubular structures. Moreover, the C5- through C10-appended amphiphiles **25e–25j** self-assembled into ribbons, whereas the C14- (**25n**) and C16-appended (**25o**) derivatives yielded fiber structures. Remarkably, the spontaneous assembly of the C11- (**25 k**), C12- (**25 l**), and C13-appended (**25 m**) PAs produced uniform nanotube structures of 56 ± 8 nm in outer diameter. Both nanotube architectures from **25 l** and **25a** are almost similar in terms of the 4 nm bilayer structures and protonation state of the lysine residues. Detailed analyses for the **25 l**- and **25a**-derived nanotubes were conducted using electron diffraction and structural simulations (Fig. 5.21). The elucidated tilt angles of the helical ribbon along the nanotube axis differ with each other, showing 11° and 27° for the **25 l** and **25a** nanotubes, respectively. A range of SSNMR technique revealed the structural difference in the confinement of the C12 and C1 alkyl chains between laminated β-sheets [44].

Click coupling reaction [45] between a range of hydrophilic azide-functionalized PEO chain and an hydrophobic alkyne-terminated tetra-phenylalanine (FFFF) or tetra-valine (VVVV) successfully afforded two classes of methoxy-terminated PEO (mPEO)-conjugated PAs, i.e., the FFFF-based **26a–26c** (Chart 5.7) and VVVV-based **27a–27c** (Chart 5.7) [46, 47]. The self-assembly of the FFFF-based PAs **26a** having sufficiently short-chain PEO segment ($mPEO_7$) in aqueous solutions resulted in the formation of nanotubes with inner diameters and wall thickness of approximately 3 nm and 7 nm, respectively (Table 5.2). Meanwhile, the VVVV-based analogues **27a–27c** self-assembled in mixture solvents of THF and water, and after dialysis to form ill-defined plate-like assemblies [46]. The nanotube self-assembly from **26a** is implicated in both the formation of antiparallel β-sheets and π–π stacking interaction

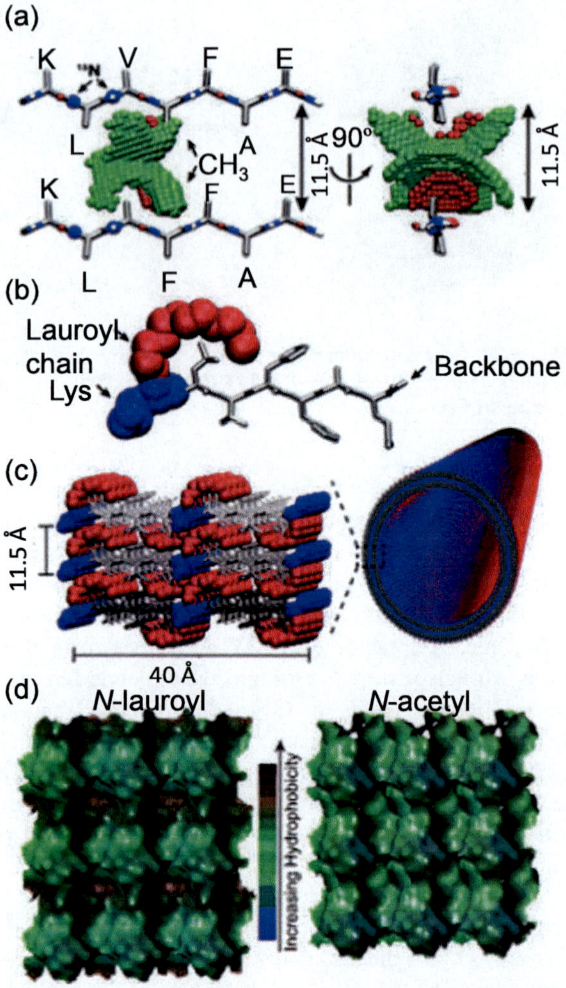

Fig. 5.21 **a** Allowable positions of the terminal $^{13}CH_3$ carbons for the C12-appended PA (**25 l**) on the basis of NMR ^{13}C–^{15}N distance measurements. **b** Model of lauroyl chain packing within the β-strand. **c** Orientation and packing of the C12-appended PA (**25 l**) within nanotube. **d** Comparison of solvent-exposed surface patterns highlighting the lipophilic potential for *N*-lauroyl **25 l** and *N*-acetyl KLVFFAE **25a**-based nanotubes (Reproduced with permission from Ref. [44], © 2012 WILEY–VCH Verlag GmbH & Co. KGaA, Weinheim)

of the FFFF segments. At higher concentrations, the nanotubes entangle with each other to hierarchically organize soft hydrogels [47].

Table 5.2 Summary of the result of varying mPEO chain length and peptide on the dominant peptide interactions and morphology adopted in aqueous solution (Reproduced with permission from Ref. [46], © 2009 American Chemical Society)

Conjugate	f_{EO}[a]	Assembly in peptide (dialysis)[b]	Morphology (dialysis)[c]	Assembly in peptide (rehydration)[b]	Morphology (rehydration)[c]
$mPEO_7$–F_4–OEt	0.32	β-sheet/π-stacking	Nanotubes	β-sheet/π-stacking	Short tapes
$mPEO_{26}$–F_4–OEt	0.62	π-stacking/β-sheet	Fibers	π-stacking/β-sheet	Fibers
$mPEO_{39}$–F_4–OEt	0.71	π-stacking	Wormlike micelles	π-stacking	Wormlike micelles
$mPEO_7$–V_4–OEt	0.35	n.d.[d]	Ill-defined aggregates	β-sheet/random coil	Ill-defined aggregates
$mPEO_{26}$–V_4–OEt	0.65	β-sheet/random coil	Ill-defined aggregates/fibers	β-sheet/random coil	Ill-defined aggregates
$mPEO_{39}$–V_4–OEt	0.73	β-sheet/random coil	Fibers	β-sheet/random coil	Ill-defined aggregates

[a]Volume fraction of ethylene oxide in the conjugate determined from the molecular weights of each component of the conjugate as measured by MALDI-TOF and the densities of the individual building blocks, as measured by helium pycnometry

[b]Determined by circular dichroism spectroscopy; in all cases, the dominant structure is listed first

[c]Determined byTEM

[d]Not determined due to poor signal arising from precipitation of conjugate

References

1. Hamley IW (2011) Self-assembly of amphiphilic peptides. Soft Matter 7:4122–4138. https://doi.org/10.1039/c0sm01218a
2. Hamley IW (2014) Peptide nanotubes. Angew Chem Int Ed 53:6866–6881. https://doi.org/10.1002/anie.201310006
3. Hamley IW (2012) The amyloid beta peptide: a chemist's perspective. Role in alzheimer's and fibrillization. Chem Rev 112:5147–5192. https://doi.org/10.1021/cr3000994
4. Krysmann MJ, Castelletto V, McKendrick JE, Clifton LA, Hamley IW, Harris PJF, King SA (2008) Self-assembly of peptide nanotubes in an organic solvent. Langmuir 24:8158–8162. https://doi.org/10.1021/la800942n
5. Madine J, Davies HA, Shaw C, Hamley IW, Middleton DA (2012) Fibrils and nanotubes assembled from a modified amyloid-beta peptide fragment differ in the packing of the same beta-sheet building blocks. Chem Commun 48:2976–2978. https://doi.org/10.1039/c2cc17118j
6. Castelletto V, Nutt DR, Hamley IW, Bucak S, Cenker C, Olsson U (2010) Structure of single-wall peptide nanotubes: in situ flow aligning X-ray diffraction. Chem Commun 46:6270–6272. https://doi.org/10.1039/c0cc00212g
7. Middleton DA, Madine J, Castelletto V, Hamley IW (2013) Insights into the molecular architecture of a peptide nanotube using FTIR and solid-state NMR spectroscopic measurements on an aligned sample. Angew Chem Int Ed 52:10537–10540. https://doi.org/10.1002/anie.201301960
8. Mehta AK, Lu K, Childers WS, Liang Y, Dublin SN, Dong J, Snyder JP, Pingali SV, Thiyagarajan P, Lynn DG (2008) Facial symmetry in protein self-assembly. J Am Chem Soc 130:9829–9835. https://doi.org/10.1021/ja801511n
9. Childers WS, Mehta AK, Ni R, Taylor JV, Lynn DG (2010) Peptides organized as bilayer membranes. Angew Chem Int Ed 49:4104–4107. https://doi.org/10.1002/anie.201000212

10. Li S, Sidorov AN, Mehta AK, Bisignano AJ, Das D, Childers WS, Schuler E, Jiang ZG, Orlando TM, Berland K, Lynn DG (2014) Neurofibrillar tangle surrogates: histone H1 binding to patterned phosphotyrosine peptide nanotubes. Biochemistry 53:4225–4227. https://doi.org/10.1021/bi500599a
11. Shimizu T, Kameta N, Ding W, Masuda M (2016) Supramolecular self-assembly into biofunctional soft nanotubes: from bilayers to monolayers. Langmuir 32:12242–12264. https://doi.org/10.1021/acs.langmuir.6b01632
12. Adamcik J, Castelletto V, Bolisetty S, Hamley IW, Mezzenga R (2011) Direct observation of time-resolved polymorphic states in the self-assembly of end-capped heptapeptides. Angew Chem Int Ed 50:5495–5498. https://doi.org/10.1002/anie.201100807
13. Morris KL, Zibaee S, Chen L, Goedert M, Sikorski P, Serpell LC (2013) The structure of cross-beta tapes and tubes formed by an octapeptide, alpha S beta 1. Angew Chem Int Ed 52:2279–2283. https://doi.org/10.1002/anie.201207699
14. Li S, Mehta AK, Sidorov AN, Orlando TM, Jiang Z, Anthony NR, Lynn DG (2016) Design of asymmetric peptide bilayer membranes. J Am Chem Soc 138:3579–3586. https://doi.org/10.1021/jacs.6b00977
15. Cheetham AG, Zhang PC, Lin Y, Lock LL, Cui H (2013) Supramolecular nanostructures formed by anticancer drug assembly. J Am Chem Soc 135:2907–2910. https://doi.org/10.1021/Ja3115983
16. Pommier Y (2006) Topoisomerase I inhibitors: camptothecins and beyond. Nat Rev Cancer 6:789–802. https://doi.org/10.1038/nrc1977
17. Ulijn RV, Smith AM (2008) Designing peptide based nanomaterials. Chem Soc Rev 37:664–675. https://doi.org/10.1039/b609047h
18. Shao H, Parquette JR (2009) Controllable peptide-dendron self-assembly: interconversion of nanotubes and fibrillar nanostructures. Angew Chem Int Ed 48:2525–2528. https://doi.org/10.1002/anie.200805010
19. Shao H, Nguyen T, Romano NC, Modarelli DA, Parquette JR (2009) Self-assembly of 1-D n-type nanostructures based on naphthalene diimide-appended dipeptides. J Am Chem Soc 131:16374–16376. https://doi.org/10.1021/ja906377q
20. Kim SH, Kaplan JA, Sun Y, Shieh A, Sun HL, Croce CM, Grinstaff MW, Parquette JR (2015) The self-assembly of anticancer camptothecin-dipeptide nanotubes: a minimalistic and high drug loading approach to increased efficacy. Chem Eur J 21:101–105. https://doi.org/10.1002/chem.201404520
21. Lee KS, Parquette JR (2015) A self-assembled nanotube for the direct aldol reaction in water. Chem Commun 51:15653–15656. https://doi.org/10.1039/c5cc06142c
22. Landschulz WH, Johnson PF, Mcknight SL (1988) The leucine zipper—a hypothetical structure common to a new class of DNA-binding proteins. Science 240:1759–1764. https://doi.org/10.1126/science.3289117
23. Kanzaki T, Horikawa Y, Makino A, Sugiyama J, Kimura S (2008) Nanotube and three-way nanotube formation with nonionic amphiphilic block peptides. Macromol Biosci 8:1026–1033. https://doi.org/10.1002/mabi.200800071
24. Shimizu T (2018) Self-assembly of discrete organic nanotubes. Bull Chem Soc Jpn 91:623–668. https://doi.org/10.1246/bcsj.20170424
25. Uesaka A, Ueda M, Makino A, Imai T, Sugiyama J, Kimura S (2012) Self-assemblies of triskelion A_2B-type amphiphilic polypeptide showing pH-responsive morphology transformation. Langmuir 28:6006–6012. https://doi.org/10.1021/la3004867
26. Uesaka A, Ueda M, Makino A, Imai T, Sugiyama J, Kimura S (2014) Morphology control between twisted ribbon, helical ribbon, and nanotube self-assemblies with his-containing helical peptides in response to pH change. Langmuir 30:1022–1028. https://doi.org/10.1021/la404784e
27. Jin H, Ding YH, Wang M, Song Y, Liao Z, Newcomb CJ, Wu X, Tang XQ, Li Z, Lin Y, Yan F, Jian T, Mu P, Chen CL (2018) Designable and dynamic single-walled stiff nanotubes assembled from sequence-defined peptoids. Nat Commun 9:270. https://doi.org/10.1038/s41467-017-02059-1

28. Morales-Rivas L, Gonzalez-Orive A, Garcia-Mateo C, Hernandez-Creus A, Caballero FG, Vazquez L (2015) Nanomechanical characterization of nanostructured bainitic steel: peak force microscopy and nanoindentation with AFM. Sci Rep 5:17164. https://doi.org/10.1038/srep17164
29. Adler-Abramovich L, Gazit E (2014) The physical properties of supramolecular peptide assemblies: from building block association to technological applications. Chem Soc Rev 43:6881–6893. https://doi.org/10.1039/c4cs00164h
30. Shimizu T, Kogiso M, Masuda M (1997) noncovalent formation of polyglycine II-type structure by hexagonal self-assembly of linear polymolecular chains. J Am Chem Soc 119:6209–6210. https://doi.org/10.1021/ja970844r
31. Kogiso M, Aoyagi M, Asakawa M, Shimizu T (2010) Highly efficient production of various organic nanotubes with different surfaces and their application to an adsorbent. Soft Matter 6:4528–4535. https://doi.org/10.1039/c0sm00077a
32. Shimizu T, Minamikawa H, Kogiso M, Aoyagi M, Kameta N, Ding W, Masuda M (2014) Self-organized nanotube materials and their application in bioengineering. Polym J 46:831–858. https://doi.org/10.1038/Pj.2014.72
33. Wakasugi A, Asakawa M, Kogiso M, Shimizu T, Sato M, Maitani Y (2011) Organic nanotubes for drug loading and cellular delivery. Int J Pharm 413:271–278. https://doi.org/10.1016/j.ijpharm.2011.04.038
34. Maitani Y, Nakamura Y, Kon M, Sanada E, Sumiyoshi K, Fujine N, Asakawa M, Kogiso M, Shimizu T (2013) Higher lung accumulation of intravenously injected organic nanotubes. Int J Nanomedicine 8:315–323. https://doi.org/10.2147/Ijn.S38462
35. Jin Q, Zhang L, Cao H, Wang T, Zhu X, Jiang J, Liu M (2011) Self-assembly of Copper(II) ion-mediated nanotube and its supramolecular chiral catalytic behavior. Langmuir 27:13847–13853. https://doi.org/10.1021/La203110z
36. Liu Y, Wang T, Li Z, Liu M (2013) Copper(II) ion selective and strong acid-tolerable hydrogels formed by an L-histidine ester terminated bolaamphiphile: from single molecular thick nanofibers to single-wall nanotubes. Chem Commun 49:4767–4769. https://doi.org/10.1039/c3cc41786g
37. Jiang J, Meng Y, Zhang L, Liu M (2016) Self-assembled single-walled metal-helical nanotube (M-HN): creation of efficient supramolecular catalysts for asymmetric reaction. J Am Chem Soc 138:15629–15635. https://doi.org/10.1021/jacs.6b08808
38. Kogiso M, Zhou Y, Shimizu T (2007) Instant preparation of self-assembled metal-complexed lipid nanotubes that act as templates to produce metal oxide nanotubes. Adv Mater 19:242–246. https://doi.org/10.1002/adma.200601117
39. Zhang W, Jin WS, Fukushima T, Ishii N, Aida T (2009) Metal-ion-coated graphitic nanotubes: controlled self-assembly of a pyridyl-appended gemini-shaped hexabenzocoronene amphiphile. Angew Chem Int Ed 48:4747–4750. https://doi.org/10.1002/anie.200900756
40. Mukai M, Aoyagi M, Minamikawa H, Asakawa M, Shimizu T, Kogiso M (2011) A simple *N*-Acyl-L-amino acid constructed metal-complexed organic nanotube having an inner diameter below 10 nm. Chem Lett 40:218–220. https://doi.org/10.1246/Cl.2011.218
41. Nishimura T, Matsuo T, Sakurai K (2011) Metal-ion induced transition from multi- to single-bilayer tubes in histidine bearing lipids and formation of monodisperse Au nanoparticles. PCCP 13:15899–15905. https://doi.org/10.1039/C1cp21065c
42. Kogiso M, Aoyagi M, Asakawa M, Shimizu T (2010) Semisolid phase synthesis of metal-complexed organic nanotubes. Chem Lett 39:822–823. https://doi.org/10.1246/Cl.2010.822
43. Hamley IW, Dehsorkhi A, Castelletto V, Furzeland S, Atkins D, Seitsonen J, Ruokolainen J (2013) Reversible helical unwinding transition of a self-assembling peptide amphiphile. Soft Matter 9:9290–9293. https://doi.org/10.1039/c3sm51725j
44. Ni R, Childers WS, Hardcastle KI, Mehta AK, Lynn DG (2012) Remodeling cross-beta nanotube surfaces with peptide/lipid chimeras. Angew Chem Int Ed 51:6635–6638. https://doi.org/10.1002/anie.201201173
45. Thirumurugan P, Matosiuk D, Jozwiak K (2013) Click chemistry for drug development and diverse chemical-biology applications. Chem Rev 113:4905–4979. https://doi.org/10.1021/cr200409f

46. Tzokova N, Fernyhough CM, Butler MF, Armes SP, Ryan AJ, Topham PD, Adams DJ (2009) The effect of PEO length on the self-assembly of Poly(Ethylene Oxide)-tetrapeptide conjugates prepared by "click" chemistry. Langmuir 25:11082–11089. https://doi.org/10.1021/la901413n
47. Tzokova N, Fernyhough CM, Topham PD, Sandon N, Adams DJ, Butler MF, Armes SP, Ryan AJ (2009) Soft hydrogels from nanotubes of Poly(Ethylene Oxide)–tetraphenylalanine conjugates prepared by click chemistry. Langmuir 25:2479–2485. https://doi.org/10.1021/la8035659

Chapter 6
Cyclic Peptide-Based and Cyclic Peptide–Polymer-Based Nanotubes

6.1 Cyclic Peptide

6.1.1 Membrane- or Sheet-Based Structure

Rationally designed cyclic peptides (CPs), if composed of even numbers (e.g., 4, 6, 8, 10, or 12) of alternating D- and L-amino acid residues, could stack one another to appear as nanotube structures through self-assembly (Fig. 6.1) [1, 2]. The critical interaction to stabilize the nanotubes is an adjacent β-sheet-like intermolecular hydrogen bonding between ring-shaped molecules with a flat conformation. Different from this stacking-based formation process in the nanotube self-assembly, Artzner and co-workers demonstrated the spontaneous assembly of nanotubes in water from synthetic CPs [3]. The dimensions of the obtained nanotubes from the therapeutical CP, Lanreotide **1a** (Chart 6.1) are highly homogeneous, featuring the outer diameter of 24.4 nm and wall thickness of approximately 1.8 nm. The amphiphilicity of the CP **1a** triggers the bilayer formation and aromatic amino acid residues segregate from aliphatic ones (Fig. 6.2). The induced β-sheet assembly, thus, drives the production of the tubular morphology. The research group fully disclosed the hierarchical system of the self-assembly, which display three successive pathways and three molecular intermediates (Fig. 6.3) [4]. The dimer formation from an initial monomer of **1a**, followed by the growth of stable open ribbons, results in the rolling-up into the hollow cylindrical structures via helical ribbons.

If the well-defined morphologies and dimensions of self-assembled nanostructures from molecules are freely tunable with high precision, the bottom-up nanotechnology for the creation of nanostructures can add further fascination and real pleasure. Notably, chemical modification of a series of tube-forming CPs **1a–1q** (Chart 6.1) enabled the fine control over the diameters of the obtained nanotubes in the range of 10–36 nm [3]. A structural model for the nanotube wall shows the presence of peptide bilayers, in which two different layers superimposed. Two aromatic amino acid residues, D-Nal^1 and D-Trp^4, are associated with the close contacts between peptides, thus playing an important role in regulating the nanotube diameters. Indeed,

T. Shimizu, *Smart Soft-Matter Nanotubes*, Nanostructure Science and Technology,
https://doi.org/10.1007/978-981-16-2685-2_6

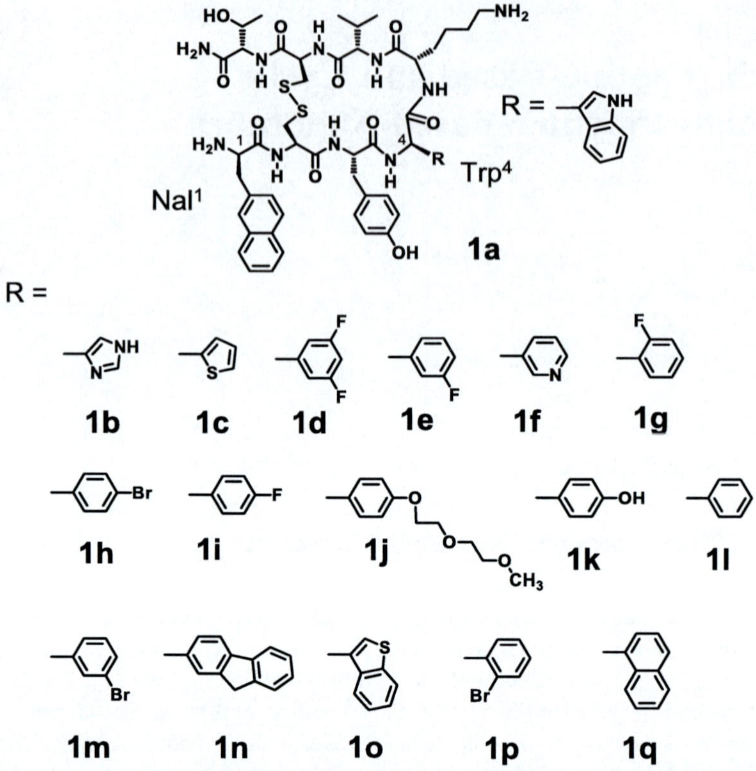

Chart 6.1 Chemical structures of **1a–1q**

the replacement of the Trp residue by 16 different kinds of aromatic side chains permitted precise size control of the nanotube diameters (Fig. 6.4) [6]. The cationic charge of the CP **1a** and the counter ion also sensitively influence on the diameters of the resultant nanotubes. Remarkably, tuning the counter ion size resulted in the fine control of the nanotube diameters ranging from 19 to 26 nm [7]. More interestingly, the replacement of the monovalent counterion with a divalent one induced the formation of double-walled peptide nanotubes without changing average diameter of the intrinsic nanotube [8].

6.1.2 Stacking-Based Structure

A number of outstanding reviews describe stacking-based self-assembly of a great deal of, structurally designed or chemically modified CPs into nanotube architectures in terms of structures, functions, and applications [2, 5, 9]. Even among them, Jolliffe, Perrier, and co-workers summarize a beneficial and comprehensive table

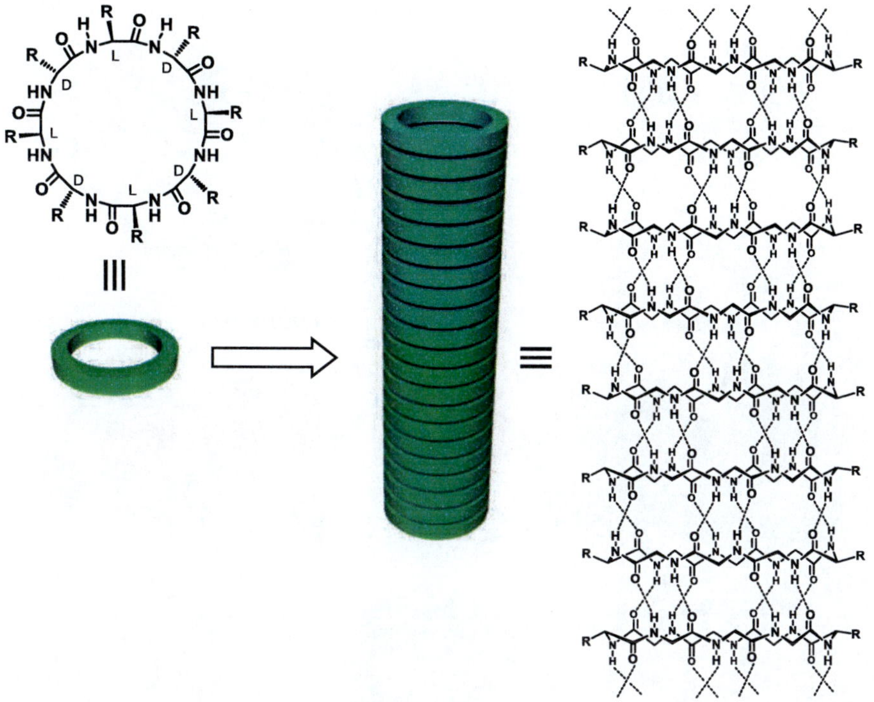

Fig. 6.1 Self-assembly of a cyclic D,L-peptide into nanotube structure. For clarity, most of the side chains of the cyclic peptide are omitted

that features representative classes of CP structures that self-assemble into stacking-based nanotubes [2]. The structures can be divided into four classes, i.e., (a) CPs consisting of alternating D- and L-α-amino acids, (b) CPs having β residues, (c) CPs having alternating α and γ residues, and **d** CPs incorporating ε-amino acids (Fig. 6.5). The stacking-based nanotube structures through β-sheet feature that the amide bonds run parallel to the long tube axis. Simultaneously, the side chains of each amino acid protrude equatorially from the external nanotube walls. Notably, the inner diameter size of the CP nanotubes is tunable, e.g., in the range of 0.7–1.3 nm by increasing the constituent numbers of amino acid residues, e.g., from an octamer to a dodecamer. The employment of β-amino acid residues as β-sheet forming building blocks can permit another type of hydrogen bonding between the ring backbones. The CPs with β residues form the ring stacking in parallel arrangement, whereas those with alternating D- and L-α residues result in antiparallel β-sheet stacking. The incorporation of cyclic γ-amino acid residues makes the ring backbone robust, exploring the possibility for the increase in the nanotube inner diameters. Moreover, the introduction of the cyclic alkane endows the interior of the nanotube channel with hydrophobicity, while maintaining the self-assembly situation.

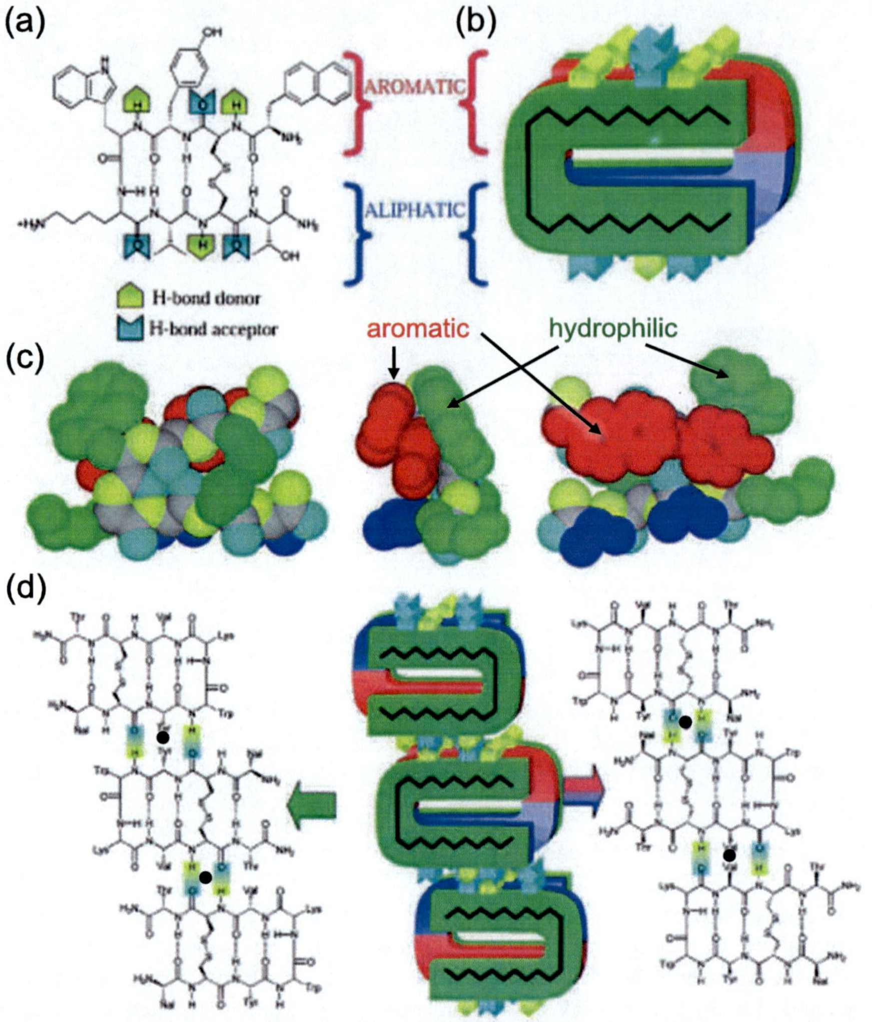

Fig. 6.2 Model for the hierarchical ordering of the cyclic peptide Lanreotide **1a** into nanotubes [3]. **a** A planar β-hairpin conformation of **1a**, which is stabilized by the turn, disulfide bridge, and intermolecular hydrogen bonds, **b** molecular packing in a dimer, and **c** CPK molecular models of the peptide that clearly shows the segregation of aromatic and hydrophilic residues. **d** Stacking of dimers in the nanotube walls, in which the black circles show C2 axes (Reproduced with permission from Ref. [5], © 2014 WILEY–VCH Verlag GmbH & Co. KGaA, Weinheim)

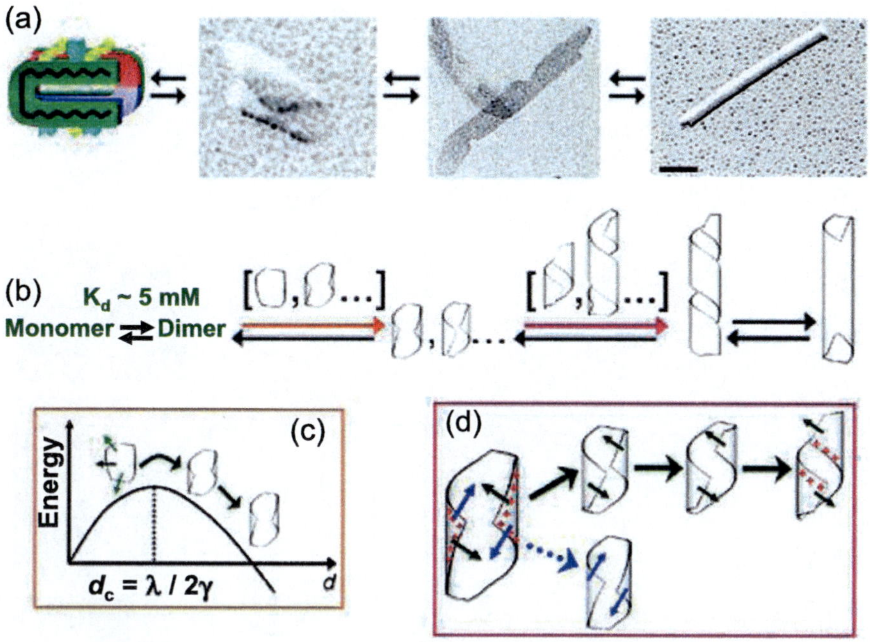

Fig. 6.3 Roadblocks and intermediates during nanotube formation from the cyclic peptide Lanreotide **1a**. **a** Schematic illustration of the dimer and TEM images of the intermediates characterized in this work. **b** Intermediates and sequence of the nanotube self-assembly process. **c** Nucleation and growth assembly and schematic evolution of the energy with the size of the nucleus that shows a typical "bell shape." **d** Helical ribbons growing with electrostatic repulsions (Reproduced with permission from Ref. [4], © 2010 American Chemical Society)

A potent advantage of CP-based nanotubes is that they have many opportunities for diverse chemical modification to the ring backbone. Such functional tailoring enables the replacement or incorporation of additional chemical functionality, reduction of toxicity, and improvement of solubility [2, 5]. Chemical modification onto the exterior walls of the CP nanotubes has drawn attention as latest trends in the self-assembly of CPs. For example, Kimura and co-workers newly prepared the polymerizable cyclic tetra-β-peptide **2** (Chart 6.2) comprising a L-β-homo-lysine residue carrying 10,12-pentacosadiynoic amide chain as well as three β-alanine residues [10]. The CP **2** self-assembles in $CDCl_3$ to form peptide nanotubes based on stacking mechanism. Upon polymerization with UV irradiation, the diacetylene moieties at the side chains successfully polymerize to result in the formation of structurally polydiacetylene-supported peptide nanotubes along the tube axis (Fig. 6.6). The research group also demonstrated the appearance of fibrous assemblies of 5–6 nm in diameters through the self-assembly of CP **3** (Chart 6.2) with a guanine residue. In the presence of K^+, the CP **3** quadruplicates to result in a G-quartet. Further stacking of the G-quartet stack to eventually give the bundle of the quadruple peptide nanotubes (Fig. 6.7) [11].

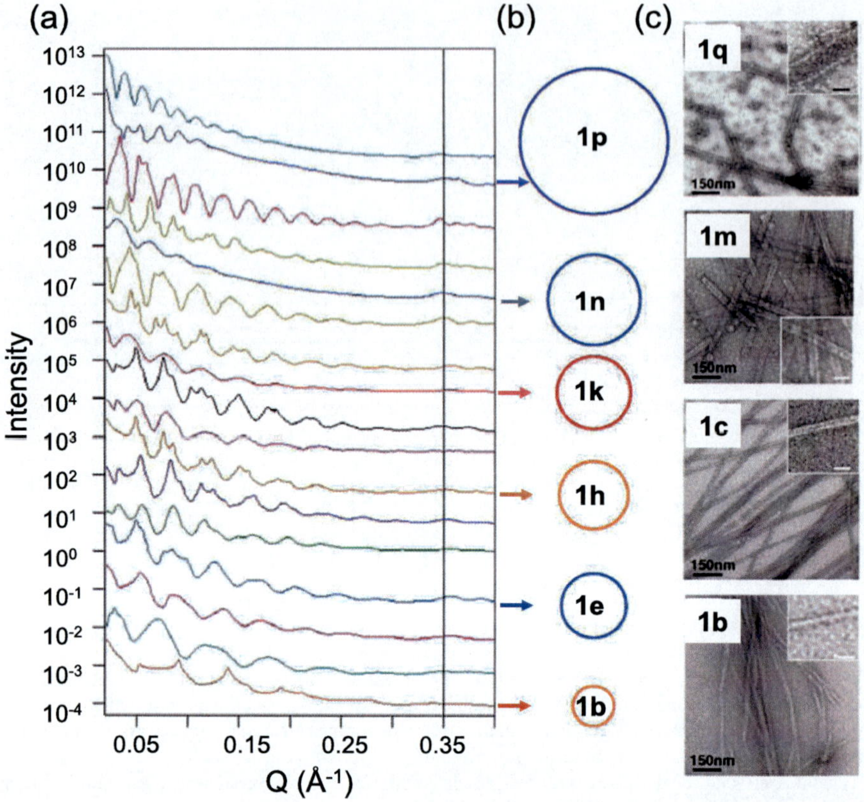

Fig. 6.4 **a** SAXS profiles plotted from the bottom to the top for all 17 derivatives. **b** Illustration of the change in the nanotube diameter (the numbers refer to the derivative numbers). **c** TEM micrographs of peptide nanotubes (the numbers refer to the derivative numbers; scale bar: 150 nm). Inset: enlarged image of each nanotube (scale bar: 50 nm) (Reproduced with permission from Ref. [6])

6.2 Cyclic Peptide (CP)–Polymer

CP–polymer hybrid nanotubes are composed of both a peptide nanotube core and polymeric shell. Pronounced chemical modification can be attained by wrapping the CPs as a core with a synthetic polymer chain as a shell. The resultant CP–polymer nanotubes give a new variety of organic nanotubes having different kinds of inner and outer functionalities. The first approach for the preparation is the chemical modification of a CP side chain with functional initiators that can induce an atom-transfer radical polymerization (ATRP) (Fig. 6.8) [12]. The self-assembled CP nanotubes with modified side chains serve as structurally well-defined templates to prepare so-called polymer-coated CP nanotubes. Meanwhile, copper-catalyzed

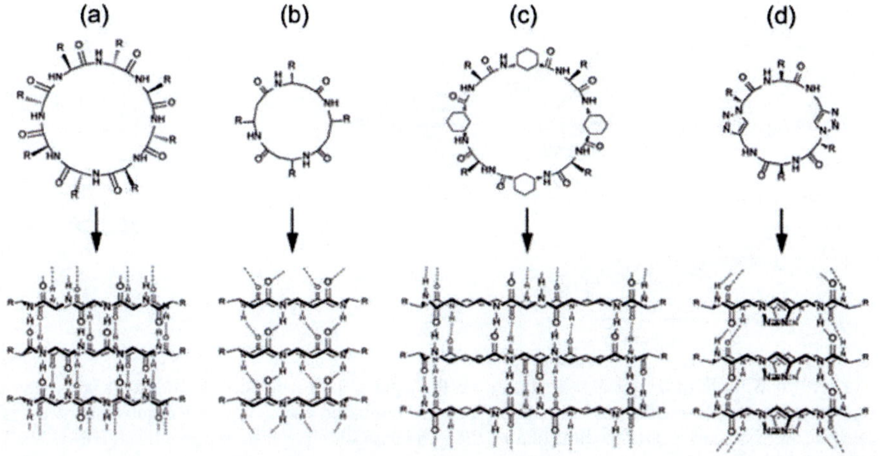

Fig. 6.5 Classification of CPs that self-assemble into nanotubes through β-sheet interactions. Dashed lines in the bottom figures represent hydrogen bonds. For clarity, side chains (R) have been omitted (Reproduced with permission from Ref. [2], © The Royal Society of Chemistry 2012)

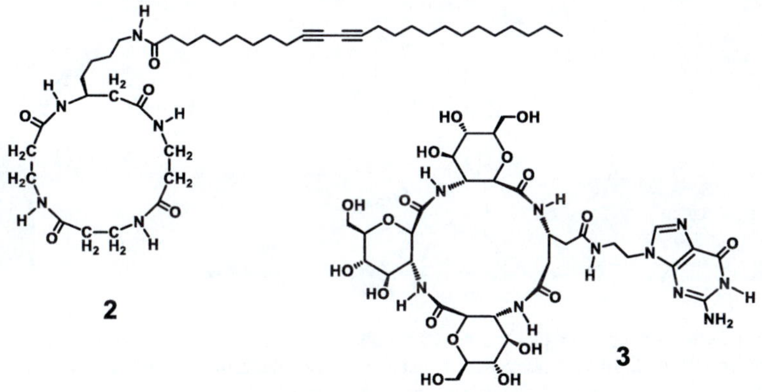

2

3

Chart 6.2 Chemical structures of **2** and **3**

azide–alkyne click reaction enables the coupling of separately synthesized functional polymers with a targeting CP in high efficiency (Fig. 6.9) [13]. This convergent (grafting-to) synthetic methodology is markedly contrast to the divergent (grafting-from) approach mentioned above. This grafting approach of macromolecular chains to the outer wall of the self-assembled CP nanotubes can dramatically improve their solubility into a variety of solvents. For example, the convergent click reaction of hydrophilic polymers onto the CPs **4a–4c** (Chart 6.3) bearing azidolysine side chains resulted in the self-assembly of water-soluble CP–polymer hybrid nanotubes [14]. In this case, the polymer shells were constructed by using alkyne modified poly

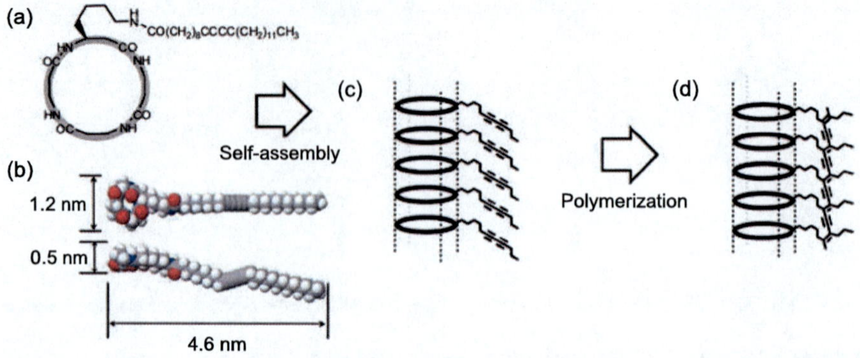

Fig. 6.6 **a** Simplified molecular structure of the cyclic tetra-β-peptide **2** and **b** its geometry-optimized structure. **c** Schematic image of peptide nanotube formation and **d** subsequent polymerization of diacetylene at the side chains (Reproduced with permission from Ref. [10], © 2012 WILEY Periodicals, Inc.)

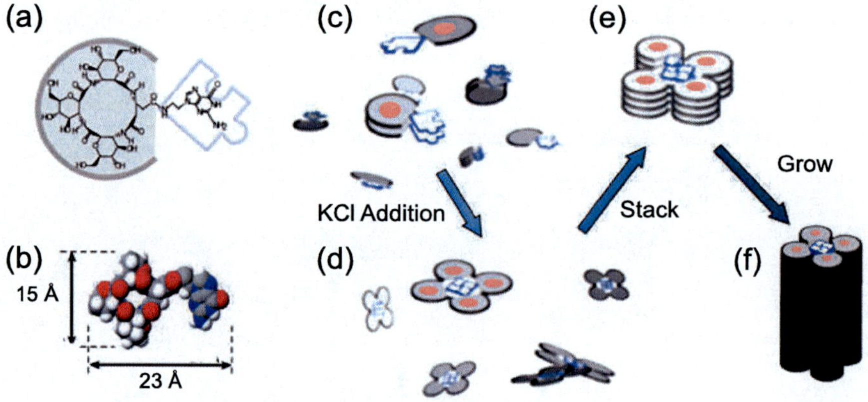

Fig. 6.7 **a** Molecular building block and **b** CPK molecular model of the cyclic peptide **3**. Schematic image of the self-assembling process **c** from dispersed molecules **d** to quadruplication via G-quartet, **e** seed formation for a quadruple peptide nanotube bundle, and **f** nanofiber assembly (Reproduced with permission from Ref. [11], © 2012 WILEY Periodicals, Inc.)

(2-hydroxyethylacrylate) (PHEA) chains **5a–5c** (Chart 6.3) having a degree of polymerization (DP) of 12, 32, and 87 or PtBA **5d** and **5e** (Chart 6.3) having a DP of 32 and 91.

Similar convergent approach led to the synthesis of the CP–polymer conjugate **6** (Chart 6.3) that carries a ruthenium-based antitumor drug (RAPTA-C) at the polymer side chains. A copolymer comprising PHEA and poly(2-chloroethyl methacrylate) (PCLEMA) forms the nanotube shell, which are involved with solubility improvement and conjugation to the CP core, respectively. As a result of

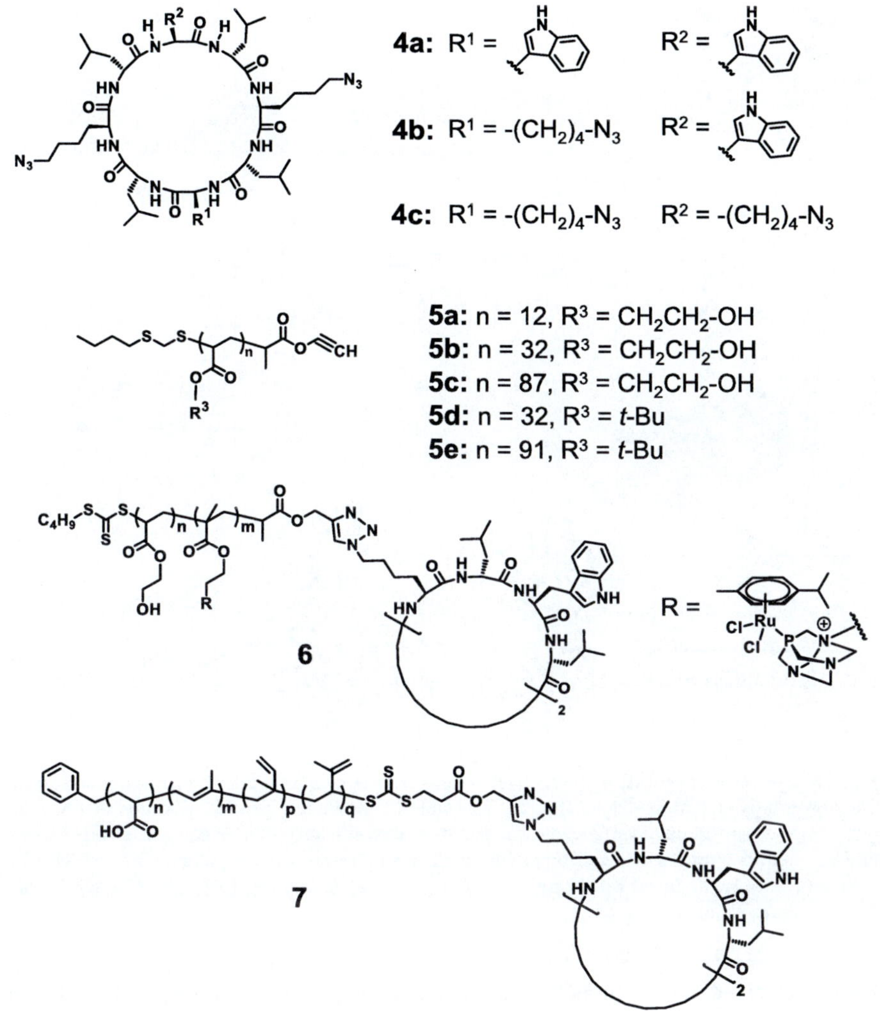

Chart 6.3 Chemical structures of **4**–**7**

successful self-assembly into nanotube structures, the obtained drug-carrying CP–polymer nanotubes displayed tenfold increase in toxicity when compared with that of free drug (Fig. 6.10) [15]. In turn, a novel class of nanotubes with multishell structures also has an outstanding tendency to self-assemble from the polymer chain conjugated cyclic peptide **7** (Chart 6.3). The core CP grafts a block copolymer chain of poly(acrylic acid) and poly(isoprene) that organize hydrophilic outer shell and hydrophobic inner shell structures, respectively [16].

Meanwhile, Jolliffe and co-workers prepared for the first time a novel type of self-assembled CP–polymer nanotubes with "de-mixed" and "mixed" corona (Fig. 6.11).

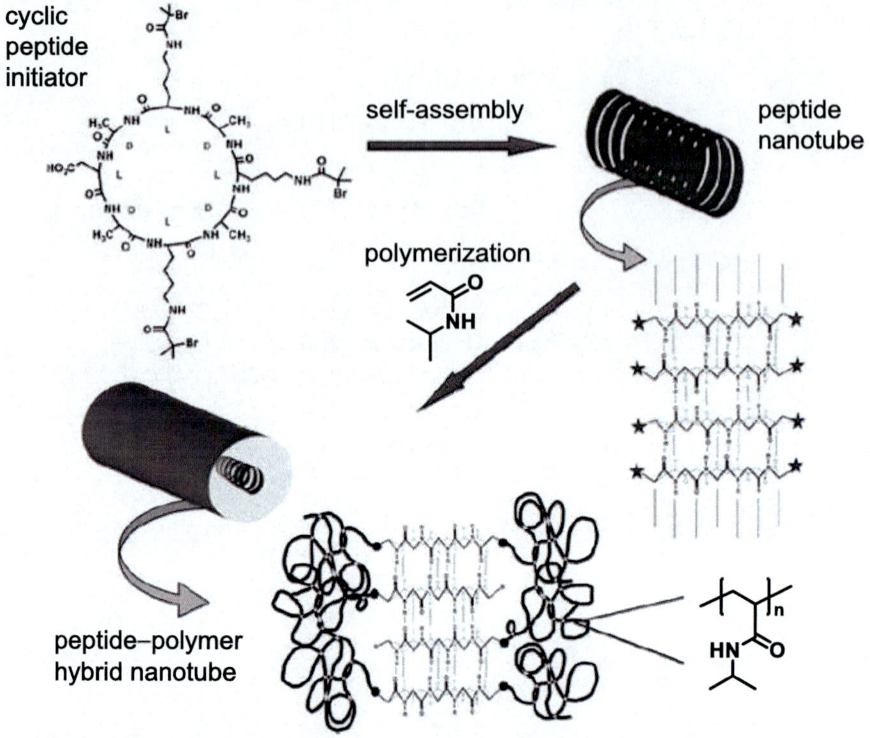

Fig. 6.8 Synthesis of peptide–polymer hybrid nanotubes. A cyclic peptide carrying polymerization initiator groups at distinct side chains self-assembles to produce a peptide nanotube that has the initiator groups on the outer surface. A subsequent surface-initiated polymerization in the presence of NIPAM monomer results in the coating of the cyclic peptide core with a covalently linked PNIPAM polymer shell (Reproduced with permission from Ref. [12], © 2005 WILEY–VCH Verlag GmbH & Co. KGaA, Weinheim)

Both "de-mixed" and "mixed" features are derived from the combination of two independently prepared polymers, i.e., the combination of microphase separating (de-mixing) polymers or that of miscible polymers. Two-step convergent reactions enabled the conjugation of the two polymer chains to the core CP [17]. Specifically, a pair of polystyrene (PS) and poly(*n*-butyl acrylate) (PBA) chains is known to cause microphase separation upon mixing in appropriate solvents. For the creation of Janus CP conjugates **8a–8d** (Chart 6.4), both polymer chains are attached to the core CP through photochemical thiol–ene reaction and succeeding copper-catalyzed azide–alkyne cycloaddition (or vice versa). Additionally, a pair of miscible PS and poly(cyclohexyl acrylate) (PCHA) chains, which transform into a single microphase upon mixing, were also integrated with the CP to afford the conjugates **8e–8 g** (Chart 6.4).

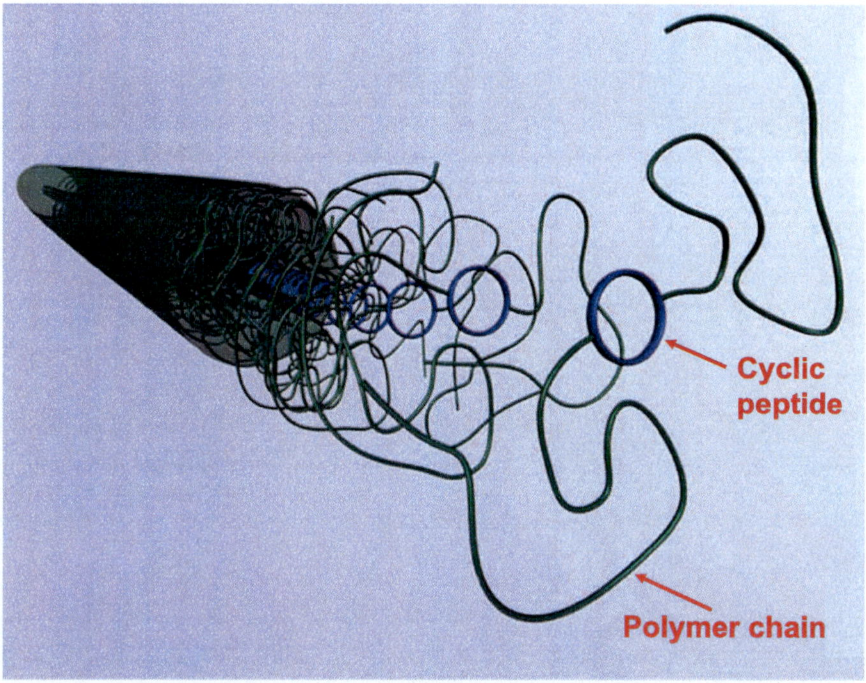

Fig. 6.9 Schematic image of a polymer nanotube that was synthesized in a convergent way. The cyclic peptides were coupled to independently prepared polymer chains (Reproduced with permission from Ref. [13], © The Royal Society of Chemistry 2011)

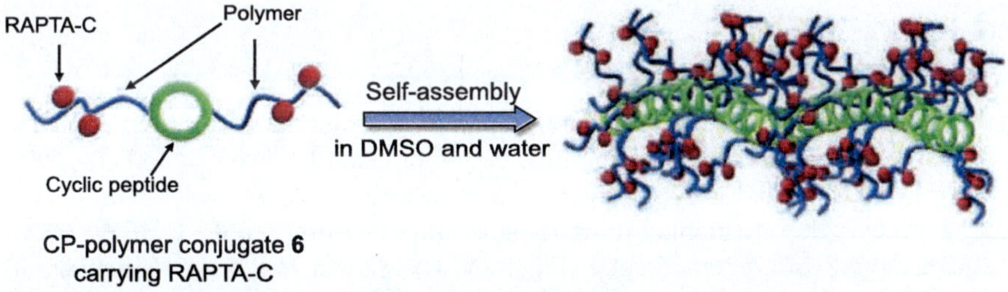

Fig. 6.10 Self-assembly of CP–polymer conjugate **6** carrying RAPTA-C into nanotube structures in DMSO and water (Reproduced with permission from Ref. [15], © 2014 WILEY–VCH Verlag GmbH & Co. KGaA, Weinheim)

Xu and co-workers demonstrated a novel synthetic strategy for the preparation of subnanometer porous thin films that possess high-aspect-ratio organic channels aligned normal to the film surface [18]. A co-assembly technique promoted the nanotube formation within cylindrical microdomains selectively formed by a block copolymer (BCP). Three different CPs **9a**, **9b**, and **9c** (Chart 6.4) with a

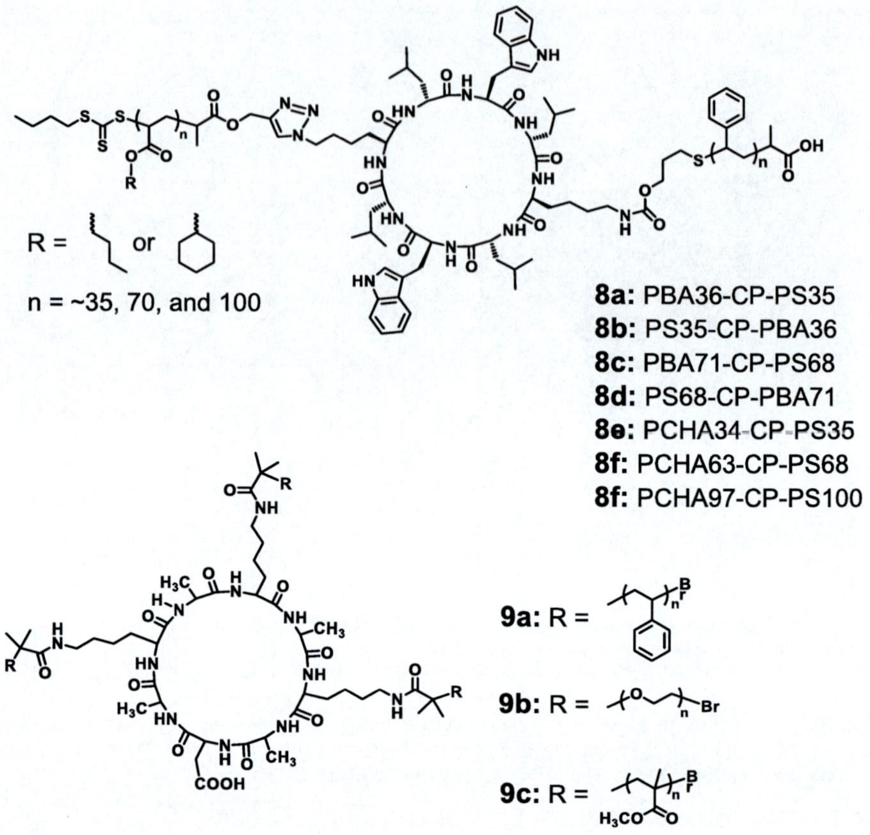

Chart 6.4 Chemical structures of **8–9**

(D-Ala-L-Lys)$_4$ peptide sequence were prepared by tethering three polymer chains via in situ surface-initiated ATRP. Thus, they fabricated PS (Mw = 3 kDa)-, PEO (Mw = 3.3 kDa)-, and Poly(methyl methacrylate) (PMMA, Mw = 5 kDa)-coated CP nanotubes. Self-assembled morphologies of two distinct diblock BCPs, PS-*b*-PMMA, having different molecular weight (Mw = 82 and 74 kDa) are lamellar and cylindrical microdomains, respectively. Mixing the BCP with the CP–polymer conjugates results in the selective segregation of the CP conjugates within a cylindrical microdomain of the BCP. Consequently, self-assembly of the CP conjugates into nanotube structures and 1D elongation that reaches the total film thickness enabled the fabrication of subnanometer porous films (Fig. 6.12).

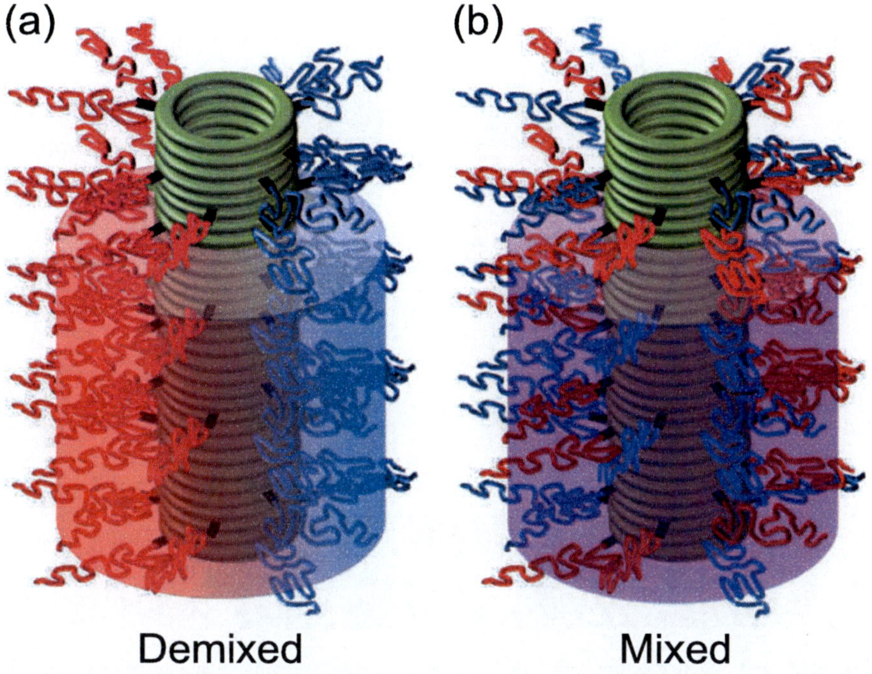

Fig. 6.11 Schematic image of the CP–polymer nanotubes with two corona configurations. **a** Janus assembly with “de-mixed” corona and **b** hybrid assembly with “mixed” corona (Reproduced with permission from Ref. [17], © 2013 Macmillan Publishers Limited)

6.3 Coiled-Coil Nanotube

On the basis of de novo design and recoding strategy, Conticell and co-workers demonstrated the uniaxial elongation of coiled-coil peptide nanotubes through self-assembly (Fig. 6.13) [19]. Employed driving force is electrostatically and structurally complementary recognition between the lower and upper interfaces of the coiled-coil nanotubes. The research group focused on the amino acid sequence of **GCN4-pAA** that forms a heptameric helix bundle in solution and a crystalline state [20]. The de novo designed peptide **7HSAP1** having a recoded sequence of **GCN4-pAA** was shown to assemble into lock-washer-type nanotube structures of 1 nm and 2.8 nm in inner and outer diameters, respectively. Coulombic attraction between the *C*- and *N*-termini of **7HSAP1** as well as additional hydrogen-bond formation promotes the self-assembly pathway into high-axial-ratio nanotubes. In addition to this example of the elongated peptide nanotubes, few examples are known as heterojunction interface-promoted elongation of nanotube length, i.e., heterogeneous nanotube from amino acid bolaamphiphile[21], uniaxial self-assembly of nanotubes from amphiphilic block peptide[22], and co-assembly of two different graphene-like nanotubes [23]. The self-assembled nanotubes from **7HSAP1** impart

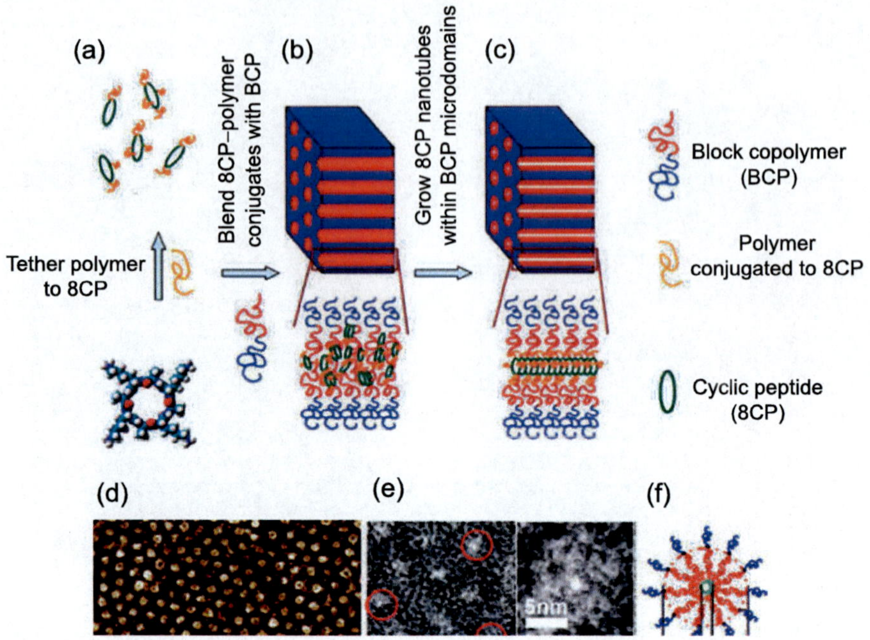

Fig. 6.12 **a–c** Schematic illustration of the process to fabricate subnanometer porous films via directed co-assembly of the cyclic peptide (8CP) and BCP that form cylindrical microdomains. **d** AFM and **e** TEM images of an approximately 32 nm thin, nanoporous polymer films. **f** Schematic image of the CP–polymer nanotube interacting with the block copolymers (Reproduced with permission from Ref. [18], © 2011 American Chemical Society)

them with surfaces of different inner and outer functionalities. The resultant peptide nanotubes can, therefore, entrap the fluorophore PRODAN **10** (Chart 6.5) that exhibit solvatochromism. Optimization of the peptide sequence through mutagenesis should expand the biological application range of selective encapsulation of small guest molecules.

6.4 Applications

6.4.1 Template and Scaffold

Aiming at the fabrication of tough organic pores with submicron diameters, Potnuru and Madhavan utilized a CP–polymer conjugate as a polymeric template [24]. The cyclic octapeptide **11** (Chart 6.5) with alternating polymerizable norbornene-modified L-serine and D-alanine residues self-assembled in solution to give bundles

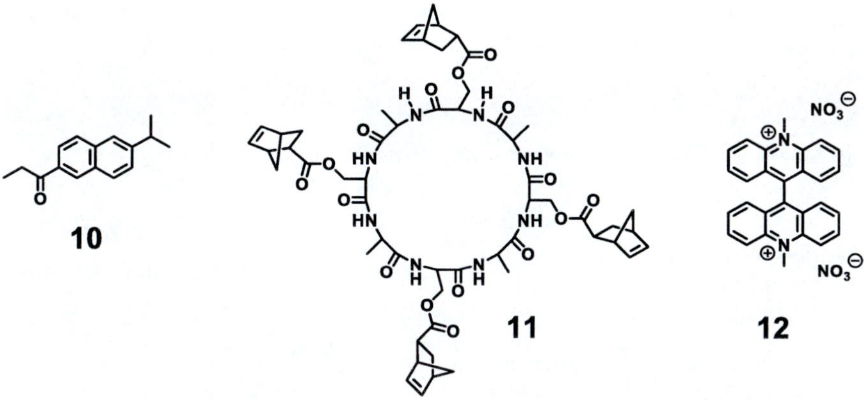

Chart 6.5 Chemical structures of **10–12**

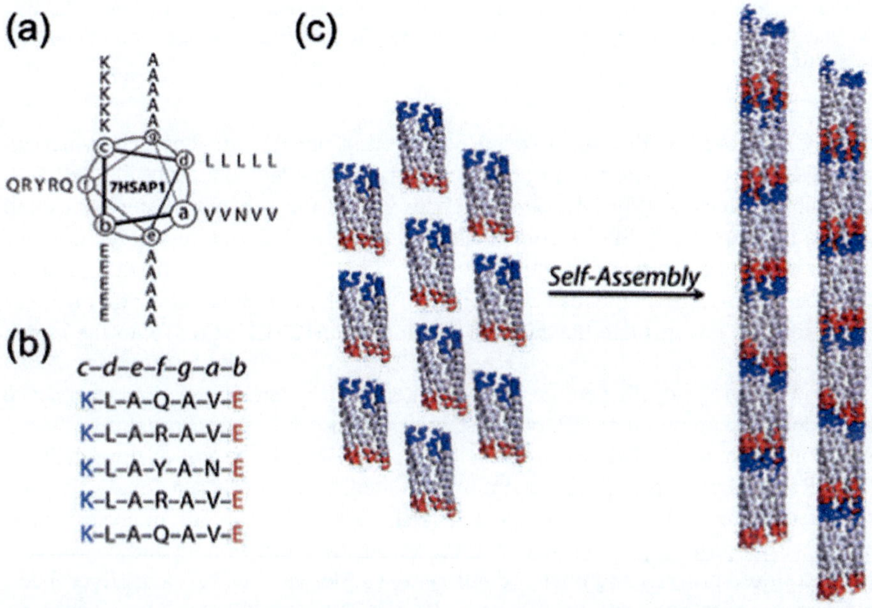

Fig. 6.13 **a** Helical wheel and **b** linear depiction of the peptide sequence of **7HSAP1**. **c** Self-assembly of **7HSAP1** into a nanotube structure via noncovalent interactions between complementary interfaces of the coiled-coil lock-washer structures (Reproduced with permission from Ref. [19], © 2013 American Chemical Society)

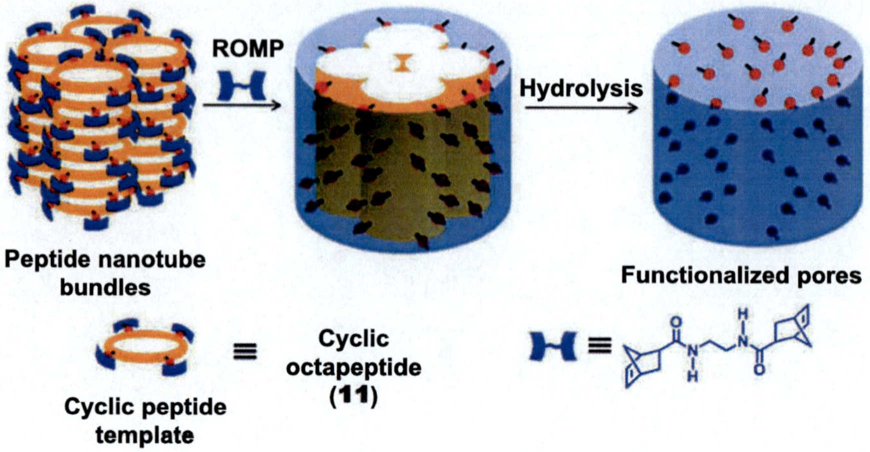

Fig. 6.14 Schematic image of the approach used to get large functionalized pores or bundles of functionalized pores (Reproduced with permission from Ref. [24], © The Royal Society of Chemistry 2016)

of the CP nanotube with diameter of ca. 36 nm. Ring-opening metathesis polymerization (ROMP) on the external norbornene monomer yielded bundles of CP–polymer nanotube with ca. 100-nm diameters. Hydrolysis of the generated ester bonds and subsequent removal of the CP core resulted in the formation of tough polymer pores having the internal carboxyl functionality (Fig. 6.14). Indeed, they verified the hollow cylindrical structures of the resultant pores. Moreover, the carboxylated polymer pore was shown to encapsulate the cationic dye lucigenin**12** (Chart 6.5) into the lumen [24].

By employing a cyclic peptide as a template, Jolliffe, Perrier, and co-workers have been demonstrating smart CP–polymer hybrid nanotubes that can exhibit diverse functions including pH-responsiveness [14], water solubility [14], or drug delivery [25]. Rational choice of functional polymer shells, which can sensitively respond to temperature [26], light [27], and pH [28], makes the CP–polymer conjugates equip with highly responsive shape and size. Furthermore, Granja and co-workers have explored ordered alignment of fullerene, single-walled CNT, and silver metal clusters on mica substrate by use of a CP or CP–polymer hybrid as a template. For example, self-assembled nanotubes from α,γ-cyclic octapeptides **13** (Chart 6.6) having fullerene side chains allow the fullerene molecules to align unidirectionally, forming 1D wire-like nanostructures of fullerene [29].

Noncovalent chemical modification provides a convincing methodology to manipulate CNTs while guaranteeing the intactness of the CNT. Montenegro, Granja, and co-workers attempted to design novel CPs that can couple them with SWCNTs

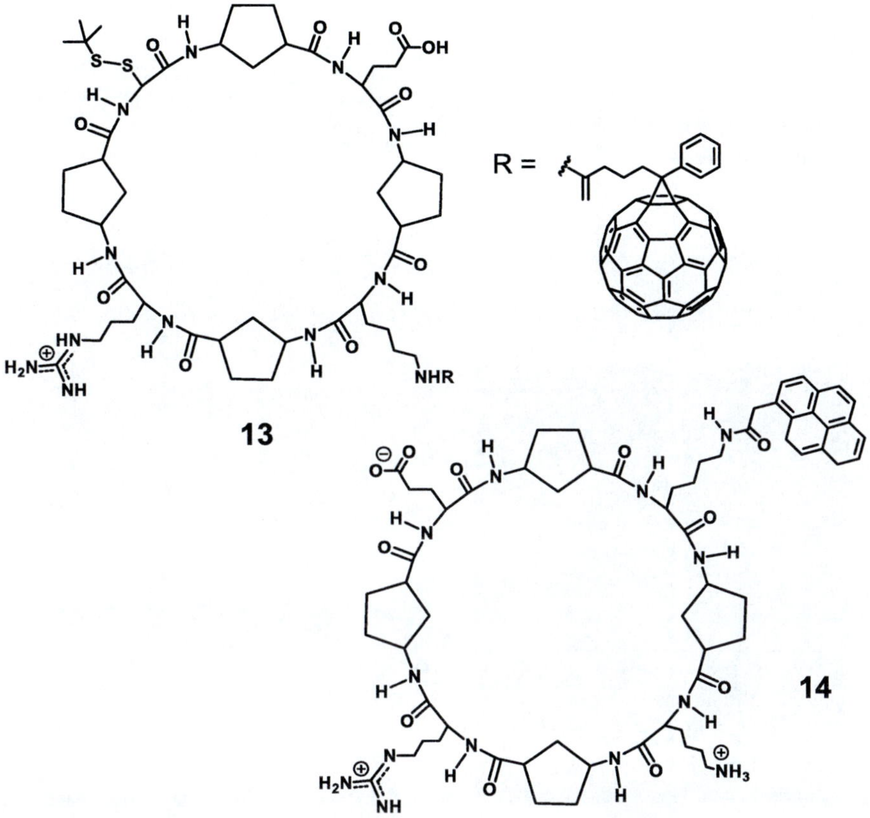

Chart 6.6 Chemical structures of **13** and **14**

through the pyrene "paddles" of the CP side chain. Similarly, the nanotubes that self-assembled from α,γ-cyclic octapeptide cyclo[(D-Arg–L-γ-Acp–D-Glu–L-γ-Acp–D-Lys(Pyr)–L-γ-Acp–D-Lys-)] **14** (Chart 6.6) where the γ-Acp residue means 3-aminocyclopentanecarboxylic acid. The CP **14** carries positively charged arginine and lysine residues, negatively charged glutamate, and pyrene-appended lysine residues. The pyrene segment enhances π–π interactions between the CP and SWCNTs in water, whereas the arginine residue commits the solubility enhancement and favorable adhesion to an anionic surface like mica substrate. Significantly, the pyrene can enwrap one side of the CNT outer surface alternatively. Thus, the binary composite system through the highly ordered noncovalent interactions between two different CP and SWCNT tubes played a critical role in immobilizing the carbon nanotubes (Fig. 6.15) [30]. Interestingly, the CPs have a tendency to recognize semiconducting CNTs more selectively than metallic one. The same CP **14**-derived nanotubes enabled the spontaneous deposition of Ag clusters on top of the nanotubes, thus, forming 1D architectures of the clusters [31].

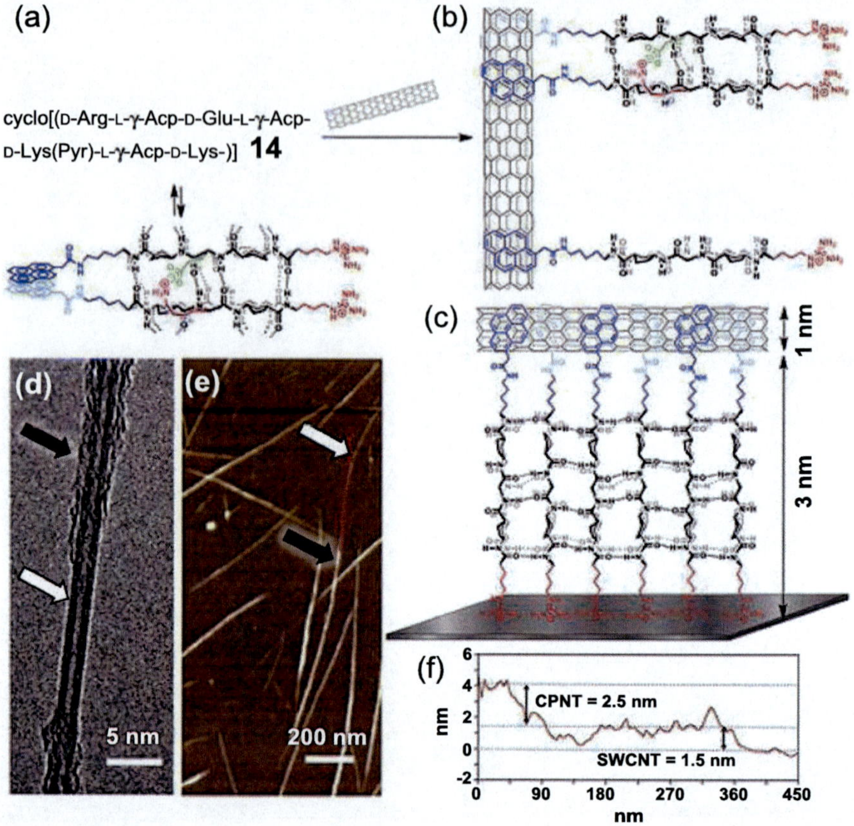

Fig. 6.15 Schematic drawing for the proposed model of SWCNT/CPNT hybrids and representative TEM and AFM images. **a** The α,γ-cyclic octapeptide **14** and its dimer structure. Plausible model of the **b** coupling of the CPNTs to SWCNTs and **c** attachment of SWCNT/CPNT hybrids to an anionic surface. **d** TEM image of a SWCNT/CPNT hybrids after deposition. The white arrow shows an CPNT-free segment of SWCNT, the black arrow a hybridized segment. **e** AFM topographic image of the SWCNT/CPNT hybrids deposited on mica, arrows as in (**d**). **f** AFM height profile along the longitudinal axis of a nanotube (Reproduced with permission from Ref. [30], © 2014 American Chemical Society)

6.4.2 Liquid Crystal Formation

Molecular building block with a mesogenic feature was conjugated with self-assembling cyclic peptides. This strategic combination resulted in the formation of columnar liquid crystals (LCs) having parallel nanometric or subnanometric channels. For example, cyclic hexapeptide was conjugated with three dendrons carrying one, two, or three C12 long hydrocarbon chains, giving three dendron-appended cyclic hexapeptides **15a–15c** (Chart 6.7) [32]. Three cyclic peptide derivatives form

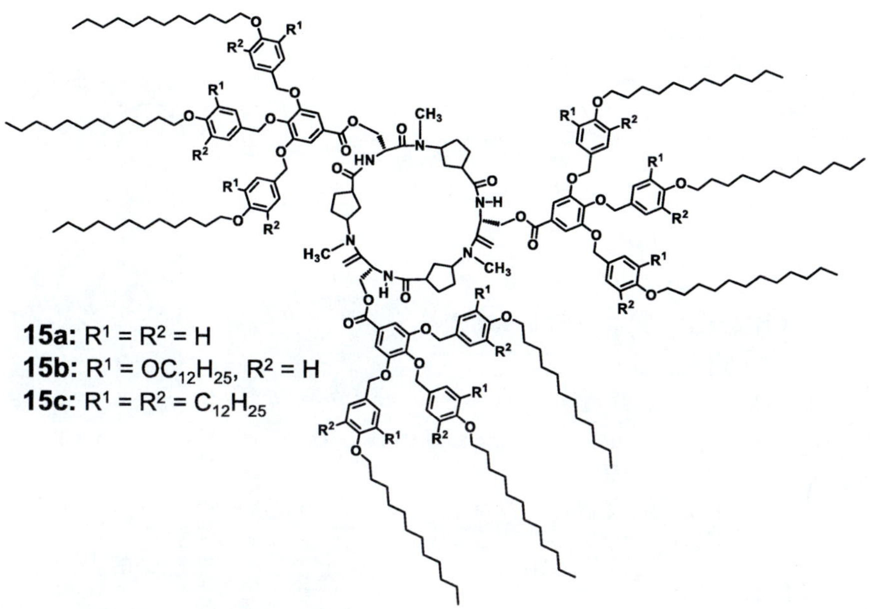

15a: $R^1 = R^2 = H$
15b: $R^1 = OC_{12}H_{25}$, $R^2 = H$
15c: $R^1 = R^2 = C_{12}H_{25}$

Chart 6.7 Chemical structure of **15**

dimers stabilized by antiparallel β-sheet interactions in chloroform. Notably, the cyclic peptide derivatives self-assemble to form hexagonal columnar mesophase for **15a** and **15b** (Fig. 6.16), whereas the CP **15c** gives homeotropic alignment of a hexagonal columnar mesophase (Table 6.1) [32]. The dimer structure plays an important role as a mesogen to drive the liquid crystal formation. Moreover, single or double nanochannels were generated along each columnar structure, depending on the number of the dendron long alkyl chains.

6.4.3 Hydrogel Formation in a Confined Nanospace

Very recently, Granja, Montenegro, and co-workers described pH-controlled hierarchical self-assembly of nanotubes from amphiphilic cyclic peptides and the resultant unique hydrogelation confined in water microdroplets [33]. They prepared the cyclic octapeptide **16** (Chart 6.8) containing three pH-responsive amino acid residues, i.e., His and Lys, and pyrene-appended Lys. Alkalinization of the acidic aqueous solutions of **16** induced nanotube self-assembly and resultant hydrogelation of the aqueous phase (Fig. 6.17). Three different morphologies including single nanotube of 3 nm wide, bundles of nanotubes of 6–10 nm wide, and entangles nanotube networks were observed for the hydrogels. To gain more insight into the hydrogelation behavior of the cyclic peptide **16** in a confined nanospace, they performed self-assembly

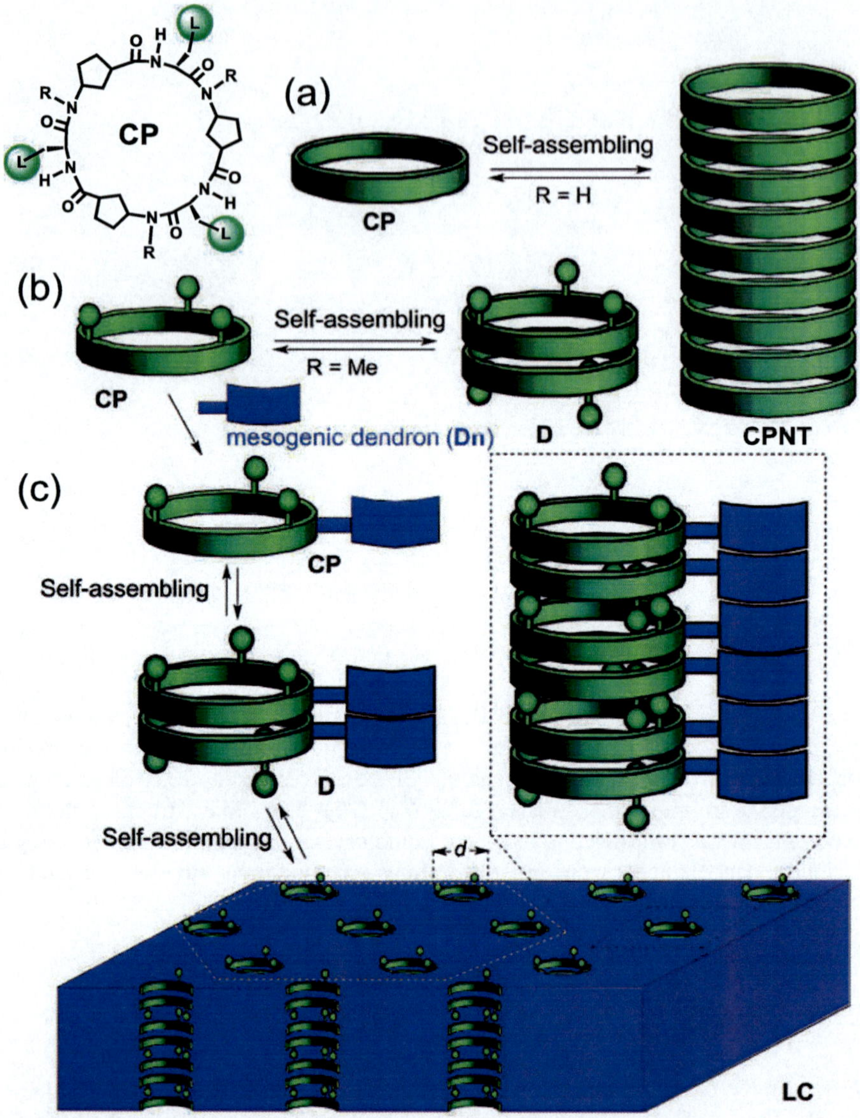

Fig. 6.16 Strategy proposed for the preparation of channeled columnar mesophases based on dimer-forming cyclic peptides (Reproduced with permission from Ref. [32], © The Royal Society of Chemistry 2014)

Table 6.1 Thermal and thermodynamic properties obtained during the second heating (h)–cooling (c) cycle (DSC experiments) (Reproduced with permission from Ref. [32], © The Royal Society of Chemistry 2014)

15a		h: Cr 158.0 Col_h 168.7 (28.7)[a] I		c: I 160.8 (23.0) Col_h[b]
15b		h: Cr 103.7 (20.3) Col_h 137.7 (2.0) I		c: I 125.4 (1.1) Col_h[b]
15c		h: Col_h 99.6 (1.8) I		c: I 91.7 (1.0) Col_h[b]

[a]The peaks corresponding to both transitions, Cr to Col_h and Col_h to I, overlapped and the enthalpy value corresponds to the integration of both peaks
[b]Crystallization was not observed on cooling down to 0 °C

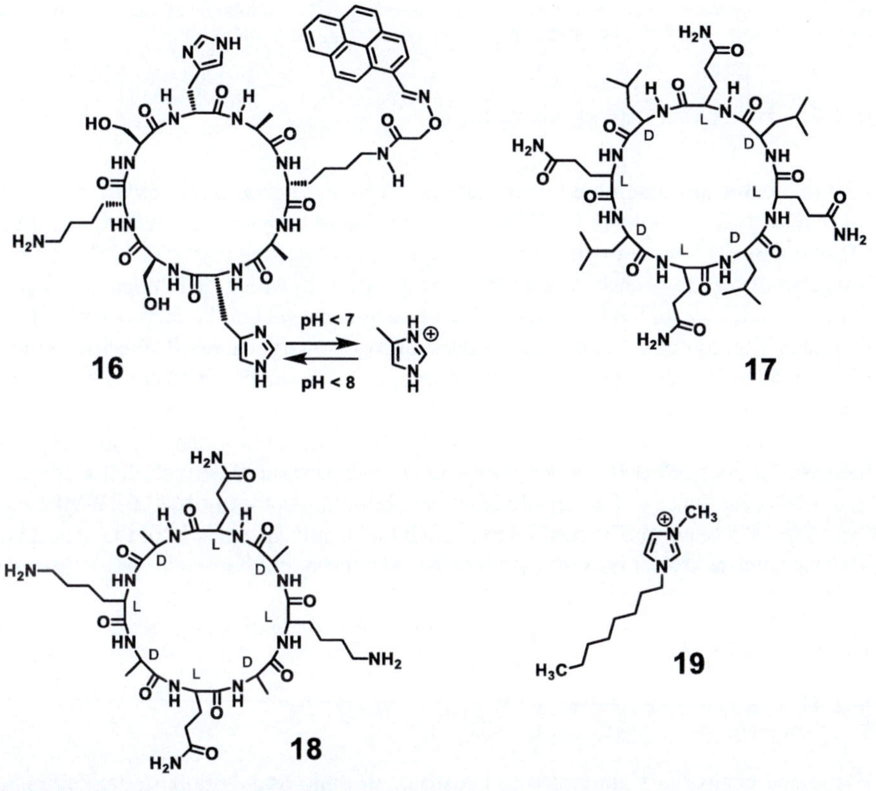

Chart 6.8 Chemical structures of **16–19**

experiments in water-in-oil droplets. Confocal microscopy revealed that the confined fibrillation really occurs and spans the shape of the microdroplet container. Finally, the shape of droplet was transformed into a distorted spherical shape. This finding gives a typical example of confined supramolecular nanotube networks, which can be achieved by using unique characteristics of the cyclic peptide nanotubes with chemical properties, external modification, and controlled diameter [34].

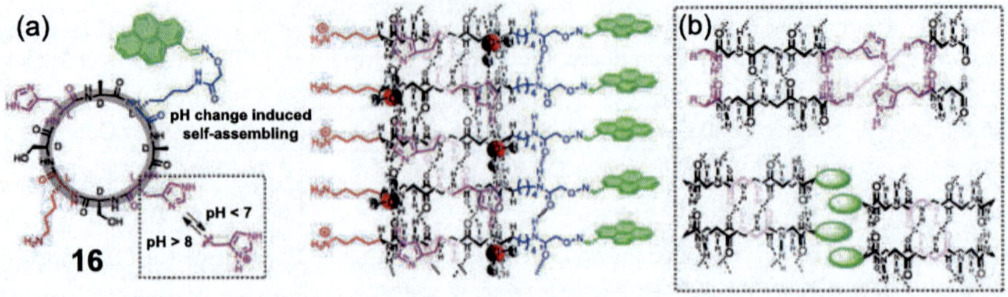

Fig. 6.17 **a** Structure and self-assembly of the cyclic octapeptide **16** into single peptide nanotubes. **b** Further aggregation of the single peptide nanotubes for hierarchical microfibrillation by histidine hydrogen-bonded networks, pyrene-associated π-stacking, and hydrophobic effects (Reproduced with permission from Ref. [33], © The Royal Society of Chemistry 2018)

6.4.4 Mechanical Reinforcement

Hierarchically self-assembled fiber materials from the nanotubes of cyclic octapeptide cyclo[(L-Gln-D-Leu)$_4$] **17** (Chart 6.8) were found to enlarge the elastic modulus of polylactide (PLA) through their incorporation as the filling materials [35]. Several image analyses are consistent with the whole aspect, in which nanotubes with 2 nm diameter, self-assembled fibers with 100 nm diameter, and bundled fibers with 1 μm diameter hierarchically assemble in this order to construct the CP fiber structures (Fig. 6.18) [36]. The utilization of this large, rod-like nanotube fibers allowed the same research group to carry out elaborated structural, mechanical, and computational studies. On the basis of nanoindentation technique, microbending studies, and molecular dynamics (MD) simulations, they evaluated an elastic modulus of 11.3 ± 3.3 GPa, hardness of 387 ± 136 MPa, and bending strength of 98 ± 19 MPa for the crystalline nanotube fibers. Notably, such mechanical properties of the nanotube fiber assemblies are equivalent to protein-based micronanofibers such as silks known as the stiffest and strongest [37].

6.4.5 Alignment of Cyclic Peptide Nanotube

Byrne and co-workers attempted to construct an ionic liquid-supported membrane having nanostructured domains [38]. To align the self-assembled CPNTs from cyclo[(L-Gln-D-Ala-L-Lys-D-Ala)$_2$] **18** (Chart 6.8), they employed the structured ionic liquid, 1-octyl-3-methylimidazolium chloride **19** (Chart 6.8). Mixing of water (ca. 5 wt%) with the ionic liquid **19** was shown to form a gel that can display liquid crystalline behavior. Thus, the ionic liquid successfully induced the alignment of the cyclic peptide nanotubes. They also constructed a flexible, robust, and self-standing membranes of polyvinylidene difluoride (PVDF) as a supported ionic liquid

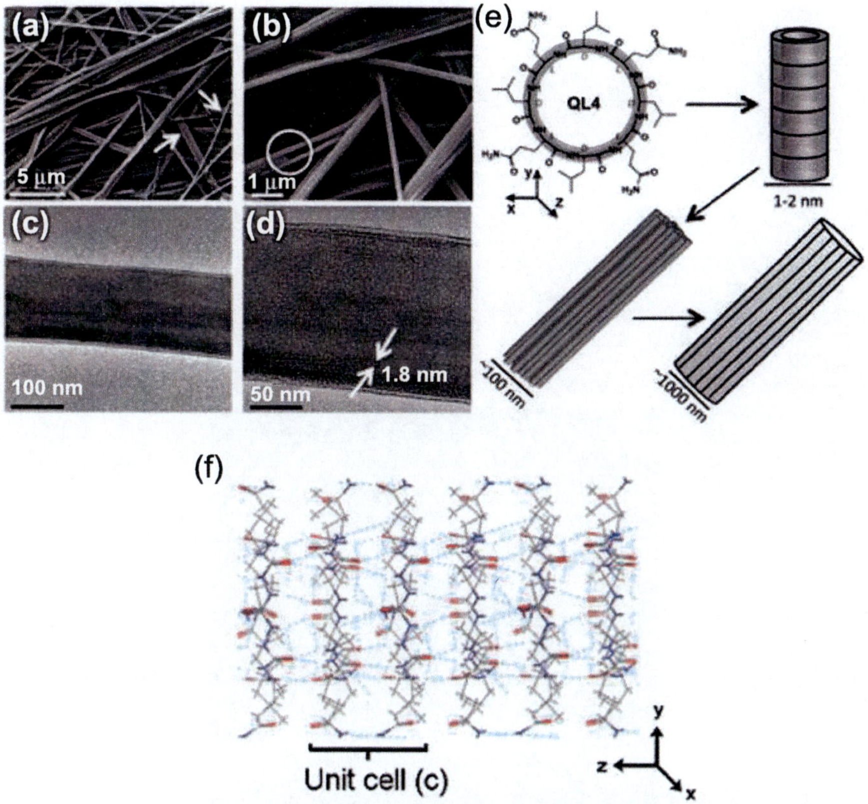

Fig. 6.18 Structure of self-assembled fibers from the cyclic octapeptide **17**. **a** and **b** SEM images of two major classes of fibers, i.e., single fibers (~ 100 nm diameter) and bundled fibers (~ 1 µm diameter), which are shown by white arrows. **c** and **d** High-resolution TEM images of single nanotubes (~ 2 nm diameter) as the building blocks of individual fibers. **e** Schematic illustration of the hierarchical self-assembly of **17** into a large, bundled fiber. **f** MD simulation of self-assembled structure of **17** (Reproduced with permission from Ref. [36], © 2015 American Chemical Society)

membrane (Fig. 6.19) [38]. The PVDF membranes include ordered domains with aligned cyclic peptide nanotubes.

6.4.6 *Transmembrane Transport*

A range of protein channels regulate the transmembrane transport of water, ion, and macromolecules elaborately, playing a crucial role in maintaining related cell viability. Particularly, the efficiency and selectivity of ion channels remarkably differ from those of passive diffusion and endo/exocytosis. Three different

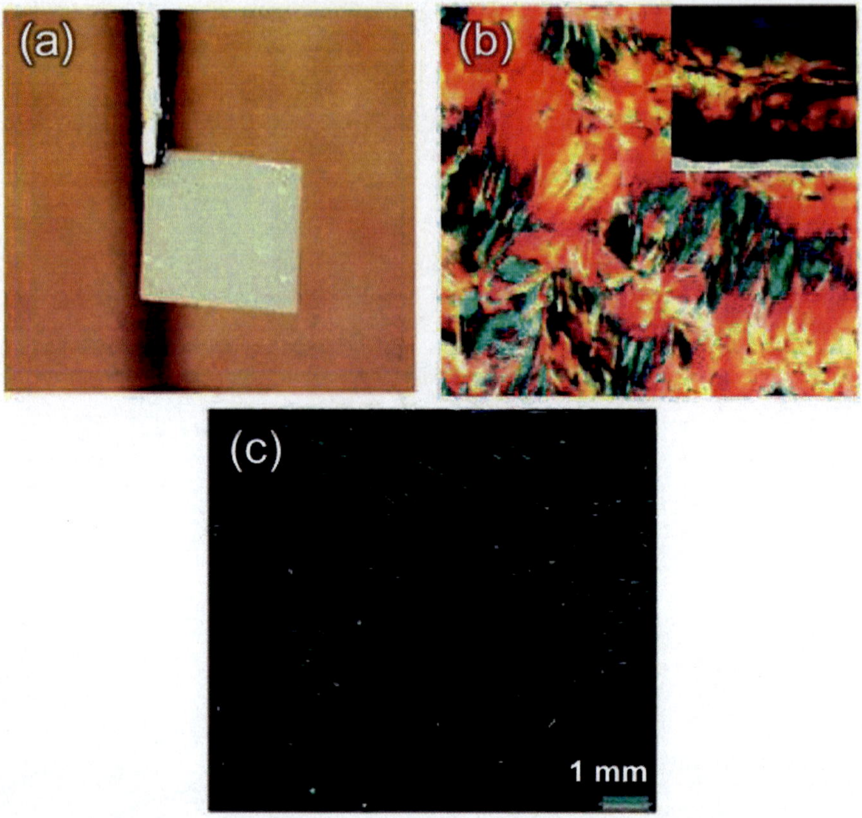

Fig. 6.19 **a** PVDF membranes swelled using ionic liquid **19** + 15 wt% H_2O, **b** polarized optical micrograph of PVDF swelled membrane (clear LC order can be observed), and **c** confocal image showing aligned CPNTs within the swelled PVDF membrane (Reproduced with permission from Ref. [38], © The Royal Society of Chemistry 2013)

cyclic peptides, i.e., hexameric α,γ-CP **20** cyclo[(L-Trp–D-γ-Ach)$_2$–L-Gln–D-γ-Ach] (Chart 6.9), octameric α,γ-CP cyclo[(L-Trp–D-γ-Ach)$_3$–L-Gln–D-γ-Ach] **21** (Chart 6.9), and octameric 3α,γ-CP cyclo[D-Trp–L-γ-Ach–D-Trp–L-Leu–D-Trp–L-γ-Ach–D-Trp–L-Gln] **22** (Chart 6.9), in which the γ-Ach residue lies every four amino acid residues, were shown to form nanotube structures in phospholipid bilayer membranes through self-assembly (Fig. 6.20). Such CPs with different cavity size and hydrophilic character were subjected to the investigation on the transport of alkali or alkaline earth cation (Na^+, K^+, Li^+, Cs^+, and Ca^{2+}) across the lipid membranes [39]. Depending on the inner diameter size of the CP nanotubes, the cyclic hexapeptide **20**-derived nanotube of ca. 0.5 nm in van der Waals inner diameter was demonstrated to accelerate bulk proton transport. In turn, the cyclic octapeptide **21**-derived nanotube exhibited outstanding selectivity for the alkaline metal cations. Relatedly, Keten and co-workers investigated how the tailoring the polarity and size of the

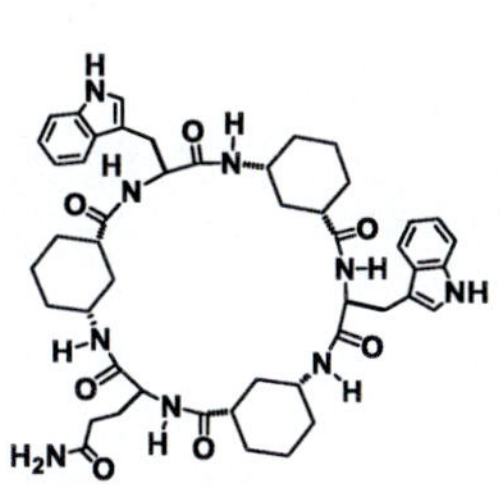

20

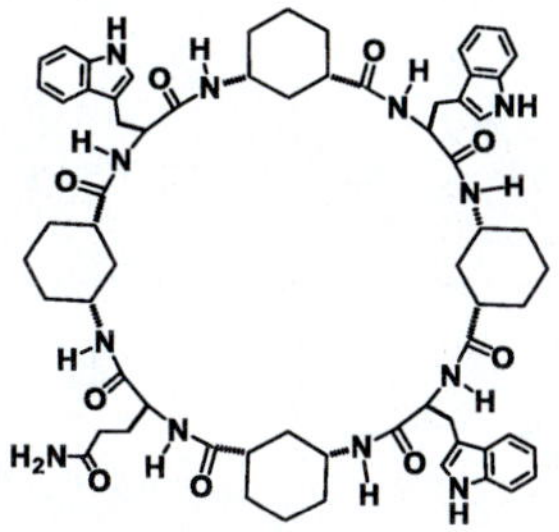

21

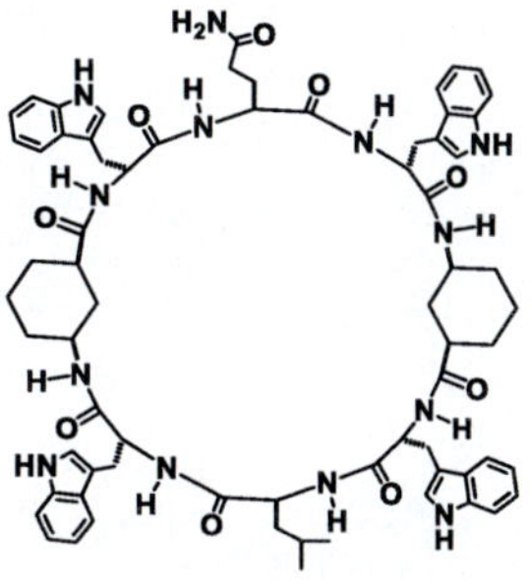

22

Chart 6.9 Chemical structures of **20–22**

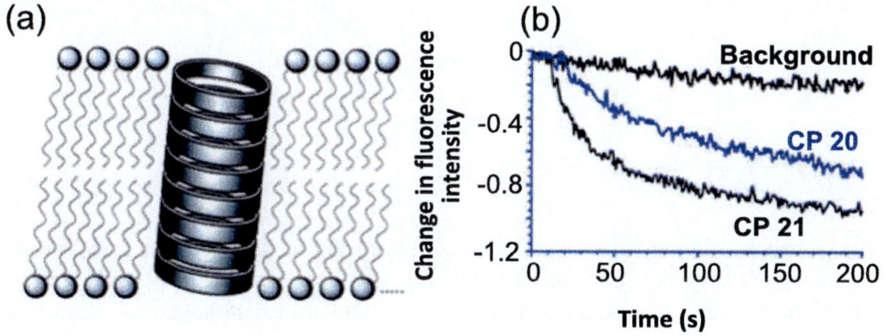

Fig. 6.20 **a** Schematic image of transmembrane ion channel formed by stacking of CPs. **b** Normalized UV fluorescence records from 5(6)-carboxyfluorescein-containing liposomes without CPs (background) or with the CP **20** or CP **21**, which show loss of fluorescence due to the breakdown of an initial transmembrane pH gradient (Reproduced with permission from Ref. [39], © The Royal Society of Chemistry 2012)

nanotube interior of CPs affect the regulation of ionic flux by utilizing nonequilibrium atomistic MD simulations [40]. The preparation and properties of artificial ion channels comprising self-assembled CP-based nanotubes with different outer surface features and inner diameters were reviewed elsewhere [34, 41, 42].

Only a few studies have reported on the influence of unprecedented amino acid configurations and nonnatural amino acid within CP sequences on the formation of nanochannel across lipid bilayers [39]. Perrier, Jolliffe, and co-workers examined the influence of amino acid constitution of six cyclic octapeptides **23–28** (Chart 6.10) consisting of alternating D- and L-α-amino acids on their self-association into transmembrane channels in phospholipid bilayers [43]. Cyclic peptides consisting of D-leucine, alternating with different constitutions of L-Trp, L-Lys, L-lysine(Alloc), and L-azidolysine, were found to self-assemble into nanochannels in phospholipid bilayers. Large unilamellar vesicle (LUV) fluorescence assay and dynamic light scattering measurement corroborated that the cyclic peptides **23** and **24**, equipping

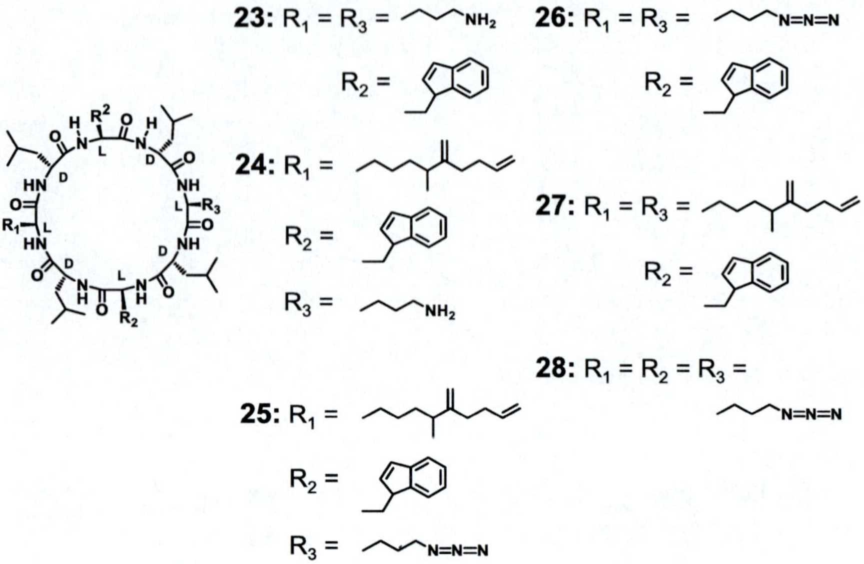

Chart 6.10 Chemical structures of **23–28**

with positively charged side chains, tend to produce a relatively larger barrel-stave-type nanopore (Fig. 6.21). In contrast to this large channel, the cyclic peptides **25–28**, bearing neutral side chains, create relatively smaller single nanopores. As mentioned in **6.2**, the same research group also fabricated a subnanometric channel from CP–polymer conjugation nanotubes with external dual functionality [17]. Remarkably, phase segregation within the Janus nanotube induces the formation of the transmembrane nanochannel on LUV.

Fu, Wu, and co-workers investigated the effect of the cyclic octapeptide nanotube self-assembled from cyclo[(L-Trp–D-Leu)$_4$–L-Gln–D-Leu] **29** (Chart 6.11) on the

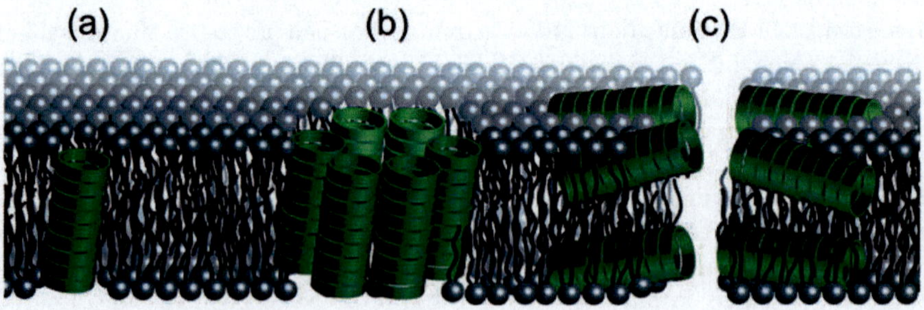

Fig. 6.21 Schematic image of the different possible modes of bilayer channel formation. Formation of cyclic peptide channels in a lipid bilayer can occur as **a** a unimeric pore, **b** a barrel-stave or **c** through a bilayer (carpet-like) disruption (Reproduced with permission from Ref. [43], © The Royal Society of Chemistry 2015)

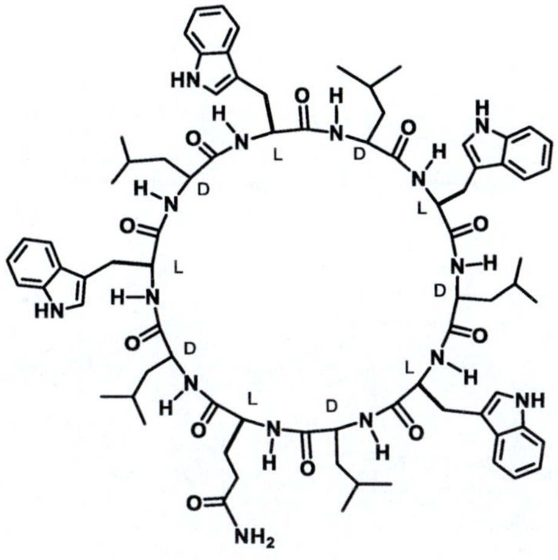

29

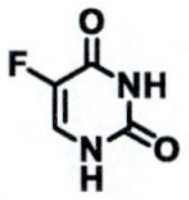

30

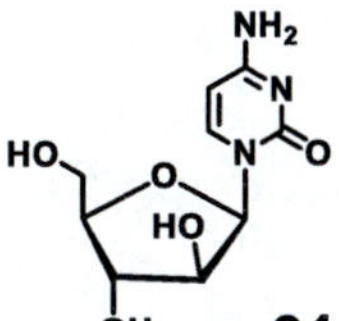

31

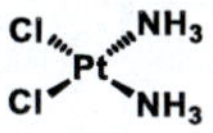

32

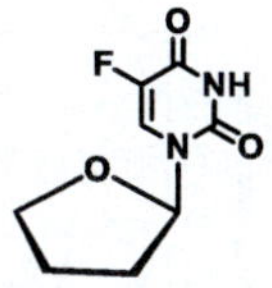

33

Chart 6.11 Chemical structures of **29–33**

transmembrane transport of the anticancer drug 5-fluorouracil (5-FU) **30** (Chart 6.11) [44]. They experimentally confirmed that in the presence of the peptide nanotube 70% of 5-FU diffuses from 5-FU-embedded liposomes. On the contrary, only 5% of 5-FU was shown to release from the drug-loaded liposomes in the absence of the nanotubes. Subsequently to the theoretical confirmation that the CP nanotubes actually mediate transmembrane transport of 5-FU [44], Subramaian et al. scrutinized the free energy involved with the transmembrane transport of 5-FU by applying steered MD and umbrella sampling simulations [45]. Furthermore, as a proof of concept test, Wu and co-workers performed the feasibility experiments for the CP **29** nanotube-assisted transport of four anticancer drugs [5-FU **30**, cytarabine **31**, cisplatin **32**, and tegafur **33**] (Chart 6.11) across cell membranes (Fig. 6.22) [46]. Among the tested drugs, the smallest molecule 5-FU in dimensions displayed the highest diffusion rate across the membrane, whereas no transportation phenomena was observed for cytarabine **31** of 1.1 nm in size.

6.4.7 Medical Applications

CP-based nanotubes as a mimic of protein channels have a potential to enhance membrane permeability. This feature changes the membrane potentials, thereby leading to the death of bacterial cells. On this hypothesis, Ghadiri and co-workers corroborated that the peptide nanotubes, assembled from cyclic hexa- and octapeptides composed of an alternating D- and L-α-amino acids, serve as selective antibiotic nanostructures against gram-positive and -negative bacteria [47]. In addition to the

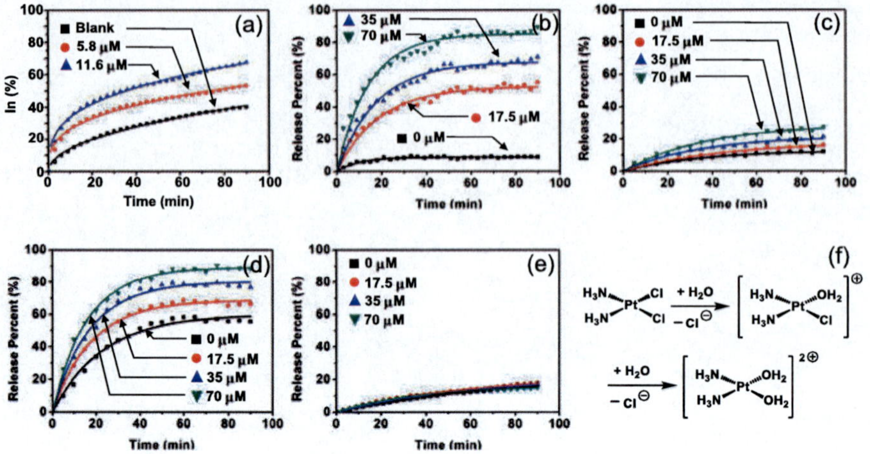

Fig. 6.22 Transmembrane transport properties of **a** H^+, **b** 5-FU **30**, **c** tegafur **33**, **d** cisplatin **32,** and **e** cytarabine **31** through self-assembly of the CP **29**-based nanotubes from liposomes at different CP levels. **a** The change in fluorescence intensity for H^+ transport and **b–e** accumulated release percentage of each drug are plotted as a function of time. **f** Possible explanation to the mechanisms of cisplatin hydration (Reproduced with permission from Ref. [46], © The Royal Society of Chemistry 2016)

aforementioned cyclic D,L-α-peptide, the research group newly designed cyclic D,L-α-glycopeptides in imitation of the glycopeptide antimicrobial agent mannopeptimcins **34** (Chart 6.12) and investigated the in vitro antimicrobial activity and membrane activity [48]. The synthesized cationic cyclic peptides contain *O*-glycosyl serine residues functionalized with either D-glucosamine (**35–37**) (Chart 6.12), D-galactose (**38** and **39**) (Chart 6.13), or D-mannose (**40** and **41**) (Chart 6.13). They also used the membrane-active amphiphilic peptide **42** (Chart 6.13) as a progenitor CP. Appropriate glycosylation maintains antimicrobial activity against gram-positive bacteria while substantially reducing mammalian cell toxicity (Table 6.2). The incorporating position of *O*-glycosyl serine residue in the cyclic sequence has a profound effect on the toxicity and selectivity by a factor of 3. The self-assembled CP nanotubes operate with a membrane permeation mode.

In succession to similar molecular design as RAPTA-C-appended nanotube from CP–polymer **6**, an organoiridium anticancer segment that exhibits profound toxicity toward widespread cancer cells was conjugated to the biocompatible polymer shell of the nanotube from CP–poly(hydroxypropyl methacrylamide) (PHPMA) **43** (Chart 6.14) [25]. The obtained drug delivery systems **44a** and **44b** (Chart 6.14) displayed equivalent or higher toxic effect on human ovarian cancer cells than the free drug. More interestingly, all the studies including in vivo assessment, pharmacokinetics, and biodistribution supported the high possibility of the CP–polymer **43** nanotubes as a novel class of drug delivery vectors [49]. One of the vital results representing the superiority of this system is that the disorganization into small segments

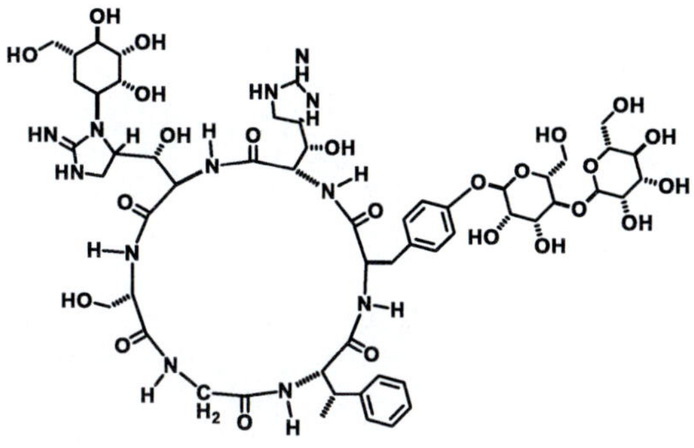

34

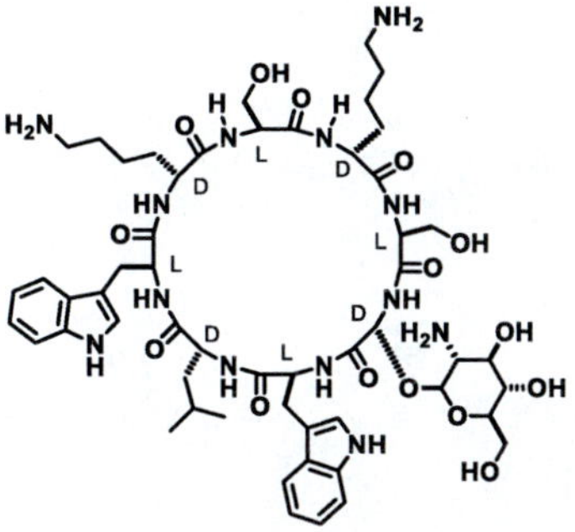

35

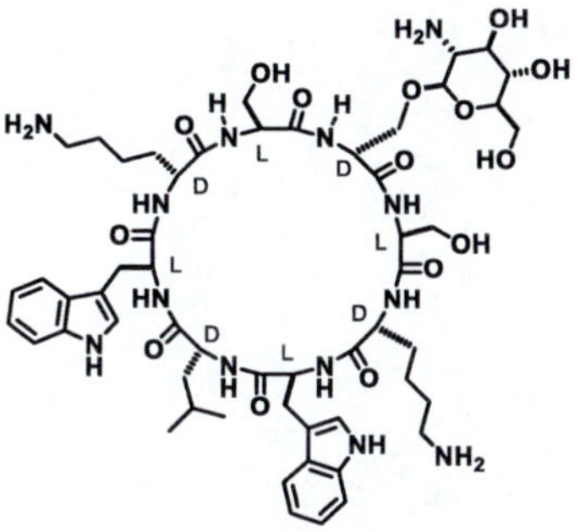

36

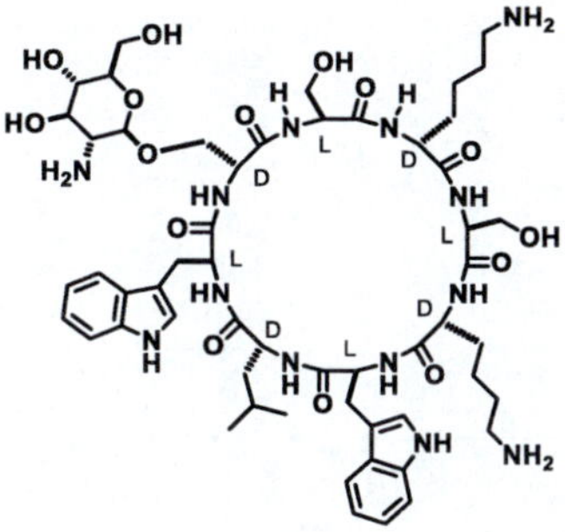

37

Chart 6.12 Chemical structures of **34**–**37**

can eliminate the conjugates efficiently and avoids their accumulation in organs (Fig. 6.23).

Similar idea as mentioned above allowed for the preparation of cationic fibers serving as a gene transfection vector [50]. A single guanidinocarbonyl pyrrole segment was introduced into a cyclic octapeptide to form the tetracationic cyclic structure **46** (Chart 6.15). The modified peptide **46** spontaneously form the tube-like fibers of micrometer long through self-assembly in buffered aqueous solution at pH 7.4. The resultant positively charged surfaces of the self-assembled fibers can bind negatively charged DNA through electrostatic interaction. The high-aspect-ratio nanostructures were, then, demonstrated to transfect genes upon nonendocytosis process for the cellular uptake (Fig. 6.24).

Meanwhile, oral or anti-apoptotic gene delivery via eye drop, using the CP nanotubes composed of cyclo(D-Trp-Tyr), was explored by Liaw and co-workers [51, 52]. They reasoned that the tyrosine- and tryptophan-containing peptide nanotubes can firmly interact with biomembranes and associate with DNA via electron-transfer interactions, respectively. The dimensions of the obtained nanotubes are 100–800 nm

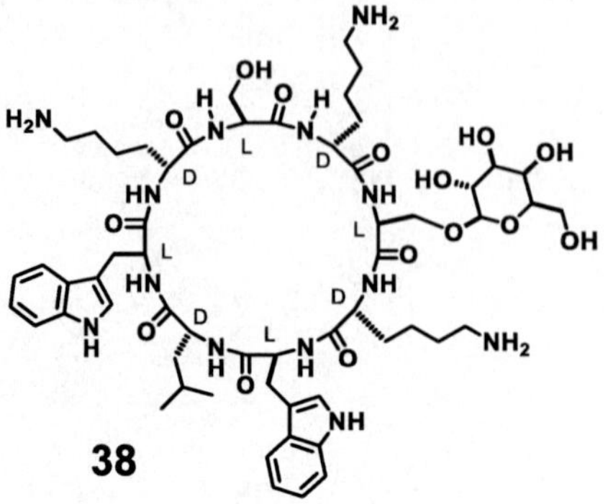

38

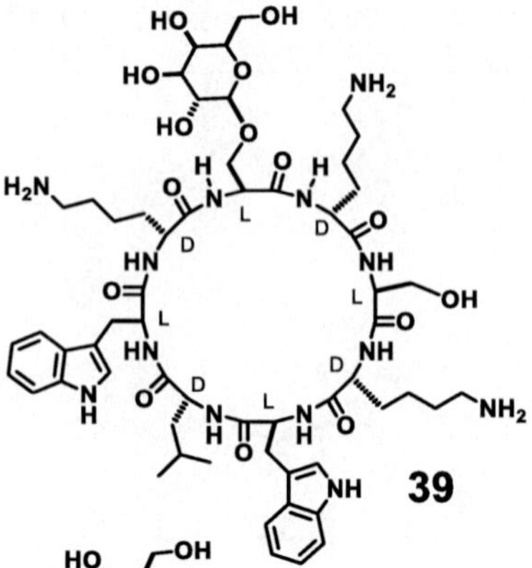

39

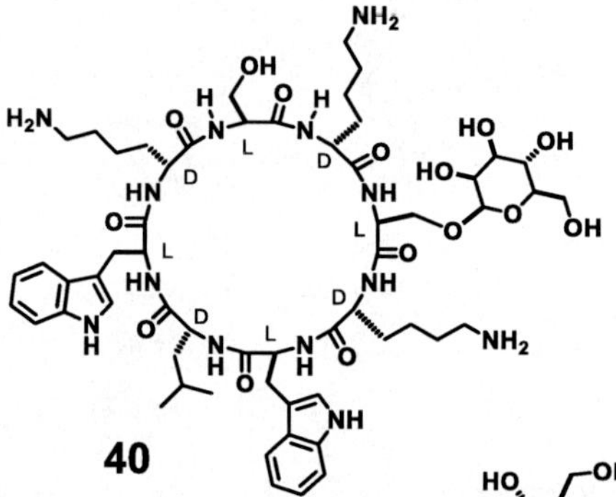

40

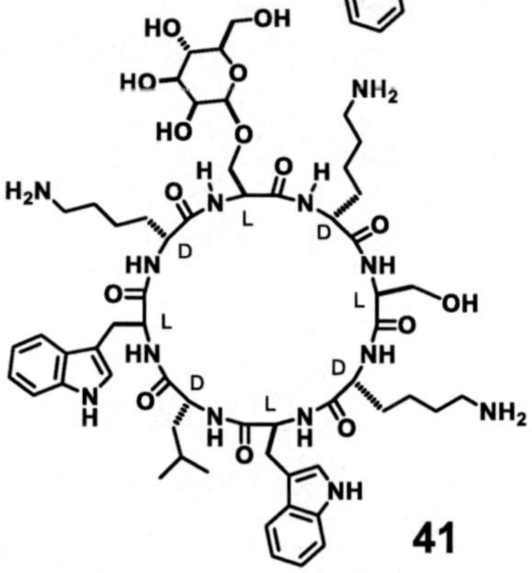

41

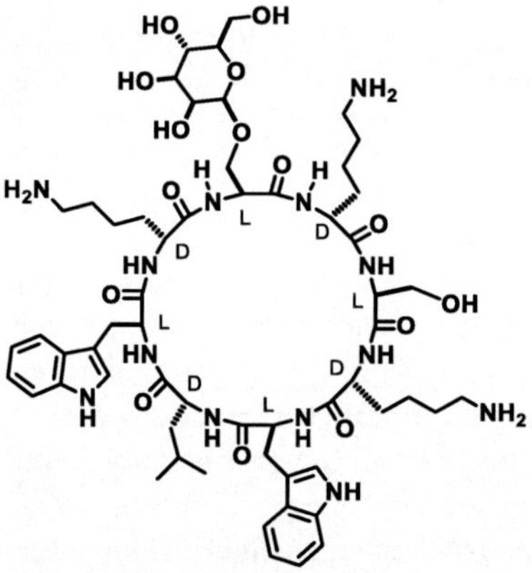

42

Chart 6.13 Chemical structures of **38–42**

wide and 1–20 μm long, suggesting that the peptide nanotubes are bundled or aggregated morphologies of single nanotubes. Thus, the in vitro study disclosed that DNA-integrated CP nanotubes increased the apparent duodenal permeability of the plasmid

Table 6.2 In vitro antibacterial, hemolytic, and cytotoxic activities of cyclic D,L-α-glycopeptides[a] (Reproduced with permission from Ref. [48], © The Royal Society of Chemistry 2009)

	MIC[b]/μM				HD50[g]/μM		LD$_{50}$/μM
Glycopeptide	MRSA[c] (MBC)[d]	MRSA[c] in FBS	*Bacillus cereus*[e]	VRE[f]	m-RBC[h]	h-RBC[i]	NIH 3T3[j]
35 [WLWKSKS**X**]	5	15	10	10	65	85	20
36[WLWKS**X**SK]	5(10)	10	10	10	90	140	50
37[WLW**X**SKSK]	5	5	15	10	80	150	20
38[WLWKSK**Z**K]	5(10)	10	10	15	150	260	50
39[WLWK**Z**KSK]	5	5	5	10	105	190	25
40[WLWKSK**U**K]	2.5(5)	5	10	10	70	120	25
41[WLWK**U**KSK]	2.5	10	5	5	50	105	15
42[WLWKSKSK]	5	5	5	10	65	120	10

[a]Peptide sequences: shown in single-letter amino acid codes with underlining representing D-residues. The bold letters indicate *O*-glycosyl serine residues with X = Ser(β-GlcNH$_2$), Z = Ser(β-Gal), and U = Ser(α-Man)
[b]Minimum inhibitory concentration
[c]Methicillin-resistant Staphylococcus aureus (ATCC33591)
[d]Minimum bactericidal concentration
[e]ATCC11778
[f]Vancomycin-resistant Enterococcus faecalis (ATCC51575)
[g]Hemolytic dose
[h]Mouse (Balb-C)
[i]Human red blood cells (erythrocytes)
[j]In vitro toxicity against mouse fibroblast cells (NIH 3T3) measured after 48 h of continued exposure to tested agents

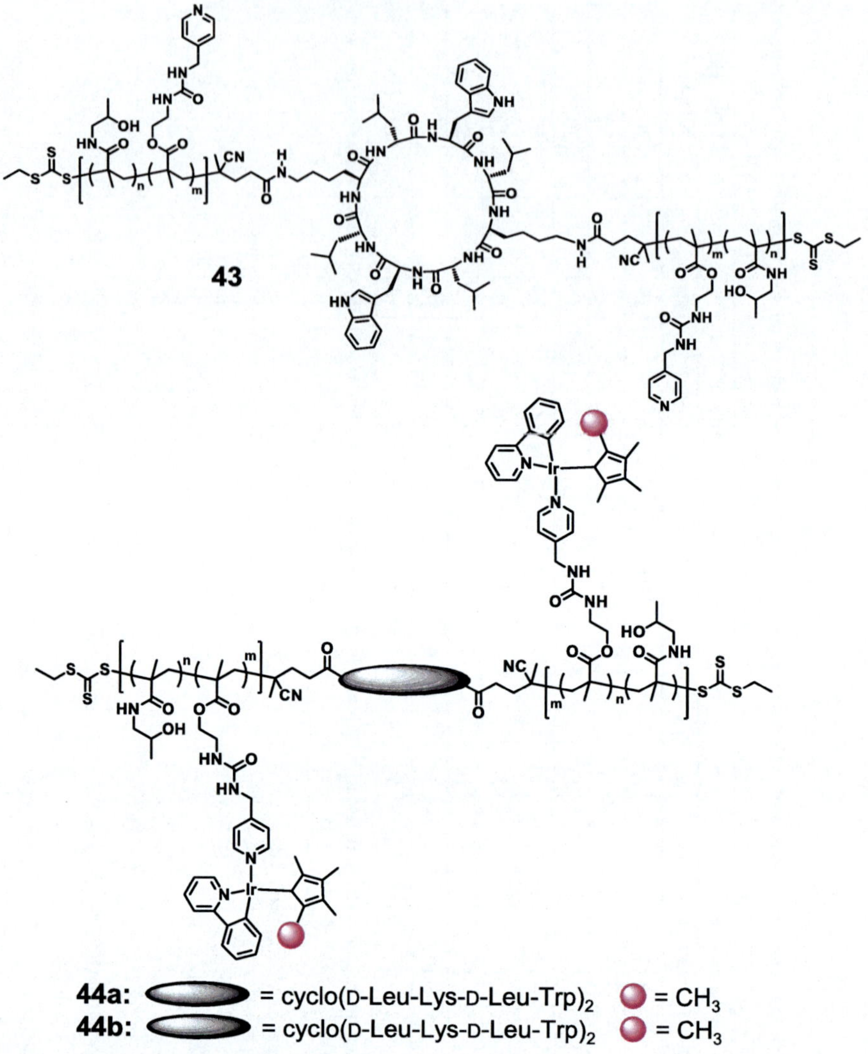

Chart 6.14 Chemical structures of **43**–**44**

DNA. The same CP nanotubes carrying caspase 3 silencing shRNA (CAP3 pRFP-C-RS) were subjected to eye drop delivery to injured cornea. They found the anti-apoptotic efficacy of gene delivery against the corneal injury, displaying that the caspase 3 activity was markedly reduced by 41% at 48 h after the first administration of the nanotube conjugates (Fig. 6.25) [52].

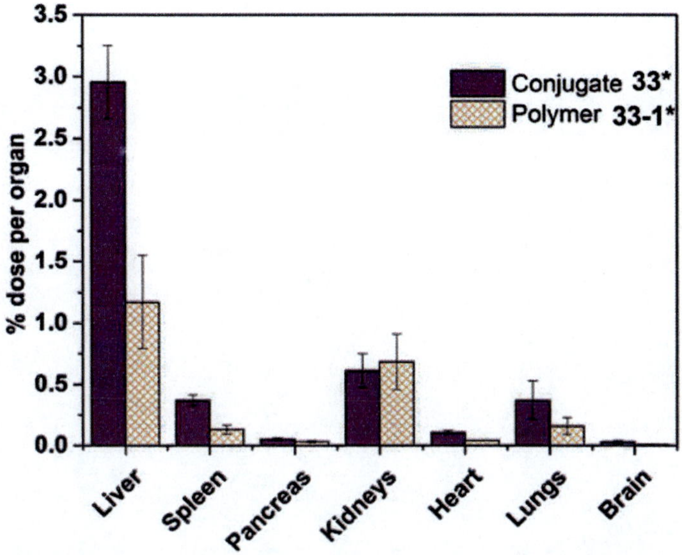

Fig. 6.23 Distribution of [^{14}C] in organs, 24 h after intravenous administration of the ^{14}C-labeled CP **45a***(Chart 6.15)–polymer nanotube and ^{14}C-labeled polymer derivative **45b***(Chart 6.15) at 12 mg kg^{-1} (mean ± SD, n = 4–5 rats) (Reproduced with permission from Ref. [49], © 2018 Elsevier Ltd.)

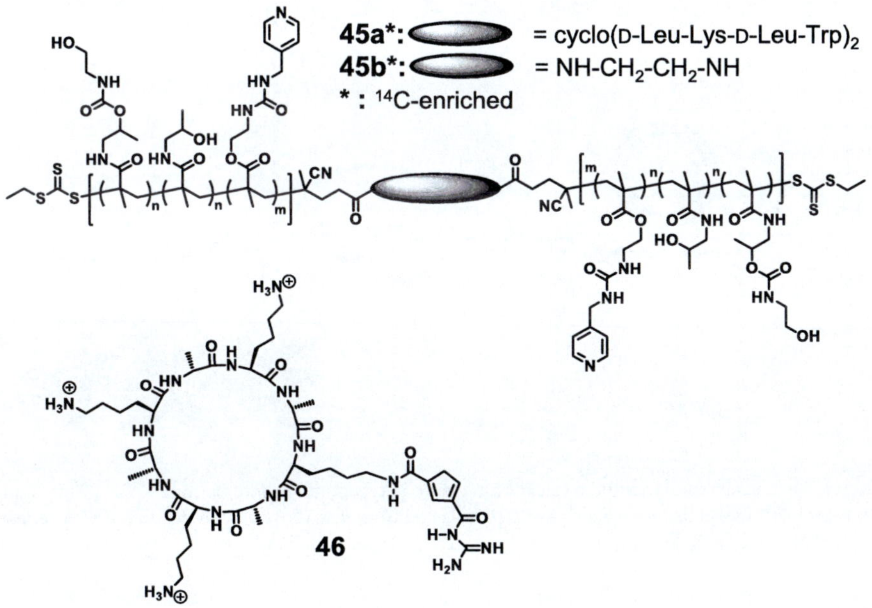

Chart 6.15 Chemical structures of **45–46**

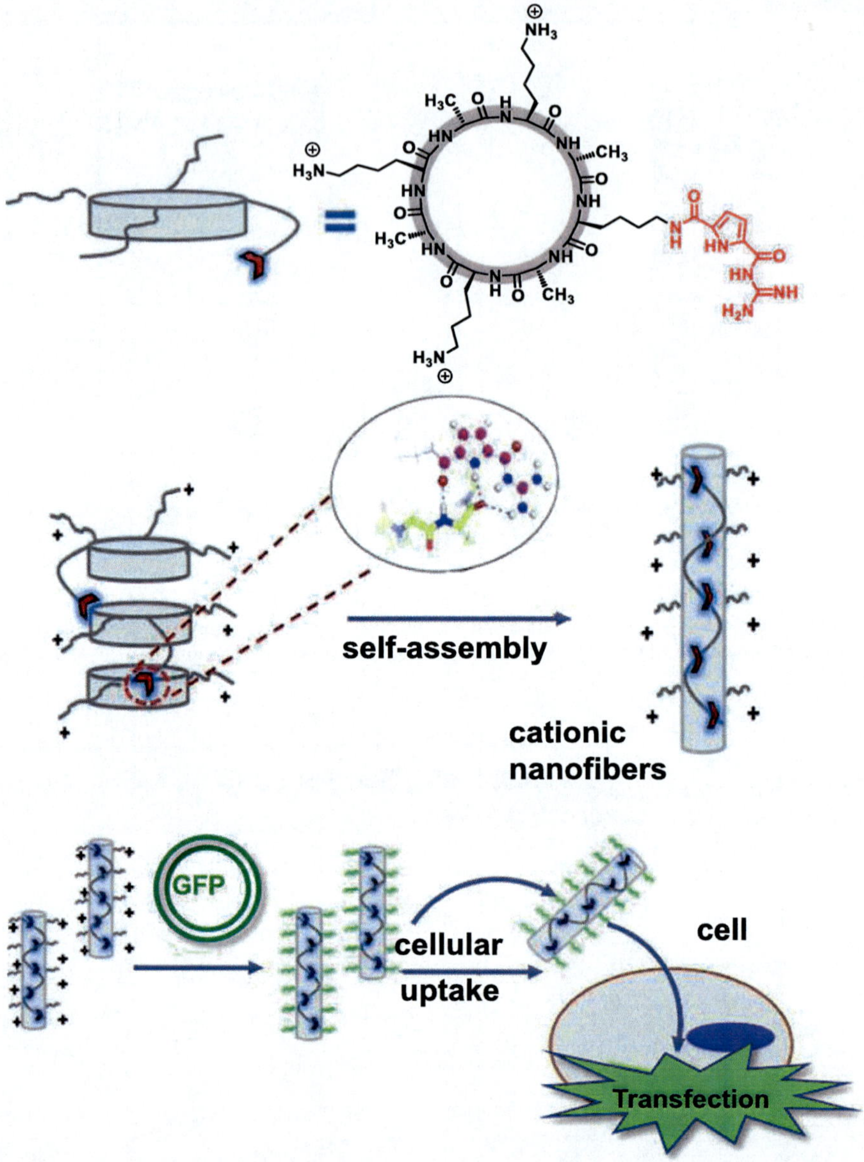

Fig. 6.24 Self-assembly of the tetracationic cyclic octapeptide **46** into cationic CPNTs that can transfer DNA into cells (Reproduced with permission from Ref. [50], © 2016 WILEY–VCH Verlag GmbH & Co. KGaA, Weinheim)

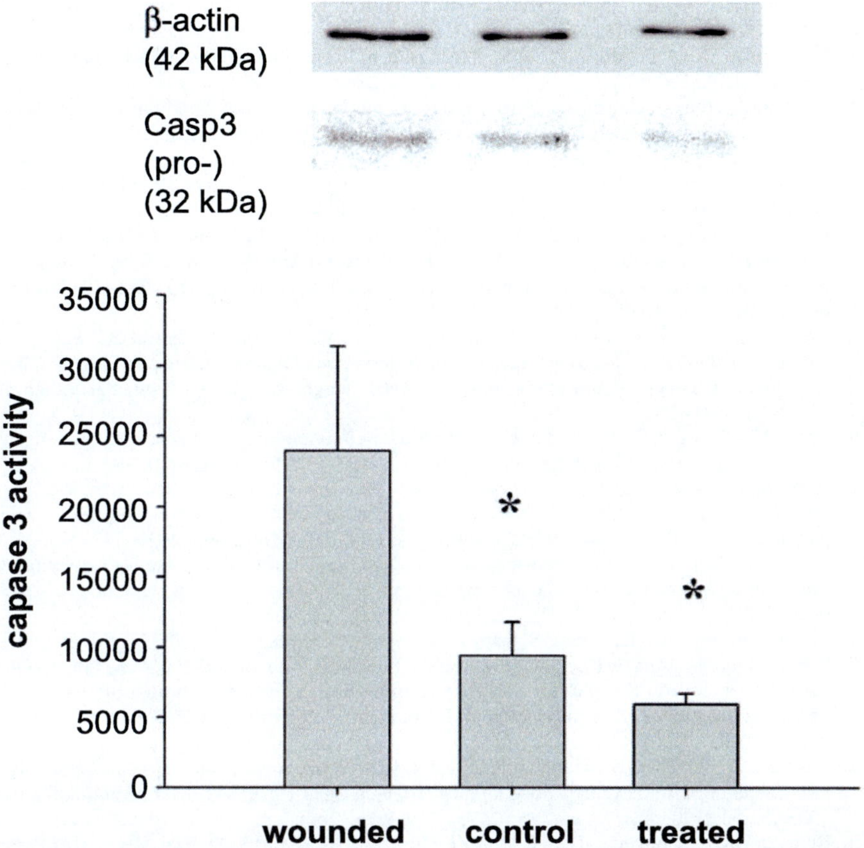

Fig. 6.25 Caspase 3 activity by the delivery of the self-assembled CPNTs from cyclo(D-Trp-Tyr), which carry CAP3 pRFP-C-RS in cornea after epithelial debridement. *Significant difference ($p < 0.05$)

References

1. Ghadiri MR, Granja JR, Milligan RA, Mcree DE, Khazanovich N (1993) Self-assembling organic nanotubes based on a cyclic peptide architecture. Nature 366:324–327. https://doi.org/10.1038/366324a0
2. Chapman R, Danial M, Koh ML, Jolliffe KA, Perrier S (2012) Design and properties of functional nanotubes from the self-assembly of cyclic peptide templates. Chem Soc Rev 41:6023–6041. https://doi.org/10.1039/C2cs35172b
3. Valery C, Paternostre M, Robert B, Gulik-Krzywicki T, Narayanan T, Dedieu JC, Keller G, Torres ML, Cherif-Cheikh R, Calvo P, Artzner F (2003) Biomimetic organization: octapeptide self-assembly into nanotubes of viral capsid-like dimension. Proc Natl Acad Sci USA 100:10258–10262. https://doi.org/10.1073/pnas.1730609100
4. Pouget E, Fay N, Dujardin E, Jamin N, Berthault P, Perrin L, Pandit A, Rose T, Valery C, Thomas D, Paternostre M, Artzner F (2010) Elucidation of the self-assembly pathway of

lanreotide octapeptide into beta-sheet nanotubes: role of two stable intermediates. J Am Chem Soc 132:4230–4241. https://doi.org/10.1021/ja9088023
5. Hamley IW (2014) Peptide nanotubes. Angew Chem Int Ed 53:6866–6881. https://doi.org/10.1002/anie.201310006
6. Tarabout C, Roux S, Gobeaux F, Fay N, Pouget E, Meriadec C, Ligeti M, Thomas D, Ijsselstijn M, Besselievre F, Buisson DA, Verbavatz JM, Petitjean M, Valery C, Perrin L, Rousseau B, Artzner F, Paternostre M, Cintrat JC (2011) Control of peptide nanotube diameter by chemical modifications of an aromatic residue involved in a single close contact. Proc Natl Acad Sci USA 108:7679–7684. https://doi.org/10.1073/pnas.1017343108
7. Gobeaux F, Fay N, Tarabout C, Meriadec C, Meneau F, Ligeti M, Buisson DA, Cintrat JC, Nguyen KMH, Perrin L, Valery C, Artzner F, Paternostre M (2012) Structural role of counterions adsorbed on self-assembled peptide nanotubes. J Am Chem Soc 134:723–733. https://doi.org/10.1021/ja210299g
8. Gobeaux F, Fay N, Tarabout C, Meneau F, Meriadec C, Delvaux C, Cintrat JC, Valery C, Artzner F, Paternostre M (2013) Experimental observation of double-walled peptide nanotubes and monodispersity modeling of the number of walls. Langmuir 29:2739–2745. https://doi.org/10.1021/la304862f
9. Bong DT, Clark TD, Granja JR, Ghadiri MR (2001) Self-assembling organic nanotubes. Angew Chem Int Ed 40:988–1011. https://doi.org/10.1002/1521-3773(20010316)40:6%3c988::AID-ANIE9880%3e3.0.CO;2-N
10. Ishihara Y, Kimura S (2012) Peptide nanotube composed of cyclic tetra-β-peptide having polydiacetylene. Biopolymers 98:155–160. https://doi.org/10.1002/Bip.22029
11. Ishihara Y, Kimura S (2013) Four-peptide-nanotube bundle formation by self-assembling of cyclic tetra-β-peptide using G-quartet motif. Pept Sci 100:141–147. https://doi.org/10.1002/bip.22177
12. Couet J, Jeyaprakash JD, Samuel S, Kopyshev A, Santer S, Biesalski M (2005) Peptide-polymer hybrid nanotubes. Angew Chem Int Ed 44:3297–3301. https://doi.org/10.1002/anie.200462993
13. Chapman R, Jolliffe KA, Perrier S (2011) Modular design for the controlled production of polymeric nanotubes from polymer/peptide conjugates. Polym Chem 2:1956–1963. https://doi.org/10.1039/c1py00202c
14. Chapman R, Warr GG, Perrier S, Jolliffe KA (2013) Water-soluble and pH-responsive polymeric nanotubes from cyclic peptide templates. Chem Eur J 19:1955–1961. https://doi.org/10.1002/chem.201203602
15. Blunden BM, Chapman R, Danial M, Lu H, Jolliffe KA, Perrier S, Stenzel MH (2014) Drug conjugation to cyclic peptide-polymer self-assembling nanotubes. Chem Eur J 20:12745–12749. https://doi.org/10.1002/chem.201403130
16. Chapman R, Jolliffe KA, Perrier S (2013) Multi-shell soft nanotubes from cyclic peptide templates. Adv Mater 25:1170–1172. https://doi.org/10.1002/adma.201204094
17. Danial M, Tran CMN, Young PG, Perrier S, Jolliffe KA (2013) Janus cyclic peptide-polymer nanotubes. Nat Commun 4:2780. https://doi.org/10.1038/ncomms3780
18. Xu T, Zhao N, Ren F, Hourani R, Lee MT, Shu JY, Mao S, Helms BA (2011) Subnanometer porous thin films by the co-assembly of nanotube subunits and block copolymers. ACS Nano 5:1376–1384. https://doi.org/10.1021/nn103083t
19. Xu C, Liu R, Mehta AK, Guerrero-Ferreira RC, Wright ER, Dunin-Horkawicz S, Morris K, Serpell LC, Zuo X, Wall JS, Conticello VP (2013) Rational design of helical nanotubes from self-assembly of coiled-coil lock washers. J Am Chem Soc 135:15565–15578. https://doi.org/10.1021/ja4074529
20. Liu J, Zheng Q, Deng Y, Cheng CS, Kallenbach NR, Lu M (2006) A seven-helix coiled coil. Proc Natl Acad Sci USA 103:15457–15462. https://doi.org/10.1073/pnas.0604871103
21. Kameta N, Ding W (2019) Direct joining of a heterogeneous pair of supramolecular nanotubes and reaction control of a guest compound by transportation in the nanochannels. Bull Chem Soc Jpn 92:1053–1059. https://doi.org/10.1246/bcsj.20190046
22. Kanzaki T, Horikawa Y, Makino A, Sugiyama J, Kimura S (2008) Nanotube and three-way nanotube formation with nonionic amphiphilic block peptides. Macromol Biosci 8:1026–1033. https://doi.org/10.1002/mabi.200800071

23. Zhang W, Jin W, Fukushima T, Saeki A, Seki S, Aida T (2011) Supramolecular linear heterojunction composed of graphite-like semiconducting nanotubular segments. Science 334:340–343. https://doi.org/10.1126/science.1210369
24. Potnuru M, Madhavan N (2016) Robust carboxylated polymer pores from a cyclic peptide template. Polym Chem 7:31–35. https://doi.org/10.1039/c5py01313e
25. Larnaudie SC, Brendel JC, Romero-Canelon I, Sanchez-Cano C, Catrouillet S, Sanchis J, Coverdale JPC, Song JI, Habtemariam A, Sadler PJ, Jolliffe KA, Perrier S (2018) Cyclic peptide-polymer nanotubes as efficient and highly potent drug delivery systems for organometallic anticancer complexes. Biomacromol 19:239–247. https://doi.org/10.1021/acs.biomac.7b01491
26. Moon HJ, Ko DY, Park MH, Joo MK, Jeong B (2012) Temperature-responsive compounds as in situ gelling biomedical materials. Chem Soc Rev 41:4860–4883. https://doi.org/10.1039/c2cs35078e
27. Jochum FD, Theato P (2013) Temperature- and light-responsive smart polymer materials. Chem Soc Rev 42:7468–7483. https://doi.org/10.1039/c2cs35191a
28. Dai S, Ravi P, Tam KC (2008) pH-responsive polymers: synthesis, properties and applications. Soft Matter 4:435–449. https://doi.org/10.1039/b714741d
29. Reiriz C, Brea RJ, Arranz R, Carrascosa JL, Garibotti A, Manning B, Valpuesta JM, Eritja R, Castedo L, Granja JR (2009) α,γ-peptide nanotube templating of one-dimensional parallel fullerene arrangements. J Am Chem Soc 131:11335–11337. https://doi.org/10.1021/ja904548q
30. Montenegro J, Vazquez-Vazquez C, Kalinin A, Geckeler KE, Granja JR (2014) Coupling of carbon and peptide nanotubes. J Am Chem Soc 136:2484–2491. https://doi.org/10.1021/ja410901r
31. Cuerva M, Garcia-Fandino R, Vazquez-Vazquez C, Lopez-Quintela MA, Montenegro J, Granja JR (2015) Self-assembly of silver metal clusters of small atomicity on cyclic peptide nanotubes. ACS Nano 9:10834–10843. https://doi.org/10.1021/acsnano.5b03445
32. Amorin M, Perez A, Barbera J, Ozores HL, Serrano JL, Granja JR, Sierra T (2014) Liquid crystal organization of self-assembling cyclic peptides. Chem Commun 50:688–690. https://doi.org/10.1039/c3cc47400c
33. Mendez-Ardoy AM, Granja JR, Montenegro J (2018) pH-triggered self-assembly and hydrogelation of cyclic peptide nanotubes confined in water micro-droplets. Nanoscale Horizons 3:391–396. https://doi.org/10.1039/c8nh00009c
34. Brea RJ, Reiriz C, Granja JR (2010) Towards functional bionanomaterials based on self-assembling cyclic peptide nanotubes. Chem Soc Rev 39:1448–1456. https://doi.org/10.1039/b805753m
35. Rubin DJ, Nia HT, Desire T, Nguyen PQ, Gevelber M, Ortiz C, Joshi NS (2013) Mechanical reinforcement of polymeric fibers through peptide nanotube incorporation. Biomacromol 14:3370–3375. https://doi.org/10.1021/bm4008293
36. Rubin DJ, Amini S, Zhou F, Su HB, Miserez A, Joshi NS (2015) Structural, nanomechanical, and computational characterization of D, L-cyclic peptide assemblies. ACS Nano 9:3360–3368. https://doi.org/10.1021/acsnano.5b00672
37. Guerette PA, Hoon S, Seow Y, Raida M, Masic A, Wong FT, Ho VHB, Kong KW, Demirel MC, Pena-Francesch A, Amini S, Tay GZ, Ding D, Miserez A (2013) Accelerating the design of biomimetic materials by integrating RNA-seq with proteomics and materials science. Nat Biotechnol 31:908–918. https://doi.org/10.1038/nbt.2671
38. Byrne N, Menzies D, Goujon N, Forsyth M (2013) Inducing alignment of cyclic peptide nanotubes through the use of structured ionic liquids. Chem Commun 49:7729–7731. https://doi.org/10.1039/c3cc43837f
39. Garcia-Fandino R, Amorin M, Castedo L, Granja JR (2012) Transmembrane ion transport by self-assembling α,γ-peptide nanotubes. Chem Sci 3:3280–3285. https://doi.org/10.1039/c2sc21068a
40. Ruiz L, Benjamin A, Sullivan M, Keten S (2015) Regulating ion transport in peptide nanotubes by tailoring the nanotube lumen chemistry. J Phys Chem Lett 6:1514–1520. https://doi.org/10.1021/acs.jpclett.5b00252

41. Montenegro J, Ghadiri MR, Granja JR (2013) Ion channel models based on self-assembling cyclic peptide nanotubes. Acc Chem Res 46:2955–2965. https://doi.org/10.1021/ar400061d
42. Rodriguez-Vazquez N, Amorin M, Granja JR (2017) Recent advances in controlling the internal and external properties of self-assembling cyclic peptide nanotubes and dimers. Org Biomol Chem 15:4490–4505. https://doi.org/10.1039/c7ob00351j
43. Danial M, Perrier S, Jolliffe KA (2015) Effect of the amino acid composition of cyclic peptides on their self-assembly in lipid bilayers. Org Biomol Chem 13:2464–2473. https://doi.org/10.1039/c4ob02041c
44. Liu H, Chen J, Shen Q, Fu W, Wu W (2010) Molecular insights on the cyclic peptide nanotube-mediated transportation of antitumor drug 5-fluorouracil. Mol Pharm 7:1985–1994. https://doi.org/10.1021/Mp100274f
45. Vijayaraj R, Van Damme S, Bultinck P, Subramanian V (2013) Theoretical studies on the transport mechanism of 5-fluorouracil through cyclic peptide based nanotubes. PCCP 15:1260–1270. https://doi.org/10.1039/c2cp42038d
46. Chen J, Zhang B, Xia F, Xie Y, Jiang S, Su R, Lu Y, Wu W (2016) Transmembrane delivery of anticancer drugs through self-assembly of cyclic peptide nanotubes. Nanoscale 8:7127–7136. https://doi.org/10.1039/c5nr06804e
47. Fernandez-Lopez S, Kim HS, Choi EC, Delgado M, Granja JR, Khasanov A, Kraehenbuehl K, Long G, Weinberger DA, Wilcoxen KM, Ghadiri MR (2001) Antibacterial agents based on the cyclic D,L-α-peptide architecture. Nature 412:452–455. https://doi.org/10.1038/35086601
48. Motiei L, Rahimipour S, Thayer DA, Wong CH, Ghadiri MR (2009) Antibacterial cyclic D,L-α-glycopeptides. Chem Commun:3693–3695. https://doi.org/10.1039/b902455g
49. Larnaudie SC, Sanchis J, Nguyen TH, Peltier R, Catrouillet S, Brendel JC, Porter CJH, Jolliffe KA, Perrier S (2018) Cyclic peptide-Poly(HPMA) nanotubes as drug delivery vectors. In vitro assessment pharmacokinetics and biodistribution. Biomaterials 178:570–582. https://doi.org/10.1016/j.biomaterials.2018.03.047
50. Li M, Ehlers M, Schlesiger S, Zellermann E, Knauer SK, Schmuck C (2016) Incorporation of a non-natural arginine analogue into a cyclic peptide leads to formation of positively charged nanofibers capable of gene transfection. Angew Chem Int Ed 55:598–601. https://doi.org/10.1002/anie.201508714
51. Hsieh WH, Chang SF, Chen HM, Chen JH, Liaw J (2012) Oral gene delivery with cyclo-(D-Trp-Tyr) peptide nanotubes. Mol Pharm 9:1231–1249. https://doi.org/10.1021/Mp200523n
52. Lee YH, Chang SF, Liaw J (2015) Anti-apoptotic gene delivery with cyclo-(D-Trp-Tyr) peptide nanotube via eye drop following corneal epithelial debridement. Pharmaceutics 7:122–136. https://doi.org/10.3390/pharmaceutics7030122

Chapter 7
Protein-Based Nanotubes

7.1 Template Process

One of the practical methodologies for the fabrication of organic tubular architectures involves a template technique. Potent template methods to prepare polymer nanotubes including synthetic and natural macromolecules can be divided into two approaches, i.e., *endo*-templating and *exo*-templating (Fig. 7.1) [1]. As building blocks of the nanotube components, synthetic amphiphiles [2], polymerizable monomers [3], and polymers [4] are available. Various shapes and roles of the template utilized lead to the formation of the polymer nanotubes with diverse composition of membrane walls, controlled diameters, and the number of constituent layers. Figure 7.2a illustrates an *endo*-templating with the use of polycarbonate (PC) membrane or microporous alumina. In the first stage, wetting process starts when the inner surface of the porous templates contacts with the solution containing the amphiphiles, monomers, or polymers. A thin film, thus, coats the pore walls. Wetting of the pore inner surface and complete filling of the pores depend on different time scales. The template process completes through selective removal of the template material after nanotube formation. The resultant-templated nanotube materials should display high physical and chemical stabilities. Moreover, these nanotubes, therefore, resist to harsh environments such as strong alkaline aqueous phase and dichloromethane that are employed to remove the templates.

Proteins are functional and yet structurally defined biomacromolecules that can exhibit a great deal of vital functions. Particularly, notable functions include storage, structural, mechanical, enzymatic, transport, chaperone, and immune actions. As described above, a procedure of wetting LbL template synthesis [5] in a nanoporous membrane can afford one of reliable and facile approaches to fabricate the structure-guided protein-based nanotube architectures [6]. Critical dimensions of the nanotubes including outer diameter, wall thickness, and length are accurately controllable by the pore diameter, the number of deposited layers, and the template thickness, respectively. Integration of the use of an appropriate protein as a building block with the

T. Shimizu, *Smart Soft-Matter Nanotubes*, Nanostructure Science and Technology,
https://doi.org/10.1007/978-981-16-2685-2_7

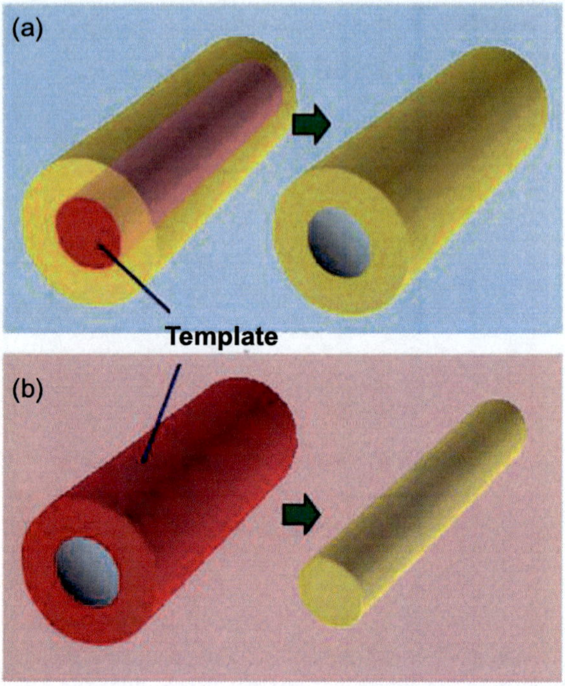

Fig. 7.1 **a** *Exo*- and **b** *endo*-templating. The templates for further synthetic process are shown in (**a**) and (**b**). Reproduced with permission from Ref. [1], © 2005 American Chemical Society

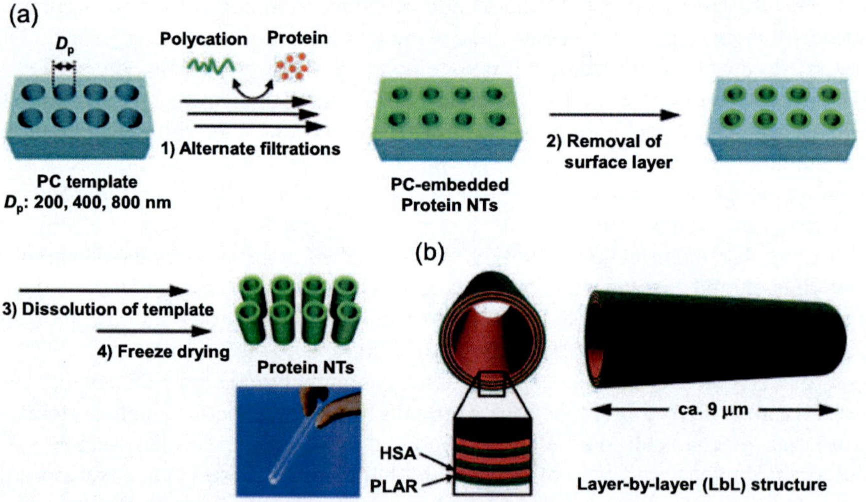

Fig. 7.2 **a** Procedure of wetting LbL template synthesis with a track-etched nanoporous PC membrane, yielding protein-based NTs. **b** Schematic image of $(PLAR/HSA)_3$ NT prepared using porous PC template (D_p = 400 nm). Reproduced with permission from Ref. [7], © The Royal Society of Chemistry 2012

alternate LbL assembly technique can, therefore, provides us with smart protein-based nanotubes. The reason why is that totally tunable disposition of desired protein components can give tailored, morphologically homogeneous polypeptide nanotubes [7].

7.2 Human Serum Albumin (HSA)–Poly-L-Arginine (PLAR)

Komatsu and co-workers developed a new LbL assembly technique using both a track-etched PC as a porous membrane (pore diameter: 400 nm) and DMF as a dissolving solvent [7]. For example, the combination of two building blocks, i.e., a positively charged PLAR and a negatively charged protein, e.g., HSA, yielded homogeneous $(PLAR/HSA)_3$ nanotubes (Fig. 7.2b) [8]. Lyophilized solid powders were obtained after alternative three depositions of the protein and polymer solution, the removal of the template membrane by DMF, and freeze-drying of the liberated nanotubes. SEM analyses revealed that the outer diameter, wall thickness, and tube length are 407 ± 13 nm, 50 ± 4 nm, and ca. 9 μm, respectively.

HSA, known as a multifunctional simple protein, is included the most in plasma. Particularly, HSA functions as a depot or carrier for many endogeneous ligands such as fatty acids and bilirubin, as well as exogeneous compounds including drugs. Unless the intrinsic 3D structure of the HSA as a constituent changes during the alternate LbL deposition process, the resultant $(PLAR/HSA)_3$ nanotubes keep the ability to capture ligand molecules as entire molecular assemblies. For example, HSA can strongly bind uranyl cation (UO_2^{2+}) that is well known as a negative staining agent for electron microscopy observation. Depending on the concentration of uranyl acetate used for the TEM observations, the resultant images remarkably differ, particularly in that the innermost HSA layer was stained darker than the outer layers [9]. This finding means that the $(PLAR/HSA)_3$ nanotubes really encapsulate the uranyl cations into the lumen and then the cations permeate from the interior to exterior surface. The $(PLAR/HSA)_3$ nanotubes also encapsulate the cationic fluorescent dye, 3,3′-diethylthiacarbocyanine iodide (DTC) **1** (Chart 7.1) that is known as a ligand for HSA (Fig. 7.3) [9]. No encapsulation phenomena was, however, observed for the cationic ligand, rhodamine 123 (R123) **2** (Chart 7.1) that is unable to bind to HSA.

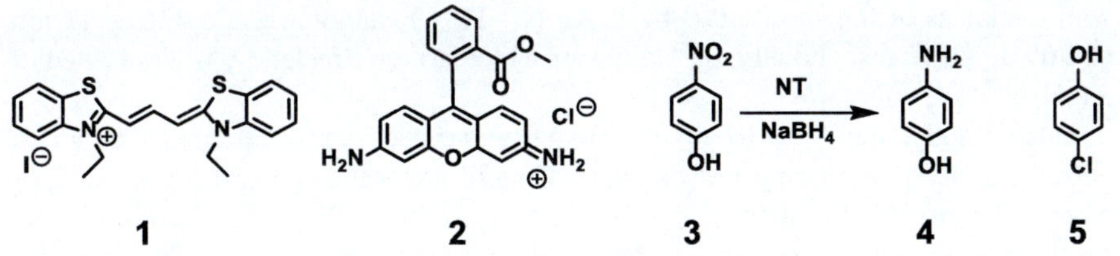

Chart 7.1 Chemical structures of **1–5**

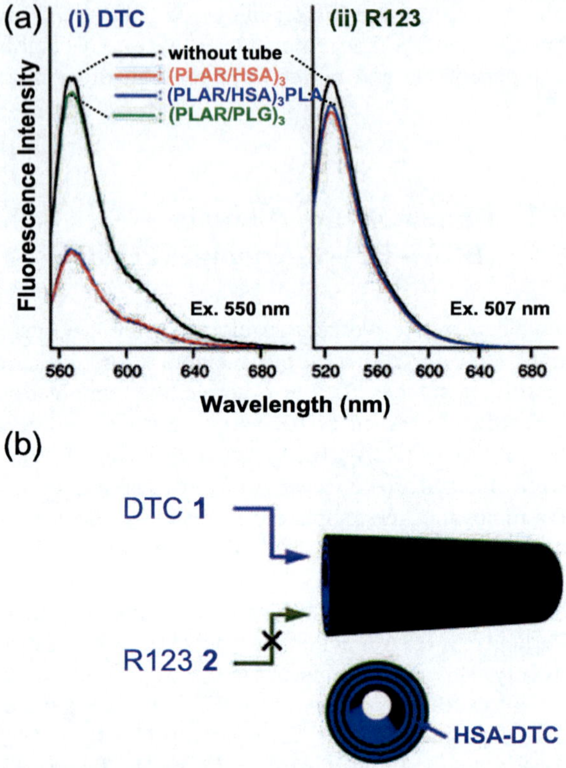

Fig. 7.3 **a** Fluorescence spectra of the sodium phosphate-buffered solution (pH 7.0, 10 mM) of (i) DTC **1** (0.2 μM) and (ii) R123 **2** (0.2 μM) after incubation with PRNTs, with subsequent centrifugation (4000 g, 10 min). **b** Schematic image of ligand capture in $(PLAR/HSA)_3$ nanotubes. The DTC **1** molecules are bound to the HSA layers of the nanotubes. Reproduced with permission from Ref. [9], © 2010 American Chemical Society

7.3 $(PLAR/Au\ NP–HSA)_3$ and $(PLAR/Ferritin)_3$

HSA has a strong tendency to bind to Au NPs, thus forming the Au NP–HSA core–shell conjugates. Fluorescence spectroscopy showed that a binding constant of HSA to Au NPs is consistent with the K value of $1.25 \times 10^9\ M^{-1}$. TEM images showed that the obtained Au NP–HSA conjugates with dimensions of 20.1 ± 2.9 nm have a bright exterior shell (ca. 3 nm) that differs from the exterior boundary of the intrinsic Au NPs. By using the similar alternating LbL technique as mentioned before, Komatsu and co-workers carried out the fabrication of functional protein nanotubes, in which Au NPs are embedded in the protein layer wall (Fig. 7.4) [10]. The outer diameter and wall thickness of the resultant $(PLAR/Au\ NP–HSA)_3$ nanotubes are 426 ± 12 nm and 65 ± 7 nm, respectively, the values of which are unaffected by the presence of Au NPs.

Interestingly, the $(PLAR/Au\ NP–HSA)_3$ nanotubes were calcinated in air at 500 °C, thereby producing red powders of solid nanotubes consisting of Au NPs of ca. 13 nm [10]. Calcination process slightly reduced the nanotube dimension, resulting in the outer diameter of 195 ± 10 nm, wall thickness of 41 ± 4 nm, and tube length of ca. 4 μm. This burning process can completely consume the organic

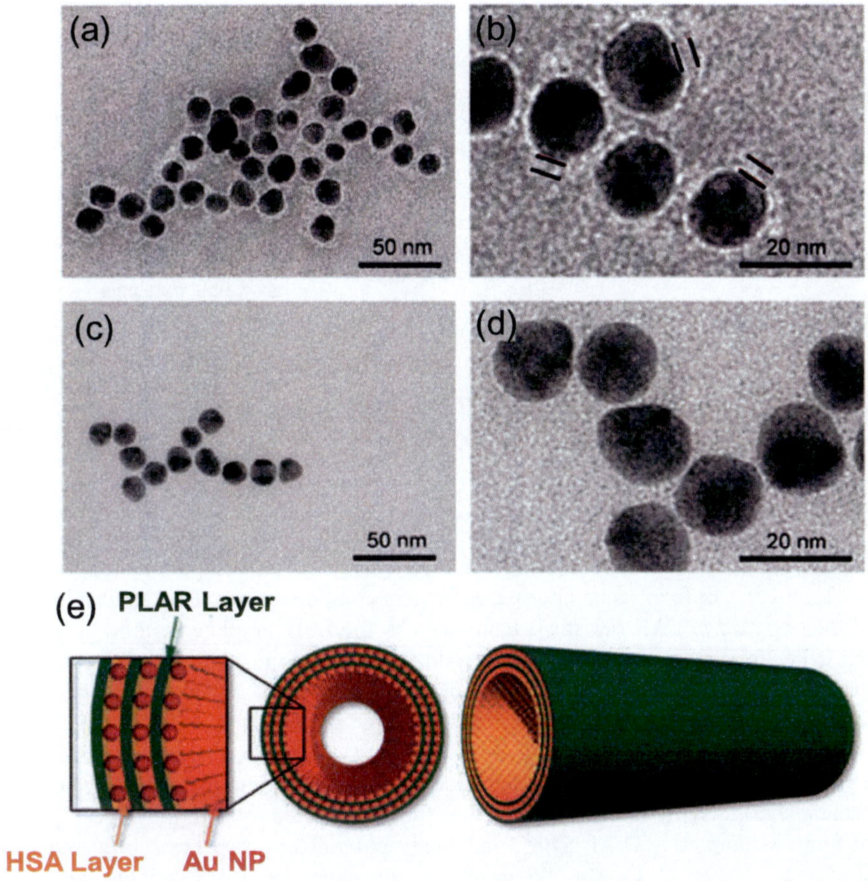

Fig. 7.4 TEM images of **a**, **b** core–shell Au NP–HSA conjugates and **c**, **d** bare Au NPs. In (**b**), the spacer between the black bars indicates the thickness of the absorbed HSAs on Au NP surface. **e** Schematic image of (PLAR/Au NP–HSA)$_3$ NT, in which numerous pieces of Au NPs are embedded in the HSA layers. Reproduced with permission from Ref. [10], © 2013 American Chemical Society

HSA and PLAR components. The resultant (PLAR/Au NP–HSA)$_3$ nanotubes and their calcinated products Au NP-based nanotubes were subjected to the demonstration experiment that catalyze the chemical reduction of 4-nitrophenol **3** (Chart 7.1) to 4-aminophenol **4** (Chart 7.1) with sodium borohydride. As a result of the conversion rate of **3**, the (PLAR/Au NP–HSA)$_3$ nanotubes exhibited more excellent catalytic performance than the Au nanotubes (Fig. 7.5) [10]. The reason for the high efficiency will be that the Au NPs are homogeneously distributed in the nanotube wall and yet the substrate molecules **3** permeate easily through the protein layers.

They also exploited the novel fabrication of iron oxide nanotubes as a solid material by using the same procedure as that used for the preparation of Au NPs-based

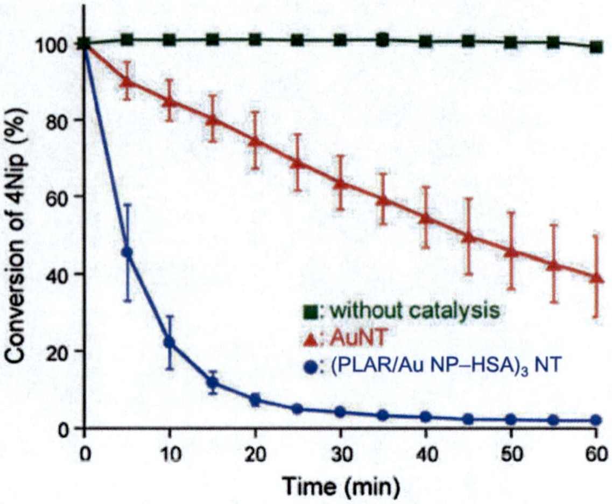

Fig. 7.5 Conversion of 4-nitrophenol **3** in aqueous dispersions of the (PLAR/Au NP–HSA)$_3$ nanotubes with $NaBH_4$ at 25 °C. Reproduced with permission from Ref. [10], © 2013 American Chemical Society

nanotubes [11]. Ferritin is an iron storage protein that has a globular protein shell composed of 24 protein subunits. Such a ferritin serves as a supply source of iron oxide and was utilized as an alternating protein component with PLAR. Calcination of the obtained (PLAR/Ferritin)$_3$ nanotubes by the LbL template method produced the α-Fe_2O_3 nanotubes colored with reddish brown (Fig. 7.6a) [11]. The tubular dimensions of the intrinsic nanotubes drastically reduced. For example, the outer diameter, wall thickness, and tube length change from 410 ± 14 to 211 ± 8 nm, from 61 ± 5 to 29 ± 3 nm, and from ca. 10 to ca. 5 μm, respectively. Burning treatment of ferritin at 500 °C in air underwent the complete disappearance of the protein shell, leaving the iron oxide cores that aggregated to form the α-Fe_2O_3 NPs of 5 nm in diameter. TEM, SEM, and X-ray photoelectron spectroscopy analyses confirmed that the α-Fe_2O_3 NPs configure the nanotube wall. Variable temperature superconducting quantum interference device measurement showed that the α-Fe_2O_3 nanotubes displayed superparamagnetism with a blocking temperature of 37 K, which is strongly different from that of bulk α-Fe_2O_3 [11]. The nanotubes also exhibited excellent photocatalytic activity for the degradation of 4-chlorophenol **5** (Chart 7.1) degradation in the presence of H_2O_2 upon irradiation of visible light as compared with the bulk α-Fe_2O_3 (Fig. 7.6b) [11].

7.4 (PLAR/HSA)$_8$PLAR/Pt Nanoparticles

The attractive feature of the wetting template synthesis for protein-integrated nanotubes is the tailoring ability for targeting functional proteins to locate at desired layer positions. The PLAR and HSA solutions were permeated into the porous PC membrane in an alternating way (totally, each eight cycles). Next again, the PLAR

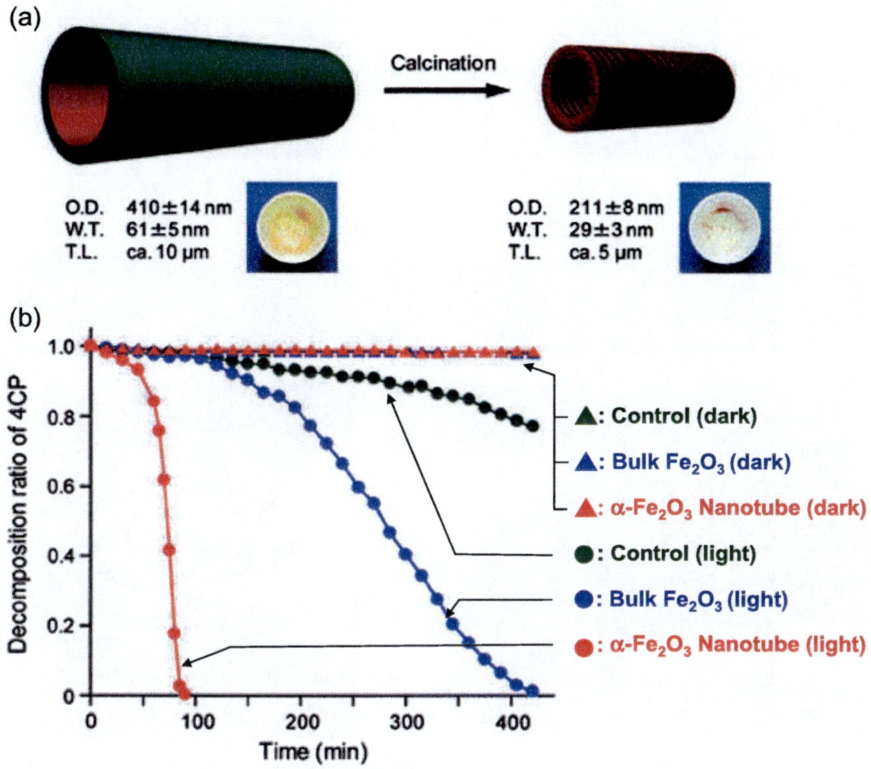

Fig. 7.6 **a** Calcination of (PLAR/Ferritin)$_3$ nanotubes to produce iron oxide nanotubes (*O.D.* outer diameter; *W.T.* wall thickness; *T.L.* tube length) and photographs of lyophilized (PLAR/Ferritin)$_3$ pale-yellow nanotubes in a crucible (left) and calcinated iron oxide nanotubes in reddish-brown color (right). **b** Photodegradation of 4-chlorophenol **5** in aqueous solution with H_2O_2 under visible light irradiation. Reproduced with permission from Ref. [11], © 2010 American Chemical Society

solution and subsequently the aqueous dispersion of citrate-capped Pt NPs (diameter: ca. 5 nm) were injected into the micropores to finish up the template-protected protein tubular structures. The template removal by DMF immersion resulted in the fabrication of multiple protein-layered (PLAR/HSA)$_8$PLAR/Pt NP microtubes with outer diameter, wall thickness, and maximum tube length of 1.16 ± 0.02 μm, 147 ± 11 nm, and ca. 23 μm, respectively [12]. Electrostatic interaction facilitates the strong binding of Pt NP layer with negative charges to the positively charged PLAR layer. The 18-layered protein hybrid nanotubes complete the innermost layer of which is covered with Pt NPs (Fig. 7.7).

Straight, hollow cylindrical structures with well-defined nano- or micrometer-scale dimensions have attracted emerging attention as a platform of an autonomous motor and smart carrier. Their rocket-like form, free installation of desired functionalities and propulsion components, and capture/deliver/release capability of targeting

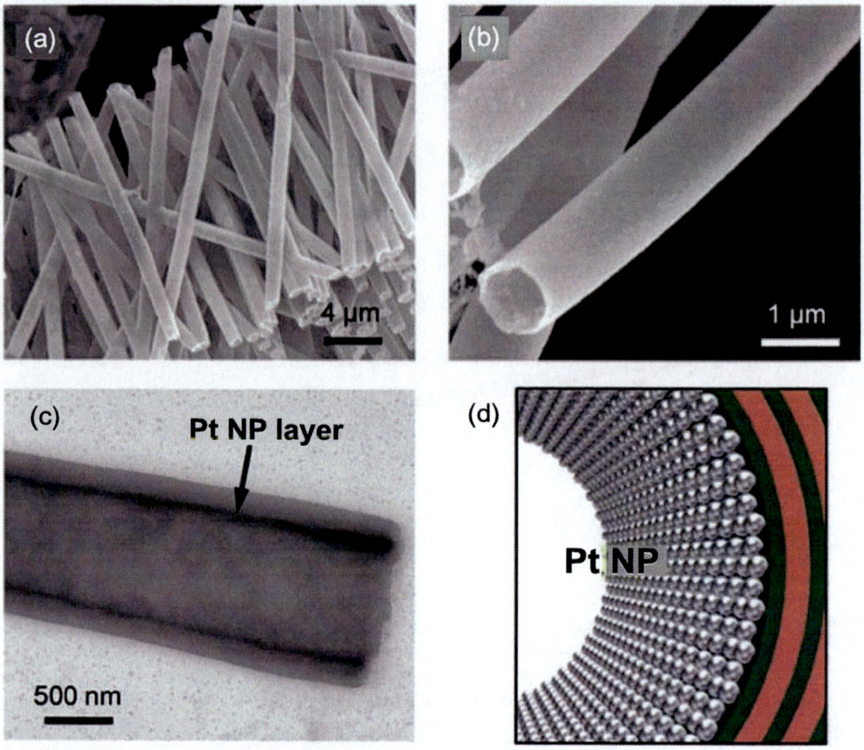

Fig. 7.7 **a**, **b** SEM images for the microtubes of HSA–Pt NPs, which were fabricated by using a 1.2 μm porous PC template. **c** TEM image of a HSA–Pt NP microtube. The Pt NP layer is clearly visible on the inner surface of the microtube. **d** Schematic image of the Pt NP layer as an inner layer of the HSA–Pt NP microtube. Reproduced with permission from Ref. [12], © 2017 WILEY-VCH Verlag GmbH & Co. KGaA, Weinheim

substances also intrigue us to explore a novel drug delivery system. In this context, He's research group reported the fabrication of self-propelled nano- and microrockets through template-assisted LbL assembly of positively charged chitosan and negatively charged sodium alginate [13]. The propulsion energy of the high-axial-ratio objects utilizes the catalytic decomposition reaction of hydrogen peroxide to water and oxygen. Komatsu and co-workers also applied the $(PLAR/HSA)_8PLAR/Pt$ NP nanotubes with Pt NPs interior surface for a self-propelled microtube motor [12]. The tubes can move in a self-propelled manner by ejecting O_2 bubbles from opened one end (Fig. 7.8). If a layer of γ-Fe_3O_4 NPs (MNP) is incorporated within the layer, the movement direction of the resultant $PLAR/HSA/MNP(PLAR/HSA)_5PLAR/Pt$ NP microtubes can be manipulated under a magnetic field [12]. Notably, the bubble-propelled protein nanotubes can also adsorb *Escherichia coli* efficiently to the positively charged exterior surfaces [12]. These approaches pertaining to microsize motor

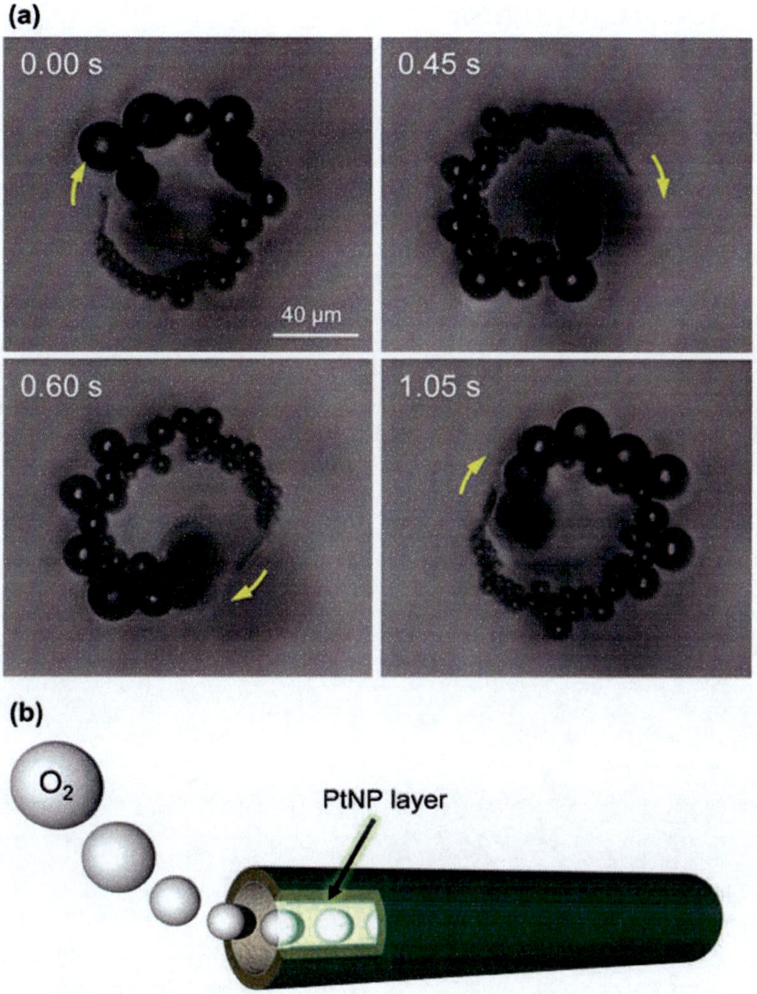

Fig. 7.8 **a** Time-lapse video images observed by inverted microscopy for the motion of self-propelled HSA–Pt NP microtubes by jetting O_2 in phosphate-buffered solution (pH 7.0, 10 mM). The microtube is moving with clockwise rotation. **b** Schematic image of the self-propulsion in aqueous H_2O_2 solution. Reproduced with permission from Ref. [12], © 2017 WILEY-VCH Verlag GmbH & Co. KGaA, Weinheim

should lead to site-targeting delivery and eventually remote-operated release of drugs in the surrounding areas of cells and tissues.

7.5 $(PLAR/HSA)_2PLAR/\alpha GluD$ and $(PLAR/HSA)_2PLAR/CalB$

Nanoscale hollow cylindrical structures are also very attractive as nanovessels for tiny channel bioreactor. Alternating LbL assembly technique in a track-etched PC membrane allowed an enzyme layer to be immobilized on the innermost layer of multilayered protein nanotubes. Martin and co-workers first prepared the protein nanotubes from GOD or hemoglobin by a LbL deposition approach using porous membrane of anodic aluminum oxide (AAO) as a template (Fig. 7.9) [6]. The deposition of glutaraldehyde solution served as a crosslinker that can tightly immobilize each protein layer. The drawback of this strategy using the AAO template was that the H_3PO_4 as a dissolving solvent for the template caused unavoidable deformation of the tubular structures under acidic conditions. However, the nanotubes were shown to exhibit higher glucose oxidation activity than that of the nanotubes left immobilized within the AAO pores [6].

To improve the serious defect of the GOD nanotubes, Komatsu and co-workers newly explored α-glucosidase (α-GluD)-embedded nanotubes into the inner most layer [14]. The αGluD enzyme catalyzes the hydrolysis of terminal nonreducing α(1

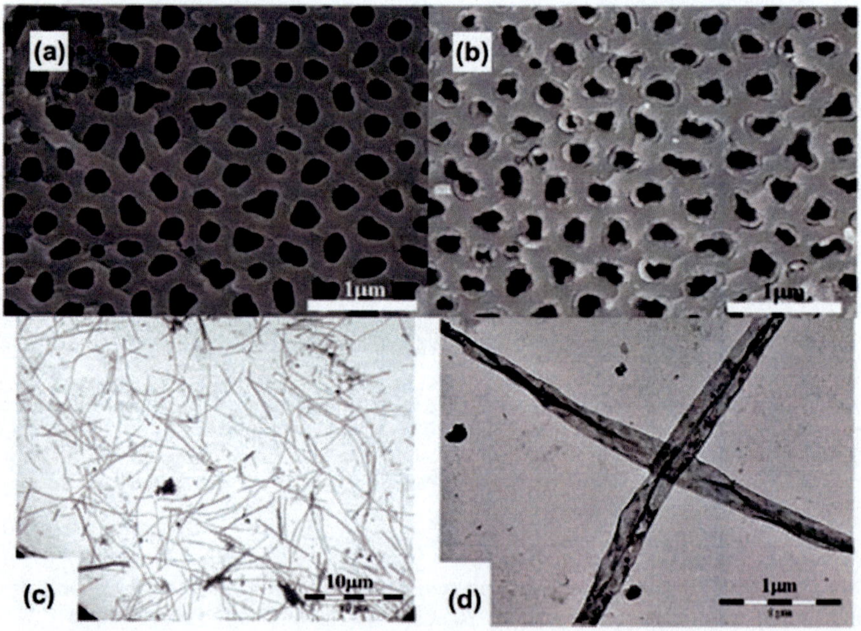

Fig. 7.9 SEM images of the surface of the alumina membrane as a template **a** before and **b** after deposition of nanotubes with six-layers GOD. **c, d** TEM images of the liberated nanotubes taken at two different magnification factors. Reproduced with permission from Ref. [6], © 2005 American Chemical Society

→ 4) glucosyl linkage, releasing a single α-glucose molecule. The wetting template synthesis eventually yielded a white powder of the αGluD-embedded protein nanotubes. The layer sequence is thus represented by $(PLAR/HSA)_2PLAR/\alpha GluD$ since the αGluD has a negative charge to strongly bind to the PLAR layer at pH 7.0. The resultant α-GluD nanotube can encapsulate the fluorescent 4-methyl-umbelliferyl-α-D-glucopyranoside (MUGlc) **6** (Fig. 7.10) as a substrate into the nanochannel space, thereby hydrolyzing the substrate to yield α-D-glucose **7** and 4-methyl-umbelliferone (MU) **8** (Fig. 7.10) [14]. Meanwhile, the biotinylated α-GluD (B-αGluD) layer is also fabricated on the interior surface of the $(PLAR/HSA)_2PLAR/PLG/Avi$ nanotubes (PLG: poly-L-glutamic acid) having an avidin (Avi) innermost layer through the Avi–biotin biospecific interaction [14]. The obtained $(PLAR/HSA)_2PLAR/PLG/Avi\text{-}B\text{-}\alpha GluD$ nanotubes was also shown to hydrolyze MUGlc, whereas no catalytic reaction was observed for the mother $(PLAR/HSA)_2PLAR/PLG/Avi$ nanotubes. Interestingly, both K_M values of the $(PLAR/HSA)_2PLAR/\alpha GluD$ and $(PLAR/HSA)_2PLAR/PLG/Avi\text{-}B\text{-}\alpha GluD$ nanotubes are roughly same as those of free αGluD and B-α-GluD, respectively. The k_{cat} values for both αGluD-coated nanotubes are, however, markedly low. Unfavorable geometries of αGluD or B-αGluD in a somewhat confined interior space should be associated with the relatively lower enzyme activity.

Enzymatic polymerization of confined monomer molecules in a nanotubular architecture has never been described. Komatsu and co-workers took notice of *Candida antarctica* lipase B (CalB) that can catalyze ester hydrolysis reaction and ROP of lactones. The alternating LbL template synthesis strategy was applied for the production of multilayered protein nanotubes containing the CalB component as

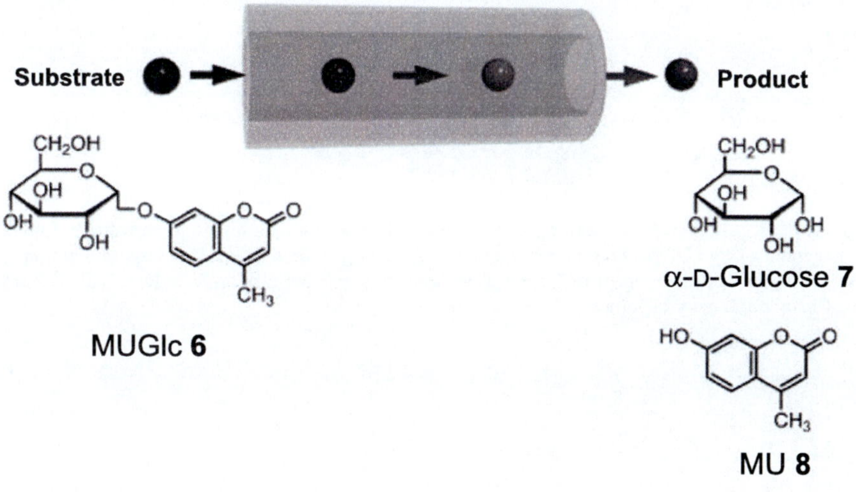

Fig. 7.10 Enzymatic exohydrolysis of the α(1 → 4) glucosyl bond of MUGlc in the 1D hollow space of protein nanotubes with an interior enzyme surface. Reproduced with permission from Ref. [14], © 2011 WILEY-VCH Verlag GmbH & Co. KGaA, Weinheim

the innermost layer [15]. They determined the K_M and k_{cat} values of the obtained $(PLAR/HSA)_2PLAR/CalB$ nanotubes for the hydrolysis of 4-methyl-umbelliferyl butyrate (MUB) **9** into MU **8** (Fig. 7.11). The obtained K_M value was nearly comparable to that of the free enzyme, whereas the k_{cat} value was considerably lower. Perhaps, the confined conditions in the nanochannel will cause the CalB protein to take unfavorable geometry in a similar way as the α-GluD nanotubes. On the contrary, the CalB protein nanotubes displayed remarkable catalytic activity for the ROP of 12-dodecanolactone (DDL) **10** (Fig. 7.11) [15]. For example, the DDL molecules can be polymerized to yield ester oligomers **11** (Fig. 7.11) with relatively higher molecular weights than those catalyzed in the presence of free CalB. The multi-layered protein nanotubes having the inner most layer decorated with enzyme proteins should lead to a novel enzymatic nanotube reactor.

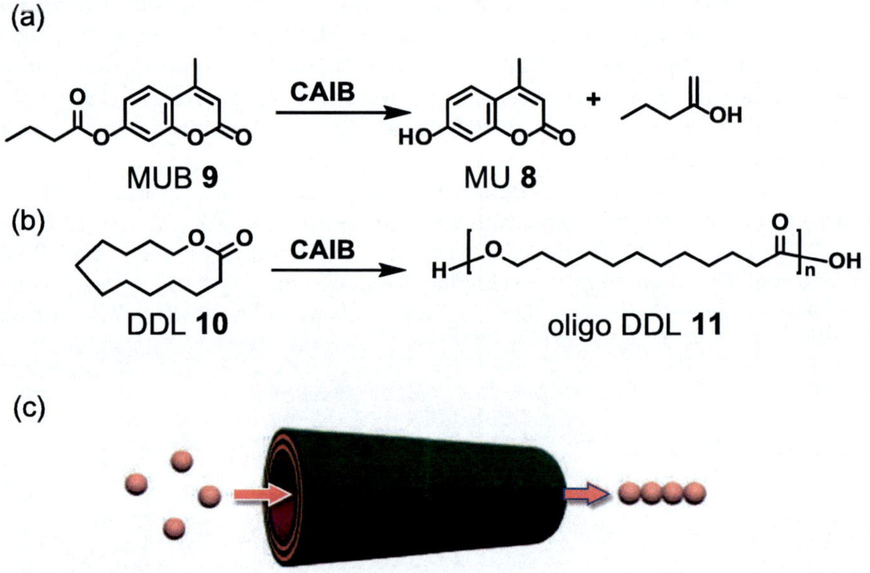

Fig. 7.11 **a** Hydrolysis reaction of MUB **9** to yield MU **8**, **b** ring-opening oligomerization of DDL **10** to generate oligo DDL **11**, and **c** schematic image of enzymatic ring-opening oligomerization of DDL **10** in the hollow cylinder of CalB NT. Reproduced with permission from Ref. [15], © 2015 The Chemical Society of Japan

7.6 $(PLAR/HSA)_5$PLAR/Fetuin and $(PLAR/HSA)_2$PLAR/PLG/Hepatitis B Surface Antigen

Let us suppose that the nanoporous PC membrane with 400-nm pore size was used for template synthesis of a protein nanotube. The diameters of the resultant swollen protein nanotubes in water are generally estimated to be ca. 200 nm. The dimensions of the lumen are therefore suitable for the encapsulation of relatively larger guest substances of 50–100 nm size. For example, corresponding target objects include metal sub-nm particle, large macromolecules, dendrimers, and viruses, which are 10–100 times larger than the small guest used in the conventional host–guest chemistry [16]. Trapping of viruses of several tens of nanometer scales is, particularly, challenging for the protein nanotubes in terms of mesoscale host–guest chemistry [17–19]. Multi-layered hollow cylindrical structures (outer and inner diameters: 400–430 and 210–280 nm, respectively), in which component protein layers are circumferentially arranged in appropriate order, were first subjected to a virus capture experiment. The particle of HBV (radius = 42 nm) is composed of both hepatitis B surface antigen (HBsAg)-embedded lipid envelope and an icosahedral nucleocapsid core including genome DNA. Komatsu and co-workers fabricated the multi-layered protein nanotubes $(PLAR/HSA)_2$PLAR/PLG/HBsAb nanotubes (AbNTs), in which anti-HBsAg antibody (HBsAb) is arranged as the innermost layer [20]. Notably, swollen AbNTs were able to trap noninfectious spherical particles self-organized from HBsAg with the binding constant of 3.5×10^7 M^{-1} in aqueous media (Fig. 7.12). Moreover, the AbNTs were shown to encapsulate the infectious HBV into the nanotube lumens.

Thereafter, they demonstrated that multi-layered, fetuin-involved protein nanotube at the innermost layer capture influenza A viruses (94 nm) through virus–receptor biospecific interactions [21]. Binding of the influenza A virus hemagglutinin to a cell surface receptor, which includes sialyloligosaccharide with *N*-acetyl neuraminic acid (Neu5Ac) terminals, triggers viral infection. Fetuin from fetal calf serum is a glycoprotein having tribranched oligosaccharides including similar Neu5Ac terminals as that of the influenza receptor. As previously described, the template process using LbL assembly allowed for free introduction of the glycoprotein receptor into the inner surface of the protein nanotube [7]. The resultant $(PLAR/HSA)_5$PLAR/fetuin nanotubes were demonstrated to encapsulate influenza A virus PR8 particles (Fig. 7.13). Programmable protein-derived nanotubes will pave the way for medical applications of smart organic nanotubes.

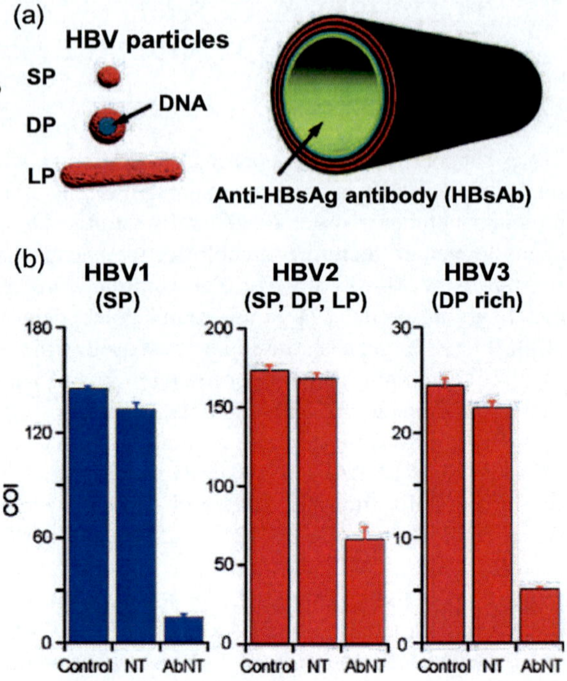

Fig. 7.12 **a** Three different HBV particles (SP, DP, and LP) and $(PLAR/HSA)_2PLAR/PLG/HBsAb$ nanotube (AbNT). **b** Amount of remaining HBsAg in the solutions of HBV1, HBV2, and HBV3 after incubation with AbNTs, with NTs, and without tubes (control). The vertical axis is defined as the cutoff index. Reproduced with permission from Ref. [20], © 2011 American Chemical Society

7.7 Collagen-Based Nanotube

Collagen (COL) is present as a fibrillar protein in extracellular matrix, playing a vital role as the chief ingredient of connective tissue. It can also offer scaffolds for functional expression of cell, i.e., differentiation, proliferation, and attachment. Separation and collection of COL-derived nanotubes on a solid substrate is of great interest to investigate the cell–COL interaction. Demoustier-Champagne and co-workers presented the first construction of COL-based nanotubes through the template synthesis based on LbL assembly of the multi-layers of anionic poly(styrenesulfonate) (PSS) **12** (Chart 7.2) and cationic COL (Fig. 7.14) [22]. In this methodology, cationic poly(allylamine hydrochloride) (PAAH) **13** (Chart 7.2) was adsorbed onto the pore walls of a polycarbonate track-etched membrane in advance to fabricate an anchoring layer. They also confirmed that at least six depositions of PSS/COL bilayers are required to form mechanically robust nanotubular structures.

They also explored the fabrication of a novel biointerface composed of COL-based nanotubes [23]. Alternating LbL assembly of either the combination of COL and PSS or that of fluorescent PAAH (Flu-PAAH) and PSS successfully produced two different multi-layered nanotubes, i.e., $(COL/PSS)_3(Flu\text{-}PAAH/PSS)_3$ having collagen layer as external one and $(Flu\text{-}PAAH/PSS)_3(COL/PSS)_3$ having PAAH layer as external one. They applied electrophoretic deposition (EPD) not only for the

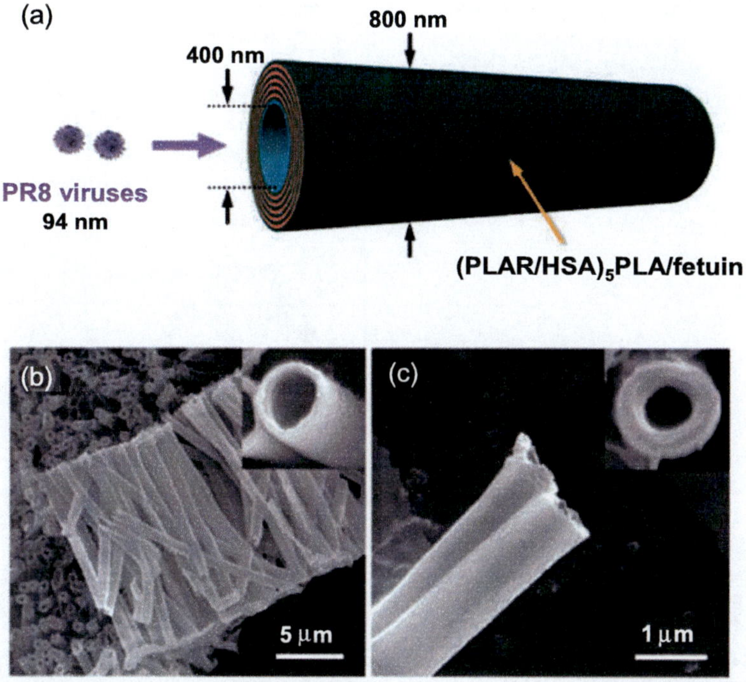

Fig. 7.13 **a** Schematic image of influenza A virus PR8 and fetuin nanotube. **b**, **c** SEM images of fetuin nanotubes; **b** dried sample and **c** lyophilized sample after swelling in water. The nanotubes were fabricated using a porous PC template (800 nm pore diameter). Reproduced with permission from Ref. [21], © 2017 The Chemical Society of Japan

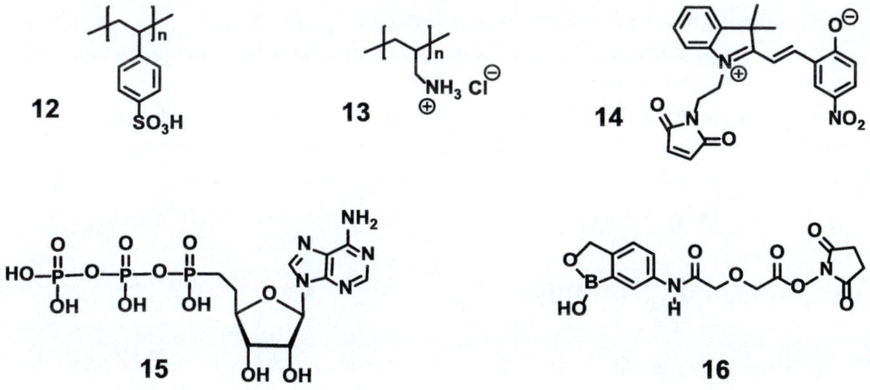

Chart 7.2 Chemical structures of **12–16**

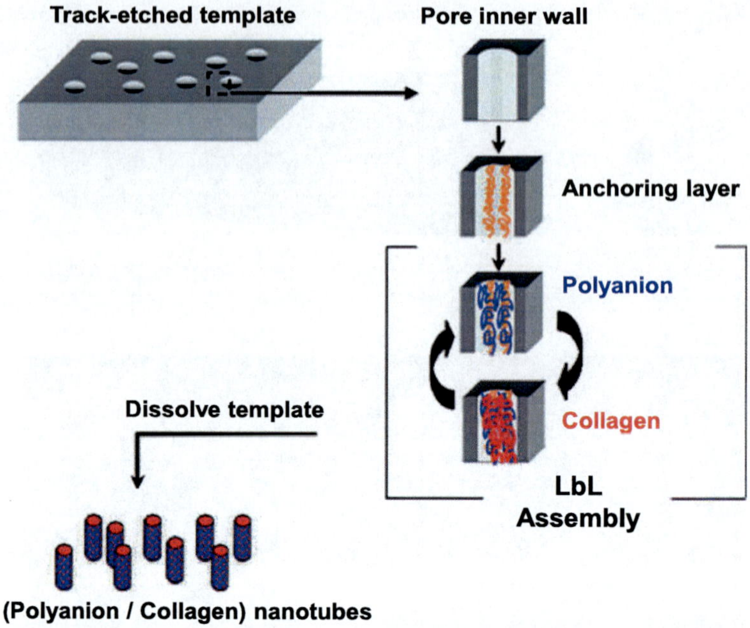

Fig. 7.14 Nanotube fabrication via the membrane-templated LbL assembly. Reproduced with permission from Ref. [22], © 2009 American Chemical Society

complete separation of the collagen-based nanotubes from the nanoporous polycarbonate template but also for the collection of the nanotubes on indium–tin oxide (ITO) glass (Fig. 7.15) [23]. A range of artificial biointerfaces coated with COL nanotubes of well-defined dimensions were, thus, produced by tuning the deposition parameters including the concentration of the deposition time, applied voltage, and nanotube concentration in the dispersion. Interestingly, the study on the interaction between cell filopodias and COL nanotubes deposited on ITO glass revealed outstanding morphological change of cell [24]. No remarkable changes were, however, observed for cell adhesion and subsequent growth on the COL nanotube-decorated ITO glass.

7.8 Enzymatic Nanotube

Microvilli that lie on the cell surfaces are associated with a variety of biological functions, e.g., adhesion, secretion, and absorption while packed in a regular way to direct the binding of other proteins. Demoustier-Champagne, Jonas, and co-workers fabricated microvilli-like structures, i.e., bioactive nanotube brushes containing β-lactamase, which stand vertically on a substrate surface (Fig. 7.16) [25]. Hard templating and LbL assembly techniques are combined to fabricate the

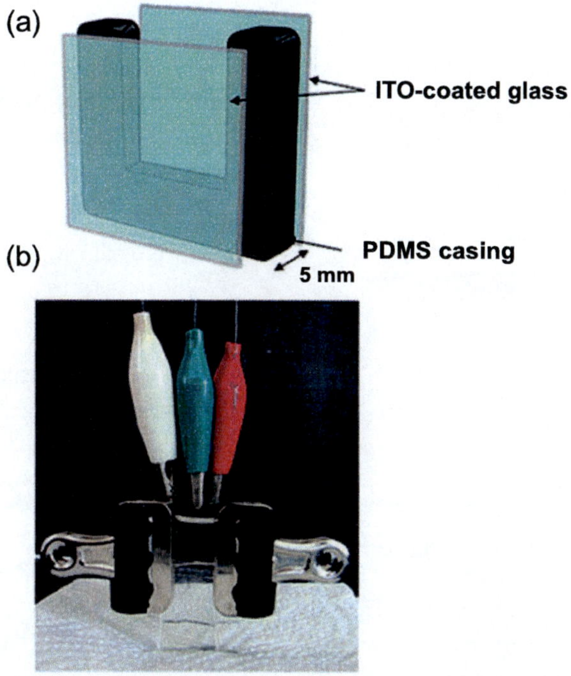

Fig. 7.15 **a** Schematic illustration and **b** picture that shows the device used for COL nanotubes collection using EPD. Reproduced with permission from Ref. [23], © 2011 American Chemical Society

nanotubes with a core–shell morphology comprising two compartments, one for mechanical robustness and the other containing β-lactamase for enzyme reaction. Thus, the enzymatic species are included either in the shell or in the core structure of the nanotubes. The nanotubes with β-lactamase in the core part exhibited a relatively longer stability of the enzymatic activity as compared the nanotubes with the enzyme in the shell [25]. In this context, the research group employed the unique technique for fabricating surface-bound nanotube arrays, which was developed by Rubner, Cohen, and co-workers [26]. They successfully constructed the arrays of reversibly swellable nanotube comprising pH-responsive multi-layers of polyelectrolytes through sacrificial porous template approach.

7.9 GroEL-Based Nanotube

A chaperonin protein of molecular chaperones, e.g., GroEL is a barrel-shaped tetradecamer that configures a double-decker architecture comprising seven subunits each. In nature, GroEL plays a vital role in facilitating proper folding and yet prevent unfavorable aggregation of denatured proteins. Kinbara, Aida, and co-workers demonstrated elaborated assembly of GroEL-based nanotubes into high-axial-ratio protein nanotubes [27]. Divalent metal cations,.e.g., Mg^{2+} was shown to accelerate

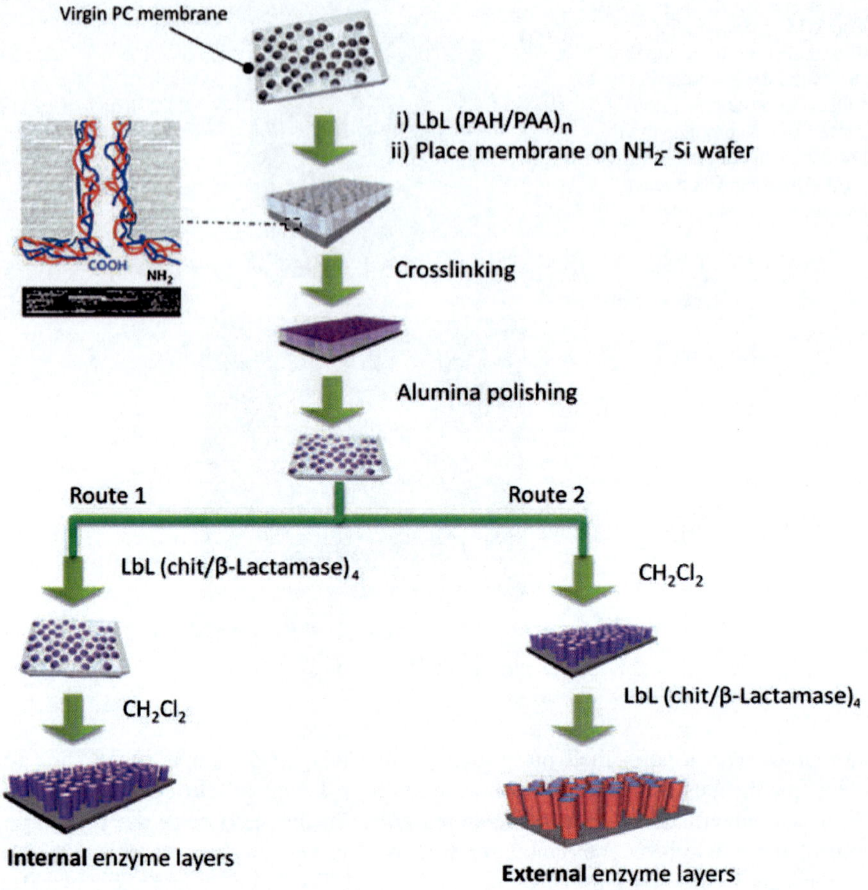

Fig. 7.16 Fabrication of nanotube brushes by LbL technique and crosslinking the template on a wafer. Depending on the order of template dissolution and (chitosan chloride/β-lactamase) LbL, the enzyme can be adsorbed onto the internal surface (Route 1) or onto the external surface of the nanotubes (Route 2). Reproduced with permission from Ref. [25], © 2017 Elsevier Inc.

supramolecular polymerization of $GroEL_{MC}$, in which a GroEL mutant was chemically modified with a range of merocyanine (MC) moieties **14** (Chart 7.2) in the vicinity of the open end (Fig. 7.17a). The connecting bridges of MC–Mg^{2+}–MC stabilize the high-aspect-ratio hollow cylindrical structures, forming the micrometer-long, discrete protein nanotubes with a 4.5 nm inner diameter. Addition of Mg^{2+} after the formation of inclusion complex of $GroEL_{MC}$ with denatured α-lactalbumin (α-$Lac_{FL/denat}$) also triggered 1D assembly of the $GroEL_{MC} \supset$ α-$Lac_{FL/denat}$ unit into the protein-based nanotubes encapsulating denatured guest proteins [27].

The GroEL-based nanotubes mentioned above can be expected to offer a novel biocontainer that exhibit both a chaperone and delivery capability for some proteins.

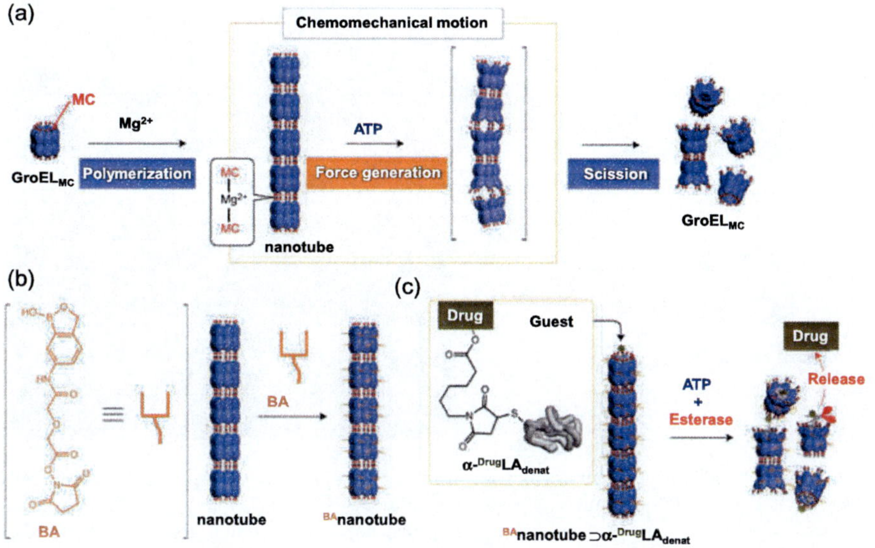

Fig. 7.17 Schematic image of the fundamental strategy for intracellular drug delivery with ATP-responsive PRNTs based on barrel-shaped chaperonin. **a** Supramolecular polymerization of $GroEL_{MC}$, ATP-triggered force generation, and the disassembly into $GroEL_{MC}$. **b** The PRNT functionalized with the boronic acid (BA) derivative. **c** Reaction of the guest-embedded PRNT (^{BA}nanotube ⊃ α-$^{Drug}LA_{denat}$) with intracellular ATP induces chemomechanical scission that can eventually lead to the release of the drugs. Reproduced with permission from Ref. [28], © 2013 Nature Publishing Group

Aida and co-workers exploited an exquisite drug delivery system driven by chemomechanical motion that sensitively responds to the existence of intracellular adenosine-5′-triphosphate (ATP) **15** (Chart 7.2) [28]. The $GroEL_{MC}$-based protein nanotubes were prepared through supramolecular polymerization triggered by the addition of Mg^{2+}. Another characteristic of the molecular design concept is to utilize the ATP-triggered generation of tiny mechanical force. The chemomechanical force induces conformational change of component protein, thus, driving the disassembly of the nanotubes into each $GroEL_{MC}$. Consequently, the appended drug is forced to release. For example, a cell-penetrable GroEL-based nanotube (^{BA}nanotube) was prepared by a surface modification with BA derivative **16** (Chart 7.2) (Fig. 7.17b). Then, a guest substance (α-$^{Drug}LA_{denat}$) comprising both irreversibly denatured α-lactalbumin as an enzymatically cleavable linker and a drug was incorporated at the apical domains of $GroEL_{MC}$ to give the drug-appended protein nanotube (^{BA}nanotube ⊃ α-$^{Drug}LA_{denat}$). Interaction of ^{BA}nanotube ⊃ α-$^{Drug}LA_{denat}$ with intracellular ATP triggers chemomechanical scission. Esterase can, thus, approach the opened lumen of the disassembled $GroEL_{MC}$ ⊃ α-$^{Drug}LA_{denat}$. As a result of a series of process, the ester bond cleaves to release the drug. Actually, the novel nanotube carrier (^{BA}nanotube ⊃ α-$^{Drug}LA_{denat}$) was demonstrated to penetrate HeLa cells, displaying the drug delivery activity in response to intracellular ATP (Fig. 7.18) [28].

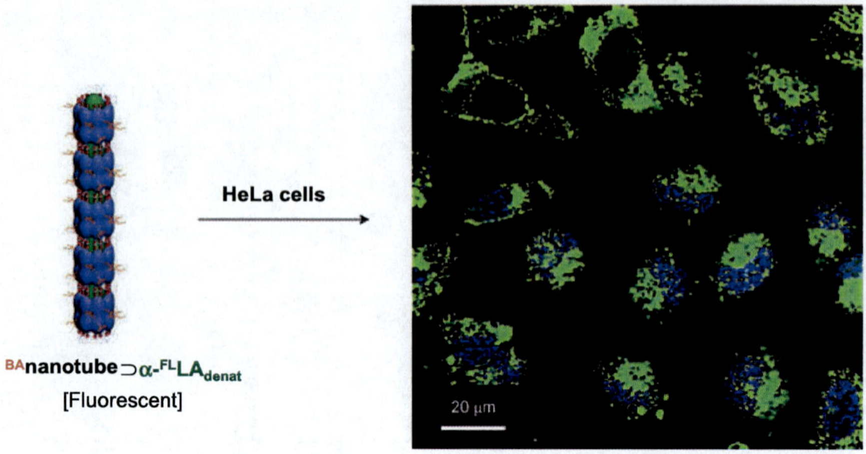

Fig. 7.18 Merged image of confocal laser scanning fluorescence micrographs of HeLa cells observed at λ_{obsd} = 500–530 nm (λ_{ext} = 488 nm) and 390–465 nm (λ_{ext} = 355 nm) after incubation at 37 °C with ^{BA}nanotube ⊃ α-^{FL}LA$_{denat}$. Reproduced with permission from Ref. [28], © 2013 Nature Publishing Group

By taking advantage of 1D array of cavities shaped by the 1D GroEL-based nanotubes, Miyajima, Aida, and co-workers fabricated micrometer-long 1D array of superparamagnetic nanoparticles (SNPs) of iron oxide [29]. A GroEL mutant was chemically decorated with multiple merocyanine (MC) segments **14** placed at its apex domains. Iron oxide SNPs, covered with functional ligand shell comprising a zwitterionic dopamine sulfonate **17** (85%) and dye-carrying hydrophobic catechol ligand **18** (15%) (Fig. 7.19b), was encapsulated in the cavity of $GroEL_{MC}$. The obtained monomer unit $GroEL_{MC}$ ⊃ SNP also polymerize through the formation of MC–Mg^{2+}–MC connecting linkage between the proteins, producing magnetically responsive protein nanotubes (Fig. 7.19). When a magnetic field of 0.5 T was applied, relatively longer nanotubes had a tendency to arrange laterally to form 1D thick bundles [29]. By contrast, turning off the magnetic field application defiberized the protein nanotube bundles into original protein component. Magnetically driven lateral assembly of superparamagnetic NT_{GroEL} ⊃ SNP was, thus, for the first time confirmed experimentally.

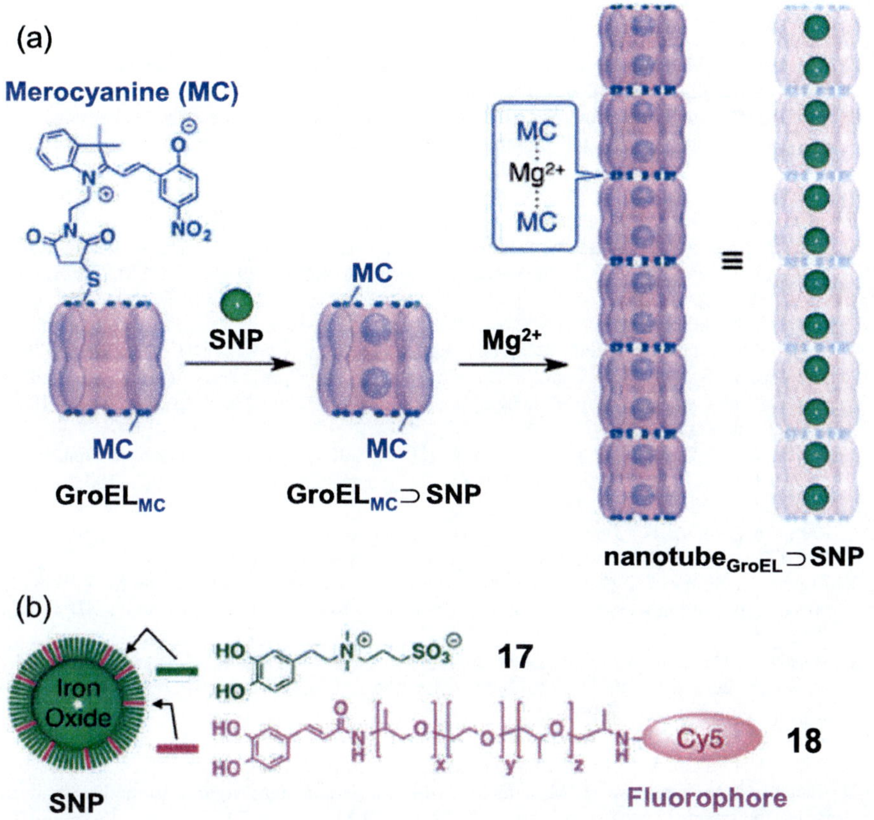

Fig. 7.19 Schematic image of **a** the preparation of 1D array of SNPs (nanotube$_{\mathrm{GroEL}}$ ⊃ SNP) through Mg^{2+}-mediated supramolecular polymerization of GroEL$_{\mathrm{MC}}$ ⊃ SNP and **b** a SNP with a dopamine sulfonate zwitterionic ligand shell containing 15% of a catechol-modified hydrophobic fluorophore Cy5. Reproduced with permission from Ref. [29], © 2015 American Chemical Society

References

1. Shimizu T, Masuda M, Minamikawa H (2005) Supramolecular nanotube architectures based on amphiphilic molecules. Chem Rev 105:1401–1443. https://doi.org/10.1021/cr030072j
2. Porrata P, Goun E, Matsui H (2002) Size-controlled self-assembly of peptide nanotubes using polycarbonate membranes as templates. Chem Mater 14:4378–4381. https://doi.org/10.1021/cm0205940
3. Kim K, Jin J-I (2001) Preparation of PPV nanotubes and nanorods and carbonized products derived therefrom. Nano Lett 1:631–636. https://doi.org/10.1021/nl010055e
4. Steinhart M, Wendorff JH, Greiner A, Wehrspohn RB, Nielsch K, Schilling J, Choi J, Goesele U (2002) Polymer nanotubes by wetting of ordered porous templates. Science 296:1997. https://doi.org/10.1126/science.1071210
5. Hou SF, Harrell CC, Trofin L, Kohli P, Martin CR (2004) Layer-by-layer nanotube template synthesis. J Am Chem Soc 126:5674–5675. https://doi.org/10.1021/ja049537t

6. Hou S, Wang J, Martin CR (2005) Template-synthesized protein nanotubes. Nano Lett 5:231–234. https://doi.org/10.1021/nl048305p
7. Komatsu T (2012) Protein-based nanotubes for biomedical applications. Nanoscale 4:1910–1918. https://doi.org/10.1039/c1nr11224d
8. Qu X, Lu G, Tsuchida E, Komatsu T (2008) Protein nanotubes comparised of an alternate layer-by-layer assembly using a polycation as an electrostatic glue. Chem Eur J 14:10303–10308. https://doi.org/10.1002/chem.200800771
9. Qu X, Komatsu T (2010) Molecular capture in protein nanotubes. ACS Nano 4:563–573. https://doi.org/10.1021/Nn901474y
10. Goto S, Amano Y, Akiyama M, Bottcher C, Komatsu T (2013) Gold nanoparticle inclusion into protein nanotube as a layered wall component. Langmuir 29:14293–14300. https://doi.org/10.1021/La403283x
11. Qu X, Kobayashi N, Komatsu T (2010) Solid nanotubes comprising α-Fe_2O_3 nanoparticles prepared from ferritin protein. ACS Nano 4:1732–1738. https://doi.org/10.1021/Nn901879d
12. Kobayakawa S, Nakai Y, Akiyama M, Komatsu T (2017) Self-propelled soft protein microtubes with a Pt nanoparticle interior surface. Chem Eur J 23:5044–5050. https://doi.org/10.1002/chem.201605055
13. Wu Z, Wu Y, He W, Lin X, Sun J, He Q (2013) Self-propelled polymer-based multilayer nanorockets for transportation and drug release. Angew Chem Int Ed 52:7000–7003. https://doi.org/10.1002/anie.201301643
14. Komatsu T, Terada H, Kobayashi N (2011) Protein nanotubes with an enzyme interior surface. Chem Eur J 17:1849–1854. https://doi.org/10.1002/chem.201001937
15. Amano Y, Komatsu T (2015) Nanotube reactor with a lipase wall interior for enzymatic ring-opening oligomerization of lactone. Chem Lett 44:1646–1648. https://doi.org/10.1246/cl.150789
16. Shimizu T, Minamikawa H, Kogiso M, Aoyagi M, Kameta N, Ding W, Masuda M (2014) Self-organized nanotube materials and their application in bioengineering. Polym J 46:831–858. https://doi.org/10.1038/Pj.2014.72
17. Shimizu T (2008) Self-assembled organic nanotubes: toward attoliter chemistry. J Polym Sci Part A: Polym Chem 46:2601–2611. https://doi.org/10.1002/Pola.22652
18. Kameta N, Minamikawa H, Masuda M (2011) Supramolecular organic nanotubes: how to utilize the inner nanospace and the outer space. Soft Matter 7:4539–4561. https://doi.org/10.1039/c0sm01559h
19. Shimizu T, Ding W, Kameta N (2020) Soft-matter nanotubes: a platform for diverse functions and applications. Chem Rev 120:2347–2407. https://doi.org/10.1021/acs.chemrev.9b00509
20. Komatsu T, Qu X, Ihara H, Fujihara M, Azuma H, Ikeda H (2011) Virus trap in human serum albumin nanotube. J Am Chem Soc 133:3246–3248. https://doi.org/10.1021/Ja1096122
21. Yuge S, Akiyama M, Ishii M, Namkoong H, Yagi K, Nakai Y, Adachi R, Komatsu T (2017) Glycoprotein nanotube traps influenza virus. Chem Lett 46:95–97. https://doi.org/10.1246/cl.160805
22. Landoulsi J, Roy CJ, Dupont-Gillain C, Demoustier-Champagne S (2009) Synthesis of collagen nanotubes with highly regular dimensions through membrane-templated layer-by-layer assembly. Biomacromolecules 10:1021–1024. https://doi.org/10.1021/bm900245h
23. Kalaskar DM, Poleunis C, Dupont-Gillain C, Demoustier-Champagne S (2011) Elaboration of nanostructured biointerfaces with tunable degree of coverage by protein nanotubes using electrophoretic deposition. Biomacromolecules 12:4104–4111. https://doi.org/10.1021/bm2011592
24. Kalaskar DM, Demoustier-Champagne S, Dupont-Gillain CC (2013) Interaction of preosteoblasts with surface-immobilized collagen-based nanotubes. Colloids Surf B Biointerfaces 111:134–141. https://doi.org/10.1016/j.colsurfb.2013.05.035
25. Ramirez-Wong DG, Bonhomme C, Demoustier-Champagne S, Jonas AM (2018) Layer-by-layer assembly of brushes of vertically-standing enzymatic nanotubes. J Colloid Interface Sci 514:592–598. https://doi.org/10.1016/j.jcis.2017.12.063

26. Chia KK, Rubner MF, Cohen RE (2009) pH-Responsive reversibly swellable nanotube arrays. Langmuir 25:14044–14052. https://doi.org/10.1021/la9015959
27. Biswas S, Kinbara K, Oya N, Ishii N, Taguchi H, Aida T (2009) A tubular biocontainer: metal ion-induced 1D assembly of a molecularly engineered chaperonin. J Am Chem Soc 131:7556–7557. https://doi.org/10.1021/ja902696q
28. Biswas S, Kinbara K, Niwa T, Taguchi H, Ishii N, Watanabe S, Miyata K, Kataoka K, Aida T (2013) Biomolecular robotics for chemomechanically driven guest delivery fuelled by intracellular ATP. Nat Chem 5:613–620. https://doi.org/10.1038/Nchem.1681
29. Sim S, Miyajima D, Niwa T, Taguchi H, Aida T (2015) Tailoring micrometer-long high-integrity 1D array of superparamagnetic nanoparticles in a nanotubular protein jacket and its lateral magnetic assembling behavior. J Am Chem Soc 137:4658–4661. https://doi.org/10.1021/jacs.5b02144

Chapter 8
Bottlebrush Copolymer-Based Nanotubes

8.1 Molecular Sculpting

In early 2000s, the fabrication technique called molecular sculpting of crosslinked nanofibers was shown to yield organic nanotubes from a triblock copolymer [1, 2]. This approach utilizes the amphiphilicity of the polymers. Representatively, the process starts with the formation of coaxial cylindrical micelles of a triblock copolymer (PI-*block*-PCEMA-*block*-PtBA) (PI-*b*-PCEMA-*b*-PtBA) **1** (Chart 8.1) via self-assembly. The nanotube formation is completed by the photocrosslinking of the PCEMA as a middle layer and subsequent removal of the PI inner core by ozonolysis (Fig. 8.1). The PtBA corona chain provides the nanotube with solvent dispersibility [1]. Positive evidence for the existence of the nanotube hollowness was given by the following facts; a light stripe was observed in the center part of each nanotube, and rhodamine B was encapsulated into the hollow cylinder structure [1, 2]. In this chapter, we firstly describe excellent works on the self-assembly of micro- or nanotube structures in the dawn of copolymer-based tiny tubes. For example, amphiphilic polymers including rod–coil and coil–coil block copolymers are involved with the micro- and nanotube formation. Thereafter, BBCNTs and MONNs are discussed in connection with their specific functions and applications. These nanotubes have emerged on the basis of the molecular sculpting.

8.2 Amphiphilic Rod–Coil Block Copolymer

Block copolymers are involved with the formation of many phase-separated supramolecular structures [3–5]. Especially, rod–coil block copolymers comprising both rigid rod-like and flexible coil-like polymer segments can form nanometer-sized supramolecular structures in solution due to the crystalline domain and anisotropic intermolecular interactions of the rod block [3]. For example, depending on the mixing ratio of trifluoroacetic acid (TFA) and dichloromethane as a solvent,

T. Shimizu, *Smart Soft-Matter Nanotubes*, Nanostructure Science and Technology,
https://doi.org/10.1007/978-981-16-2685-2_8

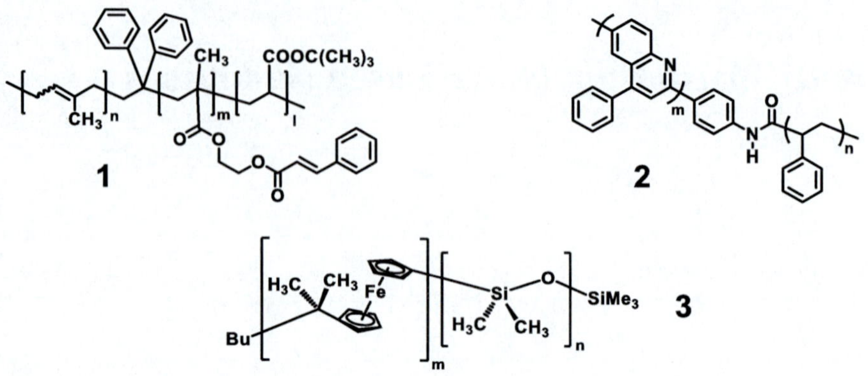

Chart 8.1 Chemical structures of **1–3**

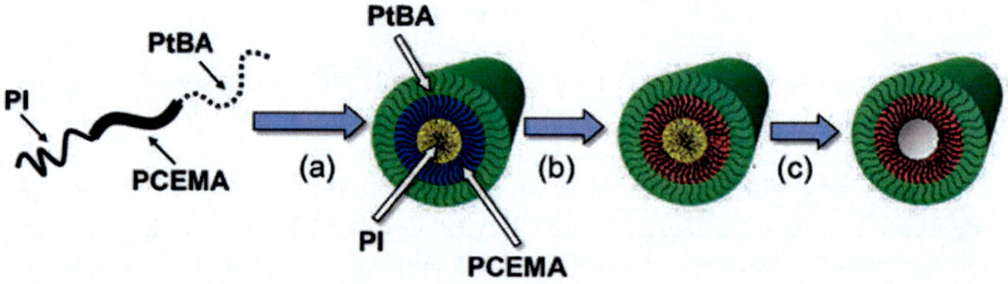

Fig. 8.1 Nanotube formation from the triblock copolymer **1** through a molecular sculpting process, which includes **a** self-assembly, **b** crosslinking of the shell, and **c** decomposition of the core by ozonolysis

poly(phenylquinoline)-*block*-polystyrene (PPQ-*b*-PS) **2** (Chart 8.1) undergoes the self-assembly into spheres, vesicles, microtubes, and lamellae (Fig. 8.2) [6]. The evaporation rate of the solvent also has an influence on the self-assembled morphologies. If the solvent composition of TFA is over 90%, microtube structures with closed

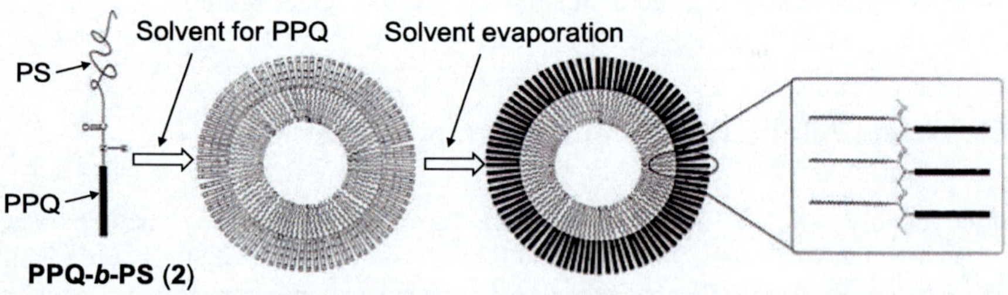

Fig. 8.2 Self-assembly of PPQ-*b*-PS **2** rod–coil diblock copolymers into hollow aggregates. Reproduced with permission from Ref. [6], © 1998 The American Association for the Advancement of Science

ends and outer diameters of 1–3 μm become dominant among the existing self-assembled structures. The PPQ block as a rod segment tends to be more miscible with TFA than the flexible PS block, thereby residing in the microtube exterior. The size scale of the self-assemblies decreases with the decrease in the fraction of the rod block of the copolymer. Notably, the resultant self-assemblies exhibit high thermal stability even at 200 °C although the glass transition temperature of the PS block is 100 °C. The block of the PS chain can trap C60 molecules in the hollow cavity of the self-assemblies. The existence of such a spherical molecule as C60 restrains the self-assembly of nonspherical structures.

The rod–coil block copolymer poly(ferrocenyldimethylsilane)-block-poly(dimethylsiloxane) (PFS-*b*-PDMS), e.g., PFS40-*b*-PDMS480 (m = 40, n = 480) and PFS80-*b*-PDMS960 (m = 80, n = 960) copolymers **3** (Chart 8.1) self-assemble in *n*-hexane and *n*-decane to give nanotube structures (Fig. 8.3). Both solvents serve as good media for the flexible PDMS chains. The obtained organometallic nanotubes are featured by the widths of 29–40 nm and their redox activity derived from the relatively stiff ferrocenyl segment [7, 8]. The length ratio of the PDMS/PFS block copolymers has a great effect on the resultant self-assembled morphology. For example, when the PDMS/PFS ratio is between 1:12 and 1:18, high-aspect-ratio hollow nanotubular structures appear [7, 8]. Meanwhile, the copolymers with a block ratio of 1:6 self-assemble to yield cylindrical micelles [9]. Interestingly, the addition of water to hexane as self-assembly solvent causes the change in aggregate state from a discrete nanotube to bundling state of the nanotubes.

As long as the rod/coil block ratio of the copolymers **3** keeps constant, the wall thickness and diameter of the resultant nanotubes are not influenced by the molecular

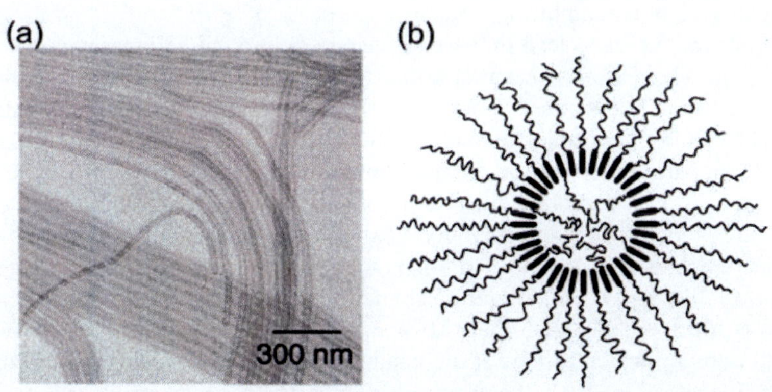

Fig. 8.3 **a** TEM micrographs of PFS80-*b*-PDMS960 **3** assemblies that are formed in *n*-decane at 61 °C, cooled to 23 °C over 2 h, and allowed to age for 1 day at room temperature. **b** Schematic drawing of a cross section of the nanotube. The PDMS chains are shown by the coils, and the PFS shell by the dark rods. Reproduced with permission from Ref. [8], © 2002 American Chemical Society

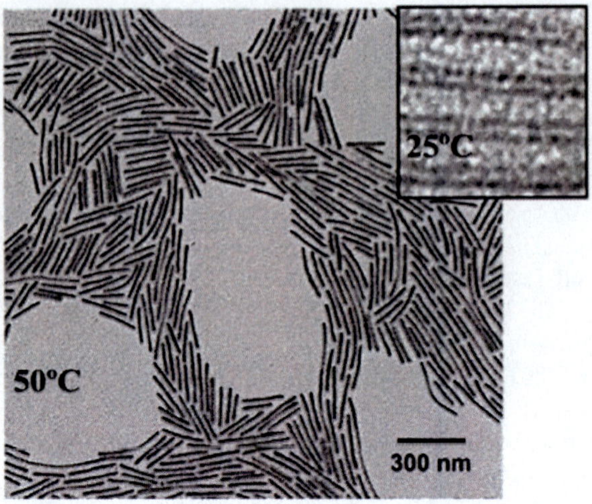

Fig. 8.4 TEM image of PFS40-*b*-PDMS480 **3** assemblies formed in *n*-decane at 50 °C. The inset shows nanotubes at 25 °C. Reproduced with permission from Ref. [10], © 2003 American Chemical Society

weight of the copolymers [8]. The nanotube wall thickness observed for PFS40-*b*-PDMS480 and PFS80-*b*-PDMS960 was shown to have almost uniform dimension, i.e., 7 nm (Fig. 8.4). If the PFS chains adopt a folding conformation, four and eight folds are required in average to maintain this shell thickness for PFS40-*b*-PDMS480 and PFS80-*b*-PDMS960, respectively. This self-assembly system of the copolymers exhibits a reversible tube-to-rod transition that responds to a temperature change [10]. For instance, the nanotubes stabilized at 23 °C transform into rod-like micelles upon heating of the aqueous solution to 50 °C. These self-assembled morphologies are, therefore, in an equilibrium state similar to the case in the aggregation of low-molecular-weight amphiphiles.

The driving forces to lead to the rearrangement into the polymer nanotubes are not yet clear. The reason is because the theory for micelles of coil-crystalline diblock copolymers mainly discusses about a lamellar structure created by the crystalline block [11]. Only a very small amount of soluble PDMS block may percolate into the internal compartment of hollow tubular structures. Strongly disfavored crystalline packing of the PFS rod block may compromise with the space filling need of the PDMS coil block. More interestingly, changing the solvent from *n*-hexane to *n*-dodecane enables the control of the nanotube length from a few micrometers to 100 μm (Table 8.1) [8]. The interfacial energy between the PFS rod block and *n*-hexane is much smaller than that between the PFS and *n*-dodecane. Exposure of the PFS domain, which resides at the nanotube end, to *n*-dodecane phase provides unfavorable situation. Thus, the nanotube growth in *n*-dodecane tends to reach high axial ratios.

Table 8.1 Dimensions of self-assembled nanotubes formed under given conditions[a]

Block copolymer		*n*-hexane		*n*-decane	
		23 °C	61 °C	61 °C	151 °C
PFS40-*b*-PDMS480	d_{cavity}[b] (nm)	9	9	7	7
	PFS wall thickness (nm)	7	7	7	7
	L_{max}[c] (nm)	0.6	4	~ 100	~ 100
PFS80-*b*-PDMS960	d_{cavity}[b] (nm)	11–12	11–12	10–21	– [d]
	PFS wall thickness (nm)	7	7	7	–
	L_{max}[c] (nm)	4	4	~ 100	–

[a]Final structures observed by TEM in our experiments. In each instance, these structures persisted for at least 1 month after the final measurements
[b]Cavity diameter
[c]Representative contour length of the nanotubes as estimated from the TEM images
[d]No tubular structures self-assembled from this sample

8.3 Amphiphilic Coil–Coil Block Copolymer

Various bilayer assemblies including rods, nanotubes, lamellae, vesicles, and large compound vesicles were prepared by the self-assembly of PS-based diblock copolymers such as PS-*block*-poly(acrylic acid) (PS-*b*-PAA) **4** (Chart 8.2) and PS-*block*-poly(ethylene oxide) (PS-*b*-PEO) **5** (Chart 8.2) [12–16]. So-called crew-cut assemblies emerge under the conditions, in which the corona-forming; i.e., soluble block is

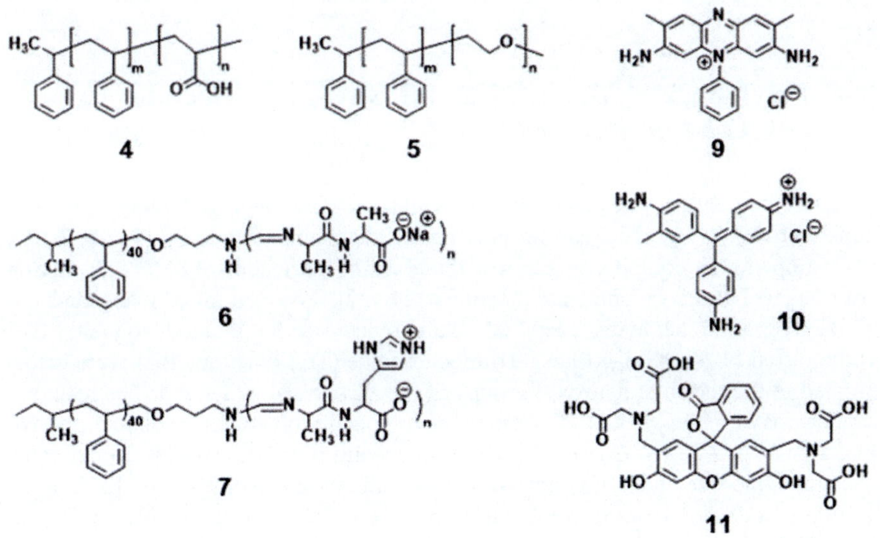

Chart 8.2 Chemical structures of **4–11**

much shorter than the core-forming, i.e., insoluble block [17]. In this context, various factors influence on the self-assembled morphology of those kinetically trapped aggregates. For example, the composition ratio of the two different coil block plays a critical role in regulating the morphology. This condition is also well-suited for the rod–coil diblock copolymers described above. The self-assembly experiment begins with addition of water dropwise, which is a poor solvent for the hydrophobic block, into the DMF solution of a copolymer. Subsequently, the self-assembly completes through dialysis to relieve the remaining DMF.

In the self-assembly of nonionizable PS-*b*-PEO diblock copolymers **5**, the composition ratio of the PS/PEO block has a decisive influence on the resultant self-assembled morphologies. Copolymers with 240/15 to 240/45 PS/PEO ratios tend to self-assemble into nanotube morphologies (Fig. 8.5) [12]. With the increase in the composition ratio of the PEO block, the self-assembled morphology converts from lamellar tape and rods to spherical polymer micelles [15, 16]. The wall thicknesses and external diameter of the nanotubes are typically $\approx$20 nm and $\approx$100 nm, respectively. The extended length of the nanotubes reaches hundreds of micrometers. The formation pathway for these tubular structures is uncertain, although unique morphology is trapped as an intermediate one during the vesicle-to-tube transition (Fig. 8.6). Probably, adhesive contact and subsequent fusion of the vesicular assemblies will be associated [12]. Although the obtained nanotubes are certainly based on bilayer structures, the chirality-derived helical feature of the amphiphilic copolymer are less significant for the nanotube formation. A superhelix structure created from chiral block copolymers **6** and **7** (Chart 8.2) is, however, of particular concerns. The self-assembled super coiled structures resembled helically coiled morphologies that are well known as an intermediate of lipid nanotube through chiral self-assembly [18].

8.4 Poly(Glycidyl Methacrylate) (PGM)-*g*-[Polylactide (PLA)-*b*-Polystyrene (PS)]

BBCs are bottlebrush-like macromolecules that comprise longer polymer chain backbone with densely grafted, shorter polymeric side chains (Fig. 8.7) [19, 20]. BBCs consisting of multicomponents can well template the fabrication of SMNTs by taking advantage of the core–shell architectures [21, 22]. For example, Huang and coworkers reported that hydrolyzed PGM chain reacted with D,L-lactide to yield PGM with grafted PLA (PGM-*g*-PLA). Trithiocarbonate (TC) functionalities were incorporated to the terminal hydroxyl groups of the PLA brush polymers. The resultant side chains of PGM-*g*-PLA-TC were copolymerized with styrene monomer to give PGM-*g*-(PLA-*b*-PS) **8** (Chart 8.3). Thus, they synthesized the BBC **8** from a PGM backbone, the side chain of which comprises a PLA core and a PS shell [23].

Importantly, interbrush, intrabrush, crosslinking reaction of the PS shell with formaldehyde dimethyl acetal and anhydrous $FeCl_3$, and complete removal of the

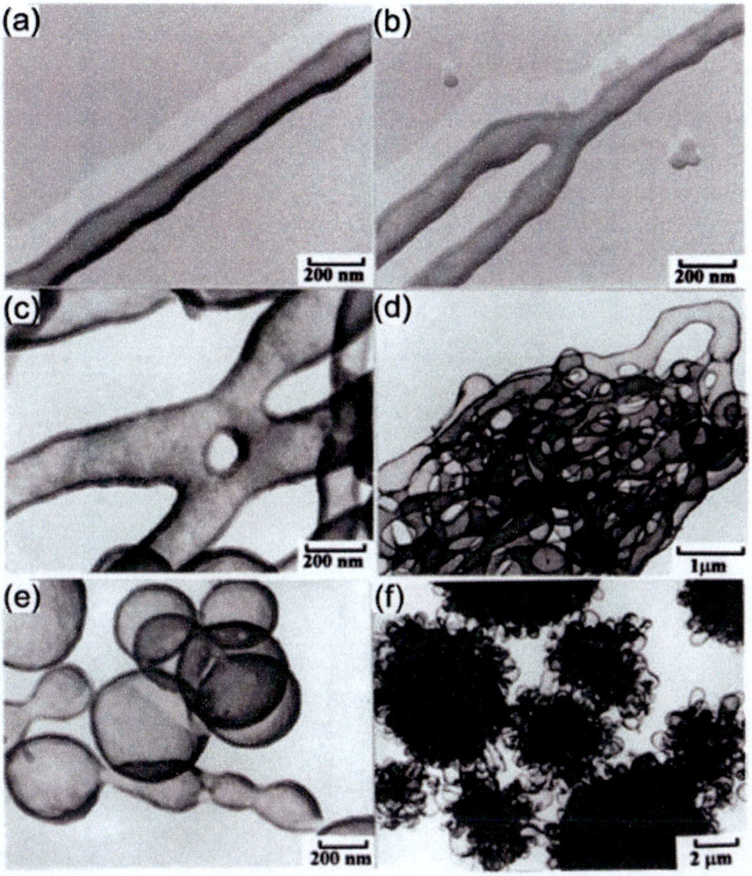

Fig. 8.5 TEM images of **a** a linear tubule, **b** a branched tubule, **c** tubules with a hole at a branch, **d** a "plumber's nightmare," **e** vesicles, and **f** large compound vesicles self-assembled from PS240-*b*-PEO15 **5** in DMF. Reproduced with permission from Ref. [12], © 1998 American Chemical Society

PLA-associated core part with acidic dioxane solution leads to the production of three different pore structures, i.e., meso/macropores, micropores, and mesopores, respectively (Fig. 8.8).

Huang and co-workers further confirmed that the obtained MONNs can selectively adsorb both charged dyes including safranine T **9** (ST, cationic), fuchsin basic **10** (FB, cationic), and calcein **11** (anionic) (Chart 8.2). Moreover, in order to investigate the size selectivity for protein adsorption, they also compared PGM210-*g*-(PLA56-*b*-PS93) having PLA-derived mesopores (ca. 7 nm) with PGM210-*g*-(PLA56-*b*-PS93) precursor and crosslinking-treated PGM210-*g*-(PLA56-*b*-PS93) having no mesopores. As a result of the UV–Vis adsorption data, the MONNs comprising PGM210-*g*-(PLA56-*b*-PS93) with mesopore can trap Cyt C (2.6 nm × 3.2 nm × 3.0 nm)

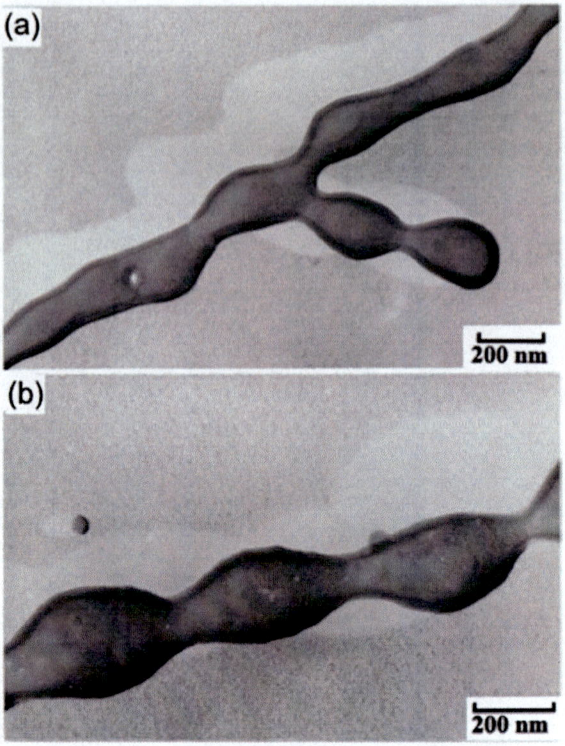

Fig. 8.6 Possible intermediates of the transition from vesicle to tubule: **a** a tubule with attached vesicles and **b** a tubule with an oscillatory perturbation in the diameter. Reproduced with permission from Ref. [12], © 1998 American Chemical Society

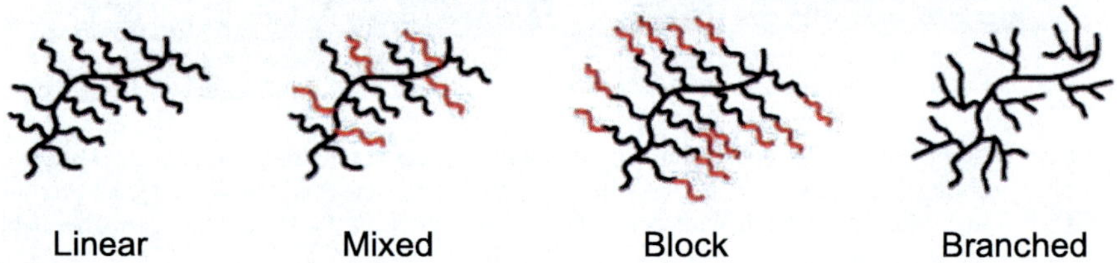

Fig. 8.7 Possible architectures of 1D polymer brushes. Reproduced with permission from Ref. [19], © 2005 Wiley Periodicals, Inc.

more efficiently than bovine serum albumin (14 nm × 3.8 nm × 3.8 nm) (Fig. 8.9). Little encapsulation of both proteins was observed for both PGM210-*g*-(PLA56-*b*-PS93) precursor and crosslinked PGM210-*g*-(PLA56-*b*-PS93) with no mesopores [23]. Uniqueness of the MONNs is that they feature large surface area, variety of chemical functionalities, rich multiporosity structures, and facile synthesis with low cost.

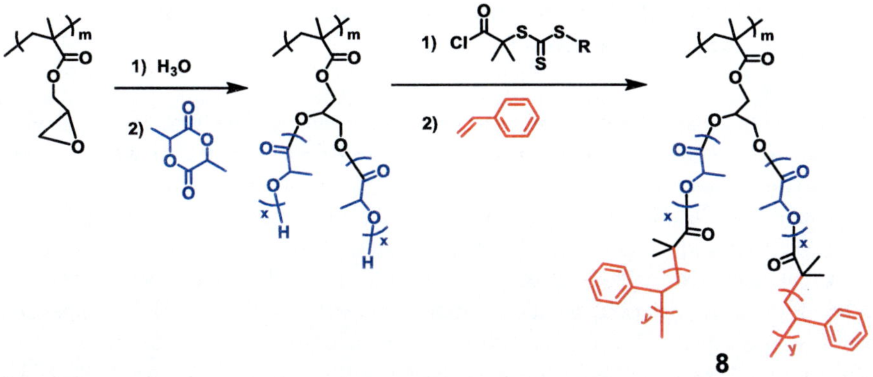

Chart 8.3 Synthesis of core–shell BBC precursor **8**

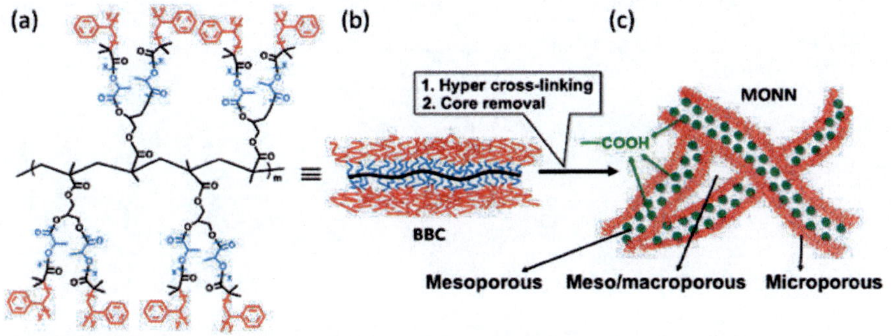

Fig. 8.8 **a** A representative molecular structure of BBC copolymer. **b**, **c** Schematic drawing of the preparation of MONNs through hyper-crosslinking and subsequent core removal of the BBC. Reproduced with permission from Ref. [24], © 2020 American Chemical Society

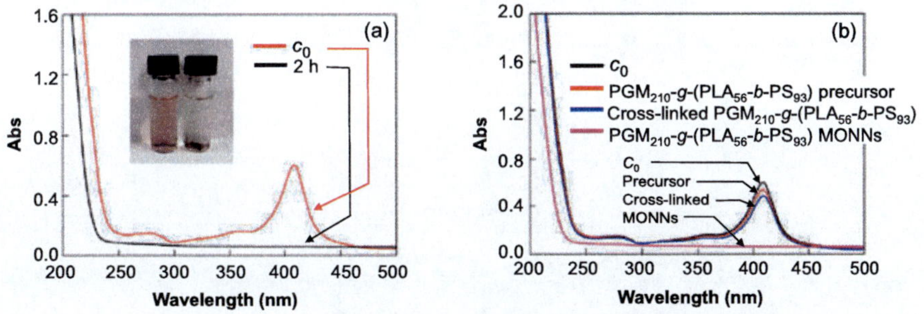

Fig. 8.9 UV–Vis absorption change of Cyt C solution **a** after the addition of MONNs PGM210-*g*-(PLA56-*b*-PS93) sample (0 and 2 h) (*Inset* Photograph of the vial taken before and after the addition of the MONNs) and **b** in 2 h after the addition of PGM210-*g*-(PLA56-*b*-PS93) precursor, crosslinked PGM210-*g*-(PLA56-*b*-PS93), or MONNs PGM210-*g*-(PLA56-*b*-PS93) samples. Reproduced with permission from Ref. [23], © Chinese Chemical Society, Institute of Chemistry, CAS, Springer-Verlag GmbH Germany 2018

8.5 Soluble Bottlebrush Copolymer-Based Nanotubes for Catalytic Systems

According to the synthetic strategy and tactics reported previously, the BBC carrying triblock terpolymer side chains P(GM-*g*-LA-*g*-VBC/BS-*g*-NIPAM) was prepared [22]. After intra-crosslinking of the poly[4-vinylbenzyl chloride-*co*-4-(3-butenyl) styrene] (PVBC/BS) middle layer and the subsequent removal of the PLA core, the benzyl chloride segment of the PVBC wall underwent transformation to yield benzyl azide-functionalized nanotubular materials [25]. The azide functional groups offer efficient reactive points that can further react with a variety of alkyne-appended catalytic molecules. The "click reaction" of the azide-functionalized inner surfaces and sodium prop-2-yne-1-sulfonate (SPS) **12**, 1-[2-(prop-2-yn-1-yloxy)ethyl]-1H-imidazole (PEI) **13**, and 2-ethynyl pyridine **14** (Fig. 8.10) gave a sulfonic acid cata-

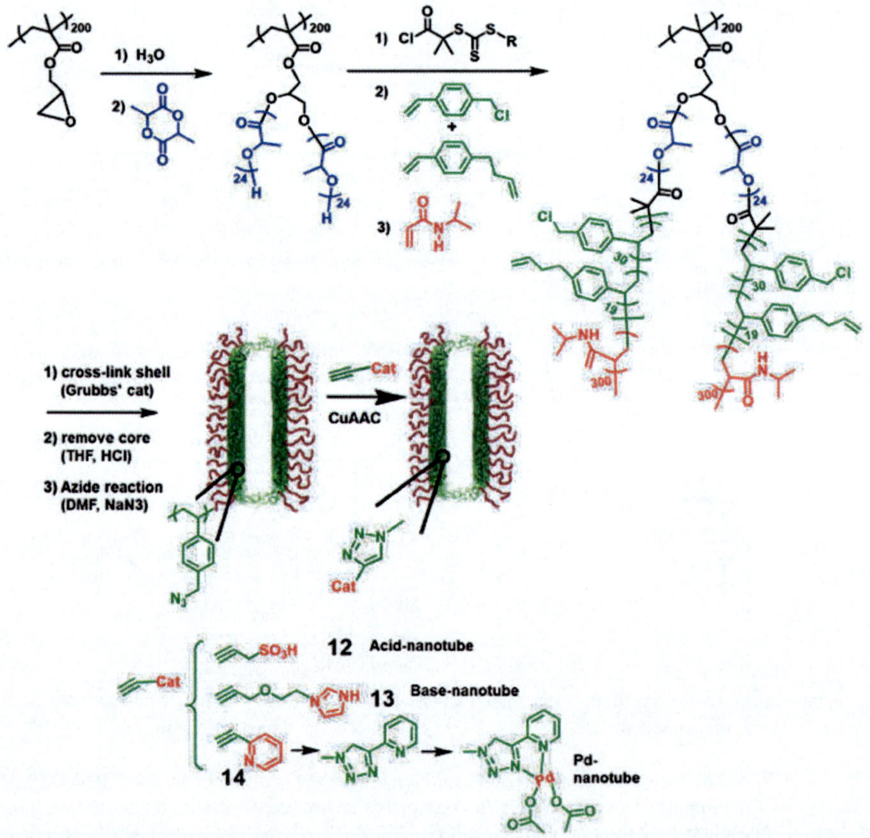

Fig. 8.10 Schematic representation of the synthesis of organic nanotube catalysts. Reproduced with permission from Ref. [25], © IOP Publishing Ltd.

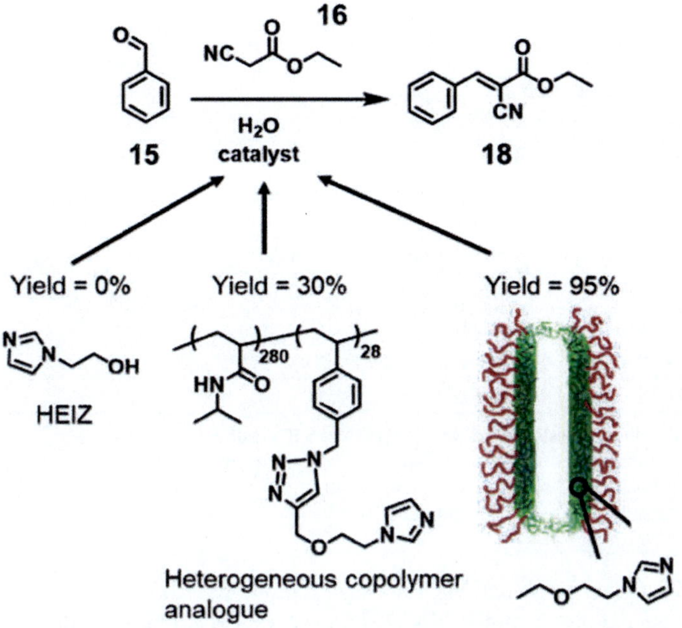

Fig. 8.11 Model Knoevenagel condensation. Reaction conditions: 1 mol% base catalysts were used. The reaction mixtures were stirring for 16 h at room temperature in H_2O. Yields are based on GC-MS measurements. Reproduced with permission from Ref. [25], © IOP Publishing Ltd.

lyst as an acid nanotube, imidazole catalyst as a base nanotube, and Pd catalyst as a Pd nanotube, respectively [25]. As a result of a series of reactions, crosslinked and hydrophobic shell layer, the inner surfaces of which were modified with catalytic functionalities, and hydrophilic corona chains entirely build up the amphiphilic organic nanotubes (Fig. 8.10).

The Knoevenagel condensation reaction between benzaldehyde **15** and ethyl cyanoacetate **16** (Fig. 8.11) in water was conducted as a typical example by using the resultant base nanotubes, homogeneous *N*-(2-hydroxyethyl)-imidazole (HEIZ), and heterogeneous copolymer analog. As a result of the product yield, the imidazole-bearing base nanotube catalyst exhibited the highest yield (95%) compared with that (30%) of the copolymer and that (0%) of HEIZ (Fig. 8.11) [25]. The reason for the excellent catalytic activity by the base nanotube is due to well distribution of the imidazole catalyst segment within the hydrophobic tube wall as well as open-ended hollow cylindrical structures.

Meanwhile, a set of acid catalyst-based SPS and base catalyst-based PEI catalyst were immobilized onto the nanotube wall to yield the acid- and base-nanotube, respectively. Catalytic performance for tandem deacetalization–Knoevenagel reactions of (dimethoxymethyl)-benzene **17** (Chart 8.4) was carried out by using the combination of the acid nanotube with the base nanotube catalysts [25]. Only the usage of both nanotube catalysts yielded the final product **18** (Chart 8.4) (100% yield)

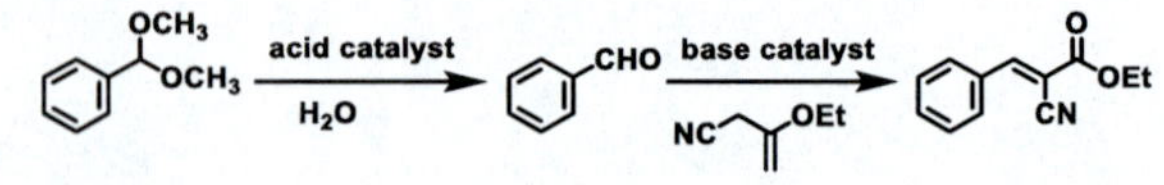

17 15 18

Chart 8.4 Tandem deacetalization–Knoevenagel reactions of (dimethoxymethyl)-benzene **17**

Table 8.2 Catalytic results for a one-pot reaction cascade using acid- and base-organic nanotube catalysts

Entry	Acid catalyst[a]	Base catalyst[a]	Conversion of A (%)[b]	Yield of C (%)[b]
1	Acid nanotube	Base nanotube	100	100
2	Acid nanotube	HEIZ	0	0
3	PTSA	Base nanotube	10	0
4	PTSA	HEIZ	0	0

Reproduced with permission from Ref. [25], © 2016 IOP Publishing Ltd.
[a]Reaction conditions: 10 mol% acid and base catalysts were used. The reaction mixtures were stirring for 12 h at 50 °C in DMF/H_2O (v/v) = (40:1)
[b]Yields are based on GC-MS measurements

(Table 8.2). Other combination of the acid nanotube/HEIZ, *p*-toluenesulfonic acid (PTSA)/base nanotube, and PTSA/HEIZ gave almost no final product **18** (0–10% yield). The low-molecular weight acid and base catalysts, PTSA and HEIZ, respectively, are able to diffuse liberally into the nanotube catalyst wall, thus deactivating the corresponding acid or base group within the nanotube by acid–base neutralization. The segregation effect of each acid and base group within the nanotubes is of critical importance to express their highly catalytic performance.

The obtained functionalized nanotube with pyridine segment was converted to Pd-coordinated nanotube catalyst (Pd nanotube) through the complexation of palladium acetate with pyridine-modified nanotube [25]. The Suzuki–Miyaura reaction of 4-iodo-4-methoxybenzene **19** with phenylboronic acid **20** (Fig. 8.12) was conducted in a mixture solvent of ethanol and water (10/1, v/v) as a test examination. The

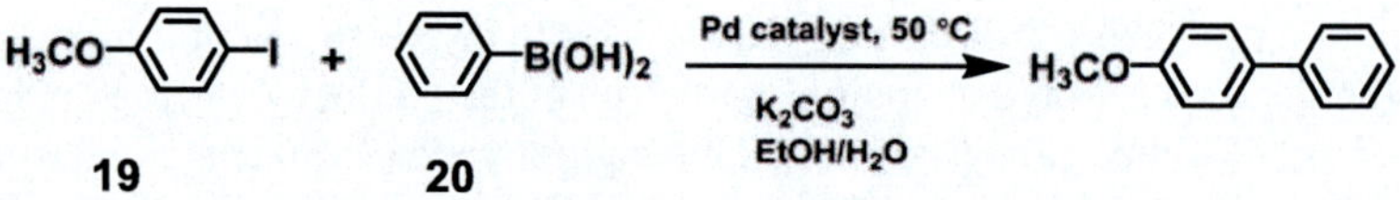

19 20

	run				
	first	second	third	fourth	fifth
Yield (%)	97	94	95	96	93

Fig. 8.12 Catalytic activity and recycle test of Pd nanotubes in the Suzuki–Miyaura crosscoupling reaction of **19** with **20**. Reaction conditions: 1 mol% Pd organic nanotube catalyst was used. The reaction mixtures were stirred for 24 h at 50 °C in EtOH/H_2O (v/v) = 10:1. Yields are based on GC-MS measurements. Reproduced with permission from Ref. [25], © IOP Publishing Ltd.

Pd nanotube showed an excellent ability for the catalysis of the Suzuki–Miyaura reaction, giving a satisfactory yield of 97%. At least the 5th cycle reaction still gave unchanged product yield (93%). No remarkable morphological changes were observed in TEM images for the Pd nanotubes even after five cycles.

8.6 Fe_3O_4-[PGM-*g*-(PLA-*b*-PS)]

By integrating magnetic nanoparticles with core–shell BBCs via in situ hyper-crosslinking reaction, Huang's research group newly prepared magnetoresponsive and repeatedly reusable MONNs [26]. Functional polymer-coated magnetic nanoparticles (Fe_3O_4@Dopa-PS) were synthesized from Fe_3O_4 magnetic particles and polystyrene with dopamine end group (Dopa-PS) through a ligand exchange reaction (Fig. 8.13). Core–shell BBC precursors PGM-*g*-(PLA-*b*-PS) were prepared in a similar way as the copolymers mentioned above [23]. Then, a Friedel–Crafts alkylation reaction between the PS side chains from the BBCs and the PS-appended polymer ligands promotes hyper-crosslinking to result in the formation of Fe_3O_4 NPs-integrated fibrous networks. Finally, selective elimination of PLA-associated core layer by degradation reaction in a 0.5 M NaOH solution affords magnetic MONNs (Fe_3O_4-MONNs) (Fig. 8.13).

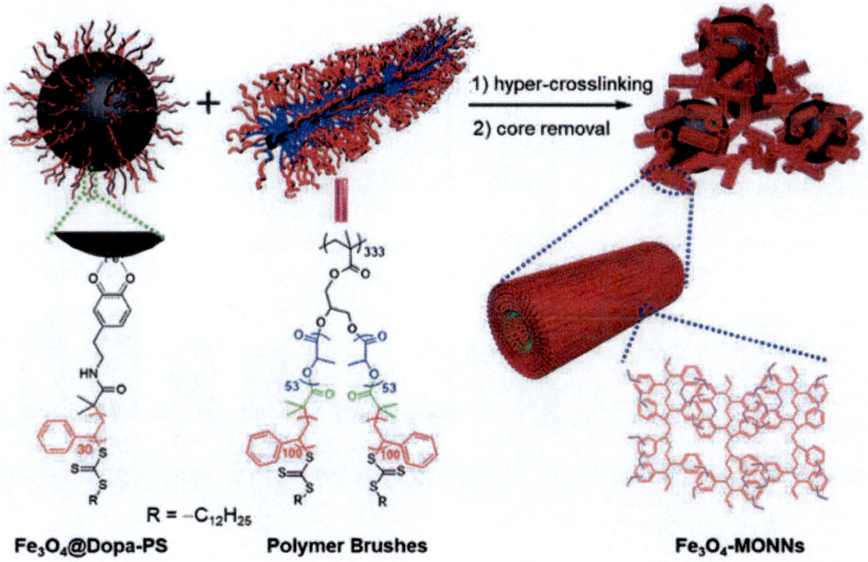

Fig. 8.13 Fabrication of magnetic microporous organic nanotube networks (Fe_3O_4-MONNs). Reproduced with permission from Ref. [26], © The Royal Society of Chemistry 2016

The obtained nanotubes have eventually rich carboxylic acid groups on the internal surface of the hollow cylinder. This means that the nanotubes can give favorable trapping field for positively charged substances. For example, the nanotubes can selectively load cationic organic dye, e.g., ST **9**, FB **10**, methylene blue (MBL) **21**, R6G **22**, and eonsin B (EB) **23** (Chart 8.5) (Fig. 8.14) [26]. Additionally, the obtained magnetic nanotubes are featured by superparamagnetism (high saturation magnetization: 19.8 emu g^{-1}), homogeneous mesopores (ca. 4 nm), and large surface area (648 m^2 g^{-1}). The magnetic nanotube hybrids are separable under an external magnetic field and therefore recyclable [26–28]. Typically, the magnetic MONNs keep the encapsulating ability of ST dye **9** even after six of cycles. Treatment of the adsorbed nanotube materials with acetic acid methanol solution (3%, v/v) can liberate the ST dye **9** from the magnetic MONNs. Notably, the desorption rate of ST dye **9** from the nanotubes keeps constant (> 95%) even after six of cycles.

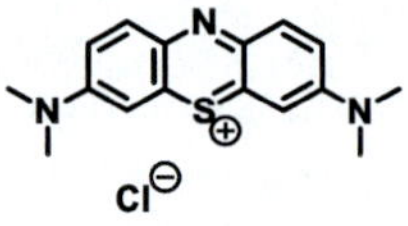

21

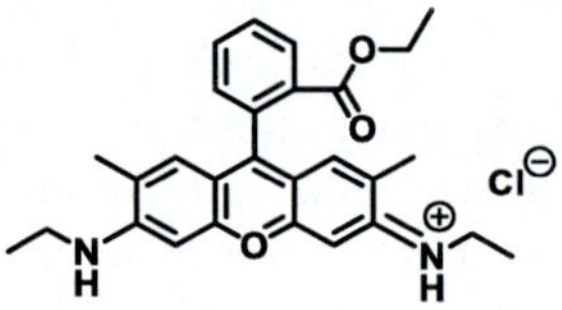

22

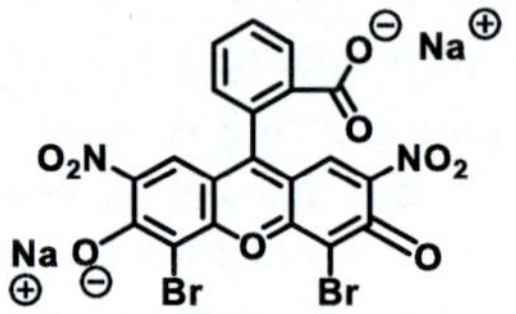

23

Chart 8.5 Chemical structures of **21–23**

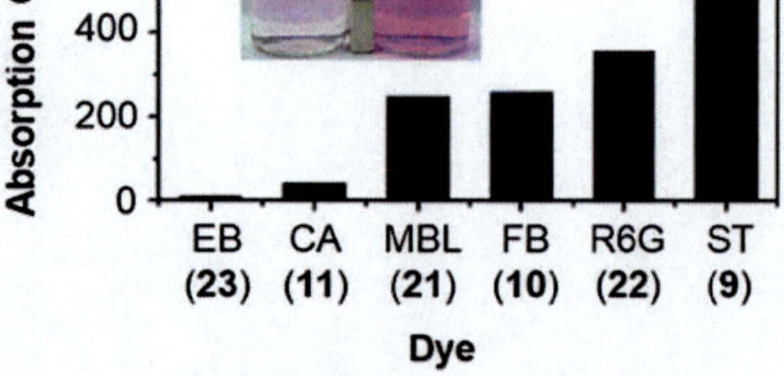

Fig. 8.14 Saturated adsorption capacities evaluated for the six hydrophilic dyes [anionic (EB **23** and CA **11**) and cationic dyes (MBL **21**, FB **10**, R6G **22**, and ST **9**)] at room temperatures. The inset represents a photograph of **a** the aqueous solution of ST **9** after adsorption by the Fe_3O_4-MONNs for 3 min and then standing under a magnetic field for 10 s and **b** initial aqueous solution (ST = 0.01 mg/mL). Reproduced with permission from Ref. [26], © The Royal Society of Chemistry 2016

8.7 SO_3H- and NH_2-Microporous Organic Nanotube Networks (MONNs)

Chemically reasonable localization of targeted functionalities into the wall and/or onto the internal surfaces of organic tubular assemblies is one of the ultimate goals for the process design of self-assembly. Huang's research group reported a novel synthetic approach for the fabrication of the MONNs. Specifically, the acidic sulfonic functional groups (–SO_3H) and basic amino functional groups (–NH_2) were tethered to the inner surfaces of the nanotube channels [29]. In order to provide the MONN substrate with acid and base catalytic functionalities, they incorporated phenyl 4-vinylbenzene sulfonate (PVBS) **24** and Boc-aminoethyl acrylamide (BAEAM) **25** (Chart 8.6) as a modification segment, respectively, between the PS shell and PLA core layers. Hyper-crosslinking of PS segments and subsequent holing treatment, similar as the procedure mentioned above, successfully resulted in the formation of PVBS- and BAEAM-protected MONNs. Finally, the deprotection of PVBS and BAEAM from the hyper-crosslinked and holed BBCs yielded targeting the SO_3H- and NH_2-MONNs, respectively (Fig. 8.15)

The obtained acidic SO_3H-anchored and basic NH_2-anchored MONNs were subjected to cascade-type one-pot reactions in order to confirm their capability as catalysts. Adopted model reaction is the combination of deacetalization with an acid catalyst and the subsequent Knoevenagel condensation with a base catalyst (Chart 8.4). As a result of the tandem reaction from acetal **17** to ethyl α-cyanocinnamate **18** via benzaldehyde **15** in the presence of ethyl cyanoacetate, the mixture of acid–base MONN catalysts exhibited excellent catalytic performance with the yield of above 99% [29]. When the same reaction was carried out by using only SO_3H-MONN, NH_2-MONN or without using any catalysts, only a little (below 5%) or no final products were yielded. Moreover, when low-molecular weight catalysts, such as benzoic acid instead of SO_3H-MONN or *n*-butylamine instead of NH_2-MONN, were employed as the combination of mixed acid–base catalysts, significantly low yields (below 4%) of the final product were observed [29]. This finding will be attributed to the deactivation of catalytic sites due to the free diffusion of the small acid or base catalyst into the nanotube channels. Complete isolation of the acid and base catalysts by their heterogeneous immobilization by the MONN substrate plays a critical role in performing the catalytic activity for the tandem reactions (Fig. 8.16).

Huang's research group further expanded the utilization of the mixed SO_3H- and NH_2-MONNs as heterogeneous catalysts into a homogeneous catalyst consisting of a single component nanotube structure [30]. By taking advantage of living radical and also ROP reported elsewhere [25], they synthesized the BBC P(GM-*g*-LA-*g*-VBC/BS-*g*-PVBS/NIPAM) **26** (Chart 8.6) as a precursor of the acid–base nanotube. Intramolecular crosslinking of the BS segments and subsequent removal of the PLA segments under basic conditions converted the precursor into nanotube structures. Through several chemical treatment including azide reaction and click reaction, the organic nanotubes come to be furnished with the functional group of PEI **13** as basic

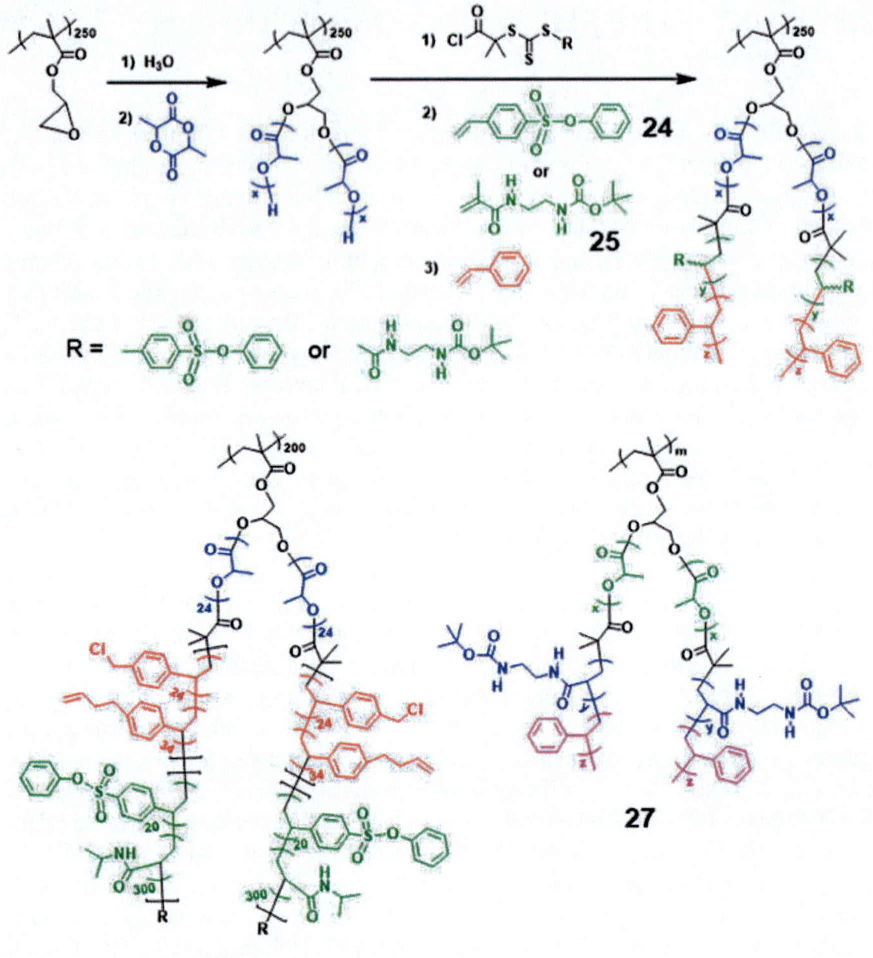

Chart 8.6 (top) Synthesis of the MONN substrate with acid and base catalytic functionalities and (bottom) chemical structures of the copolymer precursors **26** and **27**

sites on the nanotube wall and benzenesulfonic acid as acidic sites on the corona chain of nanotubes (Fig. 8.17). The resultant amphiphilic organic nanotubes serve as acid–base bifunctional catalysts. And yet, the obtained nanotubes are featured by hydrophilic corona chains, hydrophobic pore environments, and isolated placement of the acid and base catalysts. Indeed, those features resulted in the finding that the robust and stable acid–base nanotubes displayed highly catalytic performance for tandem deacetalization–Knoevenagel reactions, particularly in water (Table 8.3) [30].

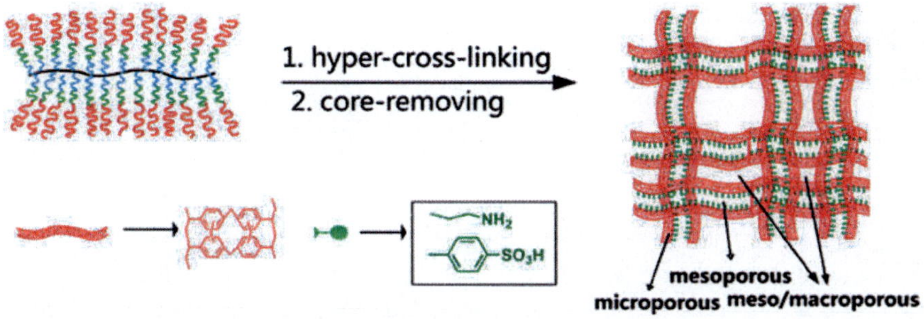

Fig. 8.15 Preparation of the acid–base-functionalized MONNs Reproduced with permission from Ref. [29], © The Royal Society of Chemistry 2016

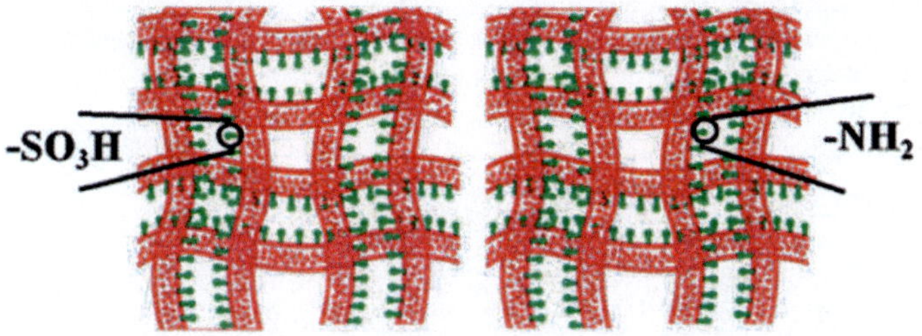

Fig. 8.16 Illustration of SO_3H-MONNs and NH_2-MONNs catalysts for the one-pot reactions. Reproduced with permission from Ref. [29], © The Royal Society of Chemistry 2016

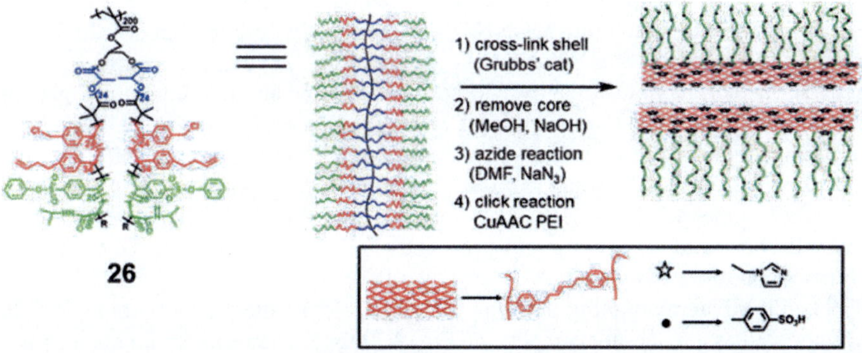

Fig. 8.17 Synthesis of acid–base bifunctional amphiphilic organic nanotubes from **26**. Reproduced with permission from Ref. [30], © The Royal Society of Chemistry and the Centre National de la Recherche Scientifique 2018

Table 8.3 Catalytic results for a one-pot deacetalization–Knoevernagel reaction using the acid–base bifunctional catalysts[a]

Entry	Catalyst	Conversion of A (%)	Yield of B (%)[b]	Yield of C (%)[b]
1	None	8	5	3
2	Acid–base nanotube	100	0	100
3	Acid nanotube	100	92	8
4	Base nanotube	13	2	11
5	Acid–base nanotube + PTSA	95	95	0
6	Acid–base nanotube + HEIZ	0	0	0

[a]Reaction conditions: **17** (0.13 mmol), about 10 mol% acid and 10 mol% base catalysts were used. The reaction mixtures were runs for 24 h at room temperature

[b]Yields are based on GC measurements

8.8 COOH-MONN and Pd@MONN

Palladium nanoparticles (Pd NPs) is a key metal catalyst that accelerates, e.g., Suzuki–Miyaura crosscoupling reaction [31]. To prevent the aggregation behavior of the Pd NPs, the particles need to be immobilized in or onto solid supports that suit a certain purpose. Huang and co-workers developed a new methodology, in which the Pd NPs could be well dispersed and immobilized onto the inner surfaces of the pores of carboxyl-group-containing MONNs [32]. As already described in **8.1**, the PGM-*g*-(PLA-*b*-PS)-derived MONNs possess abundant carboxylic functionalities on the inner surface of the hollow cavity. Moreover, additional amount of carboxyl groups can be incorporated by the insertion of PtBA chain between the PS shell and PLA core layers. Finally, in situ thermal decomposition of $Pd(OAc)_2$ was able to endow the carboxy-containing solid support MONNs with Pd NPs (Fig. 8.18). Thus, two types of Pd NPs-appended MONNs (Pd@MONNs-1 and Pd@MONNs-4) with onefold and extra threefold carboxyl groups, respectively, were prepared for the evaluation of their catalytic activity [32].

Suzuki–Miyaura coupling between *p*-iodoanisole **19** and phenylboronic acid **20** was adopted as a reaction example (see Fig. 8.12 top). The resultant kinetic profiles showed that the conversion yield reaches 100% within 3 h for the Pd@MONN-4 catalyst, whereas within 5 h for the Pd@MONN-1 catalyst (Fig. 8.19a) [32]. The relatively excellent catalytic activity of Pd@MONN-4 than that of Pd@MONN-1 will be due to the higher amount of Pd NPs within the MONN. In turn, the high catalytic behavior of Pd@MONN-1 can maintain even after eight cycles (Fig. 8.19b). Furthermore, the used Pd@MONN-1 can be easily separated and recovered by filtration treatment after individual cycle. The effect of hierarchical porosity comprising micro-, meso-, and macropores on the catalytic behavior was investigated by using

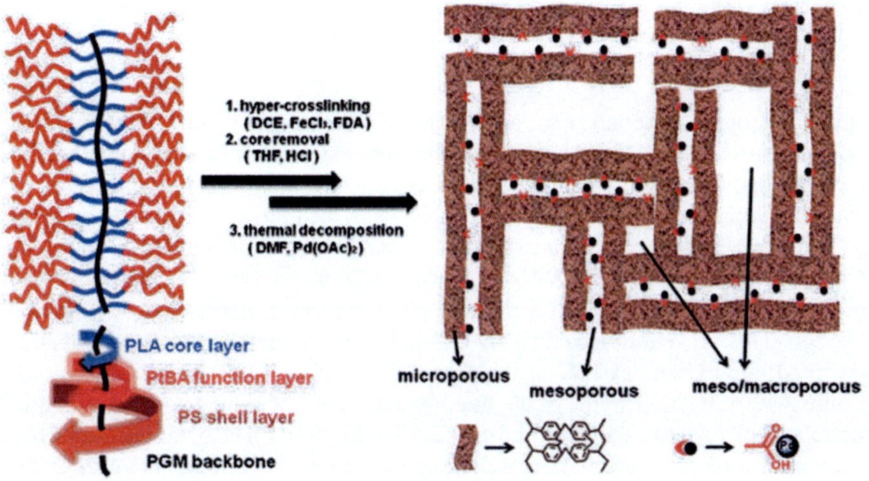

Fig. 8.18 Preparation of Pd NPs catalysts immobilized in or onto carboxyl-containing MONNs. Reproduced with permission from Ref. [32], © The Royal Society of Chemistry 2016

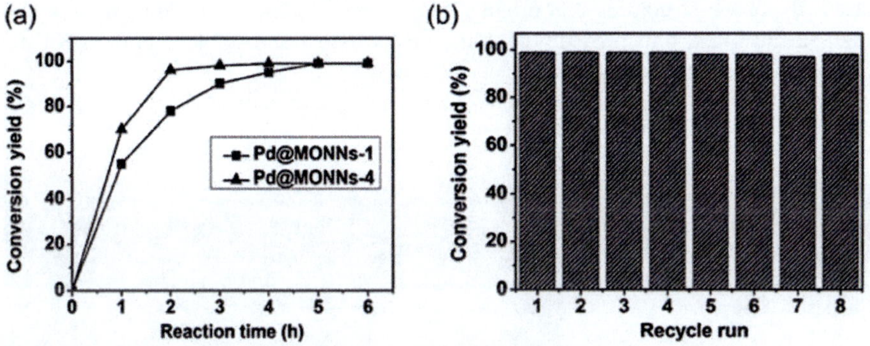

Fig. 8.19 **a** Kinetic profiles of Pd@MONNs-1 and Pd@MONNs-4 in the model Suzuki–Miyaura coupling reaction of **19** and **20**. The reaction formula is shown in Fig. 8.12. **b** Recycle performances of the typical Pd@MONNs-1 catalyst in the model reaction. The conversion yields were monitored by GC. Reproduced with permission from Ref. [32], © The Royal Society of Chemistry 2016

Pd NPs-appended microporous organic polymers (Pd@MOP) as a reference catalyst [32]. The time required for the full conversion yield took 12 h under the same conditions as Pd@MONNs. The second cycle reaction using recovered Pd@MOP exhibited significant deactivation of the catalyst. The MONN scaffold, thus, provides a platform of supporting materials for Pd catalysts.

8.9 Oxo-Vanadium-Microporous Organic Nanotube Framework

Inorganic porous materials, e.g., active carbon, silicon, and zeolite, have improved the intrinsic value of heterogeneous oxidation catalysts. Relatively low chemical acceptability for various reactions and small surface area limits the universal use as a solid supporter. Only a few reports have described on porous organic materials (POMs), on which oxo-vanadium (IV) (VO) complexes are anchored (Fig. 8.20) [33]. The copolymer precursor P(GM-*g*-PLA-*g*-PBAEAM-*g*-PS) [PBAEAM: poly(Boc-aminoethyl acrylamide)] **27** (Chart 8.6) was stirred in a mixture of HCl and 1.4-dioxane. This procedure gave NH_2-MONNs having internal amino groups that are anchored on the inner surface of the pores (Fig. 8.20) [34]. Reaction of the amino groups with salicylaldehyde yields the organic ligand of salicylaldehyde-modified microporous organic nanotube frameworks (SA-MONFs). The obtained Schiff base structure results in the formation of the oxo-vanadium (IV) complex (VO-MONFs) with the SA-MONFs (Fig. 8.21) [33]. TEM observation showed that the dimensions of the SA-MONF nanotubes are consisted of 50 ± 5 nm length, 2–3 nm shell thickness, and 3–4 nm diameter. The BET surface area and total pore volume of the SA-MONF were evaluated to be 821 $m^2\ g^{-1}$ and 1.79 $cm^3\ g^{-1}$, respectively. Hierarchical porous structures comprising micro-, meso-, and meso-/macropores and high surface area, thus provide the materials as robust and stable supporter for the heterogeneous catalysis oxidation.

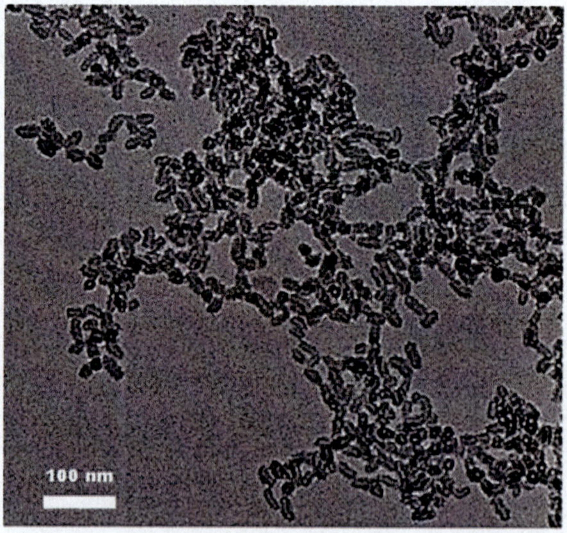

Fig. 8.20 TEM image of the NH_2-MONNs. Reproduced with permission from Ref. [34], © The Royal Society of Chemistry 2016

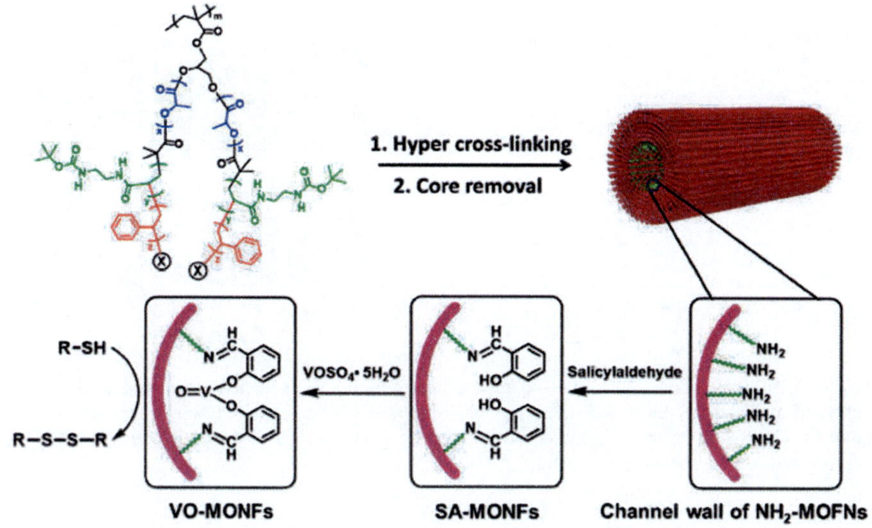

Fig. 8.21 Synthesis of VO-MONFs and the application in selective oxidation of thiols to disulfides. Reproduced with permission from Ref. [33], © 2017 Elsevier Inc.

The oxidation reaction of *p*-chlorobenzenethiol **28** to disulfides **29** was adopted as a typical example in the presence of urea hydrogen peroxide **30** as an oxidant (Chart 8.7) [33]. The oxidation catalytic activity of the VO-MONFs exhibited the highest among comparative other oxidation catalysts oxo-vanadium (II) microporous organic nanowire frameworks (VO-MONWFs) (Fig. 8.22a). The existence of the mesopores, high dispersion of vanadium catalysts, and high surface area play a critical role in expressing the high catalytic activity. Notably, the catalytic activity of the VO-MONFs maintains even after eight cycles with keeping the intrinsic hierarchically porous structures (Fig. 8.22b) [33]. And yet, the catalysts show excellent thermal and chemical stability.

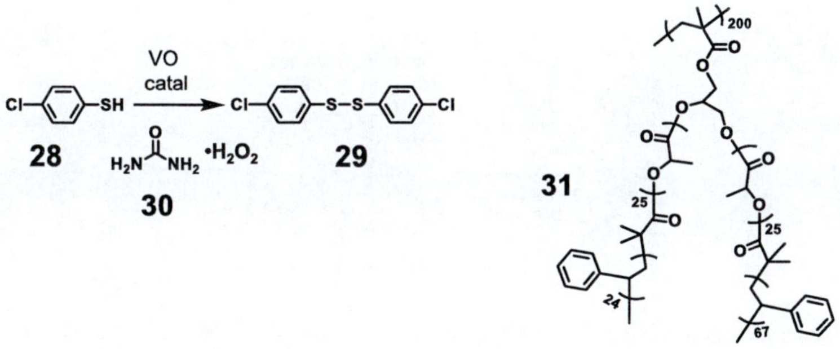

Chart 8.7 (left) Oxidation reaction of *p*-chlorobenzenethiol 28 to disulfides 29 and (right) chemical structure of the BBC precursor 31

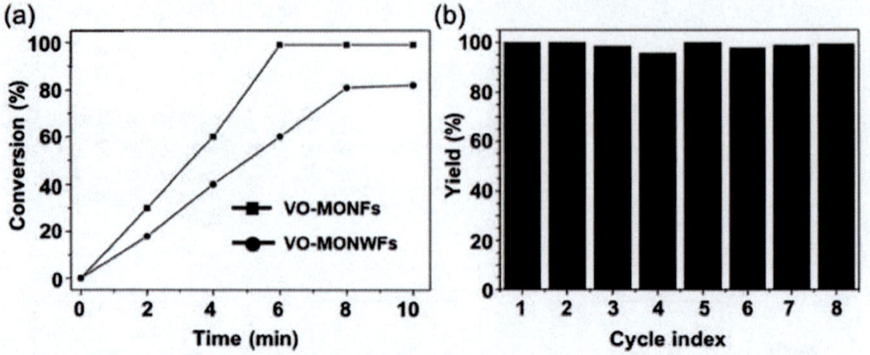

Fig. 8.22 **a** Conversion (%) plotted against time (min) for the selective oxidation of **28** to **29** catalyzed by VO-MONFs (filled square) or VO-MONWFs (filled circle). Each conversion is evaluated by GC based on starting material. **b** Catalytic performance of the recycled VO-MONFs. Reproduced with permission from Ref. [33], © 2017 Elsevier Inc.

8.10 Thiol-Functionalized Hierarchically Porous Material

MONNs were subjected to post-modification that introduce a variety of functional groups on the inner surfaces of the MONNs. Huang's research group focused on the large amount of carboxylic acid group that are left as terminal groups after the PLA degradation of the BBC precursor P(GM-*g*-PLA-*g*-PS) **31** (Chart 8.7). The combination of the incorporation of some functionalities and porous organic materials enabled the formation of Au NPs-appended thiol-functionalized MONNs in the pores [35]. Hyper-crosslinking and the subsequent removal of the PLA core resulted in the formation of hierarchically porous materials (HPMs) [35]. Condensation of the HPMs carboxyl group with cysteamine yielded the thiol-functionalized HPMs (SH-HPMs) that can bind Au NPs of ca. 3 nm prepared by in situ reduction of $HAuCl_4$ in the presence of $NaBH_4$ (Fig. 8.23). The resultant Au HPMs was employed as a heterogeneous catalyst that can accelerate the reduction reaction of

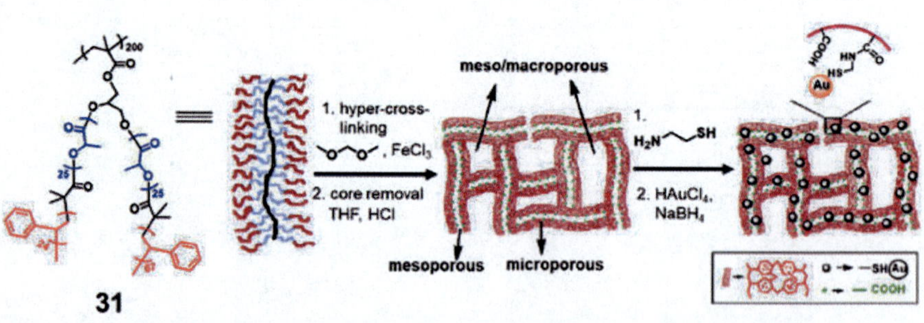

Fig. 8.23 Preparation of SH-HPMs with well-dispersed Au NPs. Reproduced with permission from Ref. [35], © 2016 Elsevier Inc.

Table 8.4 Reducibility of diverse heterogeneous Au NPs catalysts to 4-nitrophenol, which was reported in the representative literature

Entry	Au NPs catalyst support	Size of Au NPs (nm)	Capacity of Au NPs (wt%)	Dose of catalyst (mg)	[$NaBH_4$] (M)	[4-NP] (M)	Time	References
1	MOP	7.0–7.3	~ 7.0	10	0.1×10^{-3}	0.5×10^{-4}	10 min	[36]
2	COF	2.0–8.0	~ 1.2	20	0.36	1.8×10^{-4}	13 min	[37]
3	CeO_2	~ 8.0	~ 2.97	1	0.5	0.01	5 min	[38]
4	SNTs	3.0–5.0	~ 2.3	8	0.5×10^{-4}	1.2×10^{-5}	280 s	[39]
5	HGN	2.0–3.0	~ 1.0	3	30	0.03	180 s	[40]
6	HPMs	~ 3.0	~ 3.0	10	0.1×10^{-3}	0.5×10^{-4}	~ 10 s	This work

Reproduced with permission from Ref. [35], © 2016 Elsevier Inc.

4-nitrophenol to 4-aminophenol in the presence of $NaBH_4$. When compared with the catalytic activity of microporous organic polymer (MOP)-, COF-, CeO_2-, silica nanotube (SNT)-, and hollow graphene nanoshell (HGN)-supported Au NPs, that of the Au HPMs showed the most excellent values (Table 8.4) [35]. Both the catalytic activity and morphological stability keep almost unchanged even after five cycles.

References

1. Stewart S, Liu G (2000) Block copolymer nanotubes. Angew Chem Int Ed 39:340–344. https://doi.org/10.1002/(Sici)1521-3773(20000117)39:2%3c340::Aid-Anie340%3e3.0.Co;2-H
2. Li Z, Liu G (2003) Water-dispersible tetrablock copolymer synthesis, aggregation, nanotube preparation, and impregnation. Langmuir 19:10480–10486. https://doi.org/10.1021/la035263w
3. Lee M, Cho BK, Zin WC (2001) Supramolecular structures from rod-coil block copolymers. Chem Rev 101:3869–3892. https://doi.org/10.1021/cr0001131
4. Kim HC, Park SM, Hinsberg WD (2010) Block copolymer based nanostructures: materials, processes, and applications to electronics. Chem Rev 110:146–177. https://doi.org/10.1021/cr900159v
5. Cabral H, Miyata K, Osada K, Kataoka K (2018) Block copolymer micelles in nanomedicine applications. Chem Rev 118:6844–6892. https://doi.org/10.1021/acs.chemrev.8b00199
6. Jenekhe SA, Chen XL (1998) Self-assembled aggregates of rod-coil block copolymers and their solubilization and encapsulation of fullerenes. Science 279:1903–1907. https://doi.org/10.1126/science.279.5358.1903
7. Raez J, Barjovanu R, Massey JA, Winnik MA, Manners I (2000) Self-assembled organometallic block copolymer nanotubes. Angew Chem Int Ed 39:3862–3865. https://doi.org/10.1002/1521-3773(20001103)39:21%3c3862::AID-ANIE3862%3e3.0.CO;2-1

8. Raez J, Manners I, Winnik MA (2002) Nanotubes from the self-assembly of asymmetric crystalline-coil poly(ferrocenylsilane-siloxane) block copolymers. J Am Chem Soc 124:10381–10395. https://doi.org/10.1021/ja020349h
9. Massey J, Power KN, Manners I, Winnik MA (1998) Self-assembly of a novel organometallic-inorganic block copolymer in solution and the solid state: nonintrusive observation of novel wormlike poly(ferrocenyldimethylsilame)-*b*-poly(dimethylsiloxane) micelles. J Am Chem Soc 120:9533–9540. https://doi.org/10.1021/ja981803d
10. Raez J, Tomba JP, Manners I, Winnik MA (2003) A reversible tube-to-rod transition in a block copolymer micelle. J Am Chem Soc 125:9546–9547. https://doi.org/10.1021/ja030251i
11. Vilgis T, Halperin A (1991) Aggregation of coil crystalline block copolymers—equilibrium crystallization. Macromolecules 24:2090–2095. https://doi.org/10.1021/ma00008a058
12. Yu K, Eisenberg A (1998) Bilayer morphologies of self-assembled crew-cut aggregations of amphiphilic PS-*b*-PEO diblock copolymers in solution. Macromolecules 31:3509–3518. https://doi.org/10.1021/ma971419l
13. Shen H, Eisenberg A (2000) Block length dependence of morphological phase diagrams of the ternary system of PS-*b*-PAA/dioxane/H_2O. Macromolecules 33:2561–2572. https://doi.org/10.1021/ma991161u
14. Li Z-C, Liang Y-Z, Li F-M (1999) Multiple morphologies of aggregates from block copolymer containing glycopolymer segments. Chem Commun 1557–1558. https://doi.org/10.1039/A905114G
15. Zhang L, Eisenberg A (1998) Formation of crew-cut aggregates of various morphologies from amphiphilc block copolymers in solution. Polym Adv Technol 9:677–699. https://doi.org/10.1002/(SICI)1099-1581(1998100)9:10/11%3c677::AID-PAT845%3e3.0.CO;2-%23
16. Cameron NS, Corbierre MK, Eisenberg A (1999) Asymmetric amphiphilic block copolymers in solution: a morphological wonderland. Can J Chem 77:1311–1326. https://doi.org/10.1139/v99-141
17. Gao ZS, Varshney SK, Wong S, Eisenberg A (1994) Block-copolymer crew-cut micelles in water. Macromolecules 27:7923–7927. https://doi.org/10.1021/ma00104a058
18. Cornelissen JJLM, Fischer M, Sommerdijk NAJM, Nolte RJM (1998) Helical suprastructures from charged poly(styrene)-poly(isocyanodipeptide) block copolymers. Science 280:1427–1430. https://doi.org/10.1126/science.280.5368.1427
19. Zhang MF, Muller AHE (2005) Cylindrical polymer brushes. J Polym Sci Part A Polym Chem 43:3461–3481. https://doi.org/10.1002/pola.20900
20. Sheiko SS, Sumerlin BS, Matyjaszewski K (2008) Cylindrical molecular brushes: synthesis, characterization, and properties. Prog Polym Sci 33:759–785. https://doi.org/10.1016/j.progpolymsci.2008.05.001
21. Mullner M, Yuan JY, Weiss S, Walther A, Fortsch M, Drechsler M, Muller AHE (2010) Water-soluble organo-silica hybrid nanotubes templated by cylindrical polymer brushes. J Am Chem Soc 132:16587–16592. https://doi.org/10.1021/ja107132j
22. Huang K, Rzayev J (2011) Charge and size selective molecular transport by amphiphilic organic nanotubes. J Am Chem Soc 133:16726–16729. https://doi.org/10.1021/ja204296v
23. Wang TQ, Xu Y, He ZD, Zhou MH, Huang K (2018) Microporous organic nanotube networks from hyper cross-linking core-shell bottlebrush copolymers for selective adsorption study. Chin J Polym Sci 36:98–105. https://doi.org/10.1007/s10118-018-2007-0
24. Shimizu T, Ding W, Kameta N (2020) Soft-matter nanotubes: a platform for diverse functions and applications. Chem Rev 120:2347–2407. https://doi.org/10.1021/acs.chemrev.9b00509
25. Xiong L, Yang K, Zhang H, Liao X, Huang K (2016) Soluble organic nanotubes for catalytic systems. Nanotechnology 27:115603–115610. https://doi.org/10.1088/0957-4484/27/11/115603
26. Zhou M, Zhang H, Xiong L, He Z, Zhong A, Wang T, Xu Y, Huang K (2016) Synthesis of magnetic microporous organic nanotube networks for adsorption application. RSC Adv 6:87745–87752. https://doi.org/10.1039/c6ra18836b
27. Polshettiwar V, Luque R, Fihri A, Zhu HB, Bouhrara M, Basset JM (2011) Magnetically recoverable nanocatalysts. Chem Rev 111:3036–3075. https://doi.org/10.1021/cr100230z

28. Wang D, Astruc D (2014) Fast-growing field of magnetically recyclable nanocatalysts. Chem Rev 114:6949–6985. https://doi.org/10.1021/cr500134h
29. Zhang H, Xiong L, He Z, Zhong A, Wang T, Xu Y, Huang K (2016) Microporous organic nanotube network supported acid and base catalyst system for one-pot cascade reactions. New J Chem 40:7282–7285. https://doi.org/10.1039/c6nj01457g
30. Xiong L, Zhang H, He Z, Wang T, Xu Y, Zhou M, Huang K (2018) Acid-base bifunctional amphiphilic organic nanotubes as a catalyst for one-pot cascade reactions in water. New J Chem 42:1368–1372. https://doi.org/10.1039/c7nj04209d
31. Balanta A, Godard C, Claver C (2011) Pd nanoparticles for C-C coupling reactions. Chem Soc Rev 40:4973–4985. https://doi.org/10.1039/c1cs15195a
32. Xu Y, Wang T, He Z, Zhong A, Huang K (2016) Carboxyl-containing microporous organic nanotube networks as a platform for Pd catalysts. RSC Adv 6:39933–39939. https://doi.org/10.1039/c6ra05753e
33. Zhang H, Zhou M, Xiong L, He Z, Wang T, Xu Y, Huang K (2018) Oxo-vanadium (IV) complex supported by microporous organic nanotube frameworks: a high selective heterogeneous catalyst for the oxidation of thiols to disulfides. Microporous Mesoporous Mater 255:103–109. https://doi.org/10.1016/j.micromeso.2017.07.041
34. Zhang H, Xiong L, He Z, Zhong A, Wang T, Xu Y, Zhou M, Huang K (2016) Functionalized microporous organic nanotube networks as a new platform for highly efficient heterogeneous catalysis. Polym Chem 7:4975–4982. https://doi.org/10.1039/c6py01052k
35. Xu Y, Wang T, He Z, Zhong A, Huang K (2016) Well-dispersed gold nanoparticles anchored into thiol-functionalized hierarchically porous materials for catalytic applications. Microporous Mesoporous Mater 229:1–7. https://doi.org/10.1016/j.micromeso.2016.04.013
36. Lee H, Kim H, Choi TJ, Park HW, Chang JY (2015) Preparation of a microporous organic polymer by the thiol-yne addition reaction and formation of Au nanoparticles inside the polymer. Chem Commun 51:9805–9808. https://doi.org/10.1039/c5cc02269j
37. Pachfule P, Kandambeth S, Diaz DD, Banerjee R (2014) Highly stable covalent organic framework-Au nanoparticles hybrids for enhanced activity for nitrophenol reduction. Chem Commun 50:3169–3172. https://doi.org/10.1039/c3cc49176e
38. Liu BC, Yu SL, Wang Q, Hu WT, Jing P, Liu Y, Jia WJ, Liu YX, Liu LX, Zhang J (2013) Hollow mesoporous ceria nanoreactors with enhanced activity and stability for catalytic application. Chem Commun 49:3757–3759. https://doi.org/10.1039/c3cc40665b
39. Zhang ZY, Shao CL, Zou P, Zhang P, Zhang MY, Mu JB, Guo ZC, Li XH, Wang CH, Liu YC (2011) In situ assembly of well-dispersed gold nanoparticles on electrospun silica nanotubes for catalytic reduction of 4-nitrophenol. Chem Commun 47:3906–3908. https://doi.org/10.1039/c0cc05693f
40. Liu HY, Wang J, Feng ZB, Lin YM, Zhang LY, Su DS (2015) Facile synthesis of Au nanoparticles embedded in an ultrathin hollow graphene nanoshell with robust catalytic performance. Small 11:5059–5064. https://doi.org/10.1002/smll.201500635

Chapter 9
Rigid–Flexible Block Molecule-Based Nanotubes

9.1 Rigid–Flexible Block Molecule

Lee and co-workers described an intriguing review on stimuli-responsive supramolecular nanostructures and functions that can be performed and expressed through the aqueous self-assembly of rigid–flexible block molecules (RFBMs) as building blocks [1]. The RFBM is a kind of amphiphilic molecule and roughly comprises both hydrophobic, rigid rod-like segments and hydrophilic, flexible coil-like segments. For example, oligo (*p*-phenylene) and perylenediimide are employed as the rod segments, whereas PEO and poly(propylene oxide) (PPO) as the flexible chains. They synthesized RFBM **1** (Chart 9.1) that configures a macrocycle structure by the combination of a hexa-*p*-phenylene rod and a flexible chiral PEO chain. The aqueous self-assembly of the amphiphilic macrocycle **1** yielded a tubular structure having a coiled ribbon morphology [2]. The bolaamphiphilic molecule **2** (Chart 9.1) comprising a hexa (*p*-phenylene) rod block and flexible oligoether dendritic chains with chiral centers undergoes the self-assembly in water to give helical fibers [3]. The aromatic segments stack on top of each other unidirectionally with mutual rotation. Interestingly, the addition of a guest molecules induces the intercalation between the aromatic segments, resulting in the transformation from fiber structures to hollow spheres. Along these lines, researchers have devoted to exploit novel stimuli-responsive nanostructures based on tubular self-assemblies from RFBMs by using their diverse structures and abundant designability.

9.2 Perylene Diimide Derivative

9.2.1 Perylene Diimide/Hydrophobic Chain

Perylene diimide (PDI) molecules are involved in a typical class of n-type organic semiconductors [4]. Che and co-workers demonstrated that a chiral PDI-derived

T. Shimizu, *Smart Soft-Matter Nanotubes*, Nanostructure Science and Technology,
https://doi.org/10.1007/978-981-16-2685-2_9

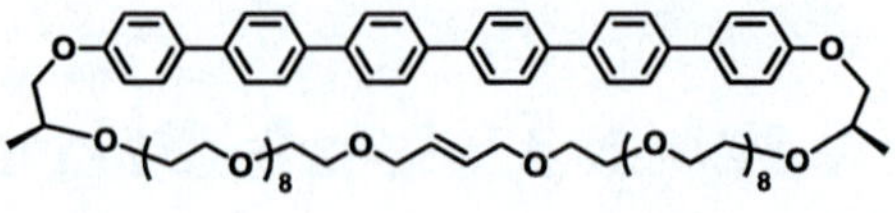

1

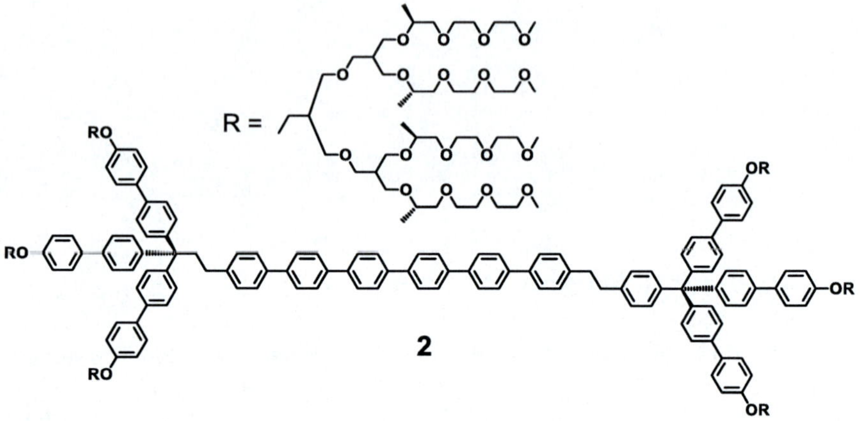

2

Chart 9.1 Chemical structures **1** and **2**

amphiphiles **3S** and **3R** (Chart 9.2) self-assemble in a mixed solvent system of chloroform and ethanol to give highly fluorescent chiral nanotubes (Fig. 9.1) [5]. The fluorescence quantum yield of the self-assembled nanotubes displayed higher values than 46%, which are extremely greater than that (ca. 25%) of the self-assembled nanocoils from other PDI-based amphiphiles [6].

Aiming at the fabrication of self-assembled nanotubes with tunable diameters and wall thickness, Che and co-workers further synthesized two families of achiral asymmetric PDI amphiphiles (**4a-4c** and **5a-5c**) (Chart 9.2) [7]. The new family of the PDI amphiphiles contains a dodecyl chain as the one side segment and bulky branched substituents at the meta- (**4a**, **4b**, and **4c**) or ortho-position (**5a**, **5b**, and **5c**) of the phenyl segment and the ethylene spacer chain as the another side segment. Self-assembly experiment was carried out by injection of the chloroform solution of each PDI amphiphile into ethanol and subsequent aging for seven days. Interestingly, the amphiphiles **4a**, **4b**, and **4c** self-assemble to construct bilayer-based nanotubes, whereas the amphiphiles **5a** and **5b** form monolayer-based nanotubes (Fig. 9.2). The bulkiness of the substituents at the meta-position has a strong effect on the diameter of the bilayer nanotubes, whereas the diameter of the monolayered nanotubes is not affected by the size of the substituents. More interestingly, all the self-assembled nanotubes exhibit highly emissive fluorescence.

Che's research group then employed the fluorescent chiral nanotubes [5] to highly sense biogenic amines (BGAs), e.g., cadaverine **6**, putrescine **7**, and trimethylamine **8** (Chart 9.2) [8]. The decomposition of meat protein expels those amine vapor.

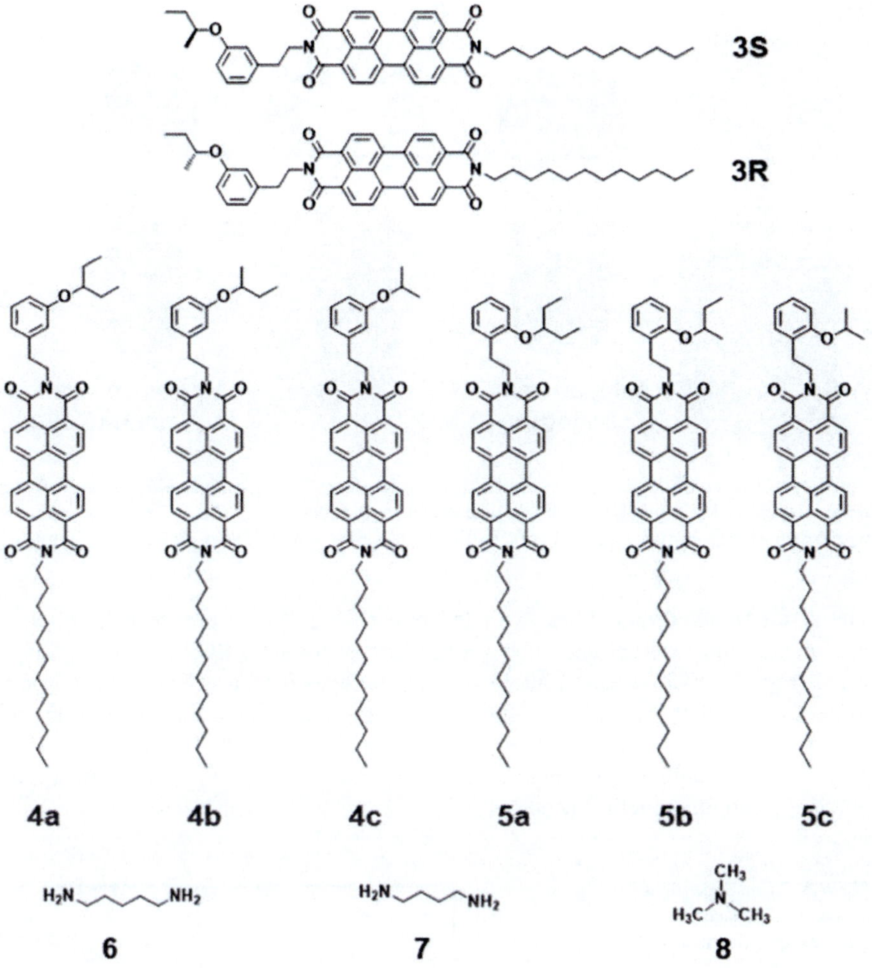

Chart 9.2 Chemical structures of **3–8**

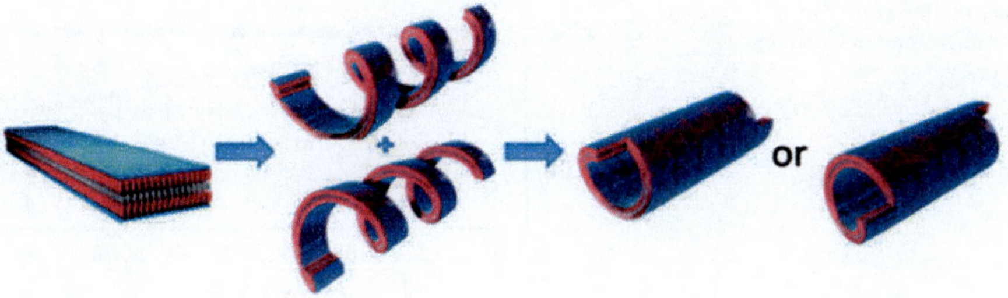

Fig. 9.1 Schematic drawing of the formation of single-handed nanotubes from **3S** or **3R**. (Reproduced with permission from Ref. [5], © The Royal Society of Chemistry 2015)

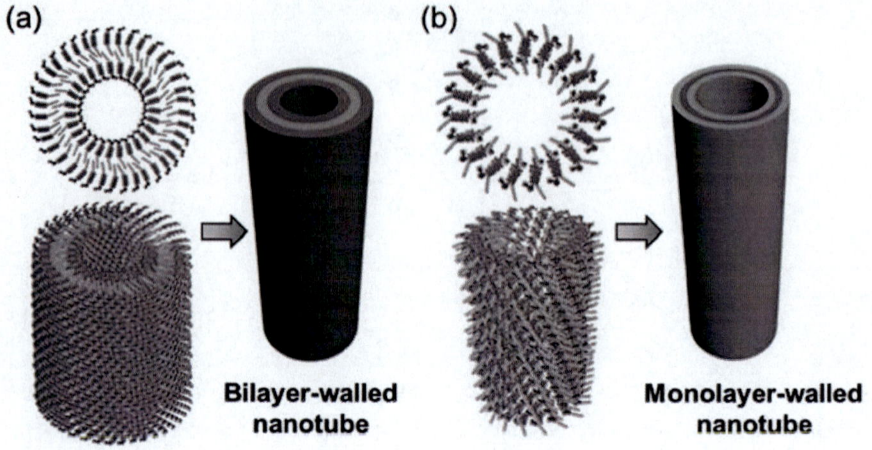

Fig. 9.2 **a** Bilayer-walled nanotube formed from asymmetric PDI derivatives (**4a**, **4b**, and **4c**) and **b** monolayer-walled nanotube formed from asymmetric PDI derivatives (**5a** and **5b**). (Reproduced with permission from Ref. [7], © 2016 WILEY–VCH Verlag GmbH & Co. KGaA, Weinheim)

The self-assembled nanotubes exhibited remarkable fluorescent quenching when the internal and external nanotube surfaces are exposed to BGAs as the spoilage vapor (Fig. 9.3). The detection limits of the nanotubes for **6** and **7** (2.6 and 1.2 ppb, respectively) excel those (>17.1 and 67.2 ppb, respectively) of the nanoribbon and nanowire analogues self-assembled from the same PDI amphiphiles. These results mean that the tubular structures are indispensable in terms of high fluorescence efficiency and inherent hollowness.

Fig. 9.3 Time dependence of the fluorescence quenching of the nanotubes self-assembled from chiral molecule **3S** or **3R** at 25 °C for 1 h. (Reproduced with permission from Ref. [8], © 2015 American Chemical Society)

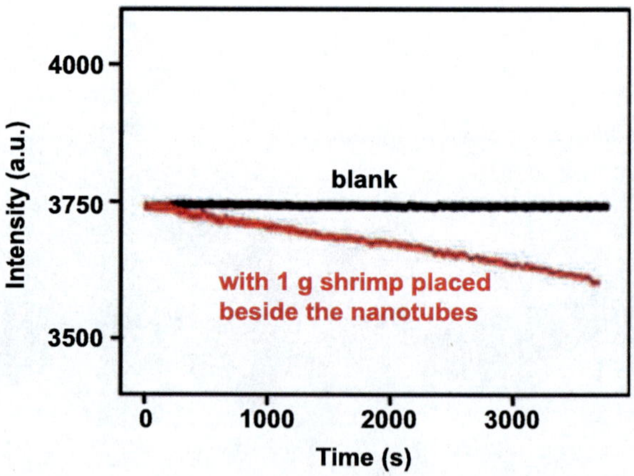

9.2.2 Perylene Diimide/Peptide Conjugate

Rybtchinski and co-workers newly designed amphiphilic RFBM (**9a** and **9b**) (Chart 9.3) that comprises PEGylated PDI/terpyridine (terpy) platinum complexes with Cys–Ala–Ala (**9a**) and Cys–Phe–Phe (**9b**) tripeptides [9]. The PEG chain serves as a flexible coil segment and the PDI–terpy–Pt complex as a rigid rod segment. Yet, the Cys–Ala–Ala and Cys–Phe–Phe segments commit the entire molecule to supramolecular chirality. The Cys–Phe–Phe sequence is relatively rigid and hydrophobic components than the Cys–Ala–Ala one. By using the PEG–PDI–terpy–Pt/peptide conjugates, they proved the viability of a noncovalent aqueous self-assembly system that is kinetically controllable in a mixture of water/THF. The self-assembly of the conjugate **9a** was conducted in water/THF mixture (80/20, v/v). After the incubation for 40 h at room temperature, the self-assembly system was diluted with water to reach the mixing ratio of the water/THF (95/5, v/v). Among several pathways examined under different conditions, such a pathway yielded hollow cylindrical nanotube structures. Cryo-TEM images revealed that the dimension of the self-assembled nanotubes is uniform, and 1.1 ± 0.2 nm in inner diameters and 1.6 ± 0.2 nm thick of tube walls [9]. Meanwhile, less selective noncovalent interactions in the conjugate **9b** resulted in the formation of somewhat diverse nanostructures including long fibers and spherical micelles (Fig. 9.4). This self-assembly feature remarkably contrasts to the pathway-derived self-assembly.

Böttcher, Hirsch, and co-workers presented the amphiphilic, tetrahedrally shaped RFBM **10** (Chart 9.3), in which both two hydrophobic PDI segments and two hydrophilic Newkome-type dendrons (first generation: G1) are covalently connected

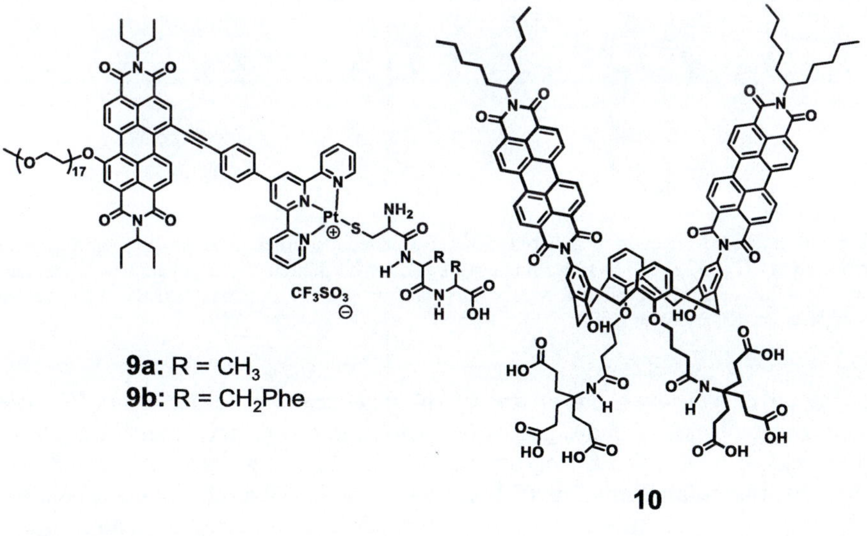

Chart 9.3 Chemical structures of **9–10**

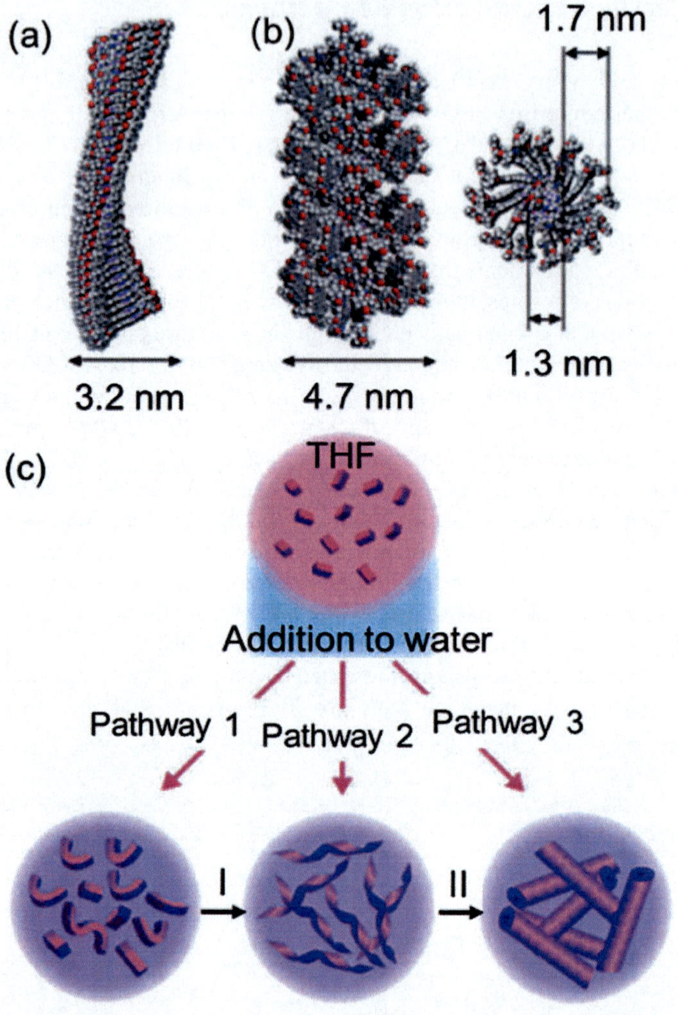

Fig. 9.4 **a** and **b** Molecular models of the self-assembled nanofibers from **9a** (PEG groups were modeled as CH_3–OCH_2CH_2–O): **a** a single stacked fiber; **b** a tubular fiber. **c** A pathway-dependent-self-assembly of **9a**. (Reproduced with permission from Ref. [9], © 2011 WILEY–VCH Verlag GmbH & Co. KGaA, Weinheim)

to the upper and lower rim of the calix[4]arene scaffold, respectively [10]. The strategy for the molecular design is based on the upward and downward arrangement of the hydrophobic rigid PDI segment and hydrophilic somewhat flexible dendron on the calix[4]arene platform. The PDI segment serves as polycyclic aromatic building block that enable effective π–π stacking interactions, whereas the G1 dendron offers adequate solubility in water. The self-assembly examination of **10** was carried out in water and in a mixture of water/THF. Cryo-TEM measurement showed that in the 1 mM solution, various curly and twisted ribbons of different width were observed as vitrified objects. Additionally, tubular assemblies were also found with diameters of 8–20 nm, length of 8–100 nm, and wall thickness of 1.8–2.5 nm (Fig. 9.5). The self-assembly of **10** in a mixture of water/THF (70/30, v/v) displayed somewhat different features. Densely arranged fiber network was only observed just after the

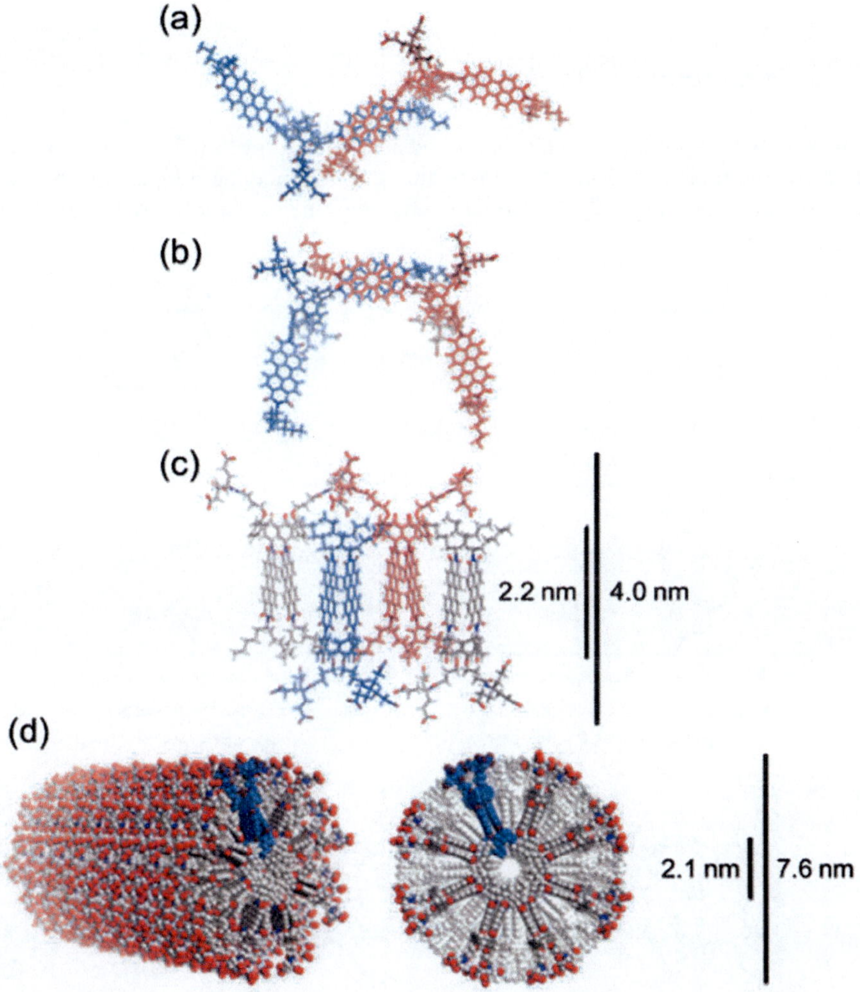

Fig. 9.5 Molecular arrangement of **10** in the aggregates formed in pure water. **a** and **b** Two different interaction profiles of the open conformers of **10**, which are derived from MD simulations and **c** π-stacked closed conformers as an alternative arrangement. **d** Perspective (left) and top view (right) of a proposed arrangement of the closed conformers of **10** leading to the fibrous aggregates. (Reproduced with permission from Ref. [10], © 2015 American Chemical Society)

dissolution, whereas after six days incubation tube-like (mean diameter: 7.63 nm) and plate-like assemblies became abundant [10]. After 26 days, crystalline patches with a lattice layer spacing of 1.95 nm appeared more clearly together with the tube-like assemblies. MD simulation studies suggested the possibility of rod-like micellar assemblies, in which monolayer membrane-based hollow cylindrical structures are formed with the hydrophilic dendrons directing outward and the hydrophobic PDI segments inward. The tiny internal lumen is filled with THF molecules [10].

9.3 Amphiphilic Porphyrin

The structure of porphin **11**, chlorin **12**, and bacteriochlorin **13** (Chart 9.4), in which four pyrrole molecules form a ring structure through methine bridge, constitutes a part of several biologically important compounds such as chlorophyll, heme, and

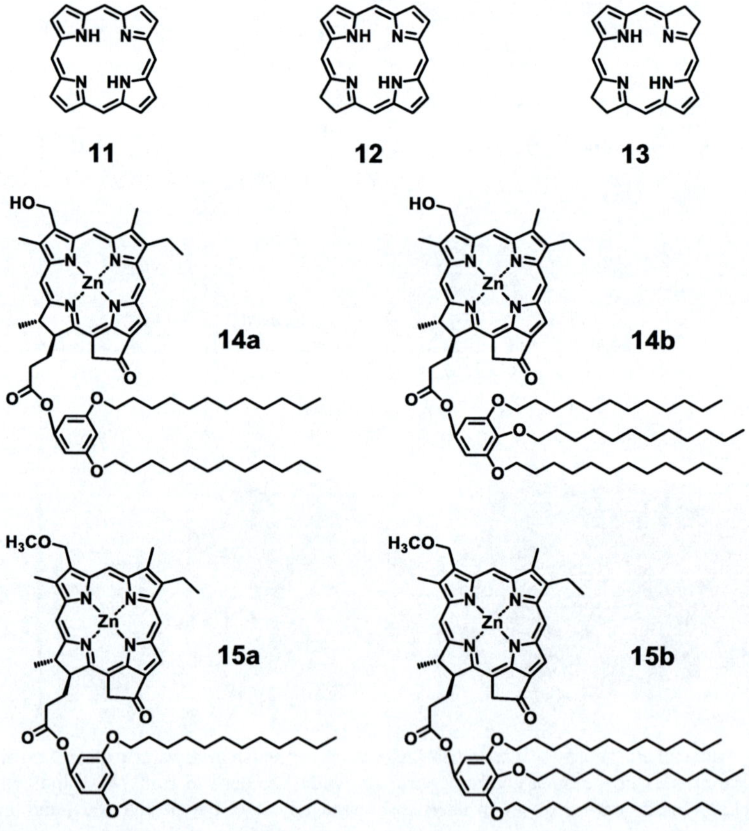

Chart 9.4 Chemical structures of **11–15**

vitamin B12. Naturally occurring chlorosomal antennae comprising bacteriochlorophyll (BChl) pigments captivate us to design 1D supramolecular assemblies that can exhibit fascinating opto-electronic and artificial photosynthetic properties. Chemical modification of the natural BChl, e.g., the incorporation of a Zn ion, tailored side chains, and 3^1-functionality, can improve the solubility and control the stacking mode. For example, hydroxy and methoxy substitution at the 3^1 position of BChl c resulted in the self-assembly of 1D tubular and 2D sheet structures, respectively [11, 12]. Würthner, Grozema, and co-workers prepared semisynthetic 3^1-hydroxy zinc chlorin (ZnChl) (**14a** and **14b**) (Chart 9.4) with two- and three-armed C12 saturated alkyl chains, respectively, as well as the 3^1-methoxy analogues (**15a** and **15b**) (Chart 9.4) [13]. Both the 3^1-hydroxy and 3^1-methoxy derivatives form a J-aggregate with slipped stacked structure of the chlorin cores. The existence of interstack hydrogen bonding between the hydroxy group and the functional 13^1-keto group drives the assembly into tubular structures. Meanwhile, the 3^1-methoxy derivative having no such a hydroxy group further assembles into a brickwork-type 2D sheets. Reflecting the self-assembled morphology (tube and sheet) based on different molecular packing, the charge carrier mobility displayed by the tubular structures from the hydroxy derivatives **14a** and **14b** is at most 0.07 cm^2V^{-1} s^{-1} that is lower than at most 0.28 cm^2V^{-1} s^{-1} by the sheet-like assemblies from the methoxy derivatives **15a** and **15b** (Fig. 9.6).

Instead of using hydrophobic, saturated long alkyl chains, Chi, Würthner, and co-workers incorporated hydrophilic, two-armed oligoethylene chains into the 3^1-hydroxy ZnChl core [12]. The obtained amphiphilic derivative **16** (Chart 9.5) underwent the self-assembly in aqueous solutions to give nanotube architectures (Fig. 9.7). The tube dimensions are highly uniform, featuring an outer diameter of 6 nm, an inner diameter of 2 nm, and a length of 300 nm up to 10 μm. Remarkably, these nanotubes were measured to display extraordinal conductivities of 0.48 Sm^{-1} and charge carrier mobilities of 0.03 cm^2 V^{-1} s^{-1}. The obtained much higher mobilities than those of previously reported chlorophyll derivatives strongly suggest the sufficiently suitable π–π stacking of **16** for charge carrier transport. The packing model for the nanotube is reminiscent of the tubular arrangement proposed for natural chlorosomal light-harvesting (LH) antenna system. Furthermore, the packing arrangement of the chlorin dye stacks is highly stable. For example, when compared to the aqueous solutions of natural BChl c derivatives with one side chain, no precipitation occurs even after several months due to the interfacial effect by the amphiphilicity of the oligoether chains of **16**.

9.4 Porphyrin–C60 Amphiphilic Dyad

If segregation of donor (D) and acceptor (A) nanodomains is rationally constructed through proper heterojunctions, an efficient photoinduced charge separation can be expected. In this context, Yamamoto, Fukushima, Aida, and co-workers have verified the effectiveness of strategical design of D–A dyad with elaborate heterojunctions

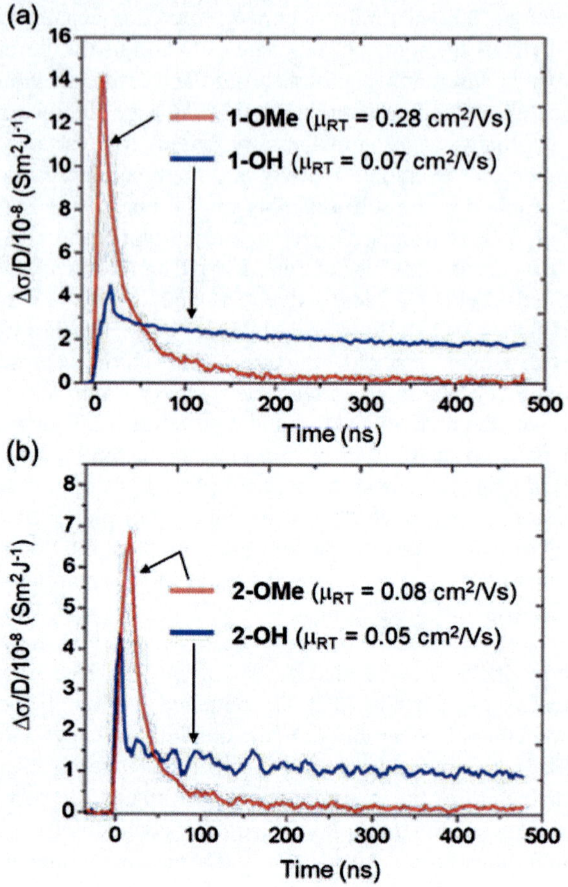

Fig. 9.6 Transient change in conductivity of ZnChl upon irradiation with a 10 ns electron pulse: **(a) 1-OH (14a)**, **1-OMe (15a)**; **(b) 2-OH (14b)**, **2-OMe (15b)**. (Reproduced with permission from Ref. [13], © 2012 American Chemical Society)

on the photoconductive performance through molecular self-assembly approaches [14]. Amphiphilic zinc porphyrin (P_{Zn}) as a D component was connected to C60 as a A component through different linkers, thereby yielding the (P_{Zn})–C60 dyads **17–19** (Chart 9.5). The self-assembly of enantiomerically pure dyad **17** with respect to the chiral fullerene unit in a mixture of CH_2Cl_2/MeOH to give nanofibers [15]. In contrast, the amphiphilic dyads **18** and **19** were shown to self-assemble in a mixture of toluene/MeOH, resulting in the formation of uniform but dimensionally different nanotube structures (Fig. 9.8) [16]. The nanotubes from **18** have outer diameter of 32 nm and wall thickness of 5.5 nm, whereas the nanotubes from **19** gave outer diameter of 7–8 nm and wall thickness of 1.7–1.8 nm. The remarkable difference in the dimensions of the self-assembled nanotubes from **18** and **19** is attributable to the difference in the geometry of the D–A dyads within each nanotube wall. For example, the dyad **18** with an ester linkage spacer forms a bilayer membrane wall, in which the D and A nanodomains are forced to be segregated coaxially along the tube axis. The

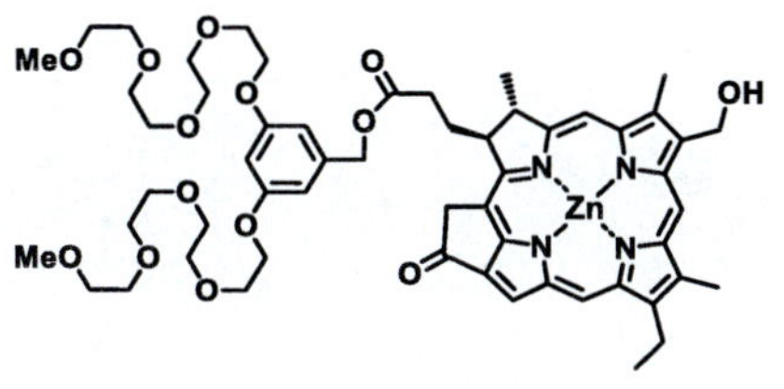

16

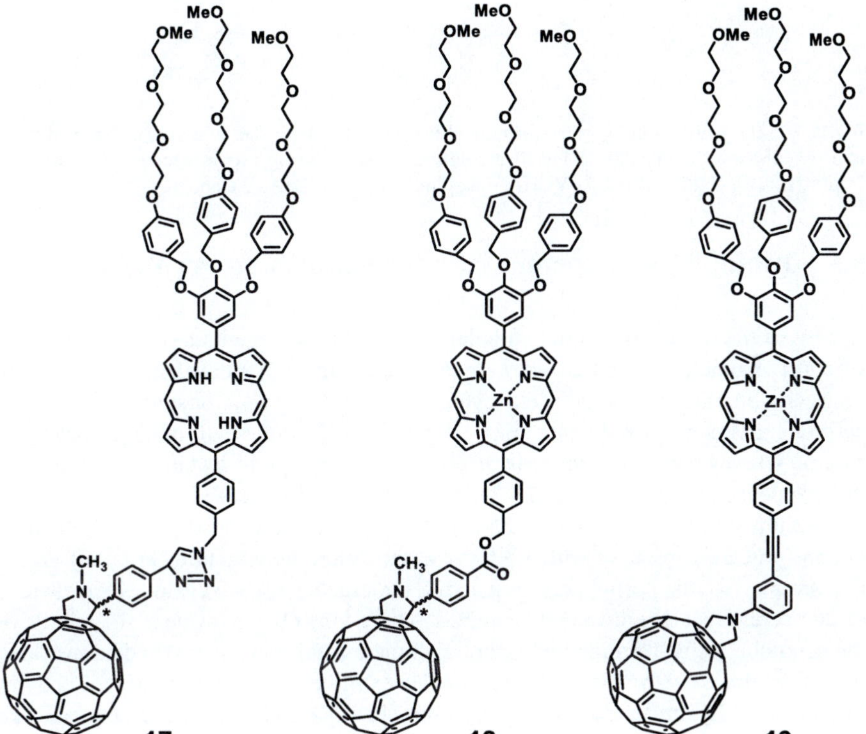

17 **18** **19**

Chart 9.5 Chemical structures of **16–19**

dyad **19** with a rigid arylacetylene spacer organizes a monolayer wall, in which the D and A dyad alternately stacked on top of each other. Noteworthy, photoconductive behavior and photovoltaic properties of each nanotube assembly strikingly depend on the relative positional relation of the D/A nanodomains [16]. Thus, the bilayer-based organic nanotube from the dyad **18** displayed much excellent photoconduction and photovoltaic properties rather than the monolayer-based nanotubes from the dyad **19**.

16 ≡

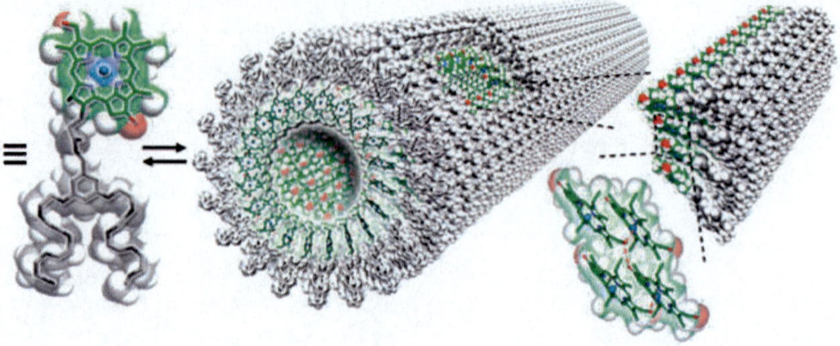

Fig. 9.7 Schematic model of the self-assembled nanotube from **16**, which were formed upon injection of a concentrated MeOH or THF solution of **16** into water (Reproduced with permission from Ref. [12], © 2012 WILEY–VCH Verlag GmbH & Co. KGaA, Weinheim)

9.5 Tubular Nanoreactor (Phthalocyanine and Porphyrin)

Aiming at the fabrication of nanotubular reactors furnished with a spatially confined reaction channel, Liu and co-workers examined strong noncovalent complexation between phthalocyanine-bridged β-cyclodextrins **20** as a rigid host and amphiphilic carboxylated porphyrin **21** as a guest (Fig. 9.9) [17]. As a result of supramolecular assembly using equimolar solution of **20** and **21**, only cylindrical tubular nanostructures appeared in aqueous media. The self-assembled nanotubes are characterized by 182 nm inner, 200 nm outer diameters and 9 nm wall thickness. The obtained dimensions are consistent with a bilayer-based nanotube structure (Fig. 9.9). In the membrane wall, the porphyrin segment of **21** interacts with two cyclodextrin cavities of **20** to construct a highly stable complex. Three long chains reside in the middle of the nanotube walls, forming hydrophobic domains. Notably, they also demonstrated that a Pd^{2+}-loaded nanotube from the complex exhibited outstanding catalytic activity in a Suzuki–Miyaura crosscoupling reaction. These RFBM-based nanotubes can take advantage of abundant number of carboxylic acid groups that are distributed around the periphery of the porphyrin. Thus, those functional groups serve as anchor spots for metal catalyst through electrostatic interaction. In the present case, Pd^{2+} was loaded onto both the inner and outer surfaces of the nanotubes.

9.6 Amphiphilic Carbocyanine Dye (Gemini-Type)

A notable tubular structure that resembles the rod-like component in the LH chlorosomes of green bacteria was formed through the self-assembly of the amphiphilic 5,5′,6,6′ -tetrachlorobenzimidacarbocyanine dye **22** (Chart 9.6) [18]. The dye **22** contains 1,1′-dioctyl group as a hydrophobic segment and 3,3′-bis(3-sulfopropyl)

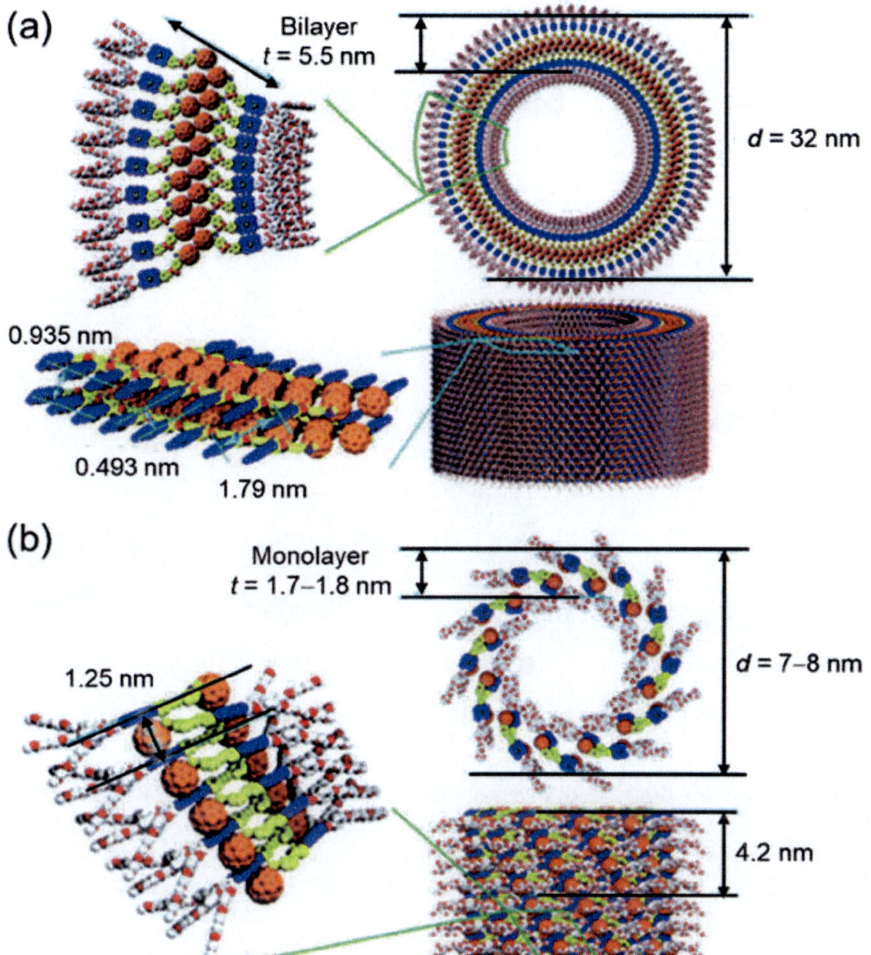

Fig. 9.8 Schematic illustration of tubular self-assemblies from the (**a**) (P_{Zn})–C60 dyad **18** and (**b**) **19** (*d* and *t* indicate nanotube diameters and wall thicknesses). (Reproduced with permission from Ref. [16], © 2012 American Chemical Society)

group as a hydrophilic moiety. In a mixture of water/MeOH (9/1, v/v), the self-assembled nanotubes have the dimensions with outer diameters of 13 ± 1 nm and lengths at most several tens μm [19]. The molecules take a bilayer arrangement to form double-walled J-aggregates with wall thicknesses of 4 nm (Fig. 9.10). The two hydrophobic octyl chains are sandwiched between the two-layer walls of the hydrophilic bis(3-sulfopropyl) groups.

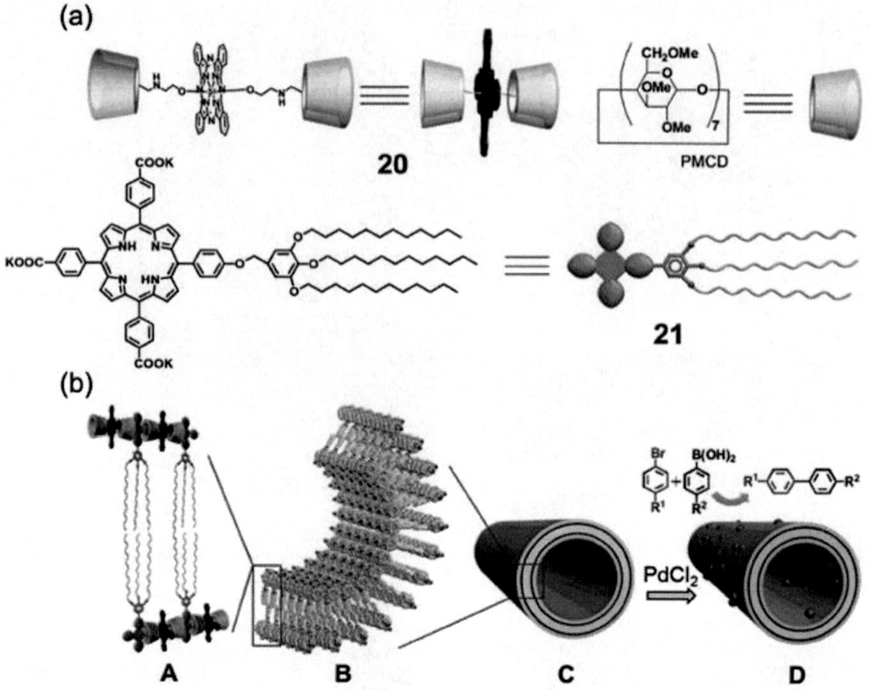

Fig. 9.9 **a** Illustration of compounds **20** and **21**, and permethyl β-cyclodextrin. **b** Schematic drawing of the tubular self-assembly. (A) Building blocks of **20** and **21**, (B) self-assembly of the nanotube wall, (C) supramolecular nanotube with a bilayer array of the **20/21** complex, and (D) test reaction catalyzed by Pd@NT. (Reproduced with permission from Ref. [17], © 2014 WILEY–VCH Verlag GmbH & Co. KGaA, Weinheim)

Chart 9.6 Chemical structures of **22** and **23**

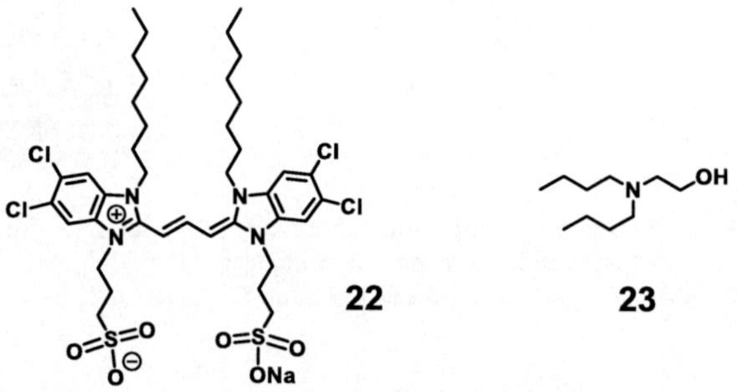

Of particular note is that the obtained J-aggregate nanotubes dissolve in aqueous solutions. In contrast to +1.675 V vs. normal hydrogen electrode measured for the monomer **22**, the nanotubes oxidize at +0.861 V. Such exclusive properties of the nanotubes arouse our interest in a new family of aqueous luminophores

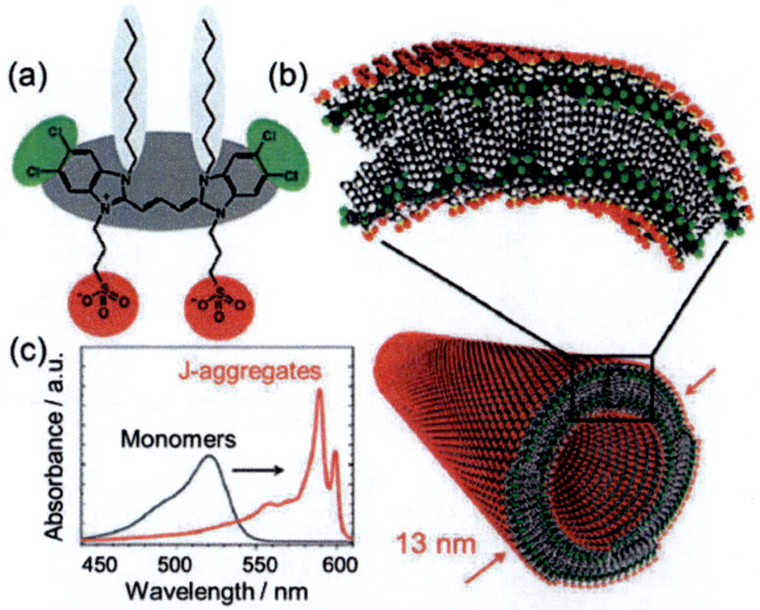

Fig. 9.10 **a** Amphiphilic cyanine dye **22**, **b** schematic illustration for the self-assembly of the dyes into well-defined nanotubular J-aggregates in water/MeOH, and **c** absorption spectra of the solution of the dye monomer and J-aggregate, which display the narrowed and red-shifted transitions typical for these tubular J-aggregates. (Reproduced with permission from Ref. [19], © 2010 American Chemical Society)

for electrogenerated chemiluminescence (ECL) applications. Aiming at the fabrication of a new class of ECL system, Bout, Stevenson, and co-workers immobilized the self-assembled tubular J-aggregates from the amphiphilic carbocyanine dye **22** on glassy carbon electrodes in the coexistence of oxidative-reductive coreactant 2-(dibutylamino)-ethanol (DBAE) **23** (Chart 9.6) [20]. As a result of the simultaneous cyclic voltammogram and ECL transient studies, the immobilized J-aggregate nanotubes were shown to display outstanding ECL signal in presence of **23** (Fig. 9.11). Direct oxidation of DBAE at the electrode as well as the catalytic oxidation of DBAE by the J-aggregate contributes to the expression of the strongest ECL signal under the conditions with ca. 17 mM **23** and at pH 2.85.

The unique cylindrical hollow channels of the nanotubular J-aggregates of **22** serve as excellent templates to enable the photoinitiated growth of silver nanowires of 6.4 nm in diameter (Fig. 9.12) [19]. The inner surfaces also function as scattered chemical reductants for Ag ions. Furthermore, the organic nanotubes have the advantage of acting as easily removable templates after the completion of the Ag nanowire growth. The dimensions of the obtained nanowires have uniform diameter of 6.4 ± 0.5 nm and length of more than one micrometer. Light exposure initiates the nucleation of Ag nanoparticle seeds and succeeding growth into the Ag nanowire in

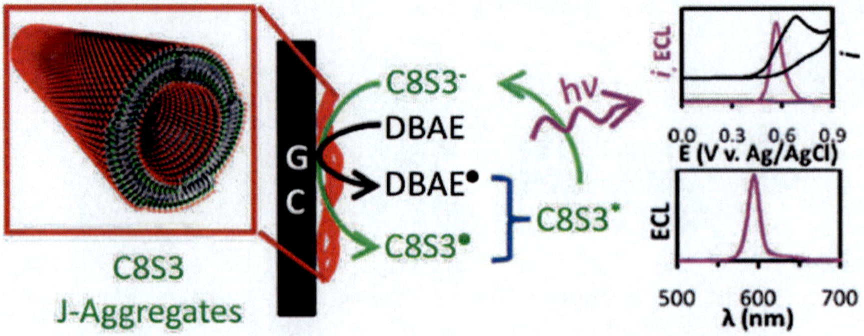

Fig. 9.11 Schematic image for aqueous electrogenerated ECL, cyclic voltammograms, and ECL transients observed for tubular J-aggregates of the amphiphilic cyanine dye **22** (C8S3). (Reproduced with permission from Ref. [20], © 2011 American Chemical Society)

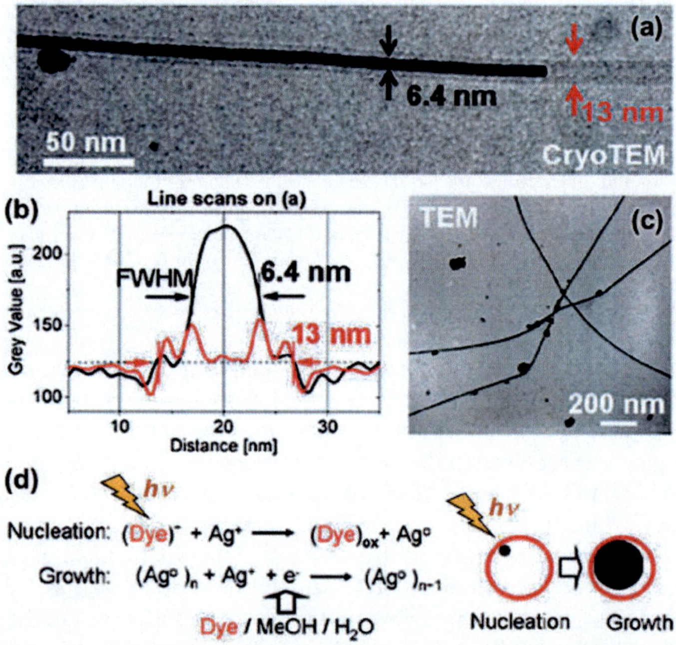

Fig. 9.12 **a** Cryo-TEM image of a silver nanowire encapsulated in a supramolecular dye nanotube of **22** after the addition of $AgNO_3$ to the solution and exposure to white light for 20 s. **b** Line scans across the template filled with silver and unfilled template. **c** TEM image of silver nanowires immobilized on a solid substrate. **d** Proposed reaction mechanism for photoinduced formation of Ag nanowires using the photochemically active dye nanotube as a template. (Reproduced with permission from Ref. [19], © 2010 American Chemical Society)

the photochemically active, hollow cylindrical template. During the growth process, methanol as a solvent is associated with the reduction of Ag ions together with the dye molecules after the formation of Ag seeds.

9.7 Hexa-*peri*-Hexabenzocoronene

Among a variety of π-conjugated molecules, polycyclic aromatic hydrocarbons (PAHs) is capable of easily forming 1D columnar structures. For example, hexa-*peri*-hexabenzocoronene (HBC) **24** (Chart 9.7) comprising 13 fused benzene rings is one of disk-shaped PAHs. The molecule **24** is, therefore, well-known as one of 2D small-sized graphene molecules. Incorporation of aliphatic side chains to HBC and subsequent π–π stacking on top of each other leads to the formation of discotic liquid crystalline phase [21]. Meanwhile, novel amphiphilic HBC derivative **25** (Chart 9.7)

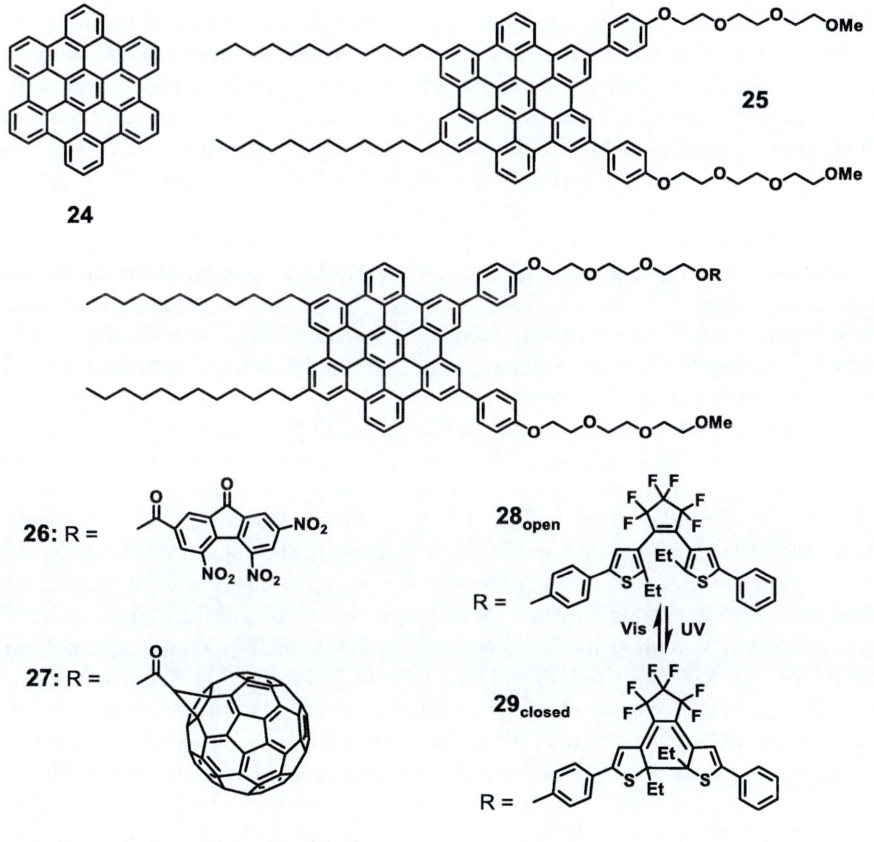

Chart 9.7 Chemical structures of **24–29**

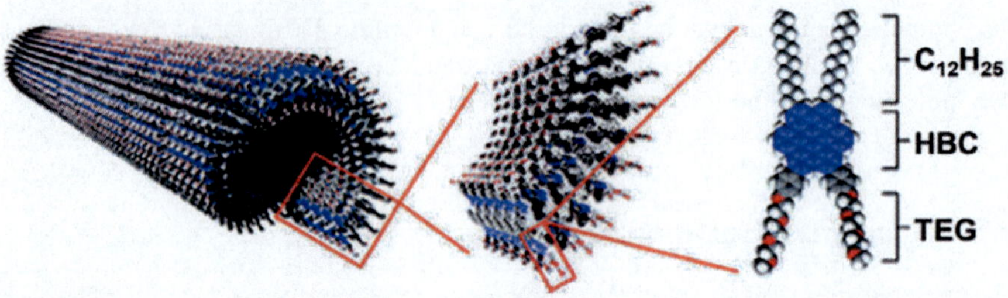

Fig. 9.13 Schematic drawing of the hierarchical structures of a self-assembled nanotube from the amphiphilic HBC derivative **25**. (Reproduced with permission from Ref. [23], © 2008 American Chemical Society)

was synthesized to carry two triethylene glycol (TEG)-appended phenyl groups on one side of the HBC core together with two dodecyl side chains on the other side. Notably, the self-assembly product from the amphiphilic building block **25** in THF was found to be uniform nanotubular structures with inner and outer diameters of 14 and 20 nm, respectively (Fig. 9.13) [22, 23]. Bilayer sheet structure stabilized by π–π stacking of the HBC core coiled up to form the hollow cylindrical morphology. In the middle part of the sheet membrane, the dodecyl side chains take an interdigitated structure. Meanwhile, the inner and outer surfaces of the nanotubes are covered with flexible and hydrophilic TEG chains. Jin, Fukushima, Aida, and co-workers investigated elaborately the structural requirement that is indispensable to stable nanotube formation from HBC-based amphiphiles. As a result of the examination for 13 different HBC derivatives, they pointed out the importance for the incorporation of two phenyl groups onto one side of the HBC core as well as the significant role of long aliphatic side chains. The presence of TEG chain is not always essential as compared to both the phenyl and aliphatic segments [23]. These results suggest that for coating of the nanotube surface with desired functional groups, the TEG chains are utilizable as a supporter molecule.

Electron acceptors including trinitrofluorenone (TNF) [24, 25], C60 [26], and dithienylethene (DTE) [27] were, thus, incorporated to one terminus of two TEG chains that were connected to the aliphatic chain-appended, electron donor HBC core. Rationally linked hole- and electron-transporting layers, i.e., p/n heterojunction constructed from D and A molecular segments are of critical importance for the excellent performance of organic photovoltaics (PVs). In this context, it is desirable that the D/A dyads undergo neither macroscopic segregation of the D/A domain nor charge-transfer complexation between the D/A components. The self-assembly of each HBC–(TEG)–TNF **26**, HBC–(TEG)–C60 **27**, HBC–(TEG)–DTE dyads $\mathbf{28}_{\mathbf{open}}$ and $\mathbf{29}_{\mathbf{closed}}$ (Chart 9.7) successfully resulted in the formation of nanotube structures, in which two electron-accepting layers sandwich an electron-donating bilayer comprising π–π stacking HBC cores [14]. Depending on the concentration of HBC–TNF **26** under exposure to MeOH vapor, the resultant self-assembled morphologies of **26** gave nanotubes with diameter of 16 nm and wall thickness of 3 nm (conc. = 0.12 mM) and nonhollow microfibers with diameter in the range of 0.2–2 μm [24].

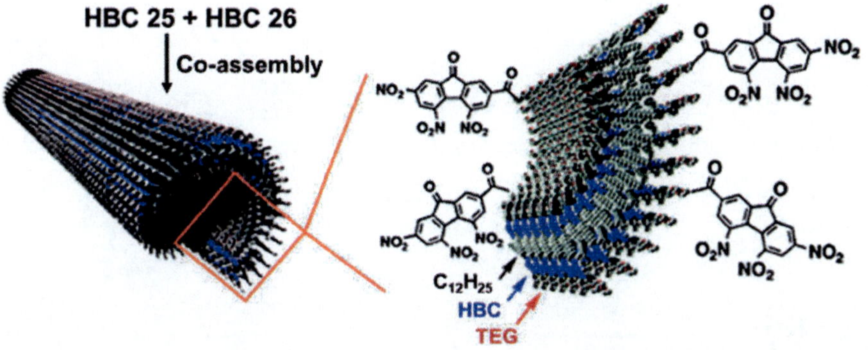

Fig. 9.14 Schematic drawing for the formation of a co-assembled graphitic nanotube from HBC **25** and **26**. (Reproduced with permission from Ref. [25], © 2007 American Chemical Society)

Irradiation of the obtained nanotube of **7** induced outstanding photocurrent generation, giving current values of over 10,000 times larger than those of dark current. Conversely, just a little bit of photocurrent was observed for the microfibers. Subsequently, the photoconductive behavior of co-assembled nanotubes from HBC–TEG **25** and HBC–(TEG)–TNF **26** was investigated under the conditions of different composition rate of TNF (Fig. 9.14) [25]. The co-assembled nanotubes gave 3 $cm^2\ V^{-1}\ s^{-1}$ as the local charge carrier mobility ($\Sigma_{\mu 1D}$) until the HBC–TNF content reaches to 75%. The $\Sigma_{\mu 1D}$ value, however, decreased to 0.7 $cm^2\ V^{-1}\ s^{-1}$ at the HBC–TNF content of 100%. On the contrary, long-range charge carrier mobility (μ) was found to be less than $1.5 \times 10^{-4}\ cm^2\ V^{-1}\ s^{-1}$ and monotonously decreased with increase in the HBC–TNF content.

Instead of TNF layer, C60 that can transport photogenerated electrons and holes were incorporated to give nanotubes from HBC–(TEG)–C_{60} D–A dyad **27** [26]. The cast film from a toluene dispersion of **27** exhibited a photovoltaic activity with an open–circuit voltage of 0.46 V. The DTE molecule undergoes an open and closed ring isomerization upon irradiation of UV and Vis light, respectively. The typical photochromic molecular switch DTE can function as an electron acceptor while taking closed form. Jin, Yamamoto, Aida, and co-workers utilized DTE as a pendant electron-accepting segment to modulate the photoconductive activity of the resultant nanotubes (Fig. 9.15) [27]. Photoconduction measurement for the nanotube from $\mathbf{29}_{\text{closed}}$ showed fivefold larger photocurrent as compared to that of $\mathbf{28}_{\text{open}}$. This finding indicates that the nanotubes from $\mathbf{29}_{\text{closed}}$ and $\mathbf{28}_{\text{open}}$ behave as distinct photoconductive nanomaterials, which are photoswitchable.

As already described, the HBC derivative **25** self-assembles into a semiconducting tubular structures [22]. Interestingly, supramolecular strategies based on stepwise co-assembly of two rationally designed amphiphilic HBC derivatives **30a** and **31a** (Chart 9.8) allowed for the direct connection of dissimilar semiconducting HBC nanotubes (***block*-$\mathbf{NT_{1\text{-}Cu}}$/ $\mathbf{NT_2}$**) (Fig. 9.16) [28]. Flash-photolysis time-resolved

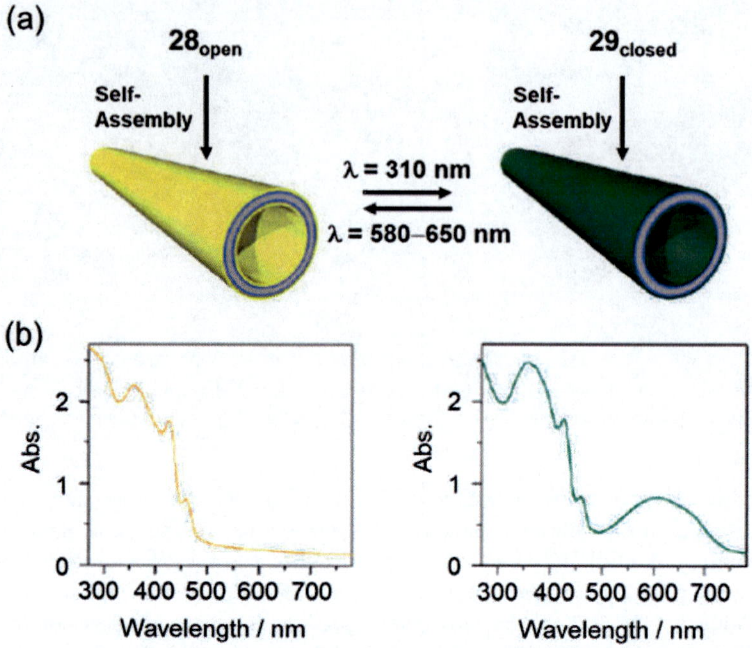

Fig. 9.15 **a** Self-assembly of the nanotubes from $\mathbf{28_{open}}$ and $\mathbf{29_{closed}}$ with a coaxial configuration, where a graphite-like bilayer of π-stacked HBC units is laminated by molecular layers of the open-form ($\mathbf{28_{open}}$) and closed-form ($\mathbf{29_{closed}}$) DTE derivatives, respectively. **b** Electronic absorption spectra of a cast film of the nanotube from $\mathbf{28_{open}}$; (left) before and (right) after exposure (30 min) to UV light at 310 nm. (Reproduced with permission from Ref. [27], © 2010 WILEY–VCH Verlag GmbH & Co. KGaA, Weinheim)

microwave conductivity method elucidated the lifetime of charge carriers generated in the connected nanotubes. The longevity of the ***block*-$\mathbf{NT_{1\text{-}Cu}}$/ $\mathbf{NT_2}$** nanotubes is verified by approximately fivefold longer lifetime ($\tau_{1/e} = 8.8 \times 10^{-6}$ s) than those of individually self-assembled nanotube blocks $\mathbf{NT_1}$ and $\mathbf{NT_2}$ ($\tau_{1/e} = 1.4 \times 10^{-6}$ and 2.5×10^{-6} s, respectively) (Fig. 9.16).

Based on the step-by-step co-assembly technique with the HBC derivatives **30a–30d** and **31a–31c**, Jin, Fukushima, Aida and coworkers demonstrated for the first time helix sense-selective noncovalent polymerization [29]. They took advantage of an elaborately designed nanotube assembly with one-handed helix sense as a seed. Such helical seeds were shown to cause the regulatory self-assembly of the amphiphilic fluorinated chiral monomers (**31b** and **31c**). The self-assembly, thus, results in the production of tubular block copolymers in which the helical sense of the employed seeds drive the helical senses (Fig. 9.17).

Sophisticated molecular engineering of HBC derivatives based on the solvophobic/solvophilic balance enabled controlled synthesis of single- and multiwalled organic nanotubes [30]. The novel HBC derivatives **32a** and **32b** (Chart 9.9) have

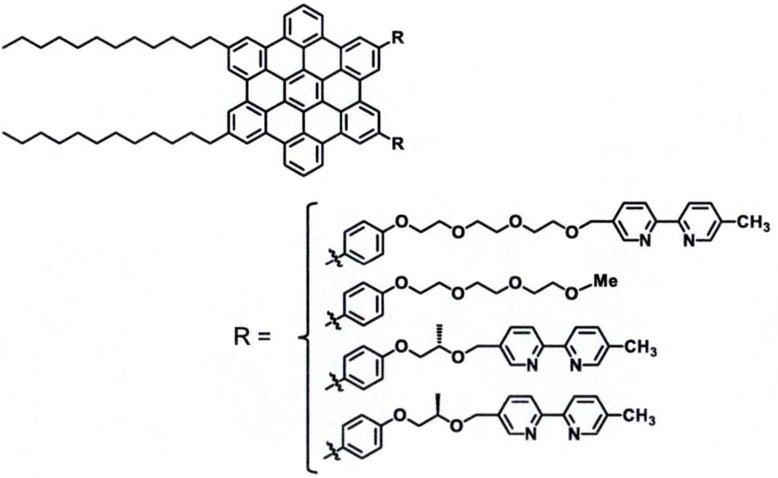

30a
30b
30c
30d

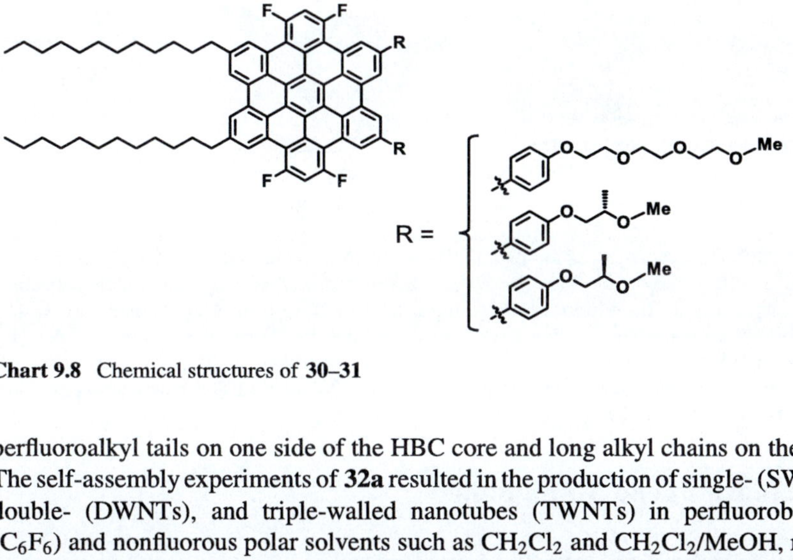

31a
31b
31c

Chart 9.8 Chemical structures of **30–31**

perfluoroalkyl tails on one side of the HBC core and long alkyl chains on the other. The self-assembly experiments of **32a** resulted in the production of single- (SWNTs), double- (DWNTs), and triple-walled nanotubes (TWNTs) in perfluorobenzene (C_6F_6) and nonfluorous polar solvents such as CH_2Cl_2 and CH_2Cl_2/MeOH, respectively (Fig. 9.18). In C_6F_6, no wall multiplication occurs due to solvophilic effect, whereas in CH_2Cl_2 solvophobic effect leads to wall multiplication through stepwise "coil-on-tube" intermediate structures. This supramolecular tactic and strategy will pave the way for the fabrication of tailored functional nanotubes comprising desired multiple layers [30].

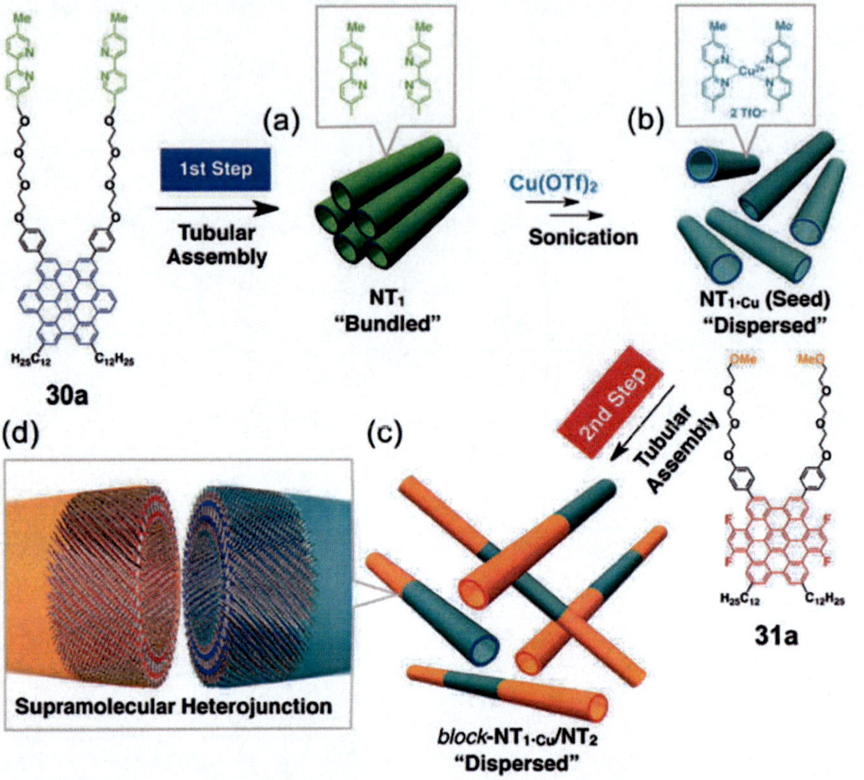

Fig. 9.16 Schematic drawing of the preparation of (**a**) $\mathbf{NT_1}$**"Bundled"** by MeOH vapor diffusion into a THF solution of **30a**, (**b**) $\mathbf{NT_{1\text{-}Cu}}$**(Seed) "Dispersed"** by post-functionalization of $\mathbf{NT_1}$ with copper(II) trifluoromethanesulfonate [$Cu(OTf)_2$] in MeOH, and (**c**) ***block*-**$\mathbf{NT_{1\text{-}Cu}/NT_2}$**"Dispersed"** by cooling a hot acetone solution of **31a** in the presence of $\mathbf{NT_{1\text{-}Cu}}$ as the seed. (**d**) Schematic image of an idealized cross section of ***block*-**$\mathbf{NT_{1\text{-}Cu}/NT_2}$ at the heterojunction interface. (Reproduced with permission from Ref. [28], © 2011 The American Association for the Advancement of Science)

9.8 Thioxanthene Amphiphile

The bolaamphiphiles **33a–33c** (Chart 9.10) are featured by two diglycine segments that are connected to the both ends of long alkyl chains. As already described in **3.5.1**, these diglycine-based bolaamphiphiles were first shown to self-assemble in water to form vesicle-encapsulated microtube structures [31, 32]. Feringa and co-workers reported that the thioxanthene-derived amphiphile **34** (Chart 9.10) as a RFBM building block self-assembles in the coexistence of the unsaturated phospholipid DOPC **35** (Chart 9.10) to yield unique nanotube structures in aqueous NaCl solution (10 mM) [33]. The amphiphile **34** consists of sterically overcrowded, thioxanthene-based ethylene that bears two hydrophilic TEG chains on the one side

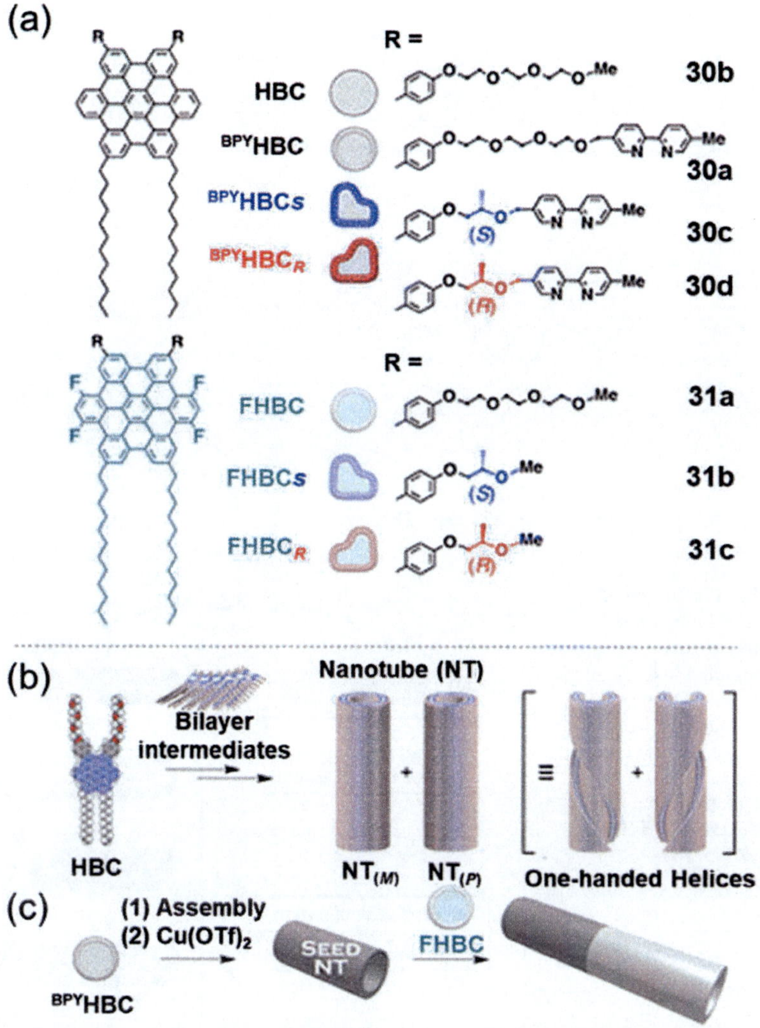

Fig. 9.17 **a** Molecular structures of HBC derivatives, **b** helical chirality of the HBC nanotubes, and **c** procedure for the formation of a nanotubular supramolecular block copolymer. (Reproduced with permission from Ref. [29], © 2015 American Chemical Society)

of the thioxanthene core together with two long alkyl chains on the other side. The RFBM unit **34** forms an interdigitated bilayer to shape robust tubular architectures with diameter of 28 nm, whereas the phospholipid **35** forms a fluid bilayer membrane to cover the end of the high-axial-ratio nanotube as a vesicle cap (Fig. 9.19). Noteworthy is the inclusion phenomena of the vesicle caps into the inner part of the hollow cylinders in response to an osmotic gradient [34]. Differences in the flexibility of the bilayer membranes between the robust nanotubes and flexible vesicles should result

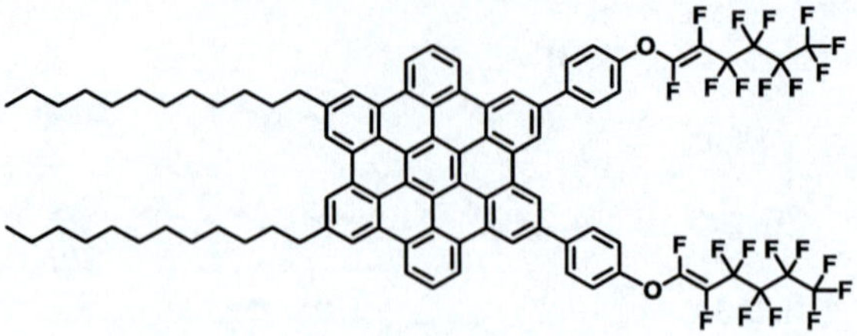

32a

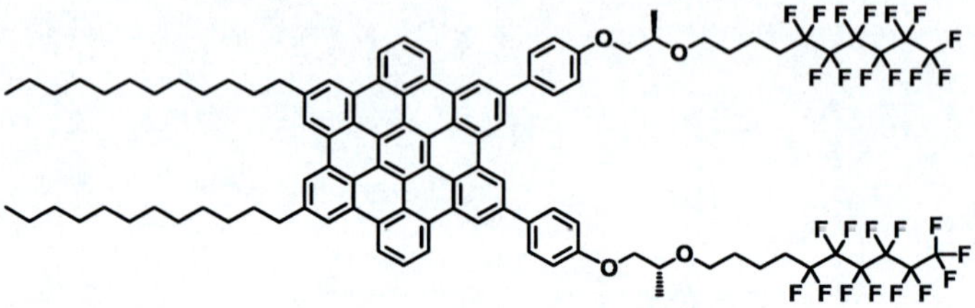

32b

Chart 9.9 Chemical structures of **32a** and **32b**

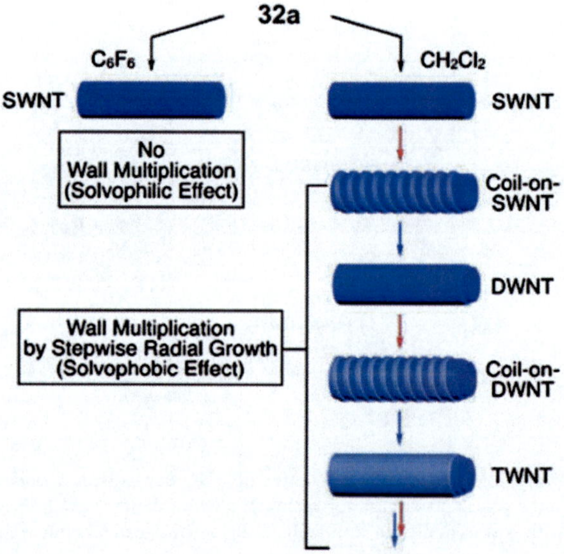

Fig. 9.18 Illustration of single- and multi-walled nanotubes of **32a** self-assembled in C_6F_6 and CH_2Cl_2, respectively, and stepwise radial growth via coil-on-tube intermediates. (Reproduced with permission from Ref. [30], © 2015 WILEY–VCH Verlag GmbH & Co. KGaA, Weinheim)

in the vesicle loading, since the osmosis-responsive loss of water volume causes the deformation of the vesicle bilayer membranes selectively. Remarkably, cationic vesicle can be encapsulated into the nanotube channels, whereas anionic vesicle caps remain intact even under hyperosmotic pressure [35].

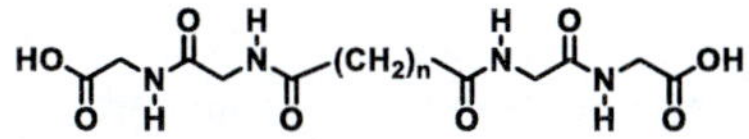

33a: n = 6
33b: n = 8
33c: n= 10

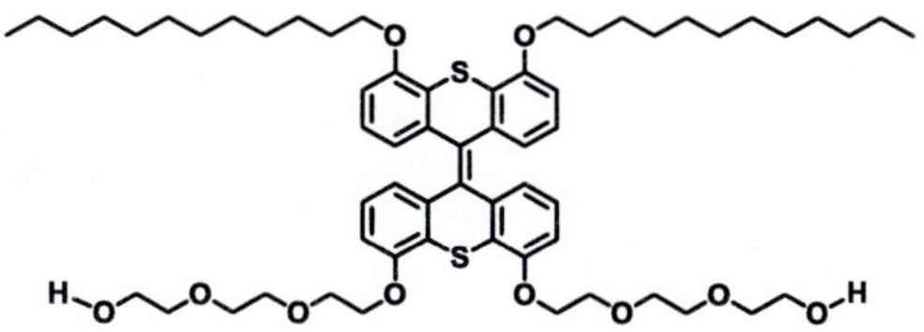

34

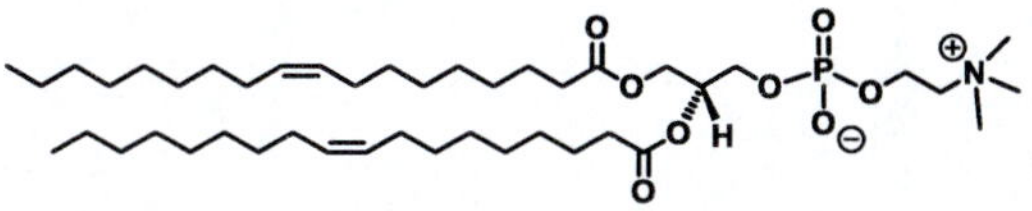

35

Chart 9.10 Chemical structures of **33**–**35**

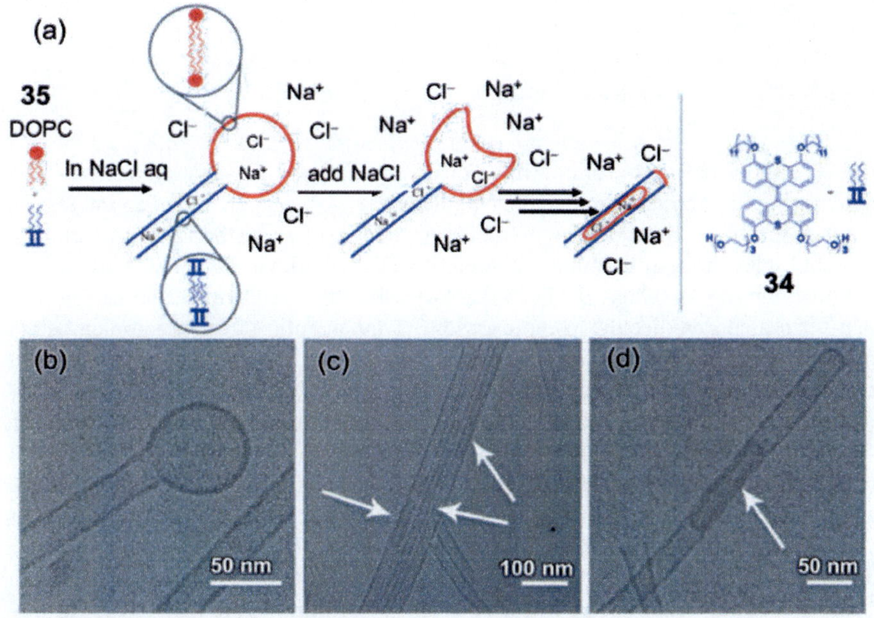

Fig. 9.19 **a** Illustration for the inclusion of vesicles into soft self-assembled nanotubes from **34**. Cryo-TEM image of (**b**) an end-capped nanotube and (**c** and **d**) inclusion of the self-assembled vesicles from **35** (arrows) under different conditions. (Reproduced with permission from Ref. [34], © 2015 WILEY–VCH Verlag GmbH & Co. KGaA, Weinheim)

9.9 Riboflavin Derivative

Riboflavin **36** (Chart 9.11), also called vitamin B_2, is a biologically active component being classified in water-soluble vitamins. The heterocyclic isoalloxazine ring as a

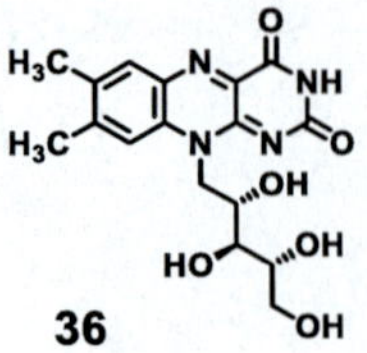

36

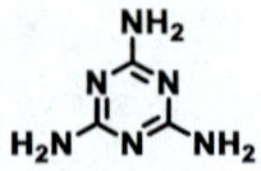

37a

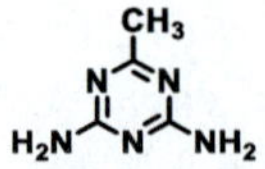

37b

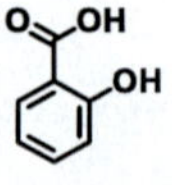

37c

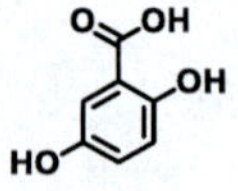

37d

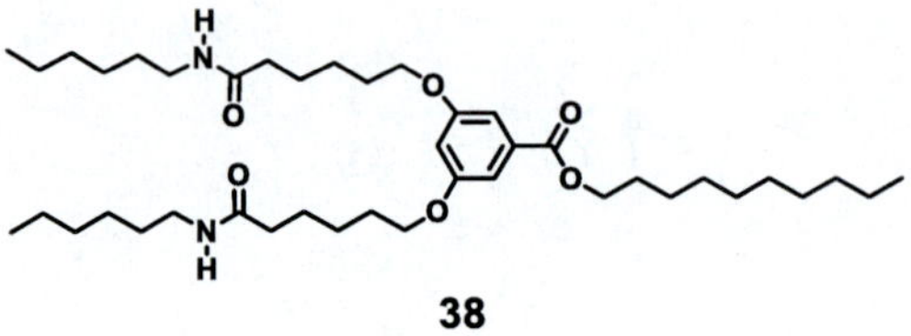

38

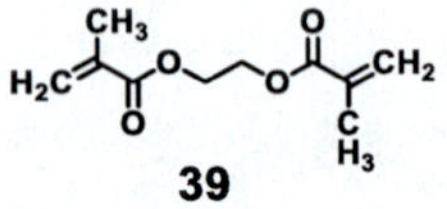

39

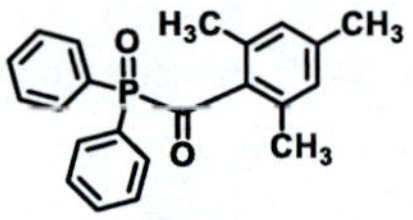

40

Chart 9.11 Chemical structures of **36–40**

rigid backbone carries the chiral ribitol chain as a hydrophilic sugar alcohol; thereby configuring the riboflavin molecule. Nandi and co-workers performed an interesting work on the effect of the riboflavin–melamine composition on the 1D self-assembled morphologies through the binary assembly of the riboflavin **36** and 1,3,5-triazine–2,4,6-triamine **37a** (Chart 9.11) [36]. Normally, the riboflavin **36** and melamine **37a** form a supramolecular complex with 3:1 molar ratio through complementary hydrogen bonding. The obtained 3:1 complex self-assembles in water to yield fibrous assemblies, resulting in the hydrogel formation (Fig. 9.20). However, detailed investigation revealed that the self-assembled 1D morphologies changed depending on the composition of the **36/37a** complexes. For example, the 4:1 (RM41) and 3:1 (RM31) complexes give helical fibers, whereas the 2:1 system produces rod-like assemblies. Interestingly, the 1:1 (RM11), 1:2 (RM12), and 1:3 (RM13) systems were observed to form hollow tubular architectures. The RM31 complex takes a nonplanar structure, in which one of three riboflavin molecules resides in the upper plane and others in somewhat lower one. This structural feature leads to the formation of 1D twisted π–stacking on top of one another, thus, resulting in the generation of helical fibers. In contrast, the RM11 complex formation takes place at one side of the melamine unit and further self-assembles to form a bent sheet-like assemblies stabilized by π–π stacking and interdigitated structure of the ribityl chains. Such a structural situation, then, produces hollow tubular assemblies [36]. Intriguingly, the new binary self-assembled hydrogels from the 1:1 **36/37b**, **36/37c**, and **36/37d** complexes (Chart 9.11) afforded hollow tubes, rod-like, and tape-like assemblies, respectively [37].

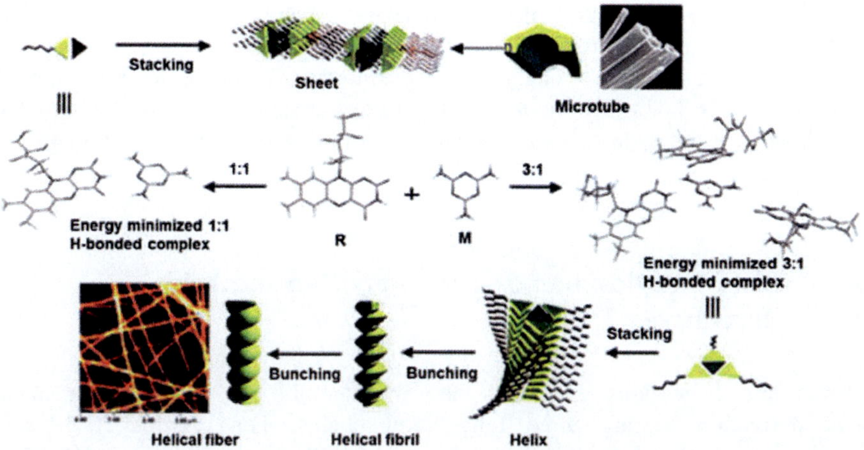

Fig. 9.20 Molecular structures of riboflavin (R) **36** and melamine (M) **37a**, and their energy minimized structure of the 1:1 and 3:1 hydrogen-bonded complex that self-assemble to form microtubes and helical bunched fibers, respectively (Reproduced with permission from Ref. [36], © The Royal Society of Chemistry 2008)

The isoalloxazine segment of riboflavin exhibits fluorescent activity and the chiral ribityl chain endow the riboflavin molecule with good water solubility. The PL efficiency of riboflavin is strongly dependent on the solvent polarity. For example, hydrogen-bond forming solvents reduce the efficiency. The PL intensity of the xerogels, self-assembled from the RM13, RM12, RM11, RM41, RM31, and 2:1 (RM21) complexes, displayed significantly lower values than that of pure riboflavin (Table 9.1) [36]. Moreover, the increase in the melamine composition augments the PL intensity. Pertaining to the riboflavin/melamine hydrogel system, the self-assembled helical fibers from the RM41 or RM31 complex, and rod-like assemblies form the RM21 complex gave almost similar PL intensity. The hollow tubes self-assembled

Table 9.1 Comparison of the fluorescence property in riboflavin (R)/melamine (M) gels and their xerogels, normalized to same riboflavin content (Reproduced with permission from ref [36], © The Royal Society of Chemistry 2008)

	Hydrogels		Xerogels		
Composition	λ_{max}/nm	Intensity ($\times 10^5$)	Morphology	λ_{max}/nm	Intensity ($\times 10^5$)
Pure R	540	0.4	–	570	9.57
RM41	565	2.5	Helical fiber	601	0.15
RM31	571	2.1	Helical fiber	612	0.067
RM21	570	2.0	Rod	610	0.064
RM11	569	3.6	Tube	600	0.34
RM12	568	3.8	Tube	601	0.4
RM13	559	9.5	Tube	600	1.3

from the RM11, RM12, and RM13 complexes, however, exhibited much higher (1.4–4.5 times compared to those of helical fibers) PL intensity [36]. The resultant tubular architectures more strongly hamper the nonradiative decay process of the π^* electrons of riboflavin than the nonhollow fibrous assemblies. The self-assembled morphology of the hydrogels was shown to affect outstandingly the PL property in the hydrogels.

9.10 Bis(5-Hexylcarbamoylpentyloxy)Benzoic Acid Derivative

Mésini and co-workers examined the self-assembly of 3,5-dihydroxybenzoic acid derivative bearing three long alkyl chains. For example, 3,5-bis(5-hexylcarbamoylpentyloxy) benzoic acid decyl ester **38** (Chart 9.11) forms organogels in organic solvents, which comprise fibrillar assemblies stabilized by hydrogen-bonding interactions [38]. The diamide compound **38** is completely soluble in $CHCl_3$ and CH_2Cl_2, whereas organogel was formed in cyclohexane after heating for dissolution and subsequent gradual cooling at concentrations of above 0.0.5 wt%. TEM observation showed that the obtained organogel from the cyclohexane solution is composed of the aggregate of hollow cylindrical tubular structure (Fig. 9.21) [38]. The dimensions of the tubes are featured by the length of the order of a micrometer, wall thickness of 4 ± 3 nm, and diameter of 25–30 nm.

Diverse molecular building blocks self-assemble to form a wide spectrum of 1D nanostructures, e.g., fibers, ribbons, helical tapes, and tubes [39]. Furthermore, the nanometric 1D structures function as a template for the fabrication of inorganic mesoporous materials [40]. Particularly, mesoporous polymer materials are superior to the inorganic ones in terms of chemical susceptibility to modification, resilient mechanical properties, and satisfactory processability. In this context, Mésini's research group demonstrated the fabrication of mesoporous polymer materials having 30 nm cylindrical pores by use of self-assembled nanotube structures as the pore template (Fig. 9.22) [41, 42]. For instance, the compound **38** can also self-assemble in polymerizable solvents such as ethylene glycol diacrylate **39** (Chart 9.11) to result in the formation of mainly nanotubes of 30 nm in outer diameters and several μm in lengths [41]. The organogel as a bulk material, which comprises polymerizable monomer solvent and nanotube templates, was photopolymerized by using diphenyl-(2,4,6-trimethylbenzoyl)-phosphine oxide **40** (Chart 9.11) as a initiator [42]. As a result of polymerization upon UV irradiation, a transparent polymer resin was successfully obtained. Organic solvents, i.e., $CHCl_3$ and CH_2Cl_2 enabled the degradation and subsequent extraction of the entire nanotubes in a polymeric matrix. The self-assembled molecules **38** of the nanotube templates are completely recoverable after the extraction process. Remarkably, the resultant polymer resin is endowed with many mesoporous channels that is just compatible with the imprints of the organic nanotubes (Fig. 9.22).

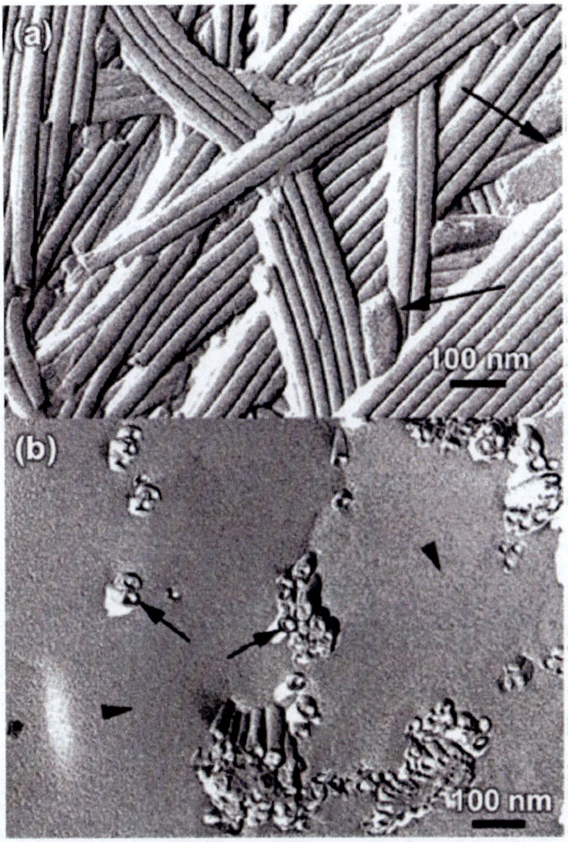

Fig. 9.21 Electron micrographs of self-assembled structures of **38** in cyclohexane. **a** Replica of a freeze fracture of **38**/cyclohexane gels (concentration: 2 wt%). Tube diameters are approximately 25–30 nm. Arrows show solvent area. **b** Same as (**a**) with a fracture plane nearly perpendicular to the main axis of the tubes. Arrows show sections of the tubes that take clear hollow cylindrical structures, whereas arrowheads amorphous solvent. (Reproduced with permission from Ref. [38], © 2005 WILEY–VCH Verlag GmbH & Co. KGaA, Weinheim)

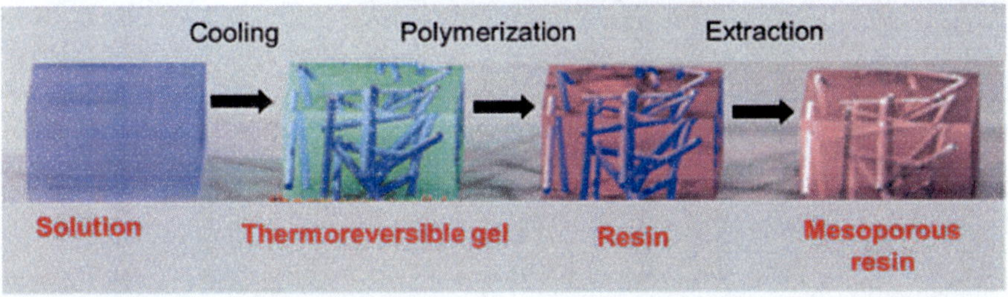

Fig. 9.22 Fabrication process for the formation of the mesoporous resin. A mixture of the compound **38** and polymerizable solvent **39** is heated to give an isotropic solution; the molecule **38** self-assembles upon cooling to form a thermoreversible gel containing the nanotubes; the photopolymerization of the matrix results in the formation of a resin embedding nanotubes; the nanotubes are disassembled and extracted to produce a porous resin. (Reproduced with permission from Ref. [42], © The Royal Society of Chemistry 2010)

9.11 Pyrimido Pyrimidine

In 1996, Lehn and co-workers first demonstrated the design and synthesis of a series of self-complementary Janus-type heterocycles **41** and **42** (Chart 9.12) that bear a unique hydrogen-bond array of DDA and AAD (A and D: hydrogen-bonding acceptor and donor sites, respectively) [43]. The ingeniously configured hydrogen-bond arrays drive their self-assembly into a hexameric supramolecular macrocycles. By taking advantage of the principles of this supramolecular design, Fenniri and co-workers carried out a pioneer work on supramolecular self-assembly of rosette-type organic nanotubes (Fig. 9.23). The designed and synthesized heteroaromatic bicyclic base is pyrimido[4,5-*d*]pyrimidine **43** (Chart 9.12) [44]. The hydrophobic bicyclic base possesses the DDA hydrogen-bond array of guanine and AAD array of cytosine. Introduction of a methyl group aims at not only minimizing the access of surrounding water but promoting thc intramolecular hydrogen-bond formation.

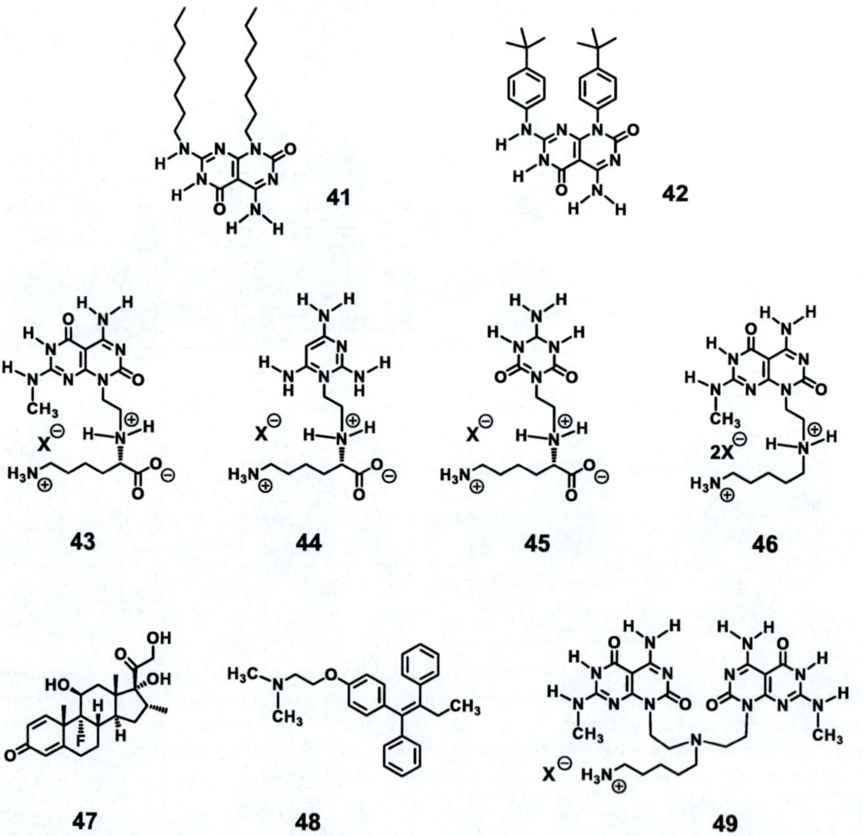

Chart 9.12 Chemical structures of **41–49**

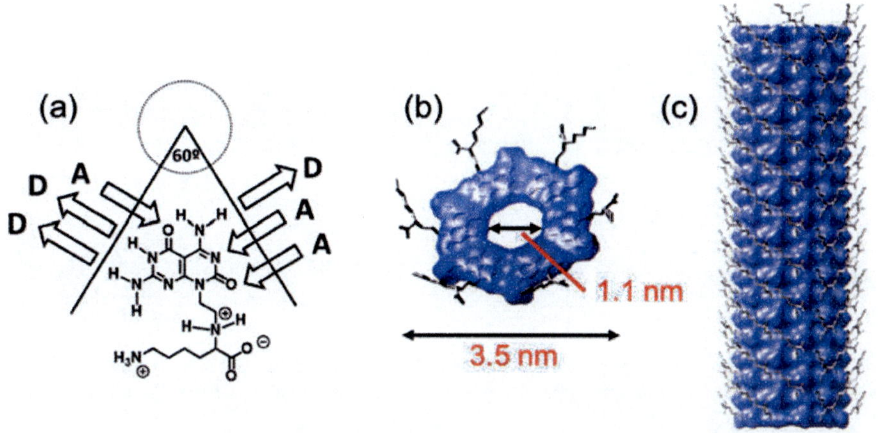

Fig. 9.23 **a** Bicyclic pyrimido[4,5-*d*]pyrimidine motif of **43** denoting the hydrogen-bond donors (D) and acceptors (A) of both guanine and cytosine. **b** Six motifs self-assemble in solution to give a hexameric rosette and **c** stacking of the rosettes produces a nanotube structure. (Reproduced with permission from Ref. [45], © The Royal Society of Chemistry 2016)

Additionally, lysine residue as an amino acid segment was linked to the base moiety to direct the expression of supramolecular chirality. To confirm the effectiveness and feasibility of the design principle, they also synthesized three other analogues **44–46** (Chart 9.12).

As a result of the self-assembly experiment in water, the bicyclic base **43** undergoes an entropically driven, hierarchical self-assembly to produce six-membered rosettes stabilized by 18 hydrogen bonds [44, 46]. Notably, following the formation of the supramacrocyclic rosettes, solvophobic effects, van der Waals interactions and π–π stacking make the rosette structure to further assemble into homogeneous tubular structures (Fig. 9.23). This rosette nanotubes roughly belong to a class of RFBM-based nanotubes. The dimensions of the obtained nanotubes are featured by the outer diameter of ca. 4.0 nm, length of around 60 nm, and inner diameter of ca. 1.1 nm (Fig. 9.24). Conversely, under the same physiological conditions, the analogues **44** and **45**, or an equimolar mixture of **44** and **45**, as well as the achiral derivative **46** are unable to perform the self-assembly. The details on the synthesis and functionalization strategies of the rosette nanotubes were reviewed elsewhere [45].

All the features of the rosette nanotubes, i.e., the hydrophobicity of the channel interior, amphiphilicity, low cytotoxicity, and biocompatibility, captivate the motivation to apply them for diverse medical applications as a drug delivery vehicle. An anti-inflammatory steroidal drug called dexamethasone (DEX) **47** (Chart 9.12) is employed to treat many rheumatic problems and skin diseases. Fenniri, Webster, and co-workers confirmed that the rosette nanotube from heteroaromatic bicyclic base **43** is able to incorporate DEX **47** into the channel interior of the nanotubes [47]. During

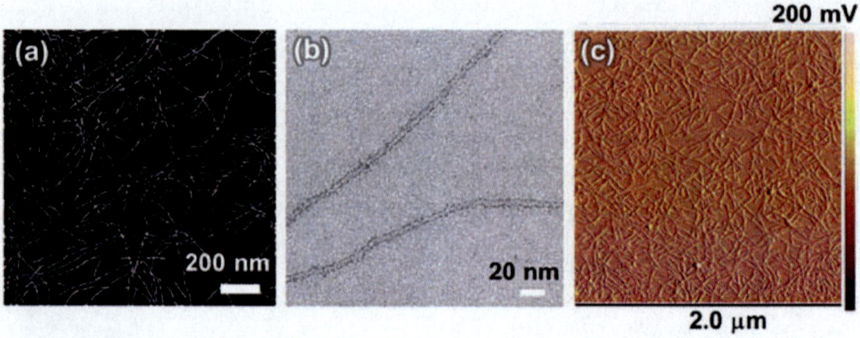

Fig. 9.24 Representative **a** SEM, **b** TEM and **c** AFM images of the self-assembled rosette nanotubes from **43**. (Reproduced with permission from Ref. [45], © The Royal Society of Chemistry 2016)

the self-assembly process of the tubular objects assisted by the hydrophobic interactions of the base units and π–π stacking, the hydrophobic channel promotes efficient loading of DEX **47**. Subsequent studies on drug–cell interactions revealed that the DEX drug is gradually released to result in the increase in the activities of osteoblast (bone-forming cell) under physiological conditions (Fig. 9.25) [47]. To further verify the potential ability of the rosette nanotubes as a drug delivery vehicle, the research group examined the loading behavior of the hydrophobic anticancer drug, tamoxifen (TAM) **48** (Chart 9.12) in physiological environments [48]. Two different building

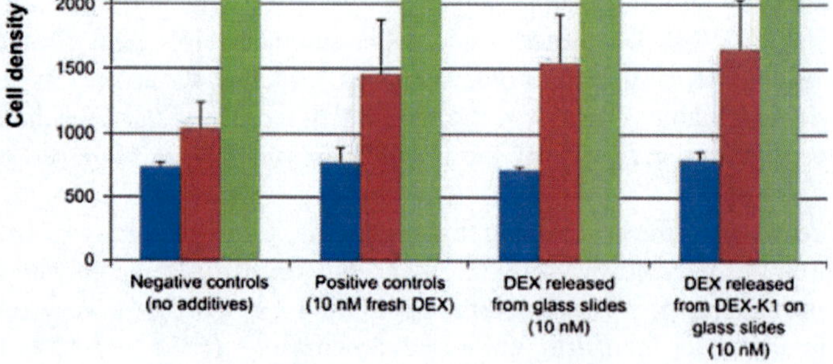

Fig. 9.25 Osteoblast density cultured with released DEX. Data show mean ± standard error of the mean (n = 9). (Reproduced with permission from Ref. [47], © 2011 Chen et al., publisher and licensee Dove Medical Press Ltd.)

blocks underwent the self-assembly from **43** and **49** (Chart 9.12) into single-base and twin-base derived nanotube architectures with almost the same dimensions (e.g., outer diameter: 3.5 and 3.8 nm, respectively). The twin-base nanotubes, stabilized more strongly by dense hydrogen bonding and π–π stacking as compared to the single-base nanotubes, exhibited higher TAM loading capacity than the single-base nanotubes. In either case, the rosette nanotubes have a strong tendency to encapsulate the hydrophobic drugs **48** in the hollow cylinder for the anticancer treatment.

9.12 Ferrocene Aromatics

Redox-active ferrocene segment was employed to drive the assembly/disassembly equilibrium between a nanoring component and nanotube assembly through redox system. The ferrocene-derived tetratopic pyridyl ligands **50a** and **50b** (Chart 9.13) are prepared by the combination of the ferrocene core, four aromatic arms with 4-pyridyl group terminus as a metal ligand, and four (or eight) TEG side chains that enhance the ligand solubility and disperse the assemblies [49]. Notably, the cyclopentadienyl rings that rotate thermally cause a dynamic change of the their tetratopic geometry in solution [50]. The Ag(I) ions complex with pyridine [Py(N)] derivatives results in the formation of a linear array of Py(N)–Ag(I)–(N)Py. Thus, three different types of assembled structures, i.e., singly bridged 2D sheet-like structure, doubly bridged 1D coordinated polymer, and doubly bridged macrocycle. In preference to the three predicted structures, the complexation of the ligands **50a** and **50b** with $AgBF_4$ in an CH_3CN solution resulted in the assembly of homogeneous metal–organic nanotube architectures with uniform large diameters of 7.5 and 14.3 nm, respectively (Fig. 9.26) [49]. The resultant diameters were, therefore, shown to depend sensitively on the

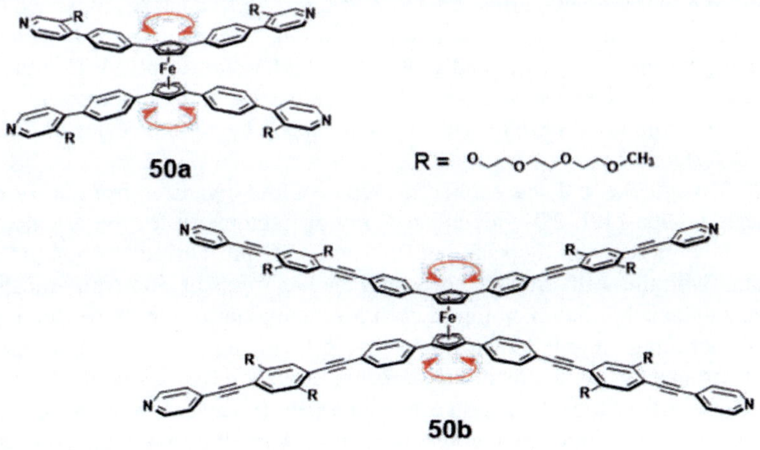

Chart 9.13 Chemical structures of **50a** and **50b**

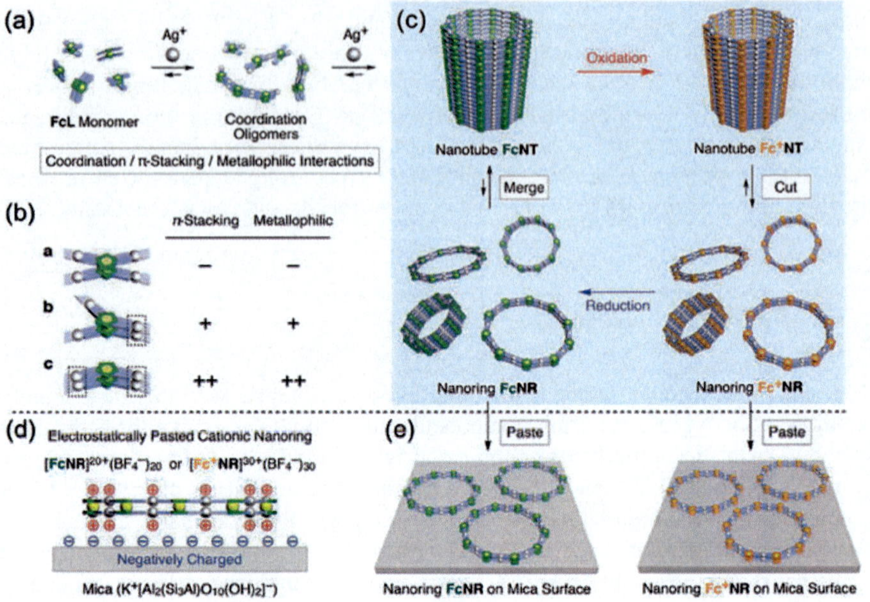

Fig. 9.26 Schematic illustration for the self-assembly of homogeneous metal–organic nanotube architectures from **50a** or **50b**. (**a**) $AgBF_4$-mediated self-assembly of **FcL** to form **FcNT**. (**b**) Geometry-dependent possible interactions in **FcL** upon complexation with Ag(I) ions. (**c**) Polymerization/depolymerization equilibrium between nanorings **FcNR** and nanotubes **FcNT** biased by redox chemistry of ferrocene. (**d**) Representation of how cationic nanoring **FcNR** is electrostatically pasted onto a negatively charged mica surface. (**e**) Cationic nanorings **FcNR** (left) and **Fc$^+$NR** (right) immobilized on mica. (Reproduced with permission from Ref. [49], © 2014 The American Association for the Advancement of Science)

length of the four aromatic arms. SAXS, AFM, and TEM analyses indicated that a self-assembled nanotube model from **50a** has 200-membered decagonal macrocycle comprising ten bent-shaped ligand molecules of **50a** and twenty Ag(I) ions. The relatively larger-diameter nanotube from **50b** possesses 360-membered decagonal cross section composed of, likewise, ten bent-shaped ligand molecules of **50b** and twenty Ag(I) ions.

A 10-T magnetic field was applied to prepare a film organized from 1D oriented nanotubes of **50a** [49]. 2D wide-angle X-ray diffractometry for the oriented film disclosed the presence of *d*-spacing of 0.70 and 0.35 nm. This diffraction profile is consistent with the wall structure, in which the nanorings of **50a** with an eclipsed geometry uniaxially pile up on top of each for the maximization of the intermolecular π–π stacking and metallophilic Ag(I)–Ag(I) interactions. The same research group envisaged that oxidation of the ferrocene core in the self-assembled nanotube enables the electrostatic attenuation of the interactions between upper nanoring and lower one to cut into individual component rings without cleaving the coordinated Ag(I)–N(Py) bonds [49]. Indeed, mild oxidation of the nanotube of **50a** with 1,1′-dichloroferrocenium tetrafluoroborate ($Cl_2Fc^+BF_4^-$) induced selective disassembly in CH_2Cl_2/CH_3CN to give toroidal objects with a homogeneous diameter of ca. 7.4 nm that corresponds to the diameter of the component nanoring. The positively charged nanorings can be successfully deposited onto negatively charged mica substrates. More interestingly, reduction of the oxidized nanorings with bis (pentamethylcyclopentadienyl) iron(II) resulted in the reconstruction of long cylindrical nanotubes [49]. Oxidative disassembly of relatively larger-diameter nanotube with the same oxidant, however, went wrong probably due to lack of robustness of the ring structure.

A mixture of the two different pyridyl ligands **50a** and **50b** was subjected to competitive assembly through coordination with $AgBF_4$. The resultant self-assemblies are not mixed-ligand nanotube products, but separate homomeric nanotube structures (Fig. 9.27) [51]. Self-sorting assembly, thus, occurred in the binary assembly of **50a** and **50b**. Remarkably, the assembled nanorings from the ligand **50a** with relatively shorter aromatic arms stacks helically. Meanwhile, the nanorings comprising relatively longer-armed ligand **50b** tend to stack without helical twist. The self-sorting feature of the ligands **50a** and **50b** should be originated from the structural difference in the favorable stacking angles.

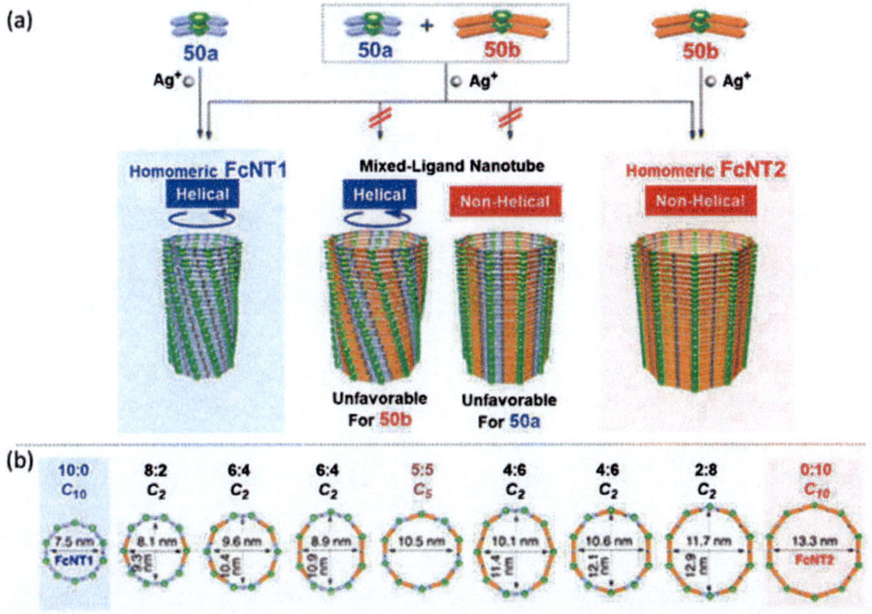

Fig. 9.27 **a** Illustration of homomeric metal–organic nanotubes **FcNT1** from **50a** and **FcNT2** from **50b**, and mixed-ligand nanotubes. The term "nonhelical geometry" neither means that the stacking angle is exactly zero degrees nor guarantees the absence of any macroscopic helical structure. **b** Schematic drawing of decagonal cross sections of possible nanotubes. (Reproduced with permission from Ref. [51], © 2016 American Chemical Society)

9.13 Bent-Shaped Aromatic Amphiphile

With a rational design principle, Lee and co-workers synthesized bent-shaped, chiral aromatic amphiphiles **51a** (S-enantiomer) and **51b** (R-enantiomer) (Chart 9.14) that consist of a rigid aromatic module with an internal angle of 120° and a four-armed hydrophilic oligoether dendron at its apex [52]. As a central core moiety, *m*-pyridine unit was employed in expectation of a dense molecular packing of the aromatic segments of adjacent molecules. As predicted, the rigid–flexible molecule **51a**, underwent the self-assembly in aqueous solutions to give hexameric macrocycles. The obtained macrocyclic rings further stack up uniaxially with their mutual rotation; thereby, resulting in the formation of chiral nanotubes with a one-handed helical sense. The uniform dimensions are featured by outer diameters of 11 nm and inner diameters of 4 nm. The hydrophilic oligoether dendrons are arranged helically on

Chart 9.14 Chemical structures of **51a** and **51b**

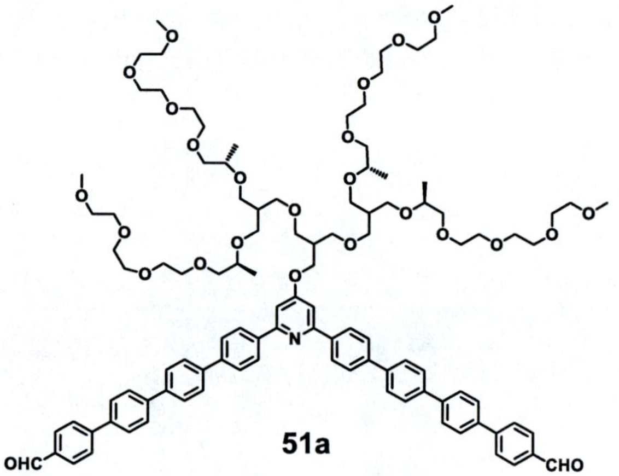

51a

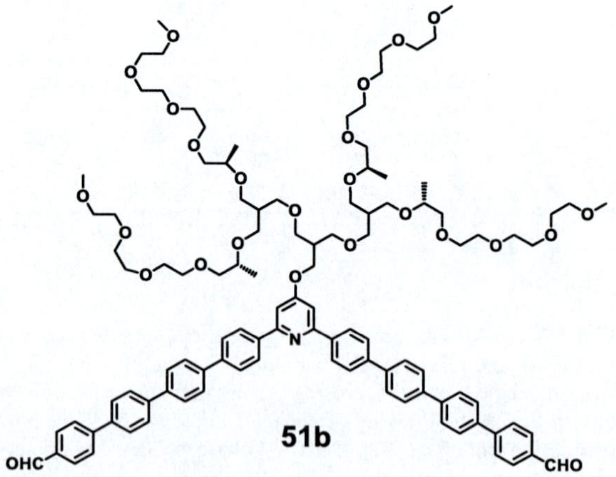

51b

the external surface, whereas the pyridine units at the valley position of the rigid aromatic segments reside on the inward side. The enantiomer **51b** exhibited the same self-assembly behavior in stacking-based nanotubes, which, but, give opposite, mirror-imaged CD signals.

Heating treatment is known to cause thermally controlled dehydration of pyridine moieties as well as ethylene oxide chains [53]. Similarly, the pyridine and ethylene oxide segments of **51a** and **51b** induce heat-responsive dehydration upon heating. Consequently, the aromatic moieties undergo a change of sliding motif from a slipped state (expansion mode) into a fully overlapped one (contraction mode) for the optimization of their π–π stacking interactions [52]. Indeed, TEM measurement revealed an outstanding contraction of 47% in cross section of the inner channel upon heating to 60 °C. Heating and cooling cycles induce reversible contraction and expansion of the channel size through an alternate sliding movement of the adjacent aromatic moieties (Fig. 9.28). Importantly, pulsating movement of the tubular structure coincide with an inversion of the helical chirality. Moreover, the nanotube structures can trap C60 molecule as a hydrophobic guest into the hydrophobic channel interior [52]. The contraction motion of the nanotube causes the partial release of the trapped C60

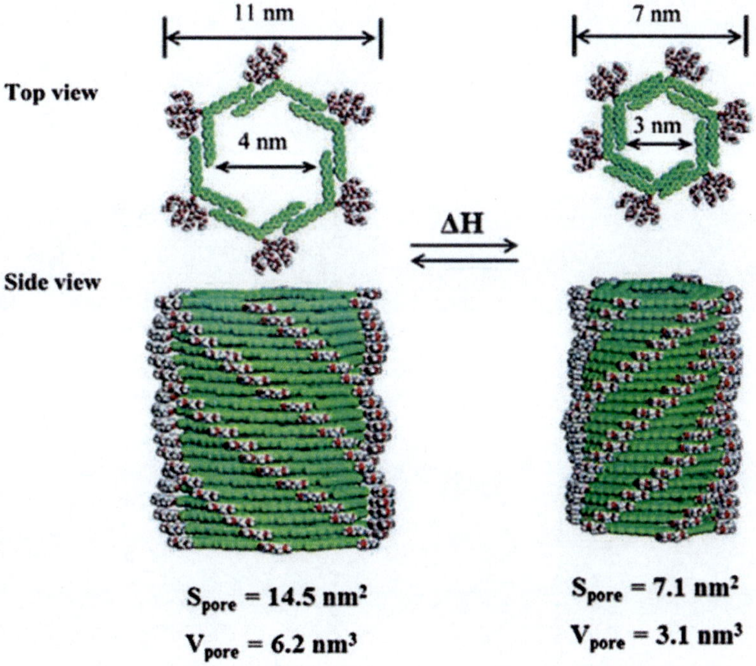

Fig. 9.28 Schematic illustration of reversible switching of the stacking-based nanotubes between expanded and contracted states with chirality inversion. S_{pore} represents the cross-sectional area of hollow pore in a slice of the macrocycles. V_{pore} means the volume of the hollow pore. (Reproduced with permission from Ref. [52], © 2012 The American Association for the Advancement of Science)

from the channel. The pulsating movement of the nanotube is, thus, involved with the reversible control of the C60–C60 interactions through loose and tight packing change.

Kim, Huang, and coworkers reported that RFBM-based supramolecular nanotubes coordinated with Pd^{2+} displayed highly catalytic activity and unprecedented selectivity for C–C and multifunctional coupling reactions, respectively [54]. The building block for the nanotube consists of the aromatic amphiphiles **52a** and **52b** (Chart 9.15) with a bent structure. The amphiphiles contain *m*-pyridine unit as a central core of the rigid aromatic module with an internal angle of 120°. The molecules also have a four-armed oligoether dendron at its apex, which provide them with hydrophilicity. Each amphiphile forms noncovalently macrocyclic structure comprising three single molecules in ethanol solution. Notably, the addition of water to the ethanol solution makes the macrocycles to stack unidirectionally, resulting in the formation of right-handed helical nanotubes and alternative stacking-based nanotubes for **52a** and **52b**, respectively (Fig. 9.29) [54]. STEM images of the nanotube from **52a** are compatible with the hollow tubes with an outer and inner diameter of 4.0 nm and 2.3 nm,

Chart 9.15 Chemical structures of **52a** and **52b**

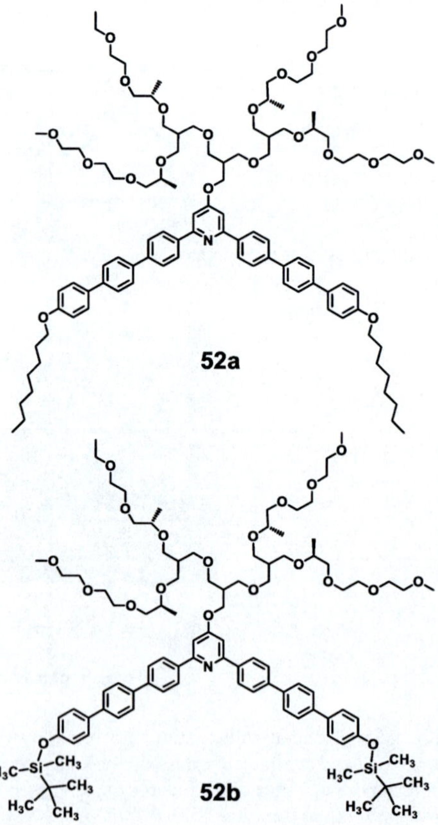

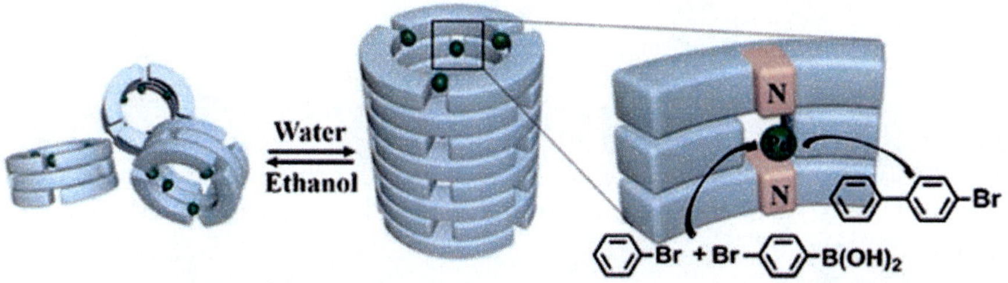

Fig. 9.29 Illustration of the nanotube regulator for the catalytic reaction based on reversible stacking of macrocycles **52b**. (Reproduced with permission from Ref. [54], © 2017 WILEY–VCH Verlag GmbH & Co. KGaA, Weinheim)

respectively. The Pd^{2+} can coordinate with pyridine units at the internal position of the aromatic amphiphile. More importantly, the Pd^{2+}-loaded catalyst of the nanotube from **52b** displayed a highly active performance for Suzuki–Miyaura crosscoupling reaction with unprecedented selectivity (Fig. 9.29). These RFBM-based nanotubes can be utilized as catalytic regulator in heterogeneous system.

Elaborated combination of hydrophobic rigid aromatic segments and hydrophilic flexible dendron leads to the self-assembly of inflatable nanofibers with stimuli-responsive functions. Lee and co-workers synthesized the amphiphilic molecules **53a** and **53b** (Chart 9.16) that comprise a short bent-shaped aromatic segment linked through *m*-pyridine core and a four-armed oligoether dendron connected at its apex [55]. They envisaged that the self-assembly of the bent-shaped amphiphiles **53a** and **53b** should yield nanofibers composed of the hydrophobic aromatic core and hydrophilic dendron shell through dimer assembly and subsequent vertical stacking. This self-assembly scheme is similar to the spontaneous assembly of nanofibers from tribranched aromatic amphiphiles [56]. Dynamic laser scattering and TEM analyzes revealed that the self-assembly of diluted aqueous solutions (0.001 wt%) of **53a** and **53b** produced micellar structures of ca. 5 nm in diameter. This finding is compatible with the dimer structure, in which two aromatic segments face each other to form dimeric aggregate with surrounding hydrophilic oligoether chains. With the increase in the concentration up to 0.01 wt%, the aggregate of **53a** elongated the length to assemble into nanofibers of 5 nm in diameter and several hundred nm in length. The nanofiber formation can be well explained by the self-assembly scheme, in which the aromatic amphiphiles **53a** aggregated into paired dimers to subsequently stack on top of each other (Fig. 9.30). Interestingly, the addition of *p*-phenylphenol **54** (Chart 9.16) as a hydrophobic guest molecule was shown to trigger the inflation of the self-assembled fibers (outer diameter: 5 nm) into nanotube structures (outer diameter: 8 nm) (Fig. 9.30) [55]. The guest **54** allowed for its hydrogen-bond formation with the pyridine nitrogen of the aromatic segments. As a result of the hydrogen bonding, the packing rearrangement in the aromatic cores from transoid dimers to cisoid macrocycles should result in the formation of the hollow tubular assemblies.

Recently, Lee and co-workers have also reported that the co-assembly of pyridine-**55a** and pyridinium-based cation molecules **55b** (Chart 9.16) (**55a/55b**, 6:4 molar

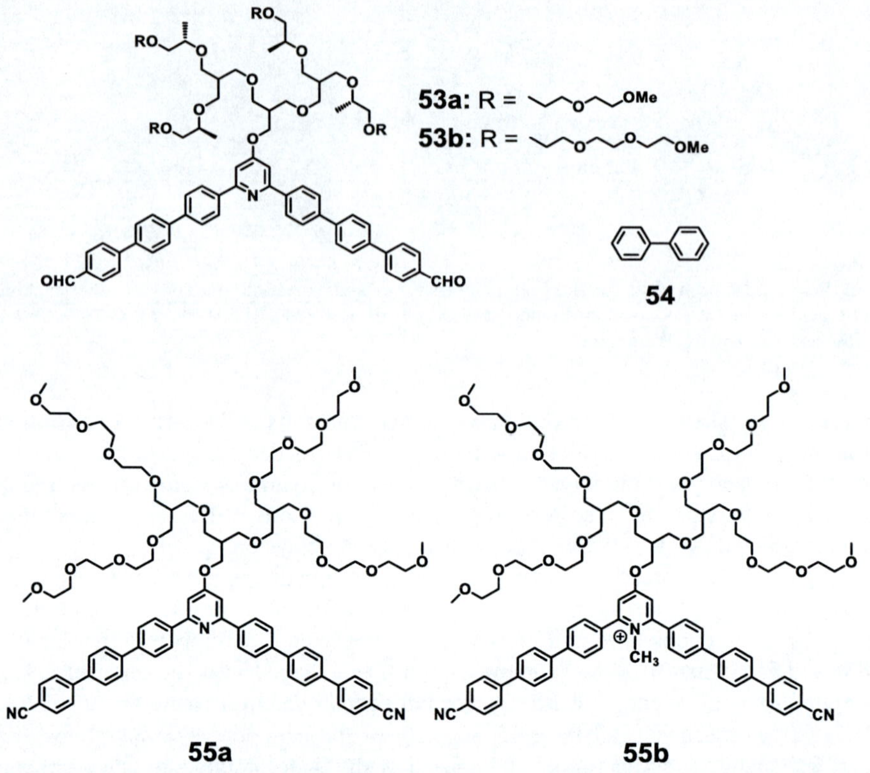

Chart 9.16 Chemical structures of **53**–**55**

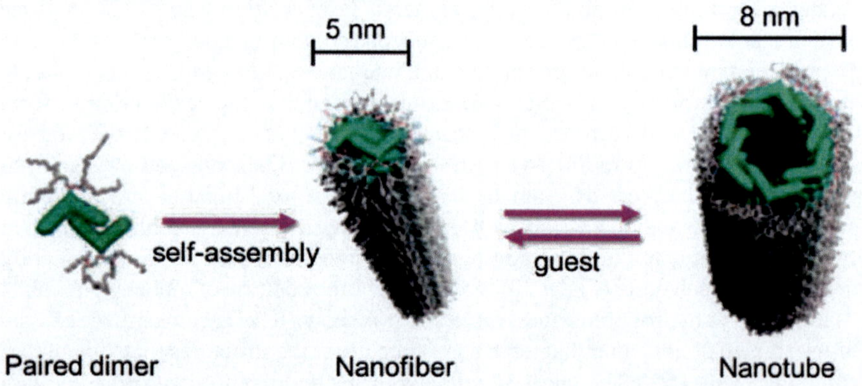

Fig. 9.30 Illustration of inflation of the self-assembled nanofiber from **53a** or **53b** into a hollow nanotube upon addition of the guest molecule **54**. (Reproduced with permission from Ref. [55], © 2014 American Chemical Society)

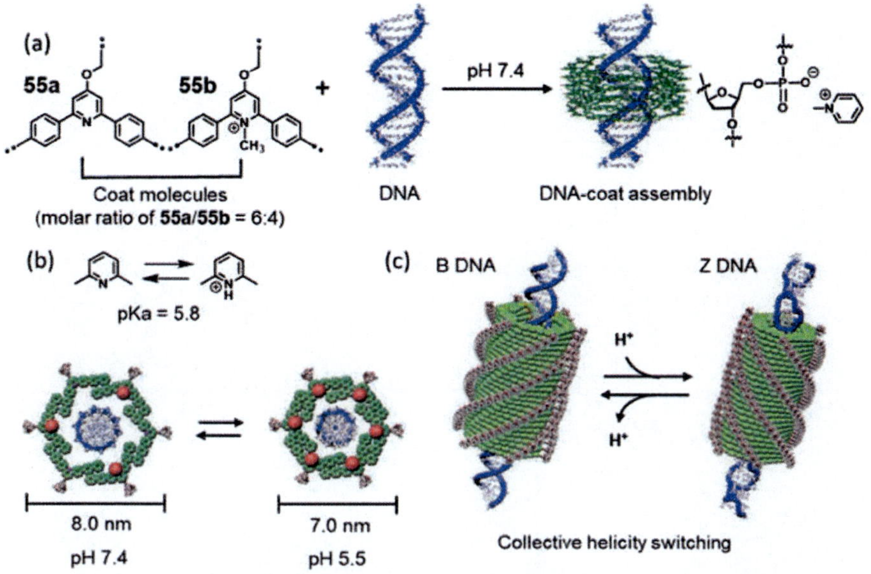

Fig. 9.31 DNA-driven self-assembly of the coat molecules **55a** and **55b**, and collective helicity switching of the DNA–coat assembly. Schematic drawing of **a** DNA–coat assembly at pH 7.4 and **b** molecular rearrangement of the coat molecules between expanded (pH 7.4) and contracted (pH 5.5) states. **c** pH-change-triggered collective motion in the helicity inversion of DNA and the synthetic coat assembly. (Reproduced with permission from Ref. [57], © 2017 Nature Publishing Group)

ratio) forms hexameric macrocycles in water, leading to the construction of TMV-like RFBM nanotubes [57]. Significantly, the nanotube assembly can capture a double-stranded DNA molecule into the hollow cylinder through electrostatic interactions (Fig. 9.31). The tubular and helical coat assembly with a diameter of 8 nm surrounds a single DNA molecule at pH 7.4. A pH change from 7.4 to 5.5 under physiological conditions induces the helicity change of both the encapsulated DNA and external tubular coat assembly. As a result, reversible collective helicity switching of the DNA and helicity inversion of the coat takes place accompanying with contraction–expansion of the diameter of the coat assembly.

9.14 Cyclic Aromatic Amphiphile

As mentioned above, amphiphilic aromatic building blocks play an important role in the self-assembly systems that exhibit a switchable function in response to some stimuli. Lee and co-workers newly synthesized amphiphilic disk-shaped aromatic macrocycles **56** (Chart 9.17) [58]. The RFBM **56** is composed of cyclohexa-*m*-phenylene as a hydrophobic disk core and four-armed oligoether dendron as a hydrophilic segment. The molecule **56** forms a hollow cylindrical nanotube with

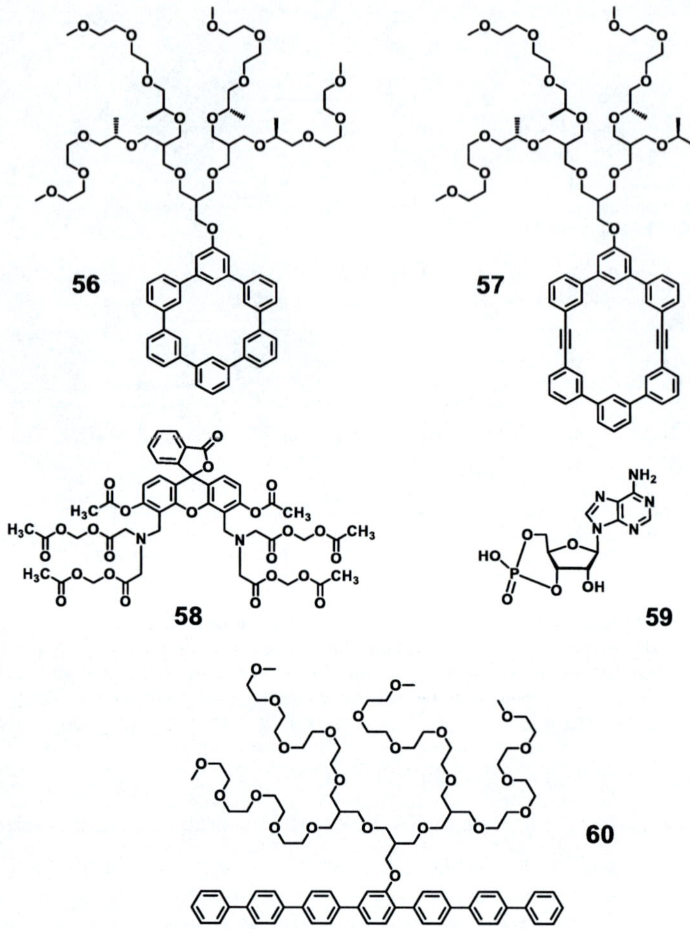

Chart 9.17 Chemical structures of **56–60**

a diameter of 9 nm through self-assembly in water. All the experiments using CryoTEM, TEM, and XRD suggested that eight pieces of self-assembled fibrils of 3.6 nm in diameter laterally associate to shape the hollow tubes with a wall thickness of 3.6 nm and inner diameter of 1.8 nm (Fig. 9.32). The self-organization of the fibril precursor was promoted by uniaxial stacking of **56** on top of one another through π–π interactions. Subsequently, the primary fibrils laterally assemble to form the tube wall comprising asymmetric bilayer molecular packing driven by mutual rotation with respect to the adjacent disks. Thus, not flat ribbon but curved bilayer structure appears to satisfy space-filling requirements of the flexible oligoether chains [58]. To corroborate this highly ordered self-assembly process, they investigated the self-assembly of the elliptical disk-shaped macrocycle amphiphile **57** (Chart 9.17). This molecule **57** requires higher energy for the mutual rotation between the neighboring

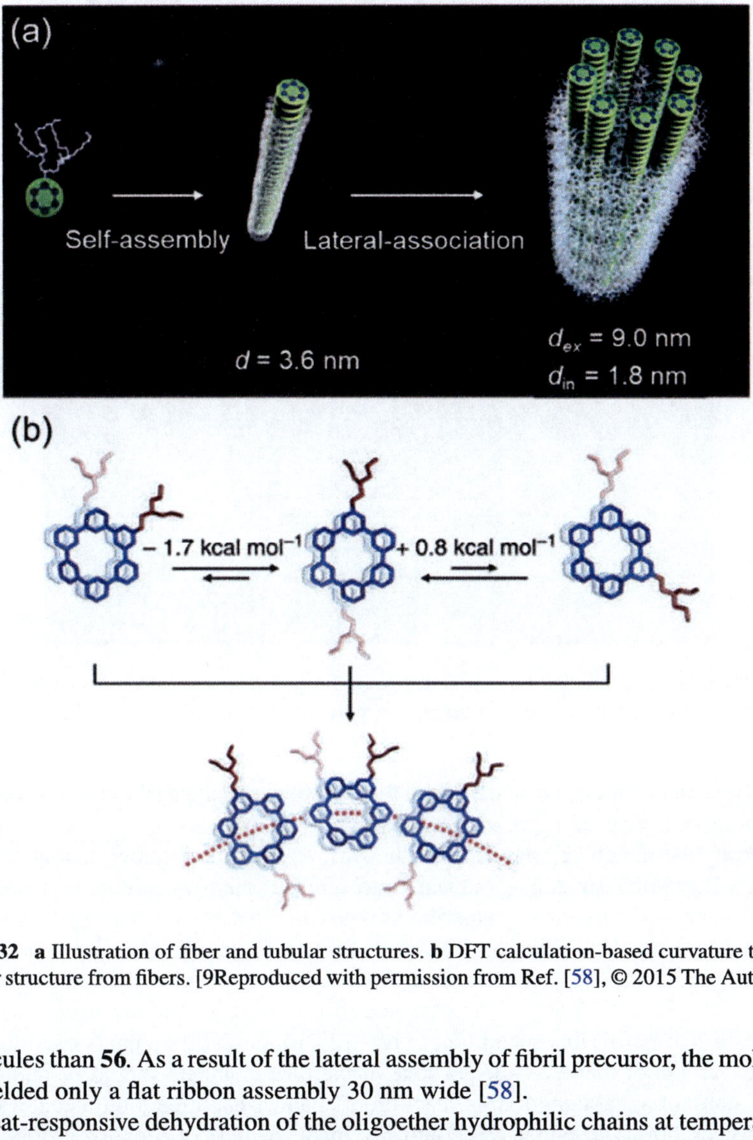

Fig. 9.32 **a** Illustration of fiber and tubular structures. **b** DFT calculation-based curvature to form tubular structure from fibers. [9Reproduced with permission from Ref. [58], © 2015 The Author(s)]

molecules than **56**. As a result of the lateral assembly of fibril precursor, the molecule **57** yielded only a flat ribbon assembly 30 nm wide [58].

Heat-responsive dehydration of the oligoether hydrophilic chains at temperatures above 40 °C undergoes a drastic change from spread open to globular conformation [59]. This outstanding conformational change of the hydrophilic chains has a significant effect on the stacking style of the aromatic macrocycles. Slipped packing mode of the macrocycle plane, thus, becomes favorable above 40 °C in water, resulting in the shrinkage of the nanotube diameter (Fig. 9.33) [58]. This closed motion of the nanotube channels contrasts with eclipsed packing arrangements of the macrocycles, which stabilize the opened hollow tubular structures at room temperature.

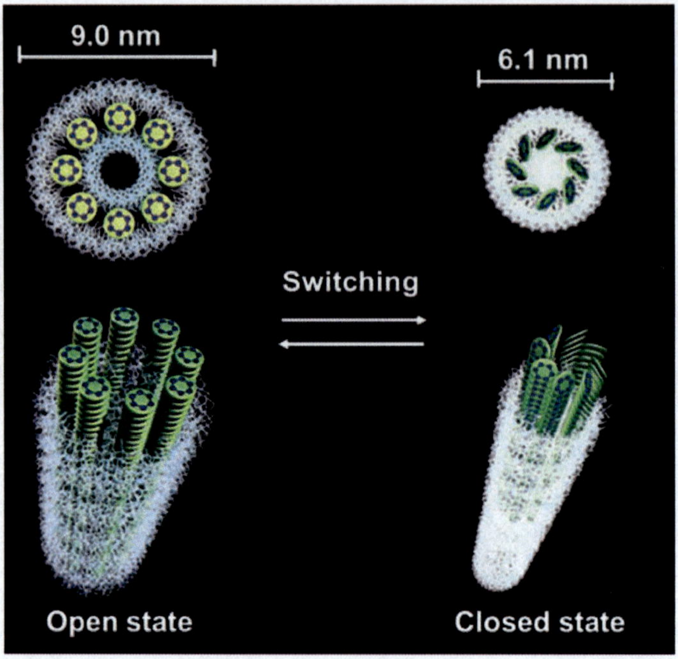

Fig. 9.33 Schematic drawing for thermally responsive and rapid open–closed switching of the nanotube channels. [Reproduced with permission from Ref. [58], © 2015 The Author(s)]

Notably, such open–closed switching of the nanotube channels takes place reversibly on cooling and succeeding heating cycles [58].

Closing the nanotube channel, i.e., the shrinkage of the internal diameter of the lumen will promote squeezing out water molecules from open ends as well as across the nanotube walls. Releasing experiment with calcein dye **58** (Chart 9.17) indicated that the large molecular dye **58** was slowly ejected from the closed nanochannels over 12 h at 45 °C [58]. At room temperature, however, the opened nanochannels released out **58** over 5 days. The obtained results suggest that the nanotube walls work as a semipermeable membrane. The Lee's research group envisioned that the interconversion of the open–closed state of the tube channels should be applicable for the nonbiological dehydration reaction of channel-captured biomolecules under aqueous conditions. The idea was confirmed by the capture of adenosine monophosphate (AMP) into the tube channel that undergoes dehydrative cyclization of AMP to give cyclic adenosine monophosphate **59** (c-AMP) (Chart 9.17) [58]. As a result of repeated heating and cooling cycles, the product amount was shown to increase successively. The presence of the rod-shaped amphiphile **60** (Chart 9.17) that can maintain the opened structure inhibited the catalytic reaction. Thus, they demonstrated that the open–closed switching of the nanotube channel successfully mediates a water-pumping catalytic reaction to yield c-AMP **59** from inactive AMP (Fig. 9.34) [58].

As already described in **2.2.2**, the binary assembly of LCA **61** and taurolithocholic acid **62** (Chart 9.18) in water starts with the formation of flat sheets that subsequently roll up spontaneously and finally zip into microtubular structures [60]. In an analogous fashion to this rolling-up and zipping assembly into microtubes, the amphiphilic cyclohexa-*m*-phenylene macrocycle **63** (Chart 9.18) was reported to give a nanotube architecture [61]. The aromatic amphiphile **63** carries a four-armed oligoether dendron at the periphery of the aromatic segment. Intriguingly, the nanotube formation in MeOH solutions is composed of a series of consecutive

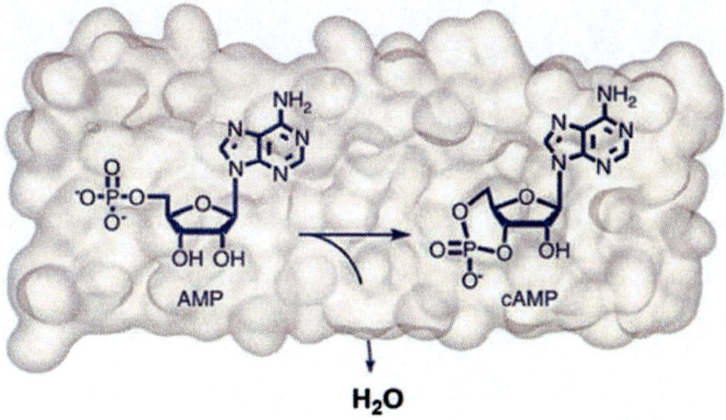

Fig. 9.34 Dehydrative cyclization of AMP to yield c-AMP **59** within the closed state of the wall membranes. [Reproduced with permission from Ref. [58], © 2015 The Author(s)]

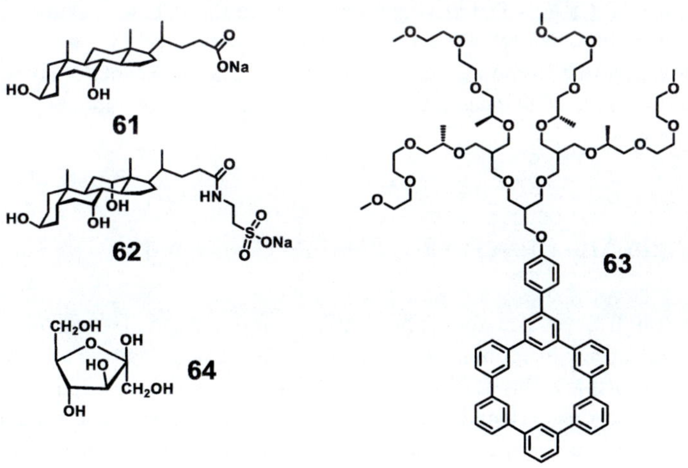

Chart 9.18 Chemical structures of **61–64**

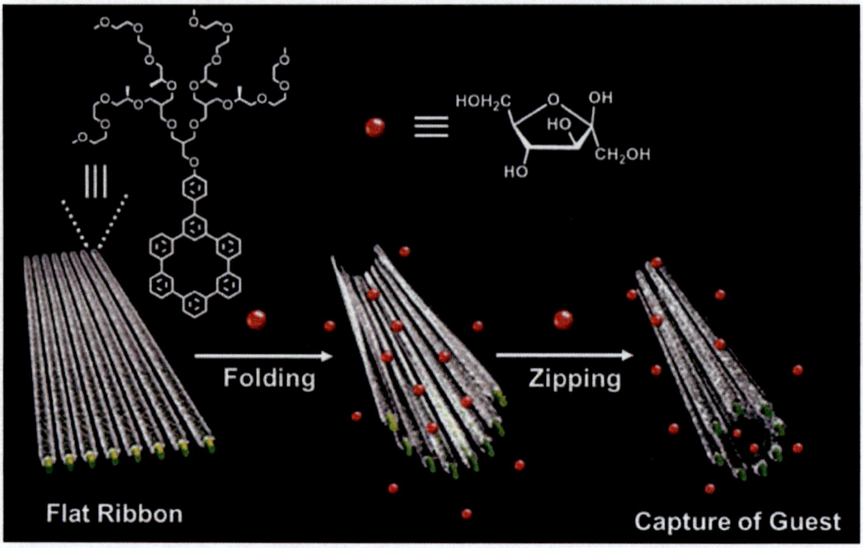

Fig. 9.35 Illustration of folding of an self-assembled flat ribbon structure from the aromatic amphiphile **63** and subsequent zipping into closed tubular structures triggered by the addition of fructose **64**. (Reproduced with permission from Ref. [61], © 2016 WILEY–VCH Verlag GmbH & Co. KGaA, Weinheim)

process, i.e., the formation of flat sheets, subsequent sheet folding, and eventual zipping into closed tubular structures (Fig. 9.35). Importantly, the addition of a guest molecule fructose **64** (Chart 9.18) is the trigger to zip two sides of the folded sheets, while the fructose is captured into the tube interiors [61]. The flat sheet comprises eight precursor fibrils that laterally associated with each other, having width of 28 nm and thickness of 3.5 nm. The driving force to curve the flat sheet into closed tubes should be originated from the closer packing of the aromatic segments upon addition of fructose **64**. More favorable interactions of fructose **64** with MeOH, as compared to the oligoether chains, enhanced the solvophobic effect, thereby inducing the packing rearrangement.

9.15 Rigid Macrocycle with Flexible Side Chains

Elaborate design and system for synthetic transmembrane channels via self-assembly of small molecules, e.g., aromatic macrocycles, oligoamides, and cyclodextrins, have resulted in the display of efficient and selective transport [62, 63]. Indeed, a class of aromatic macrocycles having both rigid oligoamide backbones and functional side chains, if stacked on top of each other, construct nanotube assemblies. The obtained nanotubes with a hardly deformable channel serve as ion channels embedded in bilayer membranes with high ion conductance [64]. For example, Yuan, Zeng, Gong, and co-workers demonstrated that aromatic oligoamide macrocycles **65a** and **65b** (Chart 9.19) with nondeformable hydrophilic nanochannel of approximately 0.85 nm stack into tubular structures that undergo further assembly into hexagonal-packing-based long fibers (Fig. 9.36) [65]. The aggregate size depends on the polarity of the

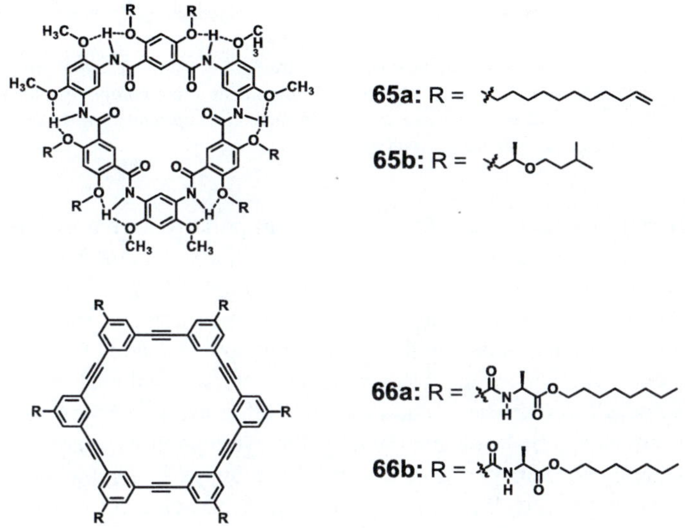

Chart 9.19 Chemical structures of **65–66**

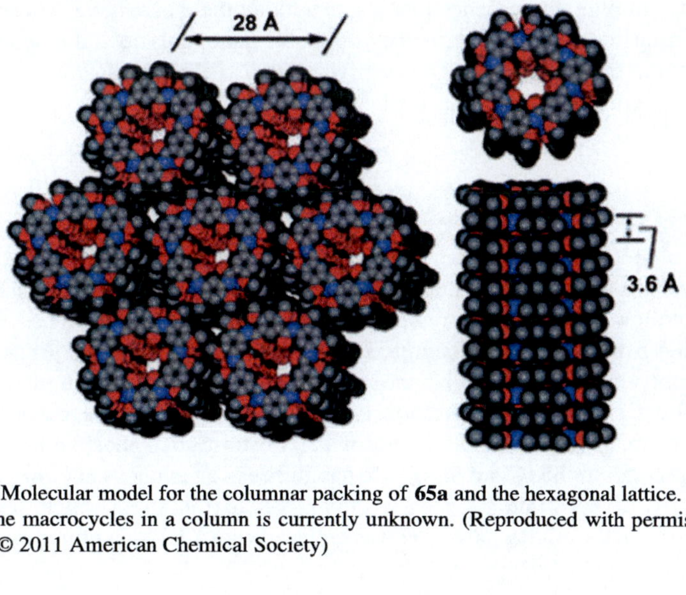

Fig. 9.36 Molecular model for the columnar packing of **65a** and the hexagonal lattice. The orientation of the macrocycles in a column is currently unknown. (Reproduced with permission from Ref. [65], © 2011 American Chemical Society)

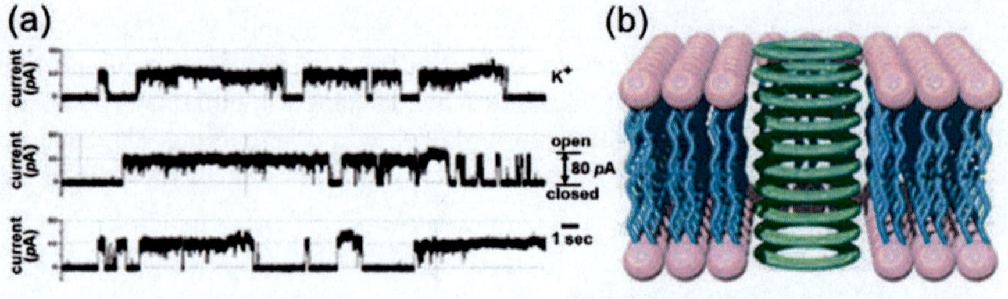

Fig. 9.37 **a** A108-s continuous K^+ single channel recording that exhibits single-step conductance changes at 50 mV with **65b** (0.32 μm). **b** A possible tubular structure comprising stacked macrocyclic molecules, which functions as a transmembrane channel. (Reproduced with permission from Ref. [64], © 2013 American Chemical Society)

solvent for self-assembly. Namely, as the solvent polarity decreases, the aggregation of **65a** promotes. Meanwhile, hydrogen-bond-derived nanotube assembly was observed for hexakis(*m*-phenylene ethylene) macrocycles **66a** and **66b** (Chart 9.19) carrying six hydrogen-bonding side chains. AFM measurement revealed the self-assembly of uniform nanofibers with a diameter of approximately 3.7 nm for **66a** or 3.2 nm for **66b**, which is compatible with that of each original macrocycle.

The macrocycle-stacked nanotube structures self-assembled from **65b** was shown to serve as a transmembrane channel in a lipid bilayer membrane. Indeed, the single tubular assembly of **65b** partition into the lipid bilayers, displaying single-step conductance changes (Fig. 9.37) [64, 65]. The same measurement with the analogue **65a** also showed the similar behavior. Meanwhile, stopped-flow kinetic measurement was performed for a carboxyfluorescein (CF)-loaded LUV, in which a macrocycle **66b**-based nanotube is embedded. The measurement exhibited a considerably faster decay of blue fluorescence than those without the macrocycle. These findings clearly demonstrate that the obtained hydrophobic nanopores not only mediate transmembrane ion transport with high selectivity but also exhibit transmembrane water permeability with high efficiency [66].

9.16 Pyrene and Phenanthrene Trimer

There are only a few reports on the self-assembly of amphiphilic pyrene derivative into hollow tubular structures [67]. Häner and coworkers designed the novel amphiphilic trimer of a 2,7-substituted pyrene **67** (Chart 9.20). As a result of the self-assembly in aqueous media, the pyrene amphiphile yielded both nanosheet and nanotube structures simultaneously. Subsequently, the same research group performed the supramolecular assembly of a 2,7-disubstituted phosphodiester-linked phenanthrene trimer **68** (Chart 9.20) into nanotubes in an aqueous environment [68]. Additionally, the substitute pyrene derivative **69** (Chart 9.20), in which the phenanthrene moiety in the middle part of **67** was replaced with a 2,7-substituted pyrene, was

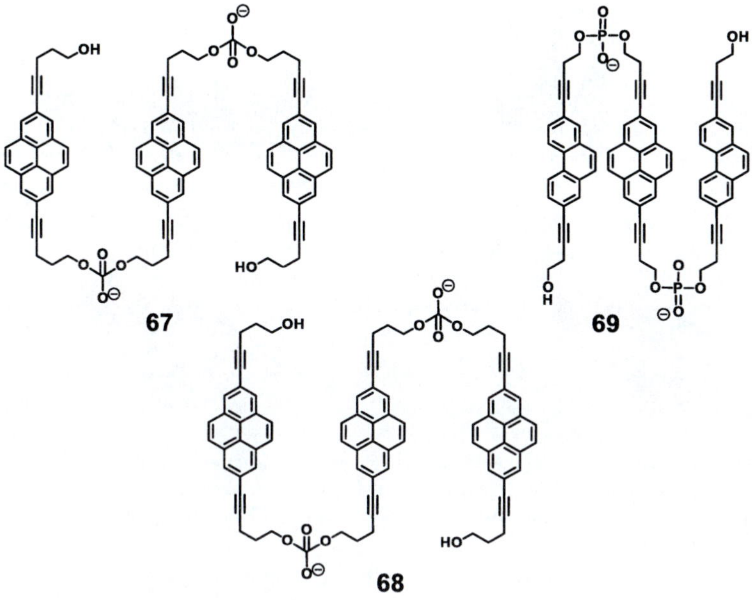

Chart 9.20 Chemical structures of **67**–**69**

prepared. Significantly, co-assembly technique using a small quantity of **69** and the matrix component **68** was shown to produce nanotubes that function as supramolecular light-harvesting antennae (Fig. 9.38) [68]. Fluorescence quantum yields reach at most 23% in the pyrene-doped nanotubes. In the membrane wall, the transfer of excitation energy occurs from phenanthrene to pyrene segments.

9.17 Rigid Block Molecule-Based Nanotube

9.17.1 Oppositely Charged Porphyrin

The fabrication of durable and well-defined porphyrin nanostructures still remains an important challenge. Ionic self-assembly technique [69] provided a powerful technique to fabricate robust porphyrin nanotubes. For example, two porphyrins, i.e., tetrakis(4-sulfonatophenyl) porphyrin diacid $(H_4TPPS_4)^{2-}$ **70** and Sn(IV) tetra(4-pyridyl) porphyrin $[Sn(OH^-)(X)TPyP]^{4+/5+}$ ($X = OH^-$ or H_2O)] **71** (Chart 9.21) with oppositely charges are mixed to assemble by electrostatic interactions in aqueous solutions at pH 2.0 [70]. As a result of ionic self-assembly, well-defined porphyrin nanotubes were formed. The dimensions of the obtained SMNTs have outer diameters of 50–70 nm, wall thickness of about 20 nm, and lengths of several μm (Fig. 9.39).

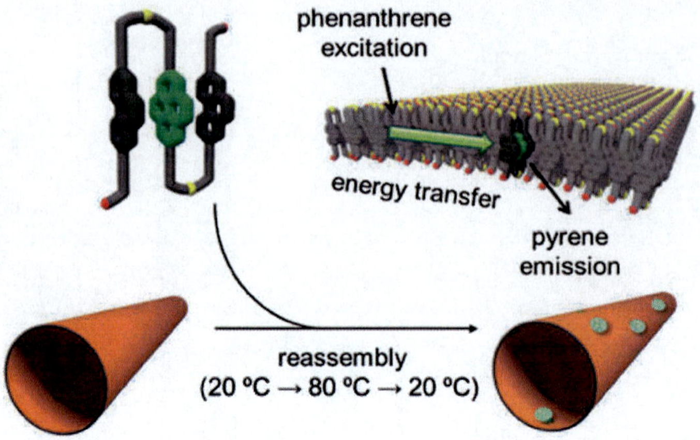

Fig. 9.38 Schematic drawing of co-assembled light-harvesting nanotubes from a small quantity of **69** and matrix component **68**, and energy transfer from excited phenanthrene to pyrene moiety as an acceptor. (Reproduced with permission from Ref. [68], © 2016 WILEY–VCH Verlag GmbH & Co. KGaA, Weinheim)

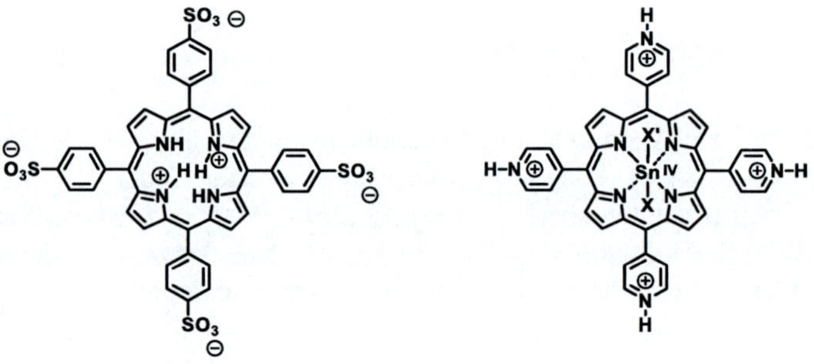

70

71

Chart 9.21 Chemical structures of **70** and **71**

Besides the electrostatic forces, several weak intermolecular interactions such as hydrogen bonding, van der Waals, and axial coordination are involved with the stabilization of the nanotube structures. The porphyrin molecules stack to take an offset J-aggregate structure in a fashion of cylindrical lamellar sheets [70]. The structure of the lamellar sheets should resemble to that proposed for the stacking mode of BChl molecules in the chlorosomal LH antenna system [71]. Resonance Raman spectroscopy indicated that the two porphyrin **70** and **71** take a separated packing arrangement [72]. The Sn-associated porphyrins **71** are, thus, not electronically involved with the J-aggregates in the nanotubes. The porphyrin **70** is entirely associated with the formation of the porphyrin nanotubes. The molar ratio of two porphyrins ($[H2TPPS_4]^{4-}$ and $[Sn(OH)_2$-TPyP] were estimated to be 2.0–2.5

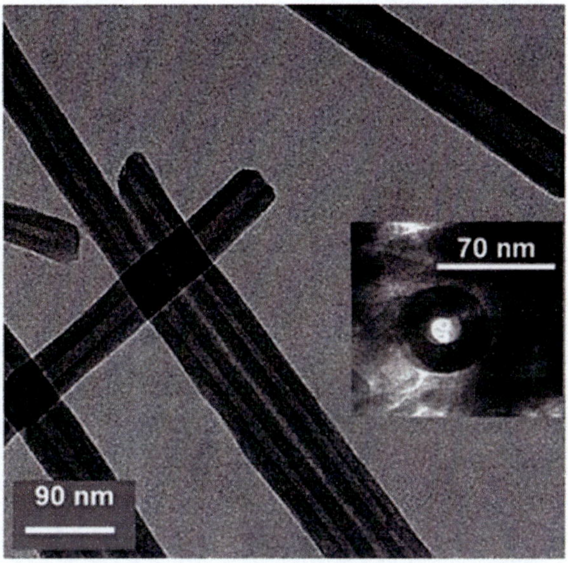

Fig. 9.39 TEM image of the porphyrin nanotubes produced via ionic self-assembly of **70** and **71** (inset: hollow cylindrical structure trapped in a vertical orientation by a thick mat of tubes. (Reproduced with permission from Ref. [70], © 2004 American Chemical Society)

on the basis of UV-vis absorption and energy-dispersive X-ray spectroscopy [70]. The dimensions of the obtained porphyrin nanotubes are tunable by changing the molecular structures.

9.17.2 Boroxine

Boroxines [$(RBO)_3$, R = aryl, alkoxy, and alkyl], i.e., boronic acid anhydrides, are the dehydrated products of organic boronic acids [$RB(OH)_2$] and eventually have tripodal molecular shapes with six-membered ring as the core motif. The boroxine ring is, therefore, generally modified with triaryl, trialkoxy, and trialkyl functional substituents. Thus, boroxine molecules serve as a useful and convenient molecular scaffold for the formation of covalent organic frameworks, hyper-branched polymers, and dendrimers [73]. Kameta and co-workers performed systematic studies on the self-assembly behavior of boroxines with ten different aromatic substituents [74]. Notably, strongly depending on the structure of the substituents, the self-assembled products of boroxines **72** and **73** (Chart 9.22) in toluene give nanofibers, whereas those of **74**, **75**, and **76** (Chart 9.22) are nanorod structures. Likewise, boroxines **77**, **78**, **79**, and **80** (Chart 9.22) form nanotape architectures through self-assembly in toluene. The boroxine molecule **81** (Chart 9.22) bearing three pyrenyl groups was, however, shown to self- assemble in toluene to yield nanotubular structures with an inner diameter of ~ 8 nm and wall thickness of ~4 nm (Fig. 9.40).

Hydrolysis of the boroxine with six-membered rings easily undergoes the structural change into original boronic acids. The research group envisioned that such

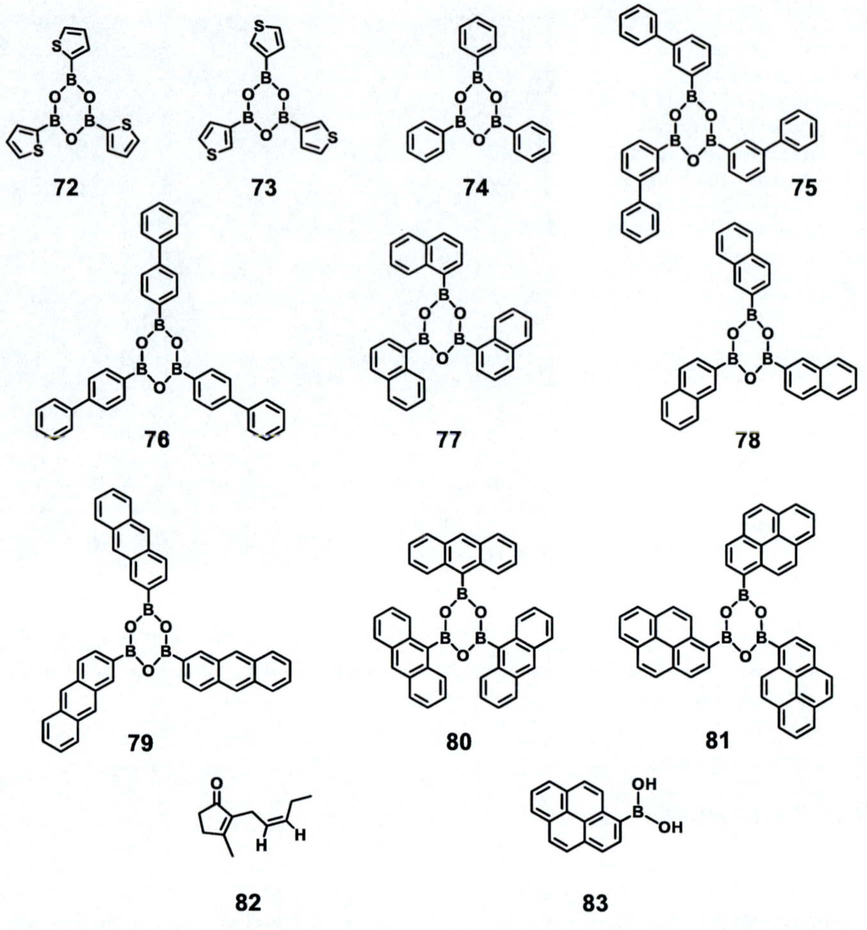

Chart 9.22 Chemical structures of **72–83**

Fig. 9.40 **a** SEM image of nanotubes self-assembled from the boroxine molecule **81** with three pyrenyl groups and **b** TEM image of nanotubes self-assembled from **81**. (Reproduced with permission from Ref. [74], © 2013 WILEY–VCH Verlag GmbH & Co. KGaA, Weinheim)

hydrolysis reactions should make a drastic change in the self-assembled tubular morphology. Indeed, humidity-responsive unzipping of the nanotube wall caused a morphological transformation from tubular to sheet structures. The reaction rate of hydrolysis sensitively depends on the humidity environment. Infrared and UV-vis spectroscopy revealed that the nanotubes consisting of J-type aggregates of boroxine were converted into sheet structures stabilized by H-type aggregates of boronic acid (Fig. 9.41). When the guest molecule cis-jasmone **82** (Chart 9.22) is encapsulated in advance in the nanotube hollow channel, such dramatical nanotube-to-sheet change should lead to active release of the guest to surrounding bulk media. They actually observed the controlled release capability of the nanotube, which strongly responds to the surrounding humidity.

Meanwhile, the chemical stability of boroxine to hydrolysis is known to enhance through the coordination of base and anion with boron atom of the boroxine as a hard Lewis acid. Addition of pyridine as a Lewis base to the boroxine nanotubes results in the formation of the boroxine/pyridine 1: 1 complex without remarkable morphological change. The presence of moisture, however, provokes the pyridine-complexed nanotube to undergo a nanotube-to-helical coil morphological change. The pyridine coordination, however, can restrain the hydrolysis reaction of boroxine with sufficiently keeping the J-aggregate molecular packing. The moisture-responsive guest release through nanotube-to-helical coil transformation exhibited much slower behavior when compared with the case by nanotube-to-sheet morphological change (Fig. 9.42).

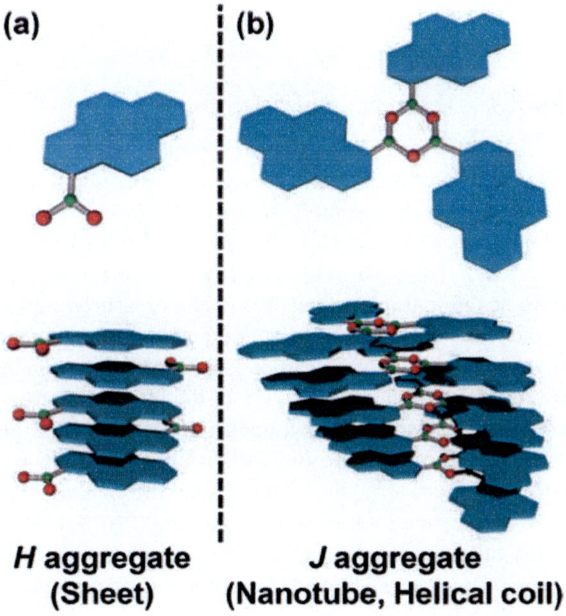

Fig. 9.41 Schematic illustration of **a** H-aggregate of the boronic acid **83** (Chart 9.22) in a sheet and **b** J-aggregate of the boroxine **81** in nanotubes and helical coils. (Reproduced with permission from Ref. [74], © 2013 WILEY–VCH Verlag GmbH & Co. KGaA, Weinheim)

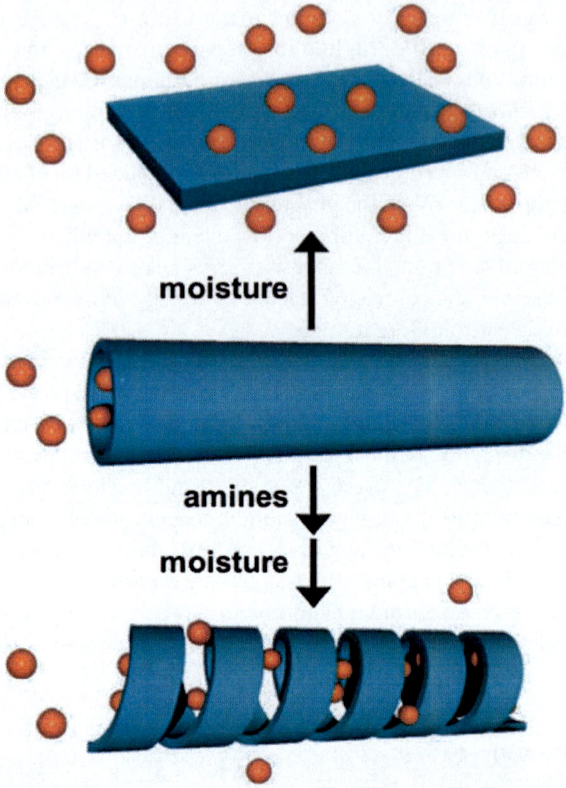

Fig. 9.42 Illustration of guest release from a sheet, a nanotube, and a helical coil. (Reproduced with permission from Ref. [74], © 2013 WILEY–VCH Verlag GmbH & Co. KGaA, Weinheim)

9.17.3 *Trimesic Acid Analogue*

Unique microtubes with a hexagonal cross section of hollow cylinder was formed through the self-assembly of trimesic amide derivatives with tripodal shape. Xu and co-workers reported that *N*,*N*′,*N*″-tris-(3-methylpyridyl)-trimesic amide **84** (Chart 9.23) self-assembled in a mixture of H_2O/THF (5:1, v/v) to form high-axial-ratio hexagonal microtubes [75]. The measured outer diameters are in the range of 1–6 μm that are relatively larger than those of the majority of organic nanotubes [76]. Intermolecular hydrogen-bond interactions between the N–H and pyridyl groups are responsible for the stabilization of the microtubes. A rosette-type primary assembly consisting of the six C_3-symmetric molecules undergoes multistep and hierarchical assembly into the microtubular structures through packing of small hexagonal 2D sheets. Remarkably, the vertical stacking of the building blocks proceeds concurrently with their lateral packing process (Fig. 9.43). The microtube formation eventually results in the formation of organogel if the mixture solvents such as H_2O/THF, H_2O/EtOH, and H_2O/MeOH compositions are used as the self-assembly solvents

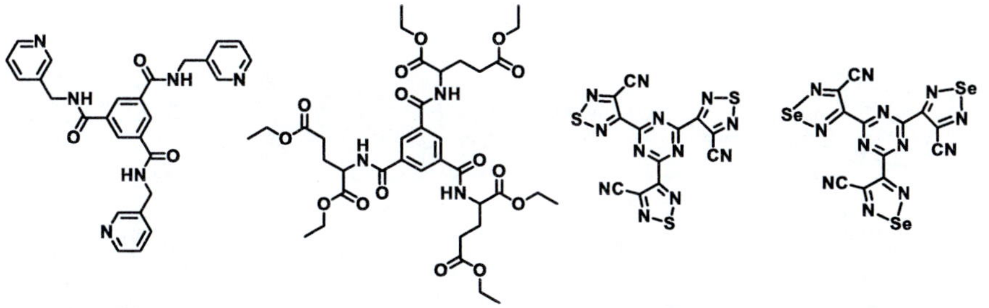

84 85 86 87

Chart 9.23 Chemical structures of **84–87**

[75]. The obtained gel system is pH-responsive, transforming into sol by acidification (pH < 3.0) and switching back to gel by neutralization (pH = 7.0).

By expanding a series of studies on the self-assembly of glutamic acid-derived amphiphiles, Liu and co-workers newly developed the C_3-symmetric, six-armed trimesic amide derivative **85** (Chart 9.23) carrying glutamic acid residue [77]. The molecule **85** undergoes the self-assembly in various organic solvents, e.g., $CHCl_3$, CH_2Cl_2, acetone, DMF, and DMSO to form hexagonal tubular structures. Depending on the employed solvents, the inner diameters are tunable in the range of 50–500 nm. Significantly, the nanotube formation is accompanied with gel formation when the solution of **85** is mixed with poorly soluble solvent, e.g., water and alkane. This means that the molecule **85** displays instant gelation capability that is inducible by the addition of anti-solvent [78]. Moreover, the instant gelation of self-assembled nanotubular structures synchronize with the capture of various guest molecules (Fig. 9.44) [77]. This encapsulation system is universal and can target biomacromolecules such as ferritin and double-stranded DNA, functional dyes such as ethidium bromide and rhodamine B, and synthetic polymers such as poly(*N*-vinylcarbazole).

Two tripodal-shaped molecules, 2,4,6-tris(4-cyano-1,2,5-thiadiazol-3-yl)-1,3,5-triazine **86** and 2,4,6-tris(4-cyano-1,2,5-selenodiazole-3-yl)-1,3,5-triazine **87** (Chart 9.23) were synthesized to explore n-type semiconductive organic nanomaterials with a 1D morphology [79]. The self-assembly proceeded by casting a diluted solution of **86** and **87** in acetone onto a Si substrate and subsequent evaporation over 10 h at 0 °C. Considerable slow evaporation of acetone and low temperature conditions are conclusive for the formation of tubular architectures. As a final process, remained acetone after evaporation serves as an etching reagent to make the self-assembled rod structures hollow (Fig. 9.45). The obtained microtubes from **86** and **87** have a similar hollow rectangular cross section, the dimensions of which are 2 μm in outer diameter and 20 μm in length. Solution process with chloroform as a solvent for the self-assembly also yielded long microwires of **86**, which function as an excellent n-type semiconductor.

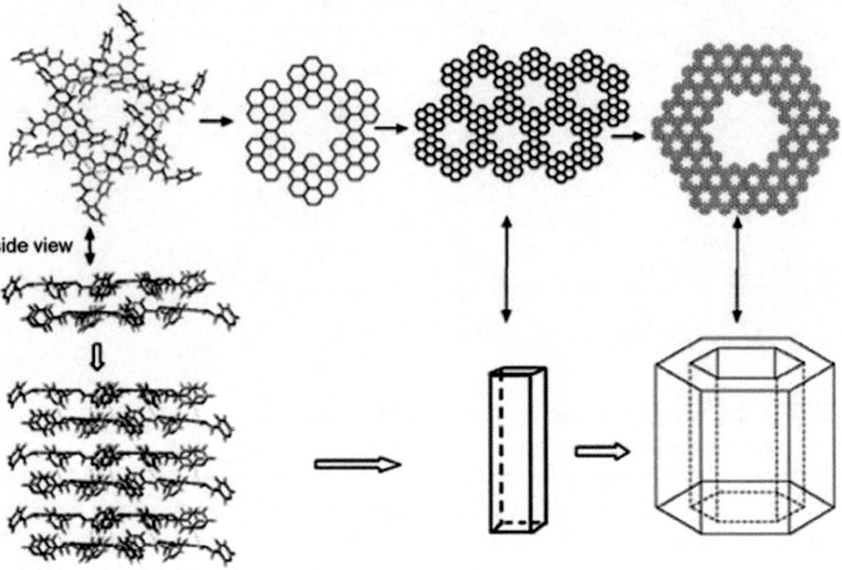

Fig. 9.43 Proposed self-assembly pathway of the trimesic amide **84** into hexagonal microtubes. This scheme does not show the definite diameter of the microtube. (Reproduced with permission from Ref. [75], © The Royal Society of Chemistry and the Center National de la Recherche Scientifique 2008)

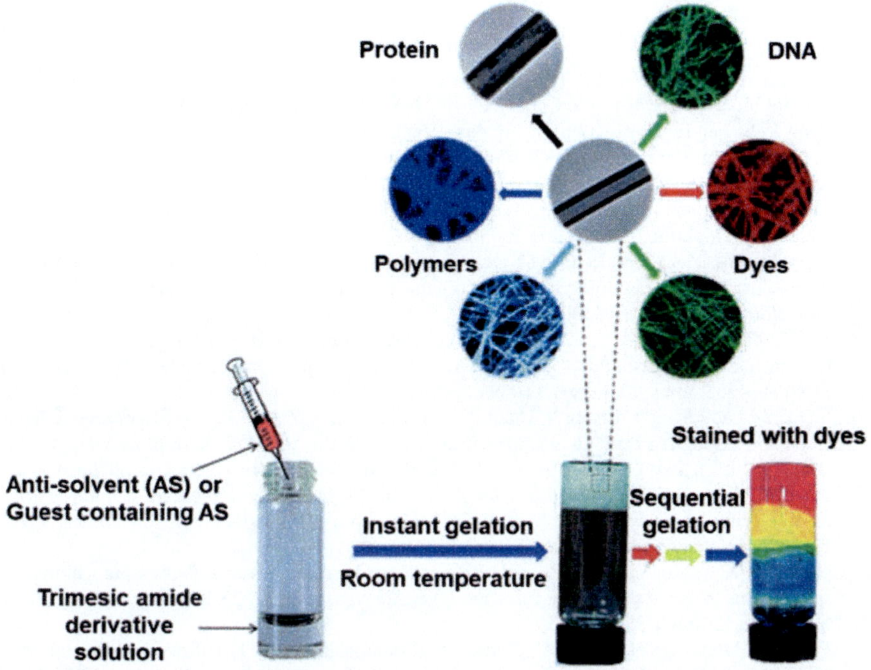

Fig. 9.44 Illustration of anti-solvent-induced instant gelation of **85** and the encapsulation of guest molecules into the nanotube when the guest molecules were pre-dissolved in either anti-solvent or the solution of trimesic amide derivative. (Reproduced with permission from Ref. [77], © 2012 WILEY–VCH Verlag GmbH & Co. KGaA, Weinheim)

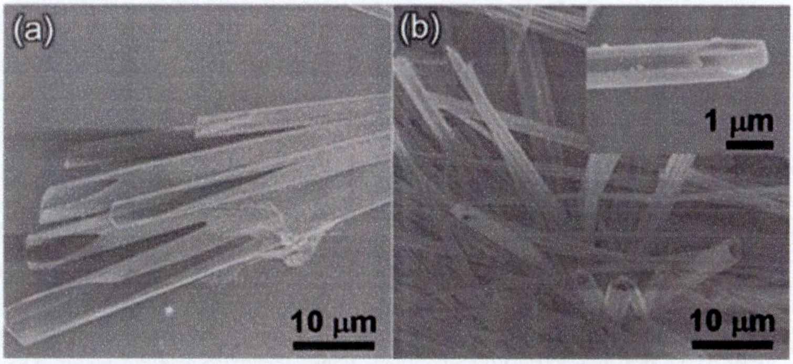

Fig. 9.45 SEM images of self-assembled microtubes from (**a**) **86** and (**b**) **87**. (Reproduced with permission from Ref. [79], © 2010 WILEY–VCH Verlag GmbH & Co. KGaA, Weinheim)

References

1. Kim HJ, Kim T, Lee M (2011) Responsive nanostructures from aqueous assembly of rigid-flexible block molecules. Acc Chem Res 44:72–82. https://doi.org/10.1021/ar100111n
2. Yang WY, Lee E, Lee M (2006) Tubular organization with coiled ribbon from amphiphilic rigid-flexible macrocycle. J Am Chem Soc 128:3484–3485. https://doi.org/10.1021/ja057823e
3. Ryu JH, Kim HJ, Huang ZG, Lee E, Lee M (2006) Self-assembling molecular dumbbells: From nanohelices to nanocapsules triggered by guest intercalation. Angew Chem Int Ed 45:5304–5307. https://doi.org/10.1002/anie.200600971
4. Zang L, Che Y, Moore JS (2008) One-dimensional self-assembly of planar π-conjugated molecules: adaptable building blocks for organic nanodevices. Acc Chem Res 41:1596–1608. https://doi.org/10.1021/ar800030w
5. Ma X, Zhang Y, Zheng Y, Zhang Y, Tao X, Che Y, Zhao J (2015) Highly fluorescent one-handed nanotubes assembled from a chiral asymmetric perylene diimide. Chem Commun 51:4231–4233. https://doi.org/10.1039/c5cc00365b
6. Yue E, Ma X, Zhang Y, Zhang Y, Duan R, Ji H, Li J, Che Y, Zhao J (2014) Fluorescent bilayer nanocoils assembled from an asymmetric perylene diimide molecule with ultrasensitivity for amine vapors. Chem Commun 50:13596–13599. https://doi.org/10.1039/c4cc06915c
7. Peng C, Zhang Y, Zhang Y, Hu Y, Che Y, Zhao J (2016) Highly fluorescent nanotubes with tunable diameter and wall thickness self-assembled from asymmetric perylene diimides. Small 12:4363–4369. https://doi.org/10.1002/smll.201601073
8. Hu Y, Ma X, Zhang Y, Che Y, Zhao J (2016) Detection of amines with fluorescent nanotubes: applications in the assessment of meat spoilage. Acs Sensors 1:22–25. https://doi.org/10.1021/acssensors.5b00040
9. Tidhar Y, Weissman H, Wolf SG, Gulino A, Rybtchinski B (2011) Pathway-dependent self-assembly of perylene diimide/peptide conjugates in aqueous medium. Chem Eur J 17:6068–6075. https://doi.org/10.1002/chem.201003419
10. Rodler F, Schade B, Jager CM, Backes S, Hampel F, Bottcher C, Clark T, Hirsch A (2015) Amphiphilic Perylene-Calix[4]arene hybrids: synthesis and tunable self-assembly. J Am Chem Soc 137:3308–3317. https://doi.org/10.1021/ja512048t
11. Huber V, Sengupta S, Wurthner F (2008) Structure-property relationships for self-assembled zinc chlorin light-harvesting dye aggregates. Chem Eur J 14:7791–7807. https://doi.org/10.1002/chem.200800764
12. Sengupta S, Ebeling D, Patwardhan S, Zhang X, von Berlepsch H, Bottcher C, Stepanenko V, Uemura S, Hentschel C, Fuchs H, Grozema FC, Siebbeles LDA, Holzwarth AR, Chi L, Wurthner F (2012) Biosupramolecular nanowires from chlorophyll dyes with exceptional charge-transport properties. Angew Chem Int Ed 51:6378–6382. https://doi.org/10.1002/anie.201201961
13. Patwardhan S, Sengupta S, Siebbeles LDA, Wurthner F, Grozema FC (2012) Efficient charge transport in semisynthetic zinc chlorin dye assemblies. J Am Chem Soc 134:16147–16150. https://doi.org/10.1021/ja3075192
14. Yamamoto Y (2011) Electroactive nanotubes from π-conjugated discotic molecules. Bull Chem Soc Jpn 84:17–25. https://doi.org/10.1246/bcsj.20100272
15. Hizume Y, Tashiro K, Charvet R, Yamamoto Y, Saeki A, Seki S, Aida T (2010) Chiroselective assembly of a chiral porphyrin-fullerene dyad: photoconductive nanofiber with a top-class ambipolar charge-carrier mobility. J Am Chem Soc 132:6628–6629. https://doi.org/10.1021/ja1014713
16. Charvet R, Yamamoto Y, Sasaki T, Kim J, Kato K, Takata M, Saeki A, Seki S, Aida T (2012) Segregated and alternately stacked donor/acceptor nanodomains in tubular morphology tailored with zinc porphyrin–C_{60} amphiphilic dyads: clear geometrical effects on photoconduction. J Am Chem Soc 134:2524–2527. https://doi.org/10.1021/Ja211334k
17. Li ZQ, Zhang YM, Chen Y, Liu Y (2014) A supramolecular tubular nanoreactor. Chem Eur J 20:8566–8570. https://doi.org/10.1002/chem.201402612

18. Didraga C, Pugzlys A, Hania PR, von Berlepsch H, Duppen K, Knoester J (2004) Structure, spectroscopy, and microscopic model of tubular carbocyanine dye aggregates. J Phys Chem B 108:14976–14985. https://doi.org/10.1021/jp048288s
19. Eisele DM, Berlepsch HV, Bottcher C, Stevenson KJ, Bout DAV, Kirstein S, Rabe JP (2010) Photoinitiated growth of sub-7 nm silver nanowires within a chemically active organic nanotubular template. J Am Chem Soc 132:2104–2105. https://doi.org/10.1021/Ja907373h
20. Walker EK, Vanden Bout DA, Stevenson KJ (2011) Aqueous electrogenerated chemiluminescence of self-assembled double-walled tubular j-aggregates of amphiphilic cyanine dyes. J Phys Chem C 115:2470–2475. https://doi.org/10.1021/Jp1108015
21. Wu J, Pisula W, Mullen K (2007) Graphenes as potential material for electronics. Chem Rev 107:718–747. https://doi.org/10.1021/cr068010r
22. Hill JP, Jin W, Kosaka A, Fukushima T, Ichihara H, Shimomura T, Ito K, Hashizume T, Ishii N, Aida T (2004) Self-assembled hexa-*peri*-hexabenzocoronene graphitic nanotube. Science 304:1481–1483. https://doi.org/10.1126/science.1097789
23. Jin W, Yamamoto Y, Fukushima T, Ishii N, Kim J, Kato K, Takata M, Aida T (2008) Systematic studies on structural parameters for nanotubular assembly of hexa-*peri*-hexabenzocoronenes. J Am Chem Soc 130:9434–9440. https://doi.org/10.1021/Ja801179e
24. Yamamoto Y, Fukushima T, Suna Y, Ishii N, Saeki A, Seki S, Tagawa S, Taniguchi M, Kawai T, Aida T (2006) Photoconductive coaxial nanotubes of molecularly connected electron donor and acceptor layers. Science 314:1761–1764. https://doi.org/10.1126/science.1134441
25. Yamamoto Y, Fukushima T, Saeki A, Seki S, Tagawa S, Ishii N, Aida T (2007) Molecular engineering of coaxial donor-acceptor heterojunction by coassembly of two different hexabenzocoronenes: graphitic nanotubes with enhanced photoconducting properties. J Am Chem Soc 129:9276–9277. https://doi.org/10.1021/Ja073577q
26. Yamamoto Y, Zhang GX, Jin WS, Fukushima T, Ishii N, Saeki A, Seki S, Tagawa S, Minari T, Tsukagoshi K, Aida T (2009) Ambipolar-transporting coaxial nanotubes with a tailored molecular graphene-fullerene heterojunction. Proc Natl Acad Sci USA 106:21051–21056. https://doi.org/10.1073/pnas.0905655106
27. He YN, Yamamoto Y, Jin WS, Fukushima T, Saeki A, Seki S, Ishii N, Aida T (2010) Hexabenzocoronene graphitic nanotube appended with dithienylethene pendants: photochromism for the modulation of photoconductivity. Adv Mater 22:829–832. https://doi.org/10.1002/adma.200902601
28. Zhang W, Jin W, Fukushima T, Saeki A, Seki S, Aida T (2011) Supramolecular linear heterojunction composed of graphite-like semiconducting nanotubular segments. Science 334:340–343. https://doi.org/10.1126/science.1210369
29. Zhang W, Jin W, Fukushima T, Mori T, Aida T (2015) Helix sense-selective supramolecular polymerization seeded by a one-handed helical polymeric assembly. J Am Chem Soc 137:13792–13795. https://doi.org/10.1021/jacs.5b09878
30. Prasanthkumar S, Zhang W, Jin W, Fukushima T, Aida T (2015) Selective synthesis of single- and multi-walled supramolecular nanotubes by using solvophobic/solvophilic controls: stepwise radial growth via "coil-on-tube" intermediates. Angew Chem Int Ed 54:11168–11172. https://doi.org/10.1002/anie.201505806
31. Shimizu T, Kogiso M, Masuda M (1996) Vesicle assembly in microtubes. Nature 383:487–488. https://doi.org/10.1038/383487b0
32. Kogiso M, Ohnishi S, Yase K, Masuda M, Shimizu T (1998) Dicarboxylic oligopeptide bolaamphiphiles: proton-triggered self-assembly of microtubes with loose solid surfaces. Langmuir 14:4978–4986. https://doi.org/10.1021/la9802419
33. Coleman AC, Beierle JM, Stuart MCA, Macia B, Caroli G, Mika JT, van Dijken DJ, Chen JW, Browne WR, Feringa BL (2011) Light-induced disassembly of self-assembled vesicle-capped nanotubes observed in real time. Nat Nanotechnol 6:547–552. https://doi.org/10.1038/Nnano.2011.120
34. Erne PM, van Bezouwen LS, Stacko P, van Dtjken DJ, Chen JW, Stuart MCA, Boekema EJ, Feringa BL (2015) Loading of vesicles into soft amphiphilic nanotubes using osmosis. Angew Chem Int Ed 54:15122–15127. https://doi.org/10.1002/anie.201506493

35. Erne PM, Stacko P, van Dijken DJ, Chen J, Stuart MCA, Feringa B (2016) End-capping of amphiphilic nanotubes with phospholipid vesicles: impact of the phospholipid on the cap formation and vesicle loading under osmotic conditions. Chem Commun 52:11697–11700. https://doi.org/10.1039/c6cc05101d
36. Saha A, Manna S, Nandi AK (2008) Hierarchical tuning of 1-D macro morphology by changing the composition of a binary hydrogel and its influence on the photoluminescence property. Chem Commun 3732–3734. https://doi.org/10.1039/b805344h
37. Saha A, Roy B, Esterrani A, Nandi AK (2011) Effect of complementary small molecules on the properties of bicomponent hydrogel of riboflavin. Org Biomol Chem 9:770–776. https://doi.org/10.1039/c0ob00670j
38. Diaz N, Simon FX, Schmutz M, Rawiso M, Decher G, Jestin J, Mésini PJ (2005) Self-assembled diamide nanotubes in organic solvents. Angew Chem Int Ed 44:3260–3264. https://doi.org/10.1002/anie.200500536
39. Shimizu T (2018) Self-assembly of discrete organic nanotubes. Bull Chem Soc Jpn 91:623–668. https://doi.org/10.1246/bcsj.20170424
40. Soler-illia GJD, Sanchez C, Lebeau B, Patarin J (2002) Chemical strategies to design textured materials: From microporous and mesoporous oxides to nanonetworks and hierarchical structures. Chem Rev 102:4093–4138. https://doi.org/10.1021/cr0200062
41. Simon FX, Khelfallah NS, Schmutz M, Diaz N, Mésini PJ (2007) Formation of helical mesopores in organic polymer matrices. J Am Chem Soc 129:3788–3789. https://doi.org/10.1021/ja067261e
42. Nguyen TTT, Simon FX, Khelfallah NS, Schmutz M, Mésini PJ (2010) Mesoporous polymeric catalysts synthesized from self-assembled organic nanotubes as templates. J Mater Chem 20:3831–3833. https://doi.org/10.1039/c000534g
43. Marsh A, Silvestri M, Lehn JM (1996) Self-complementary hydrogen bonding heterocycles designed for the enforced self-assembly into supramolecular macrocycles. Chem Commun 1527–1528. https://doi.org/10.1039/cc9960001527
44. Fenniri H, Mathivanan P, Vidale K, Sherman D, Wood K, Stwell J (2001) Helical rosette nanotubes: design, self-assembly, and characterization. J Am Chem Soc 123:3854–3855. https://doi.org/10.1021/ja005886l
45. Beingessner RL, Fan YW, Fenniri H (2016) Molecular and supramolecular chemistry of rosette nanotubes. Rsc Adv 6:75820–75838. https://doi.org/10.1039/c6ra16315g
46. Fenniri H, Deng BL, Ribbe AE, Hallenga K, Jacob J, Thiyagarajan P (2002) Entropically driven self-assembly of multichannel rosette nanotubes. Proc Natl Acad Sci USA 99:6487–6492. https://doi.org/10.1073/pnas.032527099
47. Chen YP, Song S, Yan ZM, Fenniri H, Webster TJ (2011) Self-assembled rosette nanotubes encapsulate and slowly release dexamethasone. Int J Nanomedicine 6:1035–1044. https://doi.org/10.2147/Ijn.S18755
48. Song S, Chen YP, Yan ZM, Fenniri H, Webster TJ (2011) Self-assembled rosette nanotubes for incorporating hydrophobic drugs in physiological environments. Int J Nanomedicine 6:101–107. https://doi.org/10.2147/Ijn.S11957
49. Fukino T, Joo H, Hisada Y, Obana M, Yamagishi H, Hikima T, Takata M, Fujita N, Aida T (2014) Manipulation of discrete nanostructures by selective modulation of noncovalent forces. Science 344:499–504. https://doi.org/10.1126/science.1252120
50. Muraoka T, Kinbara K, Aida T (2006) Mechanical twisting of a guest by a photoresponsive host. Nature 440:512–515. https://doi.org/10.1038/nature04635
51. Obana M, Fukino T, Hikima T, Aida T (2016) Self-sorting in the formation of metal-organic nanotubes: a crucial role of 2D cooperative interactions. J Am Chem Soc 138:9246–9250. https://doi.org/10.1021/jacs.6b04693
52. Huang Z, Kang SK, Banno M, Yamaguchi T, Lee D, Seok C, Yashima E, Lee M (2012) Pulsating tubules from noncovalent macrocycles. Science 337:1521–1526. https://doi.org/10.1126/science.1224741
53. Kim JK, Lee E, Lim Y, Lee M (2008) Supramolecular capsules with gated pores from an amphiphilic rod assembly. Angew Chem Int Ed 47:4662–4666. https://doi.org/10.1002/anie.200705863

54. Wu S, Li Y, Xie SY, Ma C, Lim J, Zhao J, Kim DS, Yang M, Yoon DK, Lee M, Kim SO, Huang Z (2017) Supramolecular nanotubules as a catalytic regulator for palladium cations: applications in selective catalysis. Angew Chem Int Ed 56:11511–11514. https://doi.org/10.1002/anie.201706373
55. Wang Y, Huang Z, Kim Y, He Y, Lee M (2014) Guest-driven inflation of self-assembled nanofibers through hollow channel formation. J Am Chem Soc 136:16152–16155. https://doi.org/10.1021/ja510182x
56. Moon KS, Kim HJ, Lee E, Lee M (2007) Self-assembly of T-Shaped aromatic amphiphiles into stimulus-responsive nanofibers. Angew Chem Int Ed 46:6807–6810. https://doi.org/10.1002/anie.200702136
57. Kim Y, Li H, He Y, Chen X, Ma X, Lee M (2017) Collective helicity switching of a DNA-coat assembly. Nat Nanotechnol 12:551–558. https://doi.org/10.1038/Nnano.2017.42
58. Kim Y, Kang J, Shen B, Wang Y, He Y, Lee M (2015) Open-closed switching of synthetic tubular pores. Nat Commun 6:8650. https://doi.org/10.1038/ncomms9650
59. Lutz JF, Weichenhan K, Akdemir O, Hoth A (2007) About the phase transitions in aqueous solutions of thermoresponsive copolymers and hydrogels based on 2-(2-methoxyethoxy)ethyl methacrylate and oligo(ethylene glycol) methacrylate. Macromolecules 40:2503–2508. https://doi.org/10.1021/ma062925q
60. Zhang X, Bera T, Liang W, Fang J (2011) Longitudinal zipping/unzipping of self-assembled organic tubes. J Phys Chem B 115:14445–14449. https://doi.org/10.1021/Jp2064276
61. Shen B, He Y, Kim Y, Wang Y, Lee M (2016) Spontaneous capture of carbohydrate guests through folding and zipping of self-assembled ribbons. Angew Chem Int Ed 55:2382–2386. https://doi.org/10.1002/anie.201509190
62. Matile S, Jentzsch AV, Montenegro J, Fin A (2011) Recent synthetic transport systems. Chem Soc Rev 40:2453–2474. https://doi.org/10.1039/c0cs00209g
63. Fyles TM (2007) Synthetic ion channels in bilayer membranes. Chem Soc Rev 36:335–347. https://doi.org/10.1039/b603256g
64. Gong B, Shao Z (2013) Self-assembling organic nanotubes with precisely defined, sub-nanometer pores: formation and mass transport characteristics. Acc Chem Res 46:2856–2866. https://doi.org/10.1021/ar400030e
65. Yang Y, Feng W, Hu J, Zou SL, Gao RZ, Yamato K, Kline M, Cai Z, Gao Y, Wang Y, Li Y, Yang Y, Yuan L, Zeng XC, Gong B (2011) Strong aggregation and directional assembly of aromatic oligoamide macrocycles. J Am Chem Soc 133:18590–18593. https://doi.org/10.1021/ja208548b
66. Zhou X, Liu G, Yamato K, Shen Y, Cheng R, Wei X, Bai W, Gao Y, Li H, Liu Y, Liu F, Czajkowsky DM, Wang J, Dabney MJ, Cai Z, Hu J, Bright FV, He L, Zeng XC, Shao Z, Gong B (2012) Self-assembling subnanometer pores with unusual mass-transport properties. Nat Commun 3:949. https://doi.org/10.1038/ncomms1949
67. Chen YL, Zhu B, Han Y, Bo ZS (2012) Self-assembly of cationic pyrene nanotubes. J Mater Chem 22:4927–4931. https://doi.org/10.1039/c2jm15997j
68. Bösch CD, Langenegger SM, Häner R (2016) Light-harvesting nanotubes formed by supramolecular assembly of aromatic oligophosphates. Angew Chem Int Ed 55:9961–9964. https://doi.org/10.1002/anie.201604508
69. Faul CFJ, Antonietti M (2003) Ionic self-assembly: facile synthesis of supramolecular materials. Adv Mater 15:673–683. https://doi.org/10.1002/adma.200300379
70. Wang Z, Medforth CJ, Shelnutt JA (2004) Porphyrin nanotubes by ionic self-assembly. J Am Chem Soc 126:15954–15955. https://doi.org/10.1021/Ja045068j
71. van Rossum BJ, Steensgaard DB, Mulder FM, Boender GJ, Schaffner K, Holzwarth AR, de Groot HJM (2001) A refined model of the chlorosomal antennae of the green bacterium Chlorobium tepidum from proton chemical shift constraints obtained with high-field 2-D and 3-D MAS NMR dipolar correlation spectroscopy. Biochemistry 40:1587–1595. https://doi.org/10.1021/bi0017529
72. Franco R, Jacobsen JL, Wang H, Wang Z, Istvan K, Schore NE, Song Y, Medforth CJ, Shelnutt JA (2010) Molecular organization in self-assembled binary porphyrin nanotubes revealed by resonance Raman spectroscopy. PCCP 12:4072–4077. https://doi.org/10.1039/b926068d

73. Korich AL, Iovine PM (2010) Boroxine chemistry and applications: a perspective. Dalton Trans 39:1423–1431. https://doi.org/10.1039/b917043j
74. Ishikawa K, Kameta N, Masuda M, Asakawa M, Shimizu T (2014) Boroxine nanotubes: moisture-sensitive morphological transformation and guest release. Adv Funct Mater 24:603–609. https://doi.org/10.1002/adfm.201302005
75. Shi N, Yin G, Li H, Han M, Xu Z (2008) Uncommon hexagonal microtubule based gel from a simple trimesic amide. New J Chem 32:2011–2015. https://doi.org/10.1039/b804455d
76. Shimizu T, Kameta N, Ding W, Masuda M (2016) Supramolecular self-assembly into biofunctional soft nanotubes: from bilayers to monolayers. Langmuir 32:12242–12264. https://doi.org/10.1021/acs.langmuir.6b01632
77. Cao H, Duan P, Zhu X, Jiang J, Liu M (2012) Self-assembled organic nanotubes through instant gelation and universal capacity for guest molecule encapsulation. Chem Eur J 18:5546–5550. https://doi.org/10.1002/chem.201103654
78. Sahoo P, Kumar DK, Raghavan SR, Dastidar P (2011) Supramolecular synthons in designing low molecular mass gelling agents: l-amino acid methyl ester cinnamate salts and their anti-solvent-induced instant gelation. Chem Asian J 6:1038–1047. https://doi.org/10.1002/asia.201000560
79. Mu X, Song W, Zhang Y, Ye K, Zhang H, Wang Y (2010) Controllable self-assembly of n-type semiconductors to microtubes and highly conductive ultralong microwires. Adv Mater 22:4905–4909. https://doi.org/10.1002/adma.201002259

Index

T. Shimizu, *Smart Soft-Matter Nanotubes*, Nanostructure Science and Technology,
https://doi.org/10.1007/978-981-16-2685-2

D

P

T